Astrophysics and Space Science Proceedings

Preface

Solar physics in India has a tradition that can be traced to the setting up of the Kodaikanal Observatory in 1899 when the Madras Observatory was relocated to a high altitude site with a view to initiate observations of the sun. This conference on *Magnetic Coupling between the Interior and the Atmosphere of the Sun* during 2–5 December 2008 was planned to coincide with centenary of the Evershed effect discovery at Kodaikanal in 1909.

The aim of this meeting was to bring to a critical focus a comprehensive understanding of the important issues pertaining to solar magnetism with particular emphasis on the various MHD processes that operate in the solar atmosphere. The current status of magnetic field measurements and their implications in the light of recent theories and numerical modeling that address the fundamental scales and processes in the highly magnetized turbulent plasma were reviewed during this meeting.

The meeting was timely for the following reasons: Space observations such as from SOHO and TRACE have provided a wealth of multiwavelength observations on processes occurring in regions of the atmosphere extending from the photosphere up to the outer corona. With the launch of Hinode and STEREO in 2006 and of SDO (Solar Dynamics Observatory) shortly, this conference provided a platform for in-depth discussions on new results from various space missions as well as a comparison with ground-based observing facilities such as the Swedish 1-m Solar Telescope. Using sophisticated image processing techniques, such telescopes routinely generate observations with a resolution better than 0.1 arcsec, thereby yielding more informative diagnostics for instance of the microstructure of flux tubes. Advanced spectropolarimetric techniques enable a more reliable determination of the magnetic field topology in internetnetwork regions, which is crucial for understanding the nature of surface magnetism. Numerical simulations are achieving a level of maturity that allows a more realistic depiction of magnetoconvection, magnetic field extrapolation, CMEs, flares, coronal heating, and so on.

Spread over four days, the meeting consisted of nine scientific sessions on the following:

1. Magnetic field generation
2. Magnetoconvection and transport
3. Local seismology
4. Magnetic field measurement

5. Sunspots and active regions
6. Quite sun and polar regions
7. Magnetic coupling through the atmosphere
8. Eruptive phenomena
9. Solar-stellar connections

The articles in these proceedings are separated out into contributions based on oral and poster presentations. In addition, short summaries have been included of the work presented at the meeting but published elsewhere. It is gratifying to note that 136 persons attended the conference, which brought together many researchers and students from all over the world, particularly several leading international experts who presented reviews on a wide range of topics.

I am grateful to the SOC for helping to put together an exciting scientific programme and also to the session chairs, speakers, and authors for a timely submission of their articles in this book. The meeting owes its success to the efforts of the LOC, to the excellent support provided by the administrative and technical staff at the Indian Institute of Astrophysics, and to the student volunteers at the Institute. In addition, I acknowledge the financial support from (a) the Department of Science and Technology, Government of India; (b) Air Force Office of Scientific Research, Asian Office of Aerospace Research and Development (AFOSR-AOARD), Japan; and (c) the Indo-US Science and Technology Forum, New Delhi. Finally, I thank Rob Rutten for volunteering to co-edit these proceedings and for his excellent efforts in their production.

Bangalore, May 2009 *Siraj Hasan*

Contents

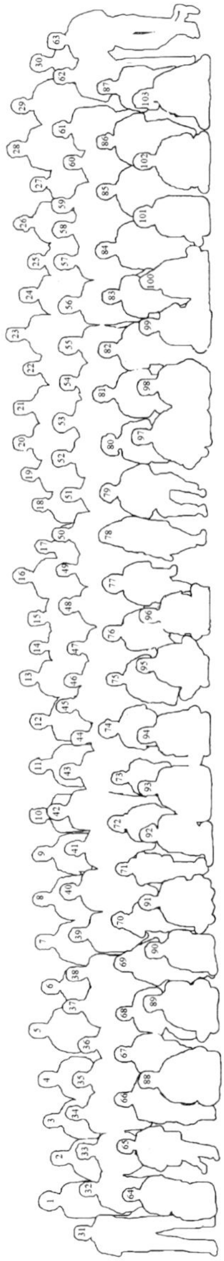

1	Robert J. Rutten	35	K. M. Hiremath	69	Vinod Krishan
2	K. B. Ramesh	36	B. S. Nagabhushana	70	Chia-Hsien Lin
3	K. Nagaraju	37	Viggo H. Hansteen	71	Sabita Mathur
4	A. Satyanarayanan	38	Bart De Pontieu	72	Dulce Isabel Gonzalez
5	Vasco Henriques	39	A. R. Choudhuri	73	Piyali Chatterjee
6	Procheta C. V. Mallik	40	John H. Thomas	74	Patrick Antolin
7	David G. Rathinavelu	41	Terry G. Forbes	75	Nandita Srivastava
8	Rohan Eugene Louis	42	Nigel O. Weiss	76	Rebecca Centeno
9	Oskar Steiner	43	S. P. Bagare	77	Ada Ortiz Carbonell
10	Gautam Narayan	44	Ramesh Chandra	78	Elena Khomenko
11	Morten Franz	45	Edwin Ebenezer	79	S. S. Hasan
12	K. Ichimoto	46	P. K. Manoharan	80	Brigitte Schmieder
13	Lokesh Bharti	47	Wahab Uddin	81	Jan Olof Stenflo
14	Girjesh R. Gupta	48	Catherine E. Fischer	82	K. N. Nagendra
15	P. Sreejith	49	Claire E. Parnell	83	Cristiana Dumitrache
16	John W. Leibacher	50	Dana W. Longcope	84	Javier Trujillo Bueno
17	Lalan Prasad	51	Eric Priest	85	Rajmal Jain
18	Ram Ajor Maurya	52	Dibyendu Nandi	86	K. R. Sivaraman
19	Dipankar Banerjee	53	S. Tsuneta	87	H. O. Vats
20	V. S. Pandey	54	Badruddin	88	Craig Deforest
21	S. K. Tiwari	55	Judith T. Karpen	89	Indrajit V. Barve
22	R. Kariyappa	56	K. Shibata	90	S. P. Rajaguru
23	James A. Klimchuk	57	K. Chandrashekhar	91	R. Ramesh
24	Brajesh Kumar	58	S. C. Tripathy	92	Davina Innes
25	A. D. Joshi	59	Bhola N. Dwivedi	93	Jagdev Singh
26	Jorge Sánchez Almeida	60	Doddamani	94	K. Sankarasubramanian
27	Shibu K. Mathew	61	S. M. Chitre	95	D. Lakshmaiah
28	Michal Sobotka	62	Ch. V. Sastry	96	Lakshmi Pradeep Chitta
29	Michal Švanda	63	Ramit Bhattachaiqryya	97	Pankaj Kumar
30	Mats Carlsson	64	Moreswhar H. Gokhale	98	Manabendra Lakhar
31	K. A. P. Singh	65	Paola Testa	99	Malali Sampoorna
32	K. P. Raju	66	P. Venkatakrishnan	100	L. S. Anusha
33	Jürgen H. M. M. Schmitt	67	Douglas O. Gough	101	Swati Routh
34	Bernhard Fleck	68	Anil Bhardwaj	102	B. B. Karak

List of Participants

Antolin, Patrick Kwasan Observatory, Kyoto University, Japan
and
Institute of Theoretical Astrophysics, University of Oslo, Norway,
antolin@kwasan.kyoto-u.ac.jp

Anusha, L. S. Indian Institute of Astrophysics, Bangalore, India,
anusha@iiap.res.in

Badruddin Aligarh Muslim University, Aligarh, India, badr_phys@yahoo.co.in

Bagare, S. P. Indian Institute of Astrophysics, Bangalore, India, bagare@iiap.res.in

Banerjee, Dipankar Indian Institute of Astrophysics, Bangalore, India,
dipu@iiap.res.in

Bellot Rubio, Luis R. Instituto de Astrofísica de Andalucia, Granada, Spain,
lbellot@iaa.es

Bhardwaj, Anil Indian Space Research Organisation, Bangalore, India,
Anil_Bhardwaj@vssc.gov.in

Bharti, Lokesh MPI für Sonnensystemforschung, Lindau, Germany,
bharti@linmpi.mpg.de

Bhattacharyya, Ramit High Altitude Observatory, Boulder, USA,
ramit@ucar.edu

Brun, A. Sacha Service d'Astrophysique, Saclay, France, sacha.brun@cea.fr

Carlsson, Mats Institute of Theoretical Astrophysics, Oslo, Norway,
mats.carlsson@astro.uio.no

S. Chakraborty Department of Theoretical Sciences, S. N. Bose Centre for Basic Sciences, Kolkata-700098, India

Chandra, Ramesh Observatoire de Paris, LESIA, Meudon, France,
Chandra.Ramesh@obspm.fr

Chandrashekhar, K. Indian Institute of Astrophysics, Bangalore, India,
kcs@iiap.res.in

Chatterjee, Piyali Department of Astronomy and Astrophysics, Tata Institute of Fundamental Research, Mumbai-400005, India, piyalic@tifr.res.in

Chia-Hsien Lin Trinity College, Dublin, Ireland, linc@tcd.ie

Chitre, S. M. University of Mumbai, Mumbai, India, kumarchitre@gmail.com

Chitta, L. Pradeep University of Hyderabad, Hyderabad, India, pradeep198621@gmail.com

Choudhuri, Arnab R. Department of Physics, Indian Institute of Science, Bangalore-560012, India, arnab@physics.iisc.ernet.in

De Pontieu, Bart Lockheed Martin, Palo Alto, USA, bdp@lmsal.com

DeForest, Craig E. Southwest Research Institute, Boulder, USA, deforest@boulder.swri.edu

Dumitrache, Cristiana Astronomy Institute, Bucharest, Romania, crisd@aira.astro.ro

Dwivedi, Bhola N. Banaras Hindu University, Varanasi, India, bholadwivedi@gmail.com

Ebenezer, Edwin Indian Institute of Astrophysics, Bangalore, India, eben@iiap.res.in

Elliott Centeno, Rebecca Instituto de Astrofísica de Canarias, La Laguna, Spain, rce@ucar.edu

Erdelyi, Robertus University of Sheffield, Sheffield, UK, robertus@sheffield.ac.uk

Fleck, Bernhard European Space Agency, GSFC, Greenbelt, USA, bfleck@esa.nascom.nasa.gov

Fischer, Catherine E. Sterrekundig Instituut, Utrecht, The Netherlands, C.E.Fischer@uu.nl

Forbes, Terry G. University of New Hampshire, Durham, USA, terry.forbes@unh.edu

Franz, Morten Kiepenheuer Institut für Sonnenphysik, Freiburg, Germany, morten@kis.uni-freiburg.de

Gokhale, M. H. D. P. Road, Kothrud, Pune, India, mhg0506@yahoo.com

Gonzalez-Gomez, Dulce I. Instituto de Geofisica, Mexico City, Mexico, dulce@geofisica.unam.mx

Gopalswamy, Nat NASA Goddard Space Flight Center, Greenbelt, Maryland, USA, Nat.Gopalswamy@nasa.gov

Gosain, Sanjay Udaipur Solar Observatory, Udaipur, India, sgosain@prl.res.in

Gough, Douglas O. Institute of Astronomy, Cambridge, UK,
douglas@ast.cam.ac.uk

Gupta, Girjesh R. Indian Institute of Astrophysics, Bangalore, India,
girjesh@iiap.res.in

Hanasoge, Shravan M. Stanford University, Palo Alto, USA,
shravan@stanford.edu

Hansteen, Viggo H. Institute of Theoretical Astrophysics, Oslo, Norway,
viggoh@astro.uio.no

Hasan, S. S. Indian Institute of Astrophysics, Bangalore, India,
hasan@iiap.res.in

Henriques, Vasco Institute for Solar Physics, Stockholm, Sweden,
vasco@astro.su.se

Hiremath, K. M. Indian Institute of Astrophysics, Bangalore, India,
hiremath@iiap.res.in

Ichimoto, Kiyoshi Kyoto University, Kyotot, Japan,
ichimoto@kwasan.kyoto-u.ac.jp

Innes, Davina MPI für Sonnensystemforschung, Lindau, Germany,
innes@mps.mpg.de

Jain, Rajmal Physical Research Laboratory, Ahmedabad, India,
rajmal@prl.res.in

Jamwal, R. Department of Physics, Barkatullah university, Bhopal, India

Javaraiah, J. Indian Institute of Astrophysics, Bangalore, India,
jj@iiap.res.in

Joshi, Anand D. Udaipur Solar Observatory, Udaipur, India,
janandd@prl.res.in

Joshi, Chandan Mohanla Sukhadia University, Udaipur, India,
chandan-joshi@hotmail.com

Karak, Bidya B. Indian Institute of Science, Bangalore, India,
bidya_karak@physics.iisc.ernet.in

Kariyappa, R. Indian Institute of Astrophysics, Bangalore, India,
rkari@iiap.res.in

Karpen, Judith T. NASA Goddard Spaceflight Center, Greenbelt, USA,
judy.karpen@nasa.gov

Kathiravan, C. Indian Institute of Astrophysics, Bangalore, India,
kathir@iiap.res.in

Khomenko, Elena Instituto de Astrofísica de Canarias, La Laguna, Spain,
khomenko@iac.es

Klimchuk, James A. NASA Goddard Space Flight Center, Greenbelt, Maryland, USA, James.A.Klimchuk@nasa.gov

Krishan, Vinod Indian Institute of Astrophysics, Bangalore, India, vinod@iiap.res.in

Kumar, Brajesh Udaipur Solar Observatory, Udaipur, India, brajesh@prl.res.in

Kumar, Pankaj Aryabhatta Research Institute of Observational Sciences, Nainital, India, pkumar@aries.ernet.in

Lahkar, Manabendra Centre for Radio Astronomy, Department of Physics, Cotton College, Guwahati, Assam, India
and
Radio Astronomy Centre, National Centre for Radio Astrophysics, Tata Institute of Fundamental Research, Mumbai, India, manab.lahkar@gmail.com

Leibacher, John W. National Solar Observatory, Tucson, USA, jleibacher@nso.edu

Longcope, Dana W. Montana State University, Bozeman, USA, dana@solar.physics.montana.edu

Louis, Rohan E. Udaipur Solar Observatory, Udaipur, India, eugene@prl.res.in

Mallik, Procheta C. V. Department of Physics and Astronomy, Glasgow, Scotland, pmallik@astro.gla.ac.uk

Manoharan, P. K. Tata Institute of Fundamental Research, Udhagamandalam (Ooty), India, mano@ncra.tifr.res.in

Mathew, Shibu K. Udaipur Solar Observatory, Udaipur, India, shibu@prl.res.in

Mathur, Savita Indian Institute of Astrophysics, Bangalore, India, smathur@iiap.res.in

Maurya, Ram A. Udaipur Solar Observatory, Udaipur, India, ramajor@prl.res.in

Moestl, Christian Space Research Institute, Graz, Austria, moestich@stud.uni-graz.at

Moradi, Hamed Monash University, Melbourne, Australia, hamed.moradi@sci.monash.edu.au

Nagabhushana, B. S. Indian Institute of Astrophysics, Bangalore, India, bhushan@iiap.res.in

Nagaraju, K. Indian Institute of Astrophysics, Bangalore, India, nagaraj@iiap.res.in

Nagendra, K. N. Indian Institute of Astrophysics, Bangalore, India, knn@iiap.res.in

Nandi, Dibyendu Indian Institute of Science Education and Research, Kolkata, India, nandi@mithra.physics.montana.edu

Narayan, Gautam Institute for Solar Physics, Stockholm, Sweden, gautam@astro.su.se

Nordlund, Åke Niels Bohr Institute, Copenhagen, Denmark, aake@nbi.dk

Ortiz Carbonell, Ada Institute of Theoretical Astrophysics, Oslo, Norway, ada@astro.uio.no

Pandey, Vinay S. Physical Research Laboratory, Ahmedabad, India, vspandey@prl.res.in

Parnell, Clare E. University of St Andrews, St Andrews, Scotland, clare@mcs.st-and.ac.uk

Prasad, L. Kumaun University, Nainital, India, lprasad@iucaa.ernet.in

Priest, Eric R. University of St Andrews, St Andrews, Scotland, eric@mcs.st-and.ac.uk

Rajaguru, S. P. Indian Institute of Astrophysics, Bangalore, India, rajaguru@iiap.res.in

Raju, K. P. Indian Institute of Astrophysics, Bangalore, India, kpr@iiap.res.in

Ramesh, K. B. Indian Institute of Astrophysics, Bangalore, India, kbramesh@iiap.res.iny

Ramesh, R. Indian Institute of Astrophysics, Bangalore, India, ramesh@iiap.res.in

Rathinavelu, David G. Gandhigram Rural University, Gandhigram, India, david_physics@yahoo.com

Ravindra, B. Indian Institute of Astrophysics, Bangalore, India, ravindra@iiap.res.in

Reddy, Eswar B. Indian Institute of Astrophysics, Bangalore, India, ereddy@iiap.res.in

Routh, Swati University of Texas, Arlington, USA, swati_astro@yahoo.co.in

Rutten, Robert J. Sterrekundig Instituut, Utrecht, The Netherlands, R.J.Rutten@uu.nl

Sampoorna, Malali Indian Institute of Astrophysics, Bangalore, India, sampoorna@iiap.res.in

Sánchez Almeida, Jorge Instituto de Astrofísica de Canarias, La Laguna, Spain, jos@iac.es

Sankarasubramanian, K. Indian Space Research Organisation, Bangalore, India, sankark@isac.gov.in

Sastry, Ch. V. Indian Institute of Astrophysics, Bangalore, India, sastrychellury@yahoo.com

Satyanarayanan, A. Indian Institute of Astrophysics, Bangalore, India, satya@iiap.res.in

Schmieder, Brigitte LESIA, Observatoire de Paris, Meudon, France, brigitte.schmieder@obspm.fr

Schmitt, Júrgen H. M. M. Hamburger Sternwarte, Hamburg, Germany, jschmitt@hs.uni-hamburg.de

Scullion, Eamon Armagh Observatory, Armagh, Northern Ireland, ems@arm.ac.uk

Shibata, Kazunari Kyoto university, Kyoto, Japan, shibata@kwasan.kyoto-u.ac.jp

Singh, Jagdev Indian Institute of Astrophysics, Bangalore, India, jsingh@iiap.res.in

Singh, K. A. P. Indian Institute of Astrophysics, Bangalore, India, alkendra@iiap.res.in

Sivaraman, K. R. Indian Institute of Astrophysics, Bangalore, India, siva@iiap.res.in

Sobotka, Michal Astronomical Institute, Ondrejov, Czech Republic, msobotka@asu.cas.cz

Sreejith, P. Indian Space Research Organisation, Bangalore, India, sreejith@isac.gov.in

Srivastava, A. K. Aryabhatta Research Institute of Observational Sciences, Nainital, India, aks@aries.ernet.in

Srivastava, Nandita Udaipur Solar Observatory, Udaipur, India, nandita@prl.ernet.in

Steiner, Oskar Kiepenheuer Institut für Sonnenphysik, Freiburg, Germany, steiner@kis.uni-freiburg.de

Stenflo, Jan Olof Institute of Astronomy, Zurich, Switzerland, stenflo@astro.phys.ethz.ch

Subramanian, K. R. Indian Institute of Astrophysics, Bangalore, India, subra@iiap.res.in

Subramanian, Prasad IISER, Pune, India, p.subramanian@iiserpune.ac.in

Švanda, Michal Astronomical Institute, Ondrejov, Czech Republic, michal@astronomie.cz

Testa, Paola Smithsonian Astrophysical Observatory, Cambridge, USA, ptesta@cfa.harvard.edu

Thomas, John H. University of Rochester, Rochester, USA, thomas@me.rochester.edu

Tiwari, Sanjiv K. Udaipur Solar Observatory, Udaipur, India, stiwari@prl.res.in

Tripathy, Sushanta C. National Solar Observatory, Tucson, USA, stripathy@nso.edu

Trujillo Bueno, Javier Instituto de Astrofísica de Canarias, La Laguna, Spain, jtb@iac.es

Tsuneta, Saku National Astronomical Observatory, Tokyo, Japan, saku.tsuneta@nao.ac.jp

Uddin, Wahab Aryabhatta Research Institute of Observational Sciences, Nainital, India, wahab@aries.ernet.in

Vats, Hari O. Udaipur Solar Observatory, Udaipur, India, vats@prl.res.in

Venkatakrishnan, P. Udaipur Solar Observatory, Udaipur, India, pvk@prl.ernet.in

Vigeesh, G. Indian Institute of Astrophysics, Bangalore, India, vigeesh@iiap.res.in

Weiss, Nigel O. DAMTP, Cambridge, UK, now@damtp.cam.ac.uk

Part I
Preamble

Republication of: Radial Movement in Sunspots

S.S. Hasan

Reproduced below is the original discovery paper on the radial motion in sunspots by John Evershed published in Kodaikanal Observatory Bulletin No. 15, 1909.

(1864 – 1956)

S.S. Hasan (✉)
Indian Institute of Astrophysics, Bangalore, India

S.S. Hasan and R.J. Rutten (eds.), *Magnetic Coupling between the Interior and Atmosphere of the Sun*, Astrophysics and Space Science Proceedings,
DOI 10.1007/978-3-642-02859-5_1, © Springer-Verlag Berlin Heidelberg 2010

Kodaikánal Observatory.

BULLETIN No. XV.

RADIAL MOVEMENT IN SUNSPOTS.

DISPLACEMENTS of the lines of hydrogen and calcium in the neighbourhood of sunspots, indicating violent motions in the line of sight, is a common characteristic of spot disturbances. Such phenomena are frequently observed during periods of active change in spot development, or during the genesis of a spot. These line-shifts rarely affect the spectra of other elements than those of the higher chromosphere. In very violent outbursts, in addition to the hydrogen and calcium lines, those of He, Mg, Na, and some of the enhanced lines of Fe are occasionally seen to be affected, but to a much less extent than those of H and Ca. The displacements may be either an increase or a decrease of wave-length, and may amount to several Ångström units, indicating movements of approach or recession of several hundred kilometers per second. These movements are seldom maintained for more than a few minutes at a time, and are usually to be found in the immediate neighbourhood of spots, rarely within the umbral area.

Recently line-shifts of quite another character have been photographed here. These are apparently permanent, affect a very large proportion of the Fraunhofer absorption lines, and are only found in the penumbræ of spots.

In the course of measurement of a large number of spot spectrum plates obtained here, it was noticed that the position of the lines of the true umbral spectrum seem usually to be almost entirely unaffected by motion in the line of sight. It is true that very slight displacements towards the violet appear to be indicated in a series of measures of some of the best plates, but the amount of the displacement (about 0·004 Å) is scarcely larger than the limits of error of measurement. In these plates the scale is about 1Å = 0·67 mm., and any movements of the solar gases of less than ⅓ km. per second would be too small for detection.

In some spectra obtained this year, in the fourth and fifth orders of a 3·2 inch grating,* in which the scale is 1Å = 1·0 to 2 mm., the displacement of the lines in the umbræ of spots is still barely measurable, but very obvious displacements are found in the penumbræ; and this shift turns out to be of a most interesting character, and is apparently a constant feature of all spots. That it should have escaped detection hitherto is most surprising, considering the high dispersion now employed in all spot spectrum work.

The first plate showing this feature clearly was exposed on January 7, 1909, using the fourth order of the grating. It covers the region λ 4650 to λ 4790, and the scale is 1Å = 1·08 mm. An unusual amount of detail is visible in the umbral spectrum, which is no doubt due to the clearness and steadiness of the air at the

* A Rowland grating of 3·2 inches ruled surface and 14,438 lines to the inch.

time of exposure. The general appearance of the Fraunhofer lines crossing the spot is represented in the accompanying diagram

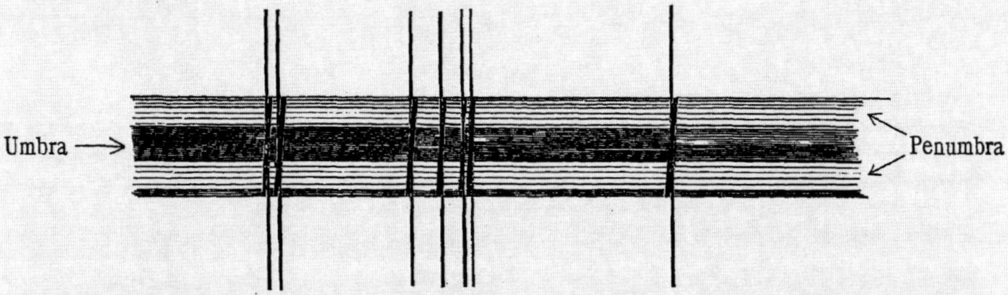

in which the displacements are somewhat exaggerated, to show their character. The lines appear to be about equally displaced, but in opposite directions, on the two sides of the spot. There are, however, considerable differences among the different lines, the stronger lines appearing less affected than the fine narrow ones.

A preliminary measurement of ten of the best defined lines gives a mean total displacement, measured at the outer edges of the penumbra, of 0·027 Å, indicating a receding velocity on the north-west side of the spot of 0·86 km. per second, and an approaching velocity of the same amount on the south-east side, the position angle of the north end of the spectrograph slit being 314°. The spot photographed was in latitude 9° N. and was 31° W of the central meridian.

The appearance of the lines in this photograph at once suggested a rotation of the absorbing gases in the spot, not a vortex movement but a rotation of the spot as a whole about a point at its centre. The lines seem quite straight over the spot, but inclined one or two degrees to the undisturbed lines.

The hypothesis of circular motion of any kind has proved, however, to be certainly untenable. From an examination of about 150 spectra obtained since January 7 and representing seven spots in the northern hemisphere and four in the southern, the following statements may be made :—

(1) All the spots examined show line-shifts of about the same order of magnitude, when at the same distance from the centre of the disc.

(2) The displacements disappear when the spot is within 10° of the centre.

(3) The displacements are most evident when the spot is between 30° and 50° from the centre of the disc, but are difficult to photograph when quite near the limb.

(4) The displacements are of opposite sign on opposite sides of the central meridian when the slit is parallel to the solar equator.

(5) The displacements are invariably towards the violet on the preceding side of a spot, towards the red on the following side, when the spot is east of the central meridian; the reverse when west.

(6) Southern spots show the same direction of movement as northern.

(7) No displacements are observed when the slit bisects a spot in a direction at right angles to a line joining the spot and the centre of the sun's disc.

A hypothesis which seems in harmony with all the facts here stated, is one which attributes the displacements to a radial movement outwards from the spot-centre. The motion must be essentially horizontal, or parallel to the sun's surface. This is shown by the total disappearance of the line-shifts when the spot is near the centre of the disc. The hypothesis of a vortex, or rotation of any kind, about an axis perpendicular to the sun's surface, is negatived by the fact stated in paragraph 7, for it is evident that for a circular movement a nodal point should be found when the slit bisects the spot in a direction passing through the centre of the sun's disc, the maximum displacement occurring in a direction at right angles to this. This direction of the node, however, differs from that actually found by about 90°.

To obtain evidence on this crucial point, advantage is taken of the rotation of the solar image due to the action of the heliostat. Thus, in the interval between 8 A.M. and 5 P.M. the position angle of the slit, which is fixed truly vertical, changes through about 100°; the north end of the slit passing from P.A. 300° to 40° and being at 360° at solar noon. Different sections through a spot can therefore be obtained by simply taking a succession of photographs at intervals throughout the day. Now it has invariably been found that the displacements diminish in amount from 8^h up to about 11^h when they generally disappear entirely, reappearing with opposite signs in the afternoon; but the time of no displacement does not occur exactly at solar noon, and the evidence so far obtained indicates that the line of nodes coincides with the direction at right angles to the line joining the spot and the centre of the disc, as it should do if the motion is radial.

The whole of the evidence obtained up to the time of writing, is presented in the accompanying table. In this all the photographs obtained are entered in the order in which they were taken, with the exception of a few duplicate exposures made on the same spot at the same time, which are omitted to save space. The third column gives the calculated position angle of the north end of the spectrograph slit; the fourth column gives the reference number of the spot; the succeeding columns give the approximate latitudes, and the longitudes reckoned east or west of the central meridian. These co-ordinates were read off from the daily spot-charts prepared at this observatory. The column headed "Shift at north side of spot" indicates by the letters V and R the direction of the displacement, whether to violet or red, and the small letters following give a rough idea of the amount of displacement, viz., 1 = large, m = medium, s = small, and tr = a just appreciable trace.

The following points are evident from the table :—

When the position angle of the north end of the slit is between 297° and 330°, eastern spots always give a violet shift on the north side, western spots a red shift.

Between 330° and 360°, the shift is either absent, or very minute in either direction.

From P.A. 20° to 70°, the shifts are, with one doubtful care, to the red for an eastern spot, and to the violet for a western.

Finally, spots near the centre show no shift at all.

Since the position angle of the slit when no displacements are observed is of fundamental importance in determining the character of the movement, it may be well to refer in detail to the photographs bearing on this point Take first the eastern spot, No. 1595, on January 27: six photographs of a central section of the spot were obtained at intervals during the day. The exposure times, position angle of the slit, and observed direction of shift are here repeated as follows :—

Exposure time.	P.A. north end of slit.	Shift at north side of spot.
H. M.	°	
8 4	299	V l
8 57	312	V s
11 27	346	R tr
11 58	352	R s
15 45	43	R m
16 7	48	R l

The change from a violet to a red shift occurs between 8^h 57^m and 11^h 27^m, when the north end of the slit was between position angles 312° and 346°. Assuming the mean angle, the change would have occurred at P.A. 329°, or 31° to the west of north. The position angle of the spot, measured from the centre of the disc, was 60°, *i.e.*, 60° to the east of north; therefore, the angle between the line of nodes and the direction of the centre of the disc would be $60° + 31° = 91°$.

Again, take the western spot No. 1591 on January 26. This was photographed four times, and the change from a red to a violet shift occurred between 12^h 2^m and 14^h 27^m, or between P.A. 353° and 25°: the mean angle between these is 9° to the east of north, and the position angle of the spot was 88° west of north, so that the angle between the line of nodes and the centre of the disc is 97°. I summarize below all the instances where the shift was observed on both sides of the line of nodes :—

Date.	Spot number.	Position angle of spot.	Position angle of node.	Angle between node and centre of disc.
		°	°	°
January 18	1588	82 E	20 W	102
Do. 18	1591	74 E	2 W	76
Do. 20	1591	60 E	23 W	83
Do. 25	1594	87 E	8 W	95
Do. 25	1591	84 W	0	84
Do. 26	1591	88 W	9 E	97
Do. 27	1595	60 E	31 W	91
				Mean 90

The estimate of the time when there is no shift must necessarily be extremely uncertain considering the small amount of the displacements, even when at their maximum, and the accordance in the above table is really remarkable, and testifies to the sensitiveness of the eye to very minute deviations from straightness in the spectrum lines.

It may be well to mention here that tests of the reality of these minute line-shifts have been applied in two ways. First, by photographing the spot spectrum in the red, in the region including the well-known group of solar and telluric lines used by Dunér and Halm in their determinations of the solar rotation. In this group, the telluric oxygen lines are very narrow and well-defined, whilst the solar iron lines are comparatively broad, and are indeed much widened in spots.* Since the displacements are much more conspicuous in the finer lines of the spectrum, the oxygen lines should show the effect strongly, if it is spurious. In the photographs, however, they are found to be absolutely straight, whilst the iron lines beside them are bent.

The second test was applied at a time when the slit was approximately in the position angle of the line of nodes for a spot which was east of the central meridian, and photographs of a central section of the spot showed no shift. Photographs were then obtained of the preceding and of the following edges of the penumbra. Here the line-shift should be at a maximum, and, as expected, the lines were found to be evenly bulged, to the violet on the preceding edge and to the red on the following edge.

It is somewhat disappointing, perhaps, that the hypothesis of a radial movement in spots, which is so strongly supported by these observations, seems entirely out of harmony with the splendid discovery of the Zeeman effect in sunspots made by Professor Hale. This seems to demand a vortex, or at any rate a circular movement in sunspots; and it was only after a considerable amount of evidence had accumulated, that the preconceived conviction that the motion must be circular was abandoned. A consoling feature of this new theory of spot movement is that it seems to explain the radial structure of the filaments in the penumbræ of spots, and the radial disposition of the calcium flocculi immediately surrounding a well-developed symmetrical spot. It also harmonizes with the well-known tendency of the principal spots in a group to separate, the leader advancing, and the follower receding.

A difficulty should also be mentioned. When the slit centrally bisects a symmetrical spot, in a direction approximating to that giving the greatest shift, the displaced lines appear quite straight, as before mentioned, and inclined to the undisturbed lines, the greatest shift occurring at the outer limits of the penumbra. This seems to imply accelerating movement, from the centre of the spot outwards, yet at the limits of the penumbra the motion apparently ceases abruptly. It is hoped that further research will throw light on this and other obscure points.

* The Fe line at 6802·709 is seen to be beautifully doubled in the third order spectrum of all spots examined by me.

KODAIKÁNAL,
3rd February 1909.

J. EVERSHED.

LINE SHIFTS IN SPOT SPECTRA.

Date.	Hour.	P.A. North end of slit.	Number of spot.	Latitude. +	Latitude. −	Longitude. E.	Longitude. W.	Shift at north side of spot.	Remarks.
1909.	H. M.	°	°	°	°	°	°		
January 5.	11 11	344	1578	7	...	...	27	R, s	Definition poor.
	33	348	—	7			27	R, s	
	16 0	348	—	7			30	V	
Do. 7.	8 47	310	1580	9	...	...	32	R, l	
	9 45	324	—	9			32	R, l	
	10 15	331	—	9			32	R, m	
	11 11	343	—	9			32	R, m	
	33	348	1578	7			58	R, m	
Do. 8.	9 5	315	1580	9	...	...	44	R, l	
	11 0	341	—	9			44	R, s	
	10 24	333	1578	8			66	R, s	
Do. 9.	8 44	310	1580	10	...	...	58	R, l	
	10 25	333	—	10			58	R, s	
Do. 11.	10 18	331	1588	10	...	20	...	—	No certain shift either way.
	45	337	—	10		20		—	
Do. 15.	9 30	321	1588	9	...	...	32	R, l	
	49	324	—	9			32	R, l	
	10 0	327	—	9			32	R, l	
	32	333	1588		2	63		V, l	
	37	335	1590		12	75		V, s	Very small spot.
Do. 16.	8 8	301	1588	...	3	50	...	V, l	
	51	311	1583	9			45	R, l	
	9 57	326	1590		13	61		V, s	Distinct, but a very small spot.
	10 1	327	1591	7		80		V, s	
	4	328	1588		3	50		V, l	
Do. 17.	8 10	301	1588	...	4	37	...	V, l	
	14	302	1583		7		58	R, l	
	20	308	1591		7	66		V, l	
	10 9	327	1590		13	48		V, s	H and K region.
	15 28	40	1588		4	35		R, s	
	32	40	1588	7			58	V, s	
	37	41	1590		14	46		—	Poor definition.
Do. 18.	7 57	298	1588	...	3	24	...	V, s	
	8 24	304	1591	7		53		V, m	
	29	306	1583	9			75	—	Spot poorly defined.
	9 3	313	1588		3	24		V, tr	
	10	315	1591	7		53		V, s	
	10 56	340	1588		3	24		—	Good definition, no shift.
	11 6	341	1583	9			75	R, l	
	15 30	40	1588		3	21		R, tr	
	34	41	1591	7		50		R, s	Poor definition.
	40	42	—	7		50		R, l	Fair definition.
	57	46	—	7		50		R, l	
	16 2	47	1588		3	21		R, tr	
	7	48	1591	7		50		R, m	
	16	51		7		53		R, m	
Do. 19.	9 12	316	1591	7	...	38	...	V, m	
	16	317	1588		3	10		—	No shift.
	52	325	1591	7		38		V, s	
	56	325	1588		3	10		V, tr	
	29	326	1591	7		38		V, s	
Do. 20.	8 16	303	1591	7	...	25	...	V, s	
	20	303	1588		3		8	—	No shift.
	29	306	1591	7		25		V	Poor focus.
	10 15	330	—	7		25		—	No shift.
	48	337	—	7		25		—	
	11 46	350	—	7		24		R, tr	
	15 11	36	—	7		23		R, tr	Spot poorly defined.
	16 23	52	—	7		23		R, tr	
Do. 21.	7 58	298	1591	7	...	14	..	V, tr	
	8 2	299	1588		4		16	R, tr	
	51	311	1591	7		14		—	No shift.
	42	309	1588		4		16	—	

2

Date.	Hour.		P.A. North end of slit.	Number of spot.	Latitude.		Longitude.		Shift at north side of spot.	Remarks.
					+	−	E.	W.		
1909.	H.	M.	°	°	°	°	°			
January 21—*cont.*	14	58	33	1591	7	...	12	...	R, tr	
	15	27	39	1588		3		17	V, s	
Do. 22.	8	4	300	1588	...	3	...	29	R, s	
		28	305	1591	10		1		—	No shift.
		42	309	1590*a*		13		20	—	} Very small spots.
		48	310	1590*b*		17		17	—	
Do. 23.	7	57	298	1588	...	4	...	41	R, m	
	8	6	300	1591	6			12	R, s	
		22	304	1590		13		32	—	No shift.
		30	306	1588		4		41	R, m	
		43	309	1593	8		73		V, m	
Do. 24.	8	39	308	1588	...	4	...	55	R, m	
		42	309	1591	6			24	R, m	
		52	311	1593	8		61		V, m	
	10	47	337	1591	6			24	R, m	
	11	13	343	1592		16	46		V, m	
Do. 25.	7	58	297	1594*a*	...	8	53	...	V, l	
		57	298	1594*b*		9	58		V, l	
	8	2	300	1591	6			37	R, l	
	9	17	317	1594*b*		8	57		V, l	Following spot.
		19	317	1594*a*		8	53		V, l	Preceding spot.
		22	318	1593	9		47		V, l	
		26	319	1595	10		66		V, l	
		56	325	1595	10		66		V, m	
		59	326	1593	10		47		V, s	
	10	4	327	1592		15	35		V, m	
		8	328	1591	6			37	R, l	
	14	20	24	1591	6			39	V, s	
		28	25	1593	9		46		R, s	
		29	25	1594*b*		9	58		R, s	Following spot.
	17	28	67	1593	9		43		R, s	
		31	68	1595	10		66		R, m	
		37	69	1594*a*		8	58		R, m	Preceding spot.
		42	70	1591	6			40	V, l	
Do. 26.	7	55	297	1594*a*	...	9	38	...	V, m	Preceding spot.
		58	298	1594*b*		9	46		V, m	Following spot.
	8	4	300	1594*a*		9	38		V, m	Preceding spot.
		7	300	1595	9		52		V, m	Plate under-exposed.
	10	23	331	1594*a*		9	38		V, l	Preceding spot.
		28	333	1594*b*		9	46		V, l	Following spot.
		31	333	1593	9		34		V, s	
		34	334	1595	9		52		V, s	
		38	335	1591	6			52	R, l	Shift greater on south side.
	11	55	352	1594*a*		9	37		V, s	Preceding spot.
		58	352	1595	9		51		R, s	
	12	2	353	1591	6			54	R, s	
	14	19	23	1595	9		50		R, m	
		27	25	1591	6			54	V, s	
	16	20	51	—	6			55	V, m	
		28	52	1595	9		50		R, m	} Spot poorly defined.
	17	26	67	—	9		49		R, l	
Do. 27.	8	4	299	1595	9	...	39	..	V, l	
		57	312	—	9		39		V, s	
		59	313	—	9		39		V, s	Following edge of penumbra.
	9	2	313	—	9		39		V, tr	Preceding edge.
		4	313	—	9		39		R, tr	Following edge.
	11	25	345	—	9		38		R	Do.
		27	346	—	9		38		R, tr	Centre of spot.
		30	346	—	9		38		V	Preceding edge of penumbra.
		57	352	—	9		38		V	Do.
		58	352	—	9		38		R, s	Centre of spot.
	12	0	353	—	9		38		R	Following edge of penumbra.
	15	45	48	—	9		36		R, m	Centre of spot.
	16	4	47	1596	10		60		R, l	
		7	48	1595	9		35		R, l	
		12	49	1591	6			70	V, l	

In the fourth column *a* and *b* refer to the preceding and following members of a group.

ADDENDUM—*February* 23.

Since these observations were made a device has been added to the spectrograph by means of which the sun's image can be rotated on the slit through 90°. The results obtained in this way leave no doubt whatever as to the position of the line of nodes, for wherever a spot may happen to be situated on the disc, outside the limit of 10° from the centre, where the shift becomes inappreciable the motion is always found to be greatest when the slit bisects a spot in the direction of the centre of the disc, and to disappear altogether when the slit is at right angles to this direction.

The greatest shift observed with the slit in the direction of the centre of the disc was on a plate exposed on February 4, representing a spot in longitude 41° west, and latitude 7° north. The spectrum includes 100 units between λλ 4782 and 4885, and all the lines in this region show the motion shift, with the probable exception of H_β, which is as usual narrowed in the spot, but seems otherwise unaffected. All the lines are not equally affected, but a large proportion are inclined over the spot a little more than one degree compared with the normal direction of the undisturbed lines. This corresponds to a maximum displacement at the outer edges of the penumbra, on each side, of 0·017 Å, indicating a velocity of approach on the following side and recession on the preceding side of the spot of 1·05 km. per second. But this is the component of velocity in the direction of the earth. The actual velocity parallel to the sun's surface is found by dividing the observed velocity by the sine of the angular distance of the spot from the centre of the disc, which was 42°. Thus, 1·05/0·67=1·57 km. per second. This represents for the majority of the lines the maximum speed attained by the projected matter on reaching the outer limits of the penumbra. Some of the lines, however, indicate a speed of over 2 km., and others under 1 km. per second, whilst it is possible, from the behaviour of H_β that the elements in the higher chromosphere do not share in the movement.

MADRAS: PRINTED BY THE SUPERINTENDENT, GOVERNMENT PRESS—1909.

Opening Remarks

E. Priest

I would like to add my warm welcome to this meeting on behalf of the Scientific Organizing Committee. I am delighted that so many of the world's leading solar physicists have agreed to attend this meeting.

My last meeting to Bangalore was in 1989, almost two solar cycles ago, together with Nigel Weiss and Terry Forbes. At exactly the same time of year, Vinod Krishan brilliantly organized an IAU Colloquium and I remember well her grace and her graciousness, qualities that are common to many Indians and something that I really appreciate about this country.

Indeed, the India I love stands for a civilized society, one in which we can all learn from each other and in which we can live at peace. Here, this week we can all play our small part in encouraging tolerance and good will, and in not allowing the violence last week in Mumbai to win.

You will realize I am sure that India has a superb international reputation in solar physics and has had a host of famous world-class scientists, including Chandrasekhar, Bappu, Kumar Chitre, and more recently Siraj Hasan and Arnab Choudhuri.

There are many natural links between the UK and India. We heard from Siraj how the British built the Kodaikanal Observatory, and indeed Lord Kelvin from Scotland had been a strong voice in favor of that decision. We also heard how Evershed came here, but what he did not tell you is that Siraj Hasan himself came to the UK for his PhD under a formidable young lady (Carol Jordan), and I remember well his visit to St. Andrews in 1986.

The Sun has many secrets, which he is challenging us to unravel here – about the intimate and subtle way that the magnetic field and plasma are coupled and the way that the solar interior and atmosphere are linked by the magnetic field. We have a cope of breakthroughs in understanding the incredible solar magnetic world that Evershed first glimpsed – but only if we do several things: if magnetic theorists and observers work together and genuinely listen to one another; if we appreciate the different gifts and experiences of our rich field – the Sun is so complex that we need each other for a full understanding, and only if the old and young communicate well

E. Priest (✉)
School of Mathematics and Statistics, St Andrews University, Scotland, UK

S.S. Hasan and R.J. Rutten (eds.), *Magnetic Coupling between the Interior and Atmosphere of the Sun*, Astrophysics and Space Science Proceedings, DOI 10.1007/978-3-642-02859-5_2, © Springer-Verlag Berlin Heidelberg 2010

with each other – so, if you are one of the more experienced solar physicists, please take time to meet the many bright young scientists here this week.

In conclusion, I look forward with great excitement and an air of anticipation to hearing marvelous talks and learning much more about our Sun's amazing magnetic field.

Solar Physics at the Kodaikanal Observatory: A Historical Perspective

S.S. Hasan, D.C.V. Mallik, S.P. Bagare, and S.P. Rajaguru

1 Background

The Kodaikanal Observatory traces its origins to the East India Company, which started an observatory in Madras "for promoting the knowledge of astronomy, geography, and navigation in India." Observations began in 1787 at the initiative of William Petrie, an officer of the Company, with the use of two 3-in achromatic telescopes, two astronomical clocks with compound pendulums, and a transit instrument. By the early nineteenth century, the Madras Observatory had already established a reputation as a leading astronomical center devoted to work on the fundamental positions of stars, and a principal source of stellar positions for most of the southern hemisphere stars. John Goldingham (1796–1805, 1812–1830), T.G. Taylor (1830–1848), W.S. Jacob (1849–1858), and Norman R. Pogson (1861–1891) were successive Government Astronomers who led the activities in Madras. Scientific highlights of the work included a catalogue of 11,000 southern stars produced by the Madras Observatory in 1844 under Taylor's direction using the new 5-ft transit instrument.

The observatory had recently acquired a transit circle by Troughton and Simms, which was mounted and ready for use in 1862. Norman Pogson, a well known astronomer whose name is associated with the modern definition of the magnitude scale and who had considerable experience with transit instruments in England, put this instrument to good use. With the help of his Indian assistants, Pogson measured accurate positions of about 50,000 stars from 1861 until his death in 1891. During this period, two total eclipses and one annular eclipse of the Sun were visible from India. Pogson led teams to all three of them. The first one of these, a total eclipse on 18 August 1868, created an enormous interest amongst European astronomers and preparations for its observation were made in England and France for many months preceding the event. Teams of professional astronomers from both countries arrived in India and established their camps at Guntoor, on the central line of the eclipse. The Madras Observatory astronomers had their camp at Masulipatam

S.S. Hasan (✉), D.C.V. Mallik, S.P. Bagare, and S.P. Rajaguru
Indian Institute of Astrophysics, Bangalore, India

S.S. Hasan and R.J. Rutten (eds.), *Magnetic Coupling between the Interior and Atmosphere of the Sun*, Astrophysics and Space Science Proceedings,
DOI 10.1007/978-3-642-02859-5_3, © Springer-Verlag Berlin Heidelberg 2010

and Vunpurthy further east. This eclipse is of great historical significance as it was the first time when spectroscopes were used during an eclipse event. A new line close to the D_2 line of sodium and to the left of it was seen in the spectrum of the chromosphere. Pogson's hand-drawing of the line is shown in Fig. 1.

The discrepancy in the wavelength of this line with the sodium line was confirmed by Norman Lockyer, who could not ascribe it to any known terrestrial element. This observation, in fact, marks the discovery of helium. During the same eclipse, observations of the hydrogen Balmer lines in the spectra of the prominences established their gaseous nature.

Historically, the eclipse of 1868 is an important landmark associated with the birth of solar physics in India. Janssen and Lockyer made effective use of the spectrograph to show that "red prominences could at any time be examined, without waiting for an eclipse at all." During the annular eclipse of 6 June 1872, which was

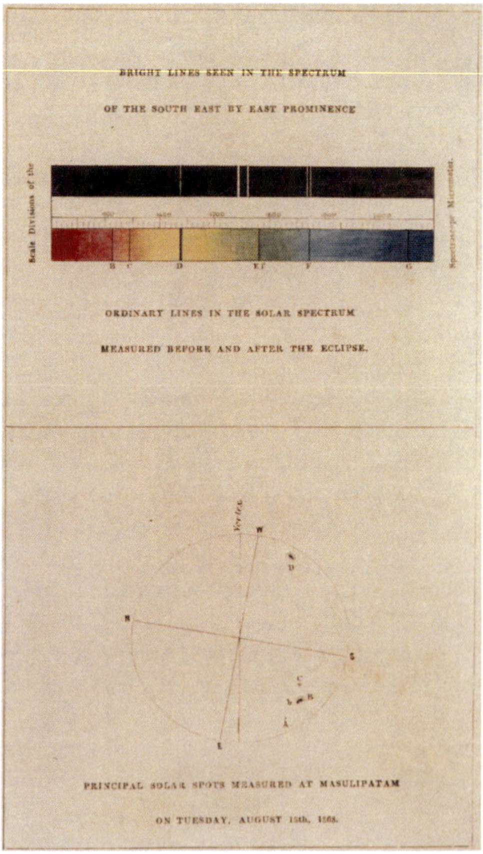

Fig. 1 Hand colored sketch of the solar spectrum recorded during the total solar eclipse at Masulipatam, India, on 18 August 1868 (from IIA archives)

visible from Madras, Pogson found the bright chromospheric spectrum flash out for a short duration on the formation and again at the breaking up of the annulus. This is the first observation on record of viewing the flash spectrum at an annular eclipse.

In May 1882, Pogson proposed a 20-in telescope to augment the needs of his observatory. The proposal had received active support both in India and Britain and the search for a suitable location in the southern highlands of India was authorized. The task was entrusted to Michie Smith, Professor of Physics at the Christian College, Madras, who had arrived in India in 1877. Michie Smith undertook a survey of the Palni and Nilgiri Hills in 1883–1885, his observations covering the twin requirements of transparency and steadiness of image during both day and night. However, there was a delay in the project being cleared by the Astronomer Royal in England as he felt "saddling Pogson with additional work connected with the new large equatorial, when he has accumulated large arrears in observations, would not be desirable".

In 1879, a Committee was appointed by the British Government, consisting of some leading astronomers in the country, to consider and advise on the methods of carrying on observations in solar physics. One of the tasks of the Committee, which came to be known as the Solar Physics Committee (hereafter, SPC), was to reduce the solar photographs taken daily since 1878 in Dehradun with a photoheliograph. The British Government in India had supported the work in the belief that a study of the Sun would help in the prediction of the monsoons, their success and failure, the latter often leading to famines that caused such a havoc. SPC also suggested to the Government of India "that photographs of the Sun should be taken frequently in order that India might assist towards securing a permanent record of the number and magnitude of the sunspots and other changes in the solar surface." At a later date, they added that special spectroscopic observations should also be taken in India in order to collect evidence that will probably throw light on the constitution of the Sun. In 1885, the Royal Society, London, constituted the Indian Observatories Committee comprised of the Astronomer Royal, and a few Fellows of the Royal and Royal Astronomical Societies.

The Committee was entrusted with the task of coordinating the work of Madras and Bombay Observatories and of advising the Secretary of State for India pertaining to the administration of these observatories. The suggestions of SPC regarding regular observations of the Sun and the thinking of the Indian Observatories Committee on the future orientation of activities of the Madras Observatory converged to pave way for the creation of a new observatory in the hills of South India. After the death of Norman Pogson on 23 June 1891, a series of initiatives were taken that eventually led to the establishment of the Solar Physics Observatory in Kodaikanal, a change in the administrative control of the observatories, and the relegation of the Madras Observatory to a secondary status. It took about 9 years to effect this transition.

2 Birth of the Kodaikanal Observatory

In October 1891, John Eliot, the Meteorological Reporter to the Government of India wrote to the Government of Madras about a rethinking of the status of the Madras Observatory by the Secretary of State, Government of India, on the advice of the Astronomer Royal. They felt it was desirable that, in place of the Madras Observatory, a large Astronomical Observatory should be maintained in India, the work at which should, as defined by the Astronomer Royal in a letter dated 14 April 1883, consist of (1) astronomical work proper, including the determination of accurate positions of the Sun, certain fundamental stars, the moon and planets, and (2) astronomical physics, more especially, daily observation of solar prominences and other solar phenomena with the spectroscope, and daily photography of the Sun. Eliot suggested that the work on the Sun, as recommended by the SPC, should get high priority and there was a need to relocate the observatory to a suitable hill-station, where the climate was nicer and more conducive to work. He further suggested that changes and additions of the European staff would be required for the observatory to carry out both solar physics and other astronomical work and changes in the control and inspection of the observatory would be necessary to ensure its working in the way desired by the Secretary of State, Government of India and the Observatories Committee in England.

A serious effort was then made to find a suitable location for an observatory. The regions chosen for further exploration were Kotagiri in the Nilgiris and Kodaikanal in the Palnis. On Pogson's death, The Government of Madras directed Michie Smith to conduct the site survey at Kodaikanal and Kotagiri. In August 1892, Michie Smith submitted to the Government of Madras a detailed report on the relative merits of Kodaikanal, Kotagiri, and Madras as sites for an astronomical observatory. He had based the report on extensive observations of (1) the Sun with respect to faculae, spots, and their spectra, (2) the Venus and Saturn, (3) the star clusters and nebulae, (4) the double stars, and (5) the measurements of apparent magnitudes of stars and (6) the photographs of star trails. The steadiness of the image of the Sun was also studied. On all counts, Kodaikanal appeared to be the best of the three sites. It also had the advantage of altitude, the spot chosen there by Smith was more than a 1,000 ft higher than the highest spot in Kotagiri.

The report was considered by the SPC in July 1892 and by the Indian Observatories Committee under the Chairmanship of Lord Kelvin in October of the same year. SPC suggested that if funds were not available for equipping a complete new observatory with two branches, the furtherance of solar physics in India would be the greater benefit to science. Continuity of the photographic registration of the sunspots in India had to be maintained as it was only with the Indian cooperation the then nearly complete record had been secured. Both the committees endorsed the appointment of Michie Smith on a more permanent basis to superintend the work at the new observatory. It was also said that the magnetic and meteorological observations could continue at Madras, which would become a branch observatory with a superintendent and a small staff. It was asserted that with these changes, the Southern India Astronomical Observatory would become the only Government Astronomical and Solar Physics Observatory in India. It would therefore be a national and not a provincial institution. The committees felt it was essential that the control of the new institution should go to the Imperial Government.

In a gazette notification dated 21 November 1893, it was stated that Kodaikanal afforded the most suitable site for the proposed Solar Physics Observatory. The Meteorological Reporter suggested that the consideration of the question of the permanent site of the Astronomical Observatory be deferred for some 5 years. In 1894, Michie Smith's position as the Government Astronomer became permanent. With the official sanction and notification, actual planning of the observatory started. Michie Smith was allowed additional duty leave to spend time in England from 17 May to 9 October 1895 to enable him to discuss with the astronomers in Europe many issues pertaining to the setting up of the observatory. On 17 July 1895, Michie Smith laid before the Indian Observatories Committee his plans for the proposed Kodaikanal Observatory and the equipment. The Committee strongly recommended acquiring a solar spectroscope. The other major equipment, already collected in Madras from various sources, that had to be shifted to Kodaikanal, consisted of a photoheliograph, a spectrograph, a 6-in Cooke Equatorial, a transit telescope and chronograph, and a 6-in Lerebour and Secretan equatorial. In October 1895, the foundation stone of the new observatory was laid by the Governor of Madras. By the end of 1896, plans and estimates of the new establishment were completed and sanctioned. In July 1897, Michie Smith visited Kodaikanal and laid out a North–South line for the observatory building. The building work was slow to begin with but picked up later. When the large spectrograph – a polar siderostat with an 11-in mirror, a 6-in lens of 40-ft focal length and a concave grating mounted on Rowland's plan – was received in 1897, it had to be temporarily housed in the library building.

On 22 January 1898, there was another total solar eclipse at Sahdol, in the princely state of Rewa. The Secretary of State invited the Astronomer Royal and Sir Norman Lockyer to visit and also report on the various Indian observatories. The Astronomer Royal, W.H.M. Christie, visited Madras early in February (1898) on his way home from his eclipse camp at Sahdol. Christie spent 2 days at the Observatory and carefully went into all questions concerning the equipment and the work that was being carried on. He then visited Kodaikanal and discussed the plans of the new buildings. Various changes which would improve the buildings were suggested and were at once adopted. Sir Norman Lockyer visited Madras but not Kodaikanal and expressed his dissenting views on the planned buildings in Kodaikanal. On his return to England, Lockyer represented to the Secretary of State for India that the buildings were too costly and too permanent, while the structures in his own South Kensington observatory were temporary and cheaper, although his observatory was the most powerful solar observatory in the world. The Secretary of State ordered stoppage of all work at Kodaikanal until the matter was sorted out with the Astronomer Royal. Whether Lockyer's protest was forwarded to the Indian Observatories Committee is not known, but the result was a delay from June to October end, after which the buildings were allowed to go on. Books and instruments were transferred from Madras to Kodaikanal and sent up the ghaut in the dry weather before end of March 1899. About 1,000 coolie loads reached Kodaikanal. No damage was detected. Michie Smith took up residence in Kodaikanal by the end of February 1899, as it was necessary for him to be there to advise the engineer in charge of arrival.

Officially, the observatory started work on 1 April 1899. After the work of construction was largely completed by December 1901, additional measures were taken to protect the Observatory from very strong winds. Progress was made in planting oak and pine trees and laying out grounds, although it was known that many years would elapse before these would take their full effect in modifying the strength of the winds to which the Observatory was exposed. In 1902, a Ca II K spectroheliograph was ordered from Cambridge Scientific Instrument Company. It arrived in Kodaikanal in 1904. The spectroheliograph consisted of a 12-in triple achromatic lens of 20-ft focal length fed by a Foucault siderostat incorporating an 18-in aperture plane silver-on-glass mirror, fitted with differential electric motors for fine adjustments in R.A. and Dec. The design of the instrument was as specified by George Ellery Hale (the discoverer of the spectroheliograph), in which the image of the Sun and the camera remain stationary while the collimating lens, the camera lens, the prisms, and slits are carried on a rigid frame which moves at right angles to the optical axis. This was the instrument that John Evershed put to effective use after his arrival in Kodaikanal in 1907. Evershed joined on 21 January 1907 as Chief Assistant to the Director. Michie Smith went on combined privilege leave and furlough for 9 months from 1 April. Naturally, Evershed took charge of the observatory as soon as he arrived. This was his first professional appointment as an astronomer.

3 John Evershed (1864–1956)

John Evershed was born at Gomshall in Surrey on 26 February 1864. He is a descendant of the ancient family of Evereshavedes in Surrey who can be traced back to before the Norman conquest (1066 A.D.). They were yeoman farmers and Evershed's grandfather was the last of the branch to carry on an unbroken family tradition as yeoman farmers in Surrey for over 600 years. Evershed's mother's side of the family was in Portsmouth where his maternal great grandfather had a shop selling nautical almanacs and charts. He was well known to Lord Nelson.

Evershed's interest in astronomy began rather early. When he was only 11, there was a partial eclipse of the Sun, visible from Surrey, and boy Evershed ran almost all the way from Gomshall to Shere to watch the eclipse using a telescope belonging to a doctor. Even before that when he was six, his imagination was spurred by a picture of German shells falling in the streets of Paris in 1868, during which Janssen, the French astronomer, escaped by balloon from the besieged city to watch the total eclipse of the Sun. It was during this eclipse that Janssen discovered the Hα line in a solar prominence and soon after obtained pictures of the same outside of an eclipse. This was the beginning of the daily spectroscopic observations of prominences without waiting for an eclipse, an activity to which Evershed had a lot to contribute when he grew up.

Evershed was, until his appointment in Kodaikanal, an accomplished amateur astronomer. In the 1890s, when Evershed was already studying solar radiation, he was induced by some great spectroscopists, including Pringsheim and Paschen, to perform experiments on heated gases and study their radiation pattern. As he says "I was able to show that colored vapors of iodine and bromine heated to the temperature of red heat glowed with a continuous spectrum, at the same time giving a discontinuous absorption spectrum by transmitted light.... I was able to show that vaporized sodium could be made to emit its characteristic D radiation by heat alone under conditions where there could be no action other than heat."

In 1890, Evershed set up a private observatory in Kenley, Surrey, and with the spectroscope, he had designed, and his 3-in telescope, he started a long series of observations of prominences and recorded their distribution in heliographic latitude and their variation with the cycle of sunspots. Between 1890 and 1905, Evershed recorded some 13,458 prominences. In 1890, he became a founder member of the British Astronomical Association and he was the director of the section on solar spectroscopy between 1893 and 1899.

The 1890s were exciting years for solar physics. In the spring of 1891, George Ellery Hale invented the spectroheliograph. Like Evershed, he had set up his own Kenwood Physical Observatory in the backyard of his house in Chicago, where he had a self-made 12-in telescope. Using the spectroheliograph and this telescope, he was soon able to see the reversals in the Ca II H and K lines. Evershed had his own way of photographing prominences in the Hβ line of hydrogen. With this technique of monochromatic photography, he was getting solar images showing the brilliant flocculi around sunspots. When he learnt about Hale's new instrument, he abandoned the idea of working on the hydrogen line and transferred his attention to the Ca II K line.

Evershed got to know Arthur Cowper Ranyard, a barrister by profession with a deep interest in astronomy, who was the editor of Knowledge. Ranyard, who knew Hale personally, introduced him to Evershed when the former visited England in 1894. This was the beginning of a long friendship between the two pioneer solar astronomers of the time, which lasted till Hale's death in 1938. Ranyard died in 1894 and left Evershed with his 18-in reflector and a small spectroheliograph. Evershed found problems with Ranyard's instruments, which produced curved spectral lines and gave distorted images of the Sun. He put the 18-in telescope at Kenley, and designed a spectroheliograph on his own using large direct vision prisms, which gave straight spectral lines and undistorted solar images.

Evershed's immediate superior, in the company he was working, was interested in science and Evershed was granted leave to join solar eclipse expeditions. His first visit to India was in connection with the eclipse of 22 January 1898. He was in the company of W.H.M. Christie, the Astronomer Royal; H.H. Turner, the Savilian Professor of Astronomy at Oxford; Sir Norman Lockyer among others. He obtained excellent flash spectra and caught the emission continuum at the head of the Balmer series extending to the ultraviolet end of the plate. He realized that this was the counterpart of the continuous absorption spectrum seen by William Huggins in stars with strong hydrogen absorption lines. At a later solar eclipse in Algeria in

1900, he repeated the same experiment and again saw the continuum emission. His results definitively proved that the flash spectrum and the Fraunhofer spectrum were of the same origin; the flash spectrum representing the higher and more diffused portion of the gases, which by their absorption gave the Fraunhofer dark line spectrum. At all eclipses, Evershed carried his own home-made instruments – prismatic spectroscopes with a long focal length camera, where only the optical parts were procured from specialized manufacturers. These prismatic spectroscopes consisting of two or more prisms are often referred to as Evershed spectroscopes.

The results of Evershed's eclipse expeditions attracted the attention of the great stellar spectroscopist Sir William Huggins, who was then the President of the Royal Society. Evershed and Huggins had an interesting correspondence in this matter. Later it was Huggins who recommended the appointment of Evershed to the position of the chief assistant to Michie Smith at the Kodaikanal Observatory.

In the original recommendation for setting up an Imperial Astronomical Observatory in the hills of southern India, the Government had indicated that the observatory would be headed by an Astronomer to be appointed by the Indian Observatories Committee in England while the Solar Physics Section would be headed by a Superintendent, either appointed by the Solar Physics Committee and the Observatories Committee, or Michie Smith, the then Government Astronomer, could be appointed to the position. Smith was deemed to have the requisite qualification and experience to run the Solar Observatory, while his suitability to head the Imperial Observatory was somewhat in doubt. It was agreed that the staff of the new observatory should consist of two Europeans and a group of Indian assistants. When the idea of establishing an Imperial Observatory was temporarily abandoned, and it was decided that the Solar Physics Observatory was the one to be immediately established in Kodaikanal, Michie Smith became the natural choice for its directorship. The European assistant to him was not immediately appointed.

The principal thrust of the work at Kodaikanal was to be solar spectroscopy. There were three existing spectroscopes, none of which appeared suitable for obtaining spectra of prominences and sunspots with good efficiency. In 1903, it was decided to build a spectroscope from one of the existing collection (a large six-prism instrument with a collimator and a telescope of 17-in focus), which was dismantled and two lenses from it were mounted with an existing diffraction grating to rig up an instrument that was then used with the 6-in Lerebour and Secretan equatorial to obtain spectra of the solar features. A three-prism Evershed spectroscope was ordered from Hilger Co. during the same year and this went into operation in November 1904. It gave excellent spectra of sunspots and prominences. Also in 1904, the Cambridge spectroheliograph arrived in Kodaikanal. Naturally there was a great need of having a person who would direct the spectroscopic work. Michie Smith visited Hale at Mt. Wilson in early 1904 and discussed in detail the kind of work Kodaikanal should undertake. Hale was also keen on developing an international network on solar research and he was particular that the work with spectroheliographs, at Catania and Yerkes Observatories, his own at Mt. Wilson, yet to be commissioned, and the newly acquired one at Kodaikanal, was done in a coordinated fashion. He was already in correspondence with Evershed and was coaxing him to join this proposed network, as Evershed had his own spectroheliograph and was producing Ca II K spectroheliograms on a regular basis. Hale narrated in detail his conversations with Michie Smith in a letter to Evershed on 2 February 1904. Evershed was quick to respond saying that he felt a cooperation in spectroscopic and spectrographic methods of observation rather than the other older methods of photographic work would indeed be of great help to the community. He called himself "an irresponsible amateur" who was not afraid to be part of any professional network.

3.1 Evershed in Kodaikanal

In 1904, the Government of India sanctioned the appointment of an European Assistant to the Director, Michie Smith. John Evershed was clearly the most eminent choice for the position. According to Evershed, Sir William Huggins greatly influenced the decision of the Government of India in making Evershed the offer, which also had the support of Gilbert Walker, the Director-General of Observatories in India. In a letter to Hale dated 3 December 1904, Evershed said that he had been offered an appointment in Kodaikanal and it would take him 6 months to decide. He was a bit depressed at the prospect of having to close down his private observatory in Kenley. He probably had in mind a personal matter, that of his marriage, before he could definitely take a final decision. He was preparing for the journey to India in 1906. On the advice of Professor Turner of Oxford, he decided to go to India not by the regular route, but by the longer Trans-Atlantic/Trans-Pacific one via America and Japan. Turner gave Evershed introductions to the leading American astronomers of the day: Pickering in Harvard, Frost at Yerkes, and Barnard and Campbell at Lick. Turner also arranged for his visits to Harvard and Princeton, and to these observatories and arranged for the India Office to sanction these visits. Evershed's main aim, of course, was to spend a month with Hale at Mt. Wilson.

At the eclipse expedition of 9 August 1896 in Norway, Evershed had first met Mary Ackworth Orr, a young and sprightly amateur astronomer and a member of the British Astronomical Association. Mary was also interested in Dante Alighieri and was to write a book titled "Dante and the Early Astronomers" in 1914. In 1906 September, John and Mary got married. In a letter to Hale on 28 June 1906, Evershed had told him about his plan to visit America on his way to India. He also wrote that before leaving England he was to be married to Mary A. Orr who would accompany him to America. On 22 September 1906, the Eversheds left England by the Anchor Liner Columbia reaching New York the end of September. The couple traveled to Williams Bay, Wisconsin, where the Yerkes Observatory is situated. Edwin Frost, the Director hosted them for a week. They arrived in Pasadena on 18 October planning to spend several weeks with Hale. However, owing to some mis-communication, Hale was away at the time from Mt. Wilson until the end of November. But the couple had a wonderful and scientifically fruitful time organized by Hale's main solar collaborator Ellerman. Evershed was deeply impressed by the instruments, the method of work, and Hale's organization of the observatory. On 21 January 1907, the Eversheds reached Kodaikanal.

Fig. 2 John Evershed working with the spectroheliograph at the Kodaikanal observatory (from IIA archives)

Evershed immediately got involved in solar observations as soon as he arrived in Kodaikanal. The Cambridge spectroheliograph had been put into operation in 1905. The building where it was housed was faulty with a leaky roof and a new one had to be constructed. There was also some problem with the setting of the second slit at the correct wavelength. So, although spectroheliograms were obtained in 1905 and 1906, the instrument did not perform to its full potential. It was left to Evershed to bring the instrument to its fine working order. By then, a new moving roof was in place in the spectroheliograph building with an addition of ruberoid sheets to prevent leaks. A new collimator slit was fitted and Evershed solved the problem with the camera slit, giving it greater stability and incorporating a new device to facilitate the setting of the slit at any desired wavelength. He also started working on an auxiliary spectroheliograph to obtain pictures of the Sun in Hα.

When Comet Daniel showed up in the sky, Evershed used his eclipse prisms (he had brought them with him to India) and made a prismatic camera fitted to the 6-in Cooke equatorial to obtain spectra of the comet. He identified the CN bands in both the nucleus and the tail of the comet. When Halley's Comet appeared in 1910, Kodaikanal geared up to observe the object both photographically and spectroscopically. Between 19 April and 16 May 1910, when the comet was most amenable to observations in the early dawn, Evershed was able to obtain spectra, which showed the CN bands and the C2 Swan bands in the head, while the tail showed bands due to CO. He was ably assisted in the observations by his wife Mary.

3.2 Evershed Effect

However, Evershed's chief work in the initial years in Kodaikanal was on the spectra of sunspots. He used the high dispersion spectrographs at the observatory to systematically study the spectra of spots whenever the weather was favorable. He noted that the sunpsot spectra were constant in character and consisted mainly of intensified Fraunhofer lines. The general absorption was continuous and was similar to the continuous absorption seen in the limb of the Sun. The pressure in the spot regions seemed to be slightly less than in the reversing layer of the photosphere. In 1909, Evershed made another "very powerful spectrograph," one he called spectrograph III. This had a parabolic silver-on-glass mirror, which formed the solar image on the slit plate and he used an excellent 6-in plane grating by Michelson, which the observatory had, as the dispersing element. He employed this instrument to solely concentrate on the spectra of sunspots.

In the early morning of 5 January 1909, on a day when the atmosphere was exceptionally steady and the sky transparency was excellent after some heavy thunderstorm activity, Evershed found two large spots and obtained their spectra. As he says, "the spectra revealed a curious twist in the lines crossing the spots which I at

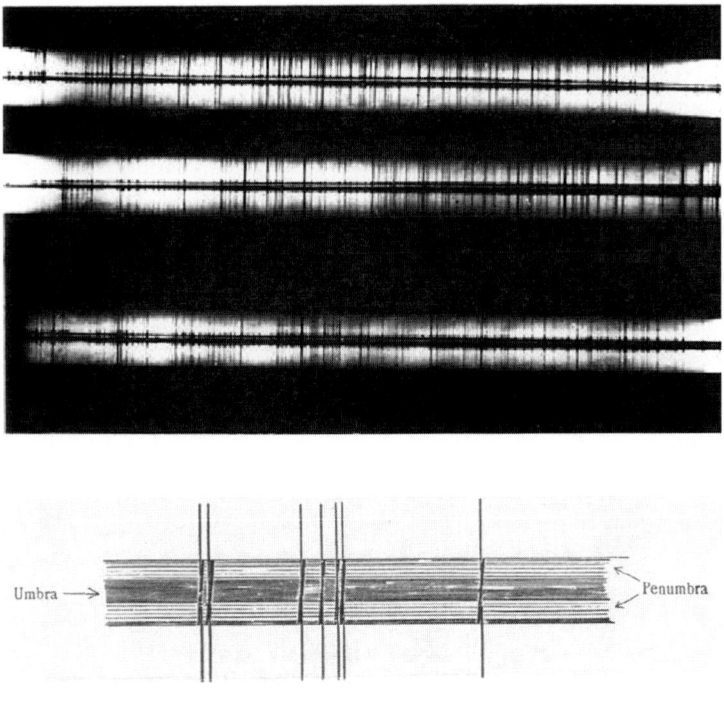

Fig. 3 *Top*: Solar spectra of a sunspot region recorded by Evershed on 5 January 1909 at Kodaikanal. *Bottom*: Line sketch of the spectrum showing the shift of the absorption line penumbra around the sunspot (from IIA archives)

once thought must indicate a rotation of gases, as required by Hale's recent discovery of strong magnetic field in spots, but it soon turned out from spectra taken with the slit placed at different angles across a spot that the displacement of the lines, if attributed to motion, could only be due to a radial accelerating motion outward from the center to the umbra. Later photographs of the Calcium H and K lines and the Hα line revealed a contrary or inward motion at the higher levels represented by these lines." This was the first time line displacements in the penumbral region were seen, indicating an outward radial flow of gases in the spots. None of the previous spectroscopic studies had revealed the dynamics of the flow of gases in the spots. Two days later, Evershed obtained more spectra that confirmed the discovery.

Earlier studies had concentrated on spots near the central meridian of the Sun. Evershed was the first to observe spots at various positions, up to 50°, on either side of the central meridian. He found that the line displacements were more pronounced in the penumbral regions of the spots, which are closer to the limb of the Sun. This established that the flow of the gases was radially outward parallel to the surface of the Sun. In a letter to Hale on 1 February 1909, Evershed described his remarkable discovery and said "I have found that there is a Doppler shift of all the lines in all spot spectra. I made a special study of it having obtained more than one hundred and fifty spectra representing seven northern spots and four southern". The complete results were published in the Kodaikanal Observatory Bulletins of 1909 and the inverse flow seen in the strong lines in a paper in MNRAS in 1910.

Hale responded soon congratulating Evershed for the discovery and stated that because of the pressure of their magnetic field work, they were unable to investigate the dynamics of gases in sunspots.

3.3 Spectral Lines and Diagnostics

The general problem of an unexplained redshift of spectral lines seen at the limb of the Sun with respect to the same lines seen from the centre had occupied the thoughts of all major observers at the time. This was first brought out by Halm at the Cape Observatory in 1907. The Mt. Wilson astronomers, who investigated the phenomenon in great detail, came to the conclusion that the shift was due to the differences in pressure in the two regions – disk centre and the limb. Evershed too was trying to understand the nature and cause of this shift. His observations, however, indicated complexity. His work further showed that the pressure interpretation was incorrect, as some of the lines supposed to be most affected by pressure were actually shifted to the violet. Evershed had developed very accurate methods of measuring extremely small line shifts and much of the spectroscopic work in Kodaikanal was concerned with the measurements of these minute shifts.

It turns out that the astronomers of the time had no idea of the actual value of pressure in the various layers of the solar atmosphere. A commonly accepted notion was that the pressure was several times the Earth's atmospheric pressure. It was also thought for some time that the weak lines formed at lower pressures than did the strong lines. Only after the appearance of Meghnad Saha's work on thermal ionization and its great success in establishing a quantitative effective temperature scale of stars were spectroscopists able to calculate the actual pressure from spectral diagnostics and it turned out that the pressure in the solar atmosphere was only about a tenth of that in the Earth's atmosphere. The solution to the problem of redshifts came from an altogether different quarter. Albert Einstein had calculated the gravitational redshift of spectral lines in his general theory of relativity. As his theory gained credibility, particularly after the solar eclipse expedition by Eddington in 1919 that gave an accurate measure of the bending of a light beam in the gravitational field of the Sun, the effect of gravitational redshift on the spectral lines in the Sun became a major target of study. Towards the end of his stay at Kodaikanal, Evershed made a comparative study of plates taken in 1914, 1921, 1922, and 1923 and found that for the weaker lines, which are supposed to form deep down in the atmosphere, the results of the redshifts were in agreement with Einstein's theory. However, there was an unexplained extra redshift detected in limb spectra whose origin was not known.

In January 1911, Michie Smith retired and Evershed became the Director of the Kodaikanal Observatory. T. Royds joined as the new Assistant Director. In the same year, Evershed put into operation a second spectroheliograph as an auxiliary to the Cambridge instrument utilizing its perfect movement and using a grating and special arrangements for getting photographs of the solar disk in Hα. As red-sensitive plates became available, he started obtaining daily spectroheliograms in Hα along with the same in the Ca II K line. Several spectacular eruptions were recorded, in some of which the speed of recession of the flying fragments was measured.

3.4 Prominences

The study of prominences had continued uninterrupted into the Kodaikanal days. Mary Evershed too developed an interest in the prominences after moving to Kodaikanal. In 1913, she published a paper in which she analyzed observations of prominences associated with sunspots made between 1908 and 1910. The number of prominences recorded at Kodaikanal during the years 1904–1914 numbered nearly 60,000 and these, supplemented by the earlier Evershed collection at Kenley, formed the subject of an exhaustive study and was jointly published by Mary and John Evershed as a Memoir of the Kodaikanal Observatory in 1917.

In 1913, the Eversheds visited Kashmir on 3 months leave and found the observing conditions in the Kashmir valley excellent for solar work. As Evershed wrote later, "The valley of Kashmir is a level plain containing a river and much wet cultivation of rice. It is 5,000 ft above sea level and is completely surrounded by high mountains. Under these conditions, the solar definition is extremely good at all times of the day, and unlike most high level stations, it is best near noon and in hot summer weather." He established a temporary observatory near Srinagar in 1915–1916 obtaining very high-quality photographs of prominences and sunspots. This experience of Evershed was later extended to various localities including one on an ocean liner in tropical waters. The view that the best solar definition is found in low level plains near the sea or on small islands surrounded by extensive sheets of water grew out of these experiences.

In 1915, Evershed was elected a Fellow of the Royal Society and he was awarded the Gold Medal of the Royal Astronomical Society in 1918. At the award ceremony, the President of the RAS, Major P.A. MacMahon, gave a full account of Evershed's work up to that time giving maximum prominence to the discovery of the radial motion in sunspots. The other contributions that he highlighted were (1) prominence observations, (2) spectra of sunspots in general, (3) investigations into the reversing layer, (4) the minute displacements of lines at all parts of the solar disk, (5) eclipse and cometary observations, and (6) the invention and perfection of instruments of observation and measurement. Major MacMahon compared Evershed with Sir William Huggins saying "There is much in our medallist's career which is a reminder of the scientific life of William Huggins. They come from the same English neighborhood, and began as amateurs of the best kind. They both possess the same kind of scientific aptitude."

John Evershed retired in 1923 and left Kodaikanal for England. As the Eversheds went down the ghaut road, a tiger darted across their path, the only time in 16 years in India that they saw a tiger! They returned to Surrey, where Evershed established his own observatory in Ewhurst near Guildford and continued with his solar observations for well over another 30 years. He made liquid-filled prisms, which revealed great possibilities for high dispersion work. He obtained Zeeman spectra of sunspots using these prisms as dispersers. He and Mary continued to take part in solar eclipse expeditions and also attended meetings of the IAU every 3 years. Mary died in 1949 at the age of 83 and John Evershed in 1956 at the age of 92.

4 Scientific Impact of the Evershed Effect

As we look back at the spectroscopic detection, through Doppler shifts, of gas motions in the penumbral photospheric layers of sunspots in 1909 by Evershed at the Kodaikanal Solar Observatory, this discovery clearly stands out as the earliest successful observation of velocity fields due to a complex magnetohydrodynamic phenomenon in action in an astrophysical setting. The most remarkable aspect of Evershed's work was his ingenuity, given the limited resources and sensitivity of the instruments that he used, in providing accurate interpretations of his observations that the gas motions in sunspot penumbrae were indeed radial outflows parallel to the solar surface. This dominant dynamical phenomenon discovered by Evershed can be clearly detected using modern observations with vastly improved spectral and spatial resolution, though additional components involving vertical and wave motions are also seen in the spectra. Yet, a century later, we still do not have a clear physical understanding of this phenomenon.

The immediate scientific impact of Evershed's discovery pertained to another momentous event in astrophysics, viz. the detection of magnetic field in sunspots

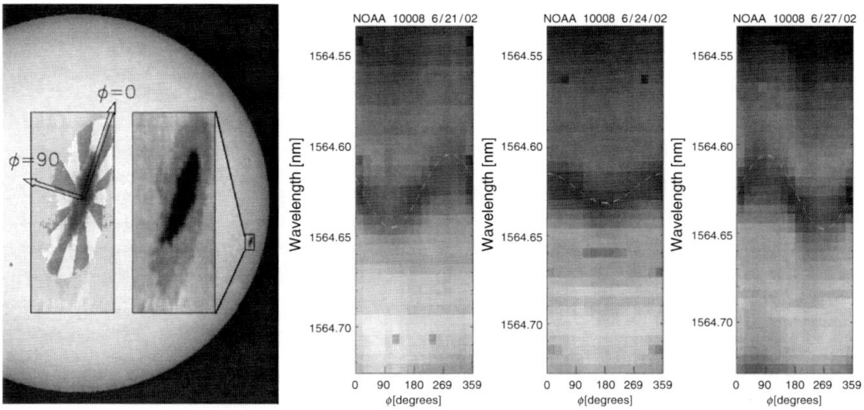

Fig. 4 A modern observation in the CN absorption line at 1546.6 nm of the Evershed effect depicting the radial outflow of material in a sunspot penumbra (observed by Penn et al. 2003)

by George Ellery Hale a year earlier in 1908 at the Mt. Wilson Observatory. This was the first ever discovery of a magnetic field outside the terrestrial environment, and was motivated by the prevailing speculations at that time that sunspots were giant vortices and that all rotating heavenly bodies harbored magnetic fields. Hale considered his detection of magnetic fields in sunspots, in turn, as an indication of swirling flow of ionized gas, which could have generated the magnetic field locally. However, as Evershed pointed out in his discovery paper of 1909, his observations of radial outflow of gas made the circular flow hypothesis untenable. It was not until the early 1930s that the local generation of magnetic field due to axisymmetric motions was shown to be impossible by Cowling through theoretical calculations, and in particular, he suggested sunspots to be composed of tubes of magnetic flux breaking through the surface.

The two momentous discoveries by Hale and Evershed occurred in an exciting era during the early twentieth century when spectroscopic techniques had begun to be used to study the Sun. An immediate follow-up of research on the Evershed effect was through intense observational investigations, particularly at the Mt. Wilson Observatory. More observational programmes followed gradually world-wide, and in general, observational studies continued over the last century. However, theoretical attempts to explain the physical causes and to understand the origin of this effect have gone through several inert periods. We are currently in a period of revived attention largely fueled by our increased numerical computational capabilities to simulate MHD processes.

Evershed's initial estimates of flow speeds were around 2 km s^{-1}, and he found increasing speed with distance from the spot center, and no tangential components for flow velocity. His later results, using chromospheric lines, indicated a radial inflow of gas at these heights, and he measured small tangential components in photospheric velocity, but considered them unreliable because they were irregular.

Detailed observational programmes started in Mt. Wilson (St. John 1913) and across several countries in Europe, most notably at Arcetri in Italy (G. Abetti 1929), at Oxford using the Oxford Solar Telescope (Kinman 1952), in France and in the Soviet Union in the early 1960s, and at Oslo during the mid 1960s. All these observational studies essentially confirmed the results of Evershed, added more details on the spatial variation of flow over the penumbra, established the filamentary structure of the flow, and tentatively identified the magnetic and flow field associations. Considerable improvements in spatial and spectral resolutions emerged in the late sixties, and with that came the identification that the Evershed flow is concentrated into dark penumbral filaments where the magnetic field is more horizontal.

Starting from the 1990s, high resolution observations, from various groups all over the world, have served to spatially resolve the flow and magnetic structures, which have the following dominant properties: the flow occurs mainly in the dark penumbral filaments where the field is more nearly horizontal than the average, the inner penumbral heads of horizontal flow structures have concentrated but bright upflows, which feed the horizontal flows.

Even though a dominant role for the magnetic field in the thermal and dynamical structures of a sunspot was envisaged early on by Biermann and Cowling, it was the work Alfvén, largely responsible for the development of the new field of magnetohydrodynamics, that provided the conceptual framework for analyzing this phenomenon. Theoretical interpretations of the Evershed flow started relatively late in 1955 when Sweet found an origin in convectively driven motions. Since then, various magneto-convective processes have been proposed – they form a large body of research on sunspot related problems.

There have been several different theories based on magneto-convective mechanisms to explain the Evershed flow, most notable ones being (1) a linear theory of convective rolls in horizontal magnetic field proposed by Danielson and its nonlinear modification by Galloway to drive a radial outflow, (2) Busse's theory of three dimensional convection in an inclined magnetic field, where Reynolds stresses drive an Evershed flow, and (3) interchange convection where a dynamical evolution of thin flux tube produces the Evershed flow. Currently, a new magnetoconvective picture of Evershed flows has emerged based on 2D and 3D MHD simulations (discussed in greater detail in this monograph), where the convective interactions between an upward hot plume and magnetic field produce the necessary ingredients to drive a horizontal Evershed flow as observed. A complete and fully consistent picture of Evershed flows with ability to match all the observed features is yet to be realized, even 100 years after its discovery. It is likely that a more comprehensive picture of the Evershed flow would emerge from the high spatial and spectral resolution spectropolarimetric observations planned in the near future. It is equally probable that the fast developing computational resources and algorithms could get complex enough to accurately reproduce the observations.

5 Kodaikanal Observatory: 1923–1960

Following John Evershed's retirement in 1923, activity in solar physics continued unabated at Kodaikanal and work progressed along the lines of the early years. The successive directors at the Observatory were T. Royds (1923–1937), A.L. Narayan (1937–1946), and A.K. Das (1946–1960). The scientific highlights of this era were (1) discovery of oxygen lines in emission in the chromosphere without the aid of an eclipse, (2) center-to-limb variations of hydrogen lines and their use to study the solar atmosphere, and (3) detailed study of the properties of the filaments, seen in Hα. Significant progress was made on the instrumentation front during this period; the new ionospheric and geomagnetic laboratory was set up in the mid-fifties and a major solar facility the solar tunnel telescope was commissioned in 1960.

5.1 Eclipse Expeditions

Eclipse studies constituted an important activity of the Observatory. Royds led expeditions to the eclipses in 1929 to Siam, and in 1936 to Japan. While the totality at Siam was lost to clouds in Japan, Royds used one of the largest ever spectrographs to record the spectra and he was the only observer to get any results. Royds carried out measurement of lines at the extreme limb of the Sun. The 1952, totality in Iraq and the 1955 one in Sri Lanka were again lost to cloudy skies.

5.2 The Saha Committee

A committee appointed by the Government of India, with Meghnad Saha as the chairman, examined in 1945 a plan for the post-war development of astronomical research and teaching at the existing observatories and the universities in the country. One of the main recommendations of this committee was aimed at improving the facilities for solar observations, especially during the first 5 year plan. A solar tower telescope, a coronagraph, and a laboratory for solar terrestrial studies in Kodaikanal were proposed. Most of these were implemented by 1960.

5.3 Ionospheric and Geomagnetic Laboratory

Another post-independence development carried out at Kodaikanal was the setting up of a magnetic and ionospheric laboratory in order to study the response of the earth's ionosphere and magnetic field to transient solar events. Two aspects of Kodaikanal greatly stimulated much of the research carried out subsequently in this area. The first one was the ready and immediate availability of information on solar phenomena observed optically with the Kodaikanal telescopes or by radio techniques. Few places in the world had readily available an ionospheric laboratory and a solar station in close proximity. The advantages in quick experimentation and inference were substantial. The second feature is Kodaikanal's location very close to the geomagnetic equator; hence several aspects of the ionosphere so vital in radio communication could be studied at Kodaikanal with much advantage.

The laboratory went into operation around 1955, just before the International Geophysical Year. There have been numerous studies of the properties of the ionosphere over low equatorial latitudes, which have improved our understanding of the phenomena governing radio propagation and its dependence on the solar radiation characteristics. Interestingly, the laboratory once recorded the geomagnetic effects of a stellar X-ray source.

5.4 Solar Instrumentation

A Hale spectrohelioscope for visual observations of the Sun was received as a gift from the Mt. Wilson Observatory during 1933–1934 for participation in an international programme for systematic monitoring of the Sun. The Hα line in the first order of the grating was used for visual observations of prominences, dark markings, and solar flare patrol. A line shift device was soon added for visual estimation of the Doppler shifts. This instrument was in regular use for the next 60 years. Transient activity first noticed with this instrument was followed up with fast sequences using the main spectroheliographs and also to alert the ionospheric and geomagnetic laboratory on campus.

A tunnel telescope by Grubb Parsons, purchased in 1958, was commissioned in 1960. It consists of a 60-cm diameter two-mirror coelostat, mounted on a 11-m high tower, that directs light via a flat mirror to a 38-cm aperture, f/90 achromat, which forms a 34-cm diameter solar image at the focal plane. A Littrow-type spectrograph with an inverse dispersion of 9 mm Å^{-1} in the fifth order of its grating and a spectroheliograph capable of giving pictures in a chosen line became available for use. These continued to be, for nearly half a century, the country's main facility for high spectral resolution observations of the Sun.

The Kodaikanal Observatory continued its programmes of cooperation with the Greenwich, Cambridge, Meudon, and the Mt. Wilson observatories, especially with regard to exchange of observations for missing days. The Observatory was responsible for communicating to the IAU material for publication of Hα flocculi and dark markings, for most part of the above period. The spectrohelioscope observations were sent regularly as inputs to the quarterly bulletins of the IAU on solar activity. Beginning 1949 May, the Observatory started sending regular messages on solar activity for the benefit of the meteorological department, geophysicists, and radio specialists in the country and abroad. The Observatory continued to publish regularly the Kodaikanal Observatory Bulletins and also the half-yearly bulletins of solar statistics.

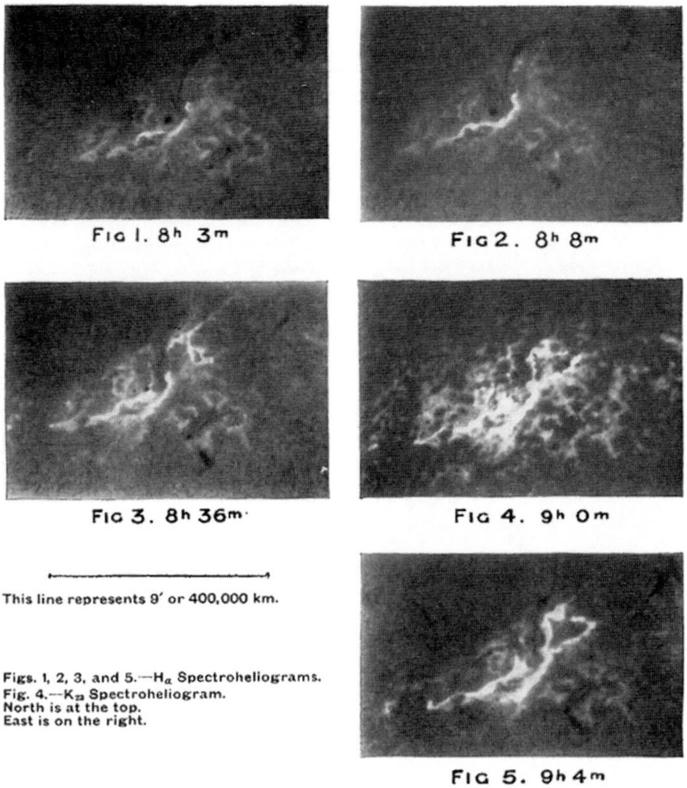

Fig. 5 Royd's flare of 22 February 1926, covering two large sunspot groups, is one of the largest ever recorded at Kodaikanal (from IIA archives)

5.5 Scientific Highlights

Royds carried out laboratory studies of spectra with a view to investigating the cause of observed displacements of solar lines with reference to lines in the spectra of terrestrial sources. Royds and Narayan later observed the centre-to-limb variations of several strong solar lines. Royds succeeded in photographing, outside an eclipse, the infrared triplet of oxygen lines at 7,771, 7,774, and 7,775 Å as emission lines in the chromosphere. These lines are normally observed as absorption lines in the Fraunhofer spectrum.

Rao and his team observed Hα, Hβ, D, and several other lines near simultaneously during the maximum phase of a solar flare. They argued that the observed line broadenings were mostly due to the Doppler effect and were not assignable either to the Zeeman effect or to the Stark effect. Narayan and co-workers studied several atoms, ions, and molecules in the laboratory and looked for their signatures in the solar spectrum. They established the presence of P_2 and studied CN spectral lines on the Sun.

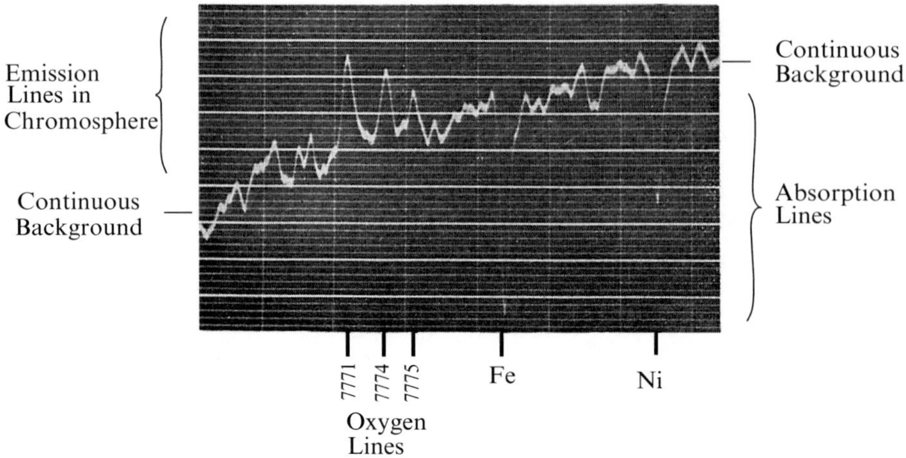

Fig. 6 Microphotometer scan of the spectrum obtained in oxygen emission lines in the chromosphere without the aid of an eclipse. Royds used a straight slit tangential to the Sun's limb to record the triplet (from IIA archives)

Das and Ramanathan observed the distribution of radiation flux across sunspots in the continuum and narrow band windows of the Ca II H and K lines. They demonstrated that certain sunspots show the presence of bright rings and that they are especially more intense in the wavelength of K_2. Statistical studies of solar features, such as sunspots, prominences, and filaments and their periodicities, longitudinal and latitudinal distributions, migration patterns, changes in orientations, and rates of occurrences were carried out.

Flare patrol and follow up observations were carried out routinely at the Observatory. All efforts were made to not miss particularly strong events. In 1926, Royds obtained extensive spectroheliogram coverage of an exceptionally strong and widespread flare (which was known for long at Kodaikanal as Royds' flare). The observations were later used for various studies. In 1949, Das and Raja Rao observed a 3+ flare followed by a great magnetic storm starting 40 h later but lasting over 2 days. Peculiarities of the storm were studied in detail.

In 1953, Das and Sethumadhavan recorded observations of an apparently quiescent prominence, not visibly associated with any sunspot, grow into a dramatic eruptive event. Spectra of the dramatic event were recorded and the associated radio noise bursts were also observed.

6 Solar Physics Research at Kodaikanal from 1960 Onwards

M.K. Vainu Bappu arrived in Kodaikanal Observatory in 1960, and by the end of 1962, he put into operation the newly installed solar instruments, especially the

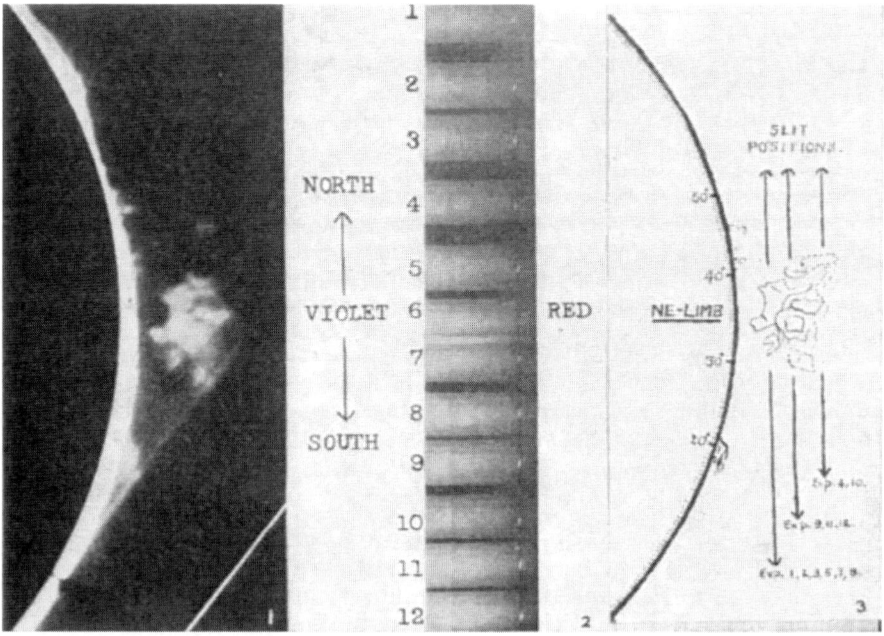

Fig. 7 An apparently quiet prominence suddenly grew to twice its height and erupted. Doppler shifts were measured from the Hα. Associated radio noise bursts were also studied (from IIA archives)

solar tunnel telescope. While solar observations continued with these facilities, new horizons were opening up in the area of night time astronomy.

Renewed vigor and productivity characterized solar research in Kodaikanal under the leadership of Bappu. The solar tower telescope was in regular operation in 1962, and by 1965 it had a spectrograph and a Babcock-type magnetograph. There were three spectroheliographs housed in the building adjacent to the solar tower. The photoheliograph of 20-cm diameter was also in operation.

6.1 Velocity Fields in the Solar Atmosphere

Bhatnagar undertook a detailed study of the Evershed effect based on observations in Zeeman insensitive lines of Ni I and Fe I. He measured sizeable tangential velocities in the sunspot penumbrae, with a maximum value of about 0.6 km s^{-1}, and studied in detail the line asymmetries, which were first observed by Evershed in 1916. Bhatnagar also performed correlative studies of continuum brightness and equivalent widths and found that darker regions of the penumbra show larger equivalent widths while an opposite correlation was found for the brighter regions. This was one of the earliest identification of flow and brightness associations that have now been well studied by high spectral and spatial resolution observations. Studies on solar 5-min oscillations were undertaken by Sivaraman during 1966–1972, culminating in a Ph.D. thesis. A comprehensive study of various properties of the solar velocity and intensity oscillations was done, and though it fell short of bringing out the resolved $k - \omega$ diagnostic diagram and thus the resonant-cavity nature of solar interior for acoustic waves, this work was among the early observational studies of solar oscillations. Temperature and velocity fields in the temperature-minimum region and the low chromosphere were studied using photoelectrically determined profiles of molecular lines of CN and C2 by Nirupama Subrahmanyam in 1965. Studies of solar rotation, by means of sunspot observations and estimation of their anchor depth in the convection zone, have formed a major research activity in more recent years and have been carried out by Sivaraman, Gupta, and Howard. The white light images recorded at Kodaikanal, from early 1900s, have been used in a number of studies, including solar surface and internal rotation, evolution of large scale bipolar regions, solar diameter, solar activity cycle dependent zonal velocity bands, optical flares, correlations between rotation of bipolar sunspots and flares.

6.2 *Magnetic Fields*

Regular observations of strong magnetic fields in sunspots began in 1963. Measurements were made by means of the compound quarter-wave-plate technique and using the 6,303 Å line, with a major objective of evaluating the spatial distribution of longitudinal magnetic field. A longitudinal magnetograph of the Babcock type was designed and brought into use by J.C. Bhattacharyya in 1965. It operated at a chopping frequency of 50 Hz and was in use for the study of weak longitudinal magnetic fields on the solar surface. Weak field in the polar regions and elsewhere were also studied by Bhattacharyya.

6.3 *Chromospheric Features and Dynamics*

Exhaustive studies of Ca II K line spectra were carried out during 1969–1976 by Bappu and Sivaraman. The analyses and results included role of the Ca II K line as a reliable diagnostic of chromospheric activity, the solar cycle variation of Ca II K line emission profiles of integrated sunlight, association between photospheric magnetic structures and Ca II K structures and their use in inferring the morphological evolution of photospheric structures, and the implications of the above for other stars. Variation of the luminosity of the Sun as a star in the Ca II K line was studied by Sivaraman, Singh, Bagare, and Gupta in the mid-eighties. During this same period, Singh and co-workers carried out a detailed study of active region contributions to Ca II K luminosity variations over a solar cycle. Intensity fluctuations in the Hα line due to chromospheric mottling were studied by Bappu in 1964. Singh and Bappu studied the supergranular network size and its variation with solar cycle. Bappu also carried out detailed observations of prominences and calcium flocculus in the early sixties. Sivaraman and Makarov observed filament migrations, and poleward migration of magnetic structures, meridional motions, and polarity reversals at the poles.

6.4 Coronal Studies

Bappu, Bhattacharyya, and Sivaraman observed the Hα and Ca II K lines during the solar eclipse of 1970. Important coronal studies, most notably on coronal wave dynamics, were carried out over several eclipse expeditions starting from 1983 to date. The recent eclipse expeditions have achieved high spatial resolution narrow band photometry of coronal structures to investigate the nature of waves in the corona. Studies of intensity oscillation in the coronal green line and red emission line have also been carried out.

6.5 Solar Terrestrial Studies

Several studies relating to the ionosphere over Kodaikanal and the electrojet have been carried out in the early sixties. Notable among them are the variation over a solar cycle of ion densities at different levels of the equatorial ionosphere, and the spatial, seasonal, and solar cycle variations in the lunar semi-diurnal oscillations in the ionospheric F-region. An analysis of magnetic crochets associated with relativistic flares was done in 1964 and the amplitude characteristics of geomagnetic sudden commencements with energetic proton events were studied.

6.6 Current Observational Facilities and Programmes

Ca II K, Hα images, and spectroheliograms continue to be obtained at Kodaikanal for studies of the chromospheric network, solar cycle variations of the background flux in the K line, and synoptic observations of solar activity. Digitization of Kodaikanal data is one major new project: a new digitizer using a sophisticated CCD camera with a uniform light source has been fabricated to digitize solar images with high spatial resolution (1 arcsec) and with high photometric accuracy.

A new indigenously designed and built spectropolarimeter has been installed and is in use at Kodaikanal for the study of active regions on the Sun. A new system to obtain K-line images of the Sun using a K-line filter has also been installed. These K-filtergrams are recorded on a 1K $\times$ 1K CCD. The system has been operational since 1996–1997.

Other research areas of study include the following:

- Oscillation in the chromospheric network
- Solar cycle variations and synoptic observations of solar activity
- Dynamics of the solar corona and coronal holes
- Sunspots and local helioseismology
- Solar interior
- Coronal mass ejections

7 Future Programmes

7.1 National Large Solar Telescope

The National Large Solar Telescope (NLST) will be a state-of-the-art 2 m class telescope for carrying out high-resolution studies of the solar atmosphere. Sites in the Himalayan region at altitudes greater than 4,000 m that have extremely low water vapor content and are unaffected by monsoons are under evaluation. This project is led by the Indian Institute of Astrophysics and has national and international partners. Its geographical location will fill the longitudinal gap between Japan and Europe and is expected to be the largest solar telescope with an aperture larger than 1.5 m till the 4 m class Advanced Technology Solar Telescope (ATST) and the European Solar Telescope (EST) come into operation.

NLST is an on-axis alt-azimuth Gregorian multi-purpose open telescope with the provision of carrying out night time stellar observations using a spectrograph at the Nasmyth focus. The telescope utilizes an innovative design with low number of reflections to achieve a high throughput and low polarization. High order adaptive optics is integrated into the design that works with a modest Fried's parameter of 7 cm to give diffraction limited performance. The telescope will be equipped with a suite of post-focus instruments, including a high-resolution spectrograph and a polarimeter. A small (20 cm) auxiliary telescope will provide full disk images.

The detailed concept design of the telescope is presently being finalized. First light is expected in 2013.

7.2 Space Coronagraph

A visible emission line coronagraph that uses an innovative design to simultaneously obtain images of the solar corona in the Fe XIV green emission line at 530.3 nm and the Fe X red line at 637.4 nm is under development. The mission is capable of taking images in the visible wavelength range covering the coronal region between 1.05 and 3 solar radii with a frequency of 4 Hz using an efficient detector. High cadence observations in the inner corona are important to understand the rapidly varying dynamics of the corona as well as to study the origin and acceleration of CMEs. There are currently no such payloads planned for the near future.

This 20 cm space coronagraph, which will be executed under the leadership of the Indian Institute of Astrophysics, is planned for launch in 2012. It will obtain simultaneous images of the solar corona in the green and red emission lines simultaneously with a field of view between 1.05 and 1.60 solar radii to (1) study the dynamics of coronal structures; (2) map the linear polarization of the inner corona; and (3) monitor the development of CME's in the inner corona by taking coronal images with high cadence up to 3 solar radii.

The large telemetry capability of the dedicated mission will permit a monitoring of CMEs for about 18 h a day. This project with several national partners has been accepted in principle by the Indian Space Research Organization.

Acknowledgment This article draws heavily on unpublished material from the IIA archives. We are grateful to Dr. Christina Birdie for her help in making the above material available to us and to Dr. Baba Varghese for his help with the figures.

References

Penn, et al. 2003, ApJ, 590, L119
St. John, C. E. 1913, ApJ, 37, 322

Vainu Bappu Memorial Lecture: What is a Sunspot?

D.O. Gough

Abstract Sunspots have been known in the West since Galileo Galilei and Thomas Harriot first used telescopes to observe the Sun nearly four centuries ago; they have been known to the Chinese for more than 2,000 years. They appear as relatively dark patches on the surface of the Sun, and are caused by concentrations of magnetism, which impede the flow of heat from deep inside the Sun up to its otherwise brilliant surface. The spots are not permanent: the total number of spots on the Sun varies cyclically in time, with a period of about 11 years, associated with which there appear to be variations in our climate. When there are many spots, it is more dangerous for spacecraft to operate. The cause of the spots is not well understood; nor is it known for sure how they die. Their structure beneath the surface of the Sun is in some dispute, although much is known about their properties at the surface, including an outward material flow, which was discovered by John Evershed observing the Sun from Kodaikanal a 100 years ago. I shall give you a glimpse of how we are striving to deepen our understanding of these fascinating features, and some of the phenomena that appear to be associated with them.

1 Introduction

Sunspots are dark blotches apparent on the surface of the Sun which, under suitable conditions, such as when the Sun is seen through a suitably thin cloud, can sometimes be seen with the naked eye. Reports from China date back more than 2,000 years, but in the West the history is less clear. It is likely that the pre-Socratic Greek philosopher Anaxagoras observed sunspots with the naked eye, and there have been scattered reports of sightings in the literature since. In 1607, Johannes Kepler tried to observe with a camera obscura a transit of Mercury that he had predicted, and did

D.O. Gough (✉)
Institute of Astronomy, University of Cambridge, UK
and
Department of Applied Mathematics and Theoretical Physics, University of Cambridge, UK

S.S. Hasan and R.J. Rutten (eds.), *Magnetic Coupling between the Interior and Atmosphere of the Sun*, Astrophysics and Space Science Proceedings,
DOI 10.1007/978-3-642-02859-5_4,

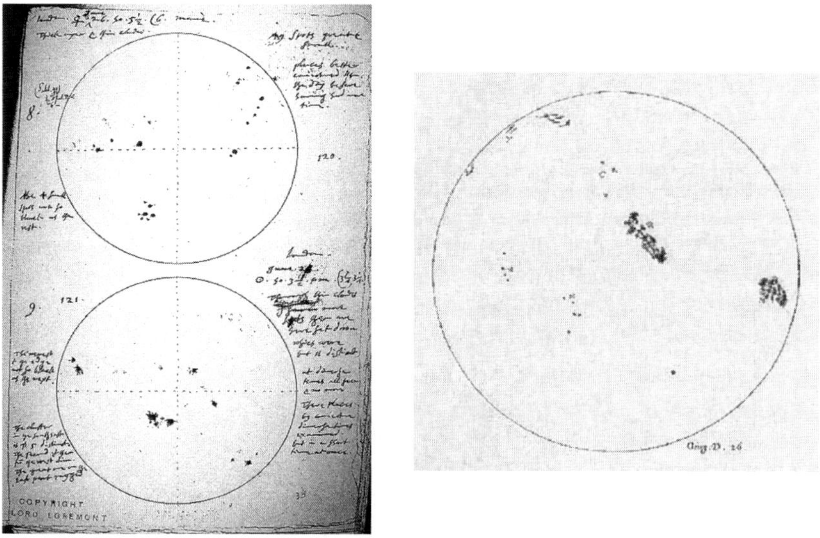

Fig. 1 On the *left* is Harriot's sunspot drawing of December 1610. On the *right* is one of a sequence of drawings by Galileo, which demonstrates the rotation of the Sun; the rotation is very clearly displayed when the drawings are projected in quick succession, as in a movie. It is then evident that the axis of rotation is diagonal in the image: from *bottom left* to *top right*. It is also evident that the sunspots lie in two latitudinal bands roughly equidistant from the equator

indeed see a dark spot that he believed to be Mercury, but it is likely that what he saw was actually a sunspot (Fig. 1).

The scientific study of sunspots began when Thomas Harriot and Galileo Galilei independently observed the Sun through telescopes late in 1610. The following year, David Fabricius, who had made the first discovery of a periodic variable star, namely Mira, together with his son Johannes, also observed spots with a telescope, and published about them in the autumn of that year. They had tracked the passage of the spots across the solar disc, and noticed their reappearance on the eastern limb a dozen or so days after they had disappeared to the west, and inferred that the Sun was rotating, a notion that had already been entertained by Giordano Bruno and Kepler. Christoph Scheiner began a serious study at that time: believing the Sun to be perfect, he attributed the spots to solar satellites, which appeared dark when they passed in front of the disc. In contrast, with the help of his protégé Benedetto Castelli, who developed the method of projecting the Sun's image onto a screen where it could be studied in great detail, Galileo inferred that the cloud-like spots were actually on the surface of the Sun, blemishes on what others believed to be a perfect object, thereby criticizing Scheiner's premise. The spots were not permanent features on the surface, nor were their lifetimes all the same. A large spot might last a rotation period or two, after which it disappears, perhaps to be replaced by a spot at a different location. Smaller spots are shorter-lived. Galileo also disagreed with

Scheiner's adherence to a geocentric cosmology, having been rightly convinced by Copernicus's cogent arguments. The two men, though civil at first, subsequently became enemies.

Scheiner published a massive book, Rosa Ursina, which became the standard work on sunspots for a century or more. By that time he had at least shed his belief in an unblemished Sun, accepting that the spots were on the Sun's surface, and by careful measurement of the motion of the spots he was able to ascertain that the axis of the Sun's rotation was inclined by about 7° to the normal to the plane of the ecliptic. But he continued to uphold his Ptolemaic viewpoint.

Further productive work was hampered by a dearth of sunspots throughout the second half of the seventeenth century, an epoch now known as the Maunder Minimum. Perhaps the most important discovery immediately after that period was by Alexander Wilson in 1769, who realized from the changing appearance of a spot as it approaches the solar limb that the central dark umbra is lower than its surroundings, a phenomenon now known as the Wilson depression.

2 Subsequent Milestones of Discovery

An extremely important milestone for the whole of astronomy is Joseph von Fraunhofer's introduction of spectroscopy, which has enabled astronomers to draw conclusions about the physical conditions and chemical composition of celestial objects, most notably the Sun, and to recognize and measure Doppler wavelength shifts to determine line-of-sight velocity. We now know from spectroscopy that sunspots are cooler than the surrounding photosphere, more of which I shall discuss later.

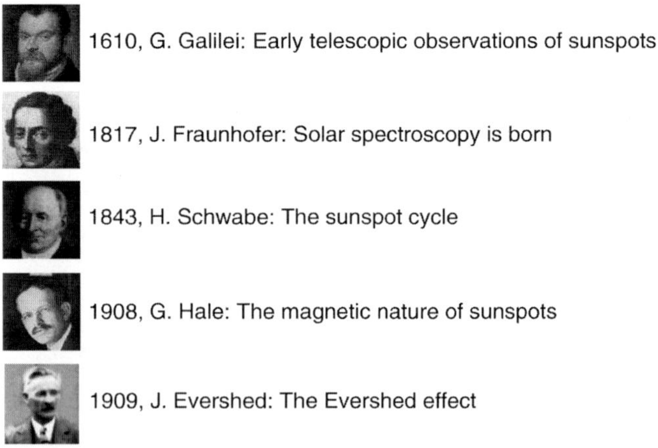

Fig. 2 Landmarks in sunspot discovery

In the few decades after the discovery of sunspots in the West, it was recognized that the number of spots varied with time. And then there was the Maunder Minimum – more than half a century with almost no spots, an epoch when the appearance of but a single spot was worthy of comment. After the reappearance of spots at the beginning of the eighteenth century, sunspot numbers were again quite variable. Nobody at the time appears to have noticed any pattern. Indeed, it was not until 1843 that the amateur astronomer Heinrich Schwabe pointed out a cyclicity, with an estimated period of about 10 years, although further work revealed that the intervals between successive maxima vary from 9 to 11.5 years, with an average of about 10.8 years.

In 1908, George Ellery Hale, the man who pioneered astrophysics as a science beyond the mere identification and plotting of stars, first observed and recognized Zeeman splitting in sunspots, and so established the magnetic nature of the spots. The vertical field is strongest in the central darkest regions of the spot, where the strength is about 3,000 G, and declines gradually outwards (Fig. 3). Why should such a field concentration come about, and what maintains it? Hale subsequently led an investigation into the polarity of sunspots: large sunspots usually occur in pairs, one leading the other as the Sun rotates, with the polarity of all leaders being the same in any hemisphere, but oppositely directed in the northern and southern hemispheres, and with that polarity changing each sunspot cycle (producing a magnetic cycle of duration about 22 years). These properties are now called Hale's polarity laws. The presence of a concentrated magnetic field is now known to be what causes the spot to exist. Precisely how the field became so concentrated is less clear.

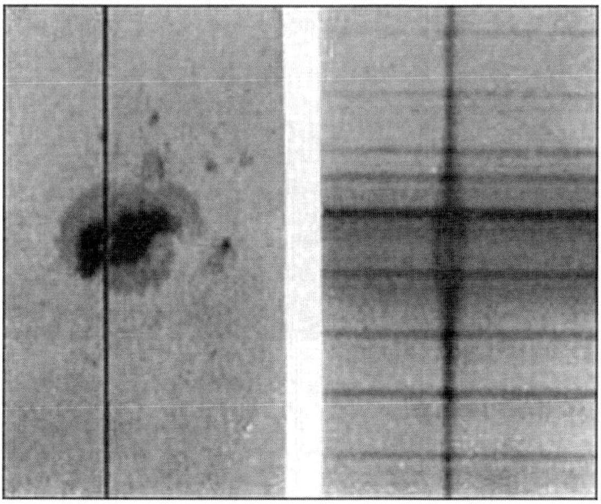

Fig. 3 The *right hand panel* is a Fraunhofer line in the spectrum of light passed through a slit lying across a sunspot, indicated in the *left-hand panel*, in a portion of the solar image not far from disc center. The line is split by the magnetic field, by an amount which is proportional to the intensity of the field. Notice that the field intensity is roughly uniform in the umbra, and then declines gradually to imperceptibility through the penumbra. This is consistent with the sketch reproduced in Fig. 9

Some obvious questions come to mind:

- How do sunspots form?
- Why are sunspots dark?
- What is their structure?
- What holds the field together?
- How long do sunspots live, and what determines the lifetime?
- What is their global effect on the Sun? ... and why?
- What causes the sunspot cycle?
- Is it predictable?

In this lecture I shall address these questions, some of them only quite cursorily (and not in the order listed), but I shall not be able to provide satisfactory answers to them all.

3 Superficial Sunspot Structure

Figure 4 is a photograph of a sunspot. There is a central very dark (in comparison with the normal photosphere) region called the umbra, which is surrounded by a less dark annulus called the penumbra. Beyond the penumbra, one can see the granulation pattern of convection in the normal photosphere. With appropriate exposure, some intensity variation is visible in the umbra: typically small bright temporally varying bright dots against a less variable darker background.

Fine structure in the penumbra is more evident. It consists mainly of light and dark filaments radiating from the umbra, apparently aligned with the magnetic

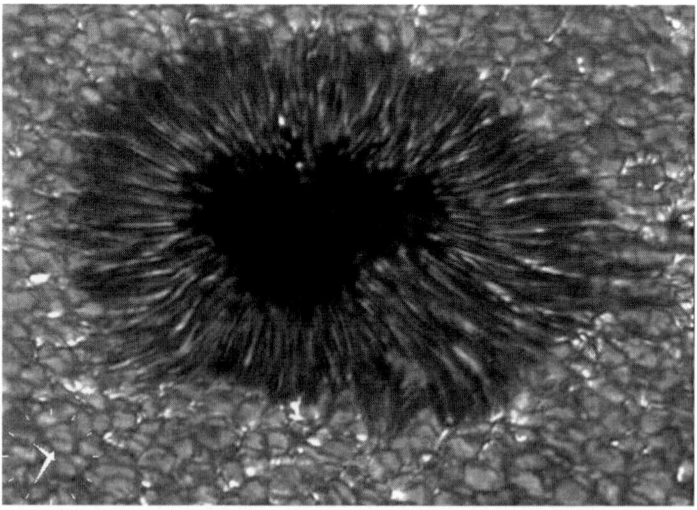

Fig. 4 Photograph of a sunspot in the G band taken through the Dutch Open Telescope

field. There are also elongated bright regions aligned with the filaments that extend through only part of the penumbra; they are called penumbral grains. Figure 4 is a single frame of a movie; when the movie is played, it can be seen that the grains move along the filaments, predominantly inwards in the inner regions of the penumbra near the umbra, predominantly outwards in the outer regions.

Doppler observations of weak photospheric spectrum lines reveal a radially outward flow in the penumbra, the velocity increasing with radius out to the sunspot boundary. This is the discovery of John Evershed, in 1909, to which this conference is dedicated. In stronger lines formed in the chromosphere above the photosphere, a reverse flow is observed.

Sunspots are to be found in a variety of sizes; a medium spot is not very different in size from the Earth (see Fig. 10).

4 The Sunspot Cycle

I have already mentioned that the sunspot number varies cyclically, with a cycle time of 10.8 ± 0.9 years. Figure 5 depicts the variation of a measure of sunspot number (area)[1] with time since the Maunder Minimum, with some pre-minimum estimates from the time of Galileo and Scheiner. There is proxy evidence that the post-minimum cycle is a continuation of similar cyclic behavior occurring before the Maunder Minimum, with some hint that phase was maintained between them to the extent that phase is maintained at all. Figure 6 illustrates not only the variation of sunspot area but also the latitudes at which the spots occur. At a typical epoch, sunspots are concentrated mainly in latitudinal belts located roughly symmetrically

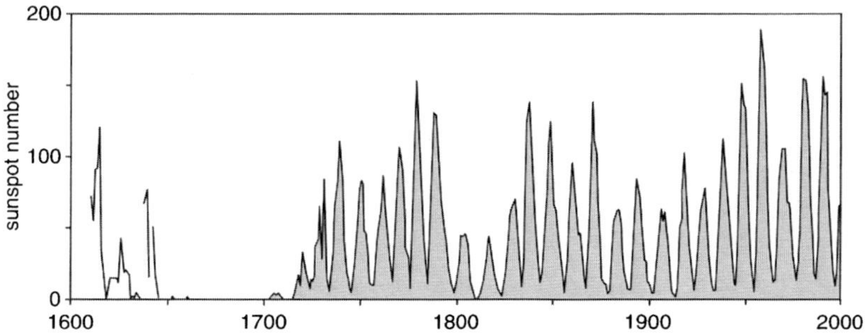

Fig. 5 Smoothed plot of sunspot numbers through the last three complete centuries

[1] Rudolf Wolf invented a measure of sunspot number, which he called "relative sunspot number," and which is now called the Wolf or Zürich, sunspot number. It is approximately proportional to an effective proportion of the area of the solar disc occupied by sunspots, and as the intensity of sunspot fields does not vary very much from one spot to another, it provides an estimate of the total (unsigned) magnetic flux emerging from sunspots.

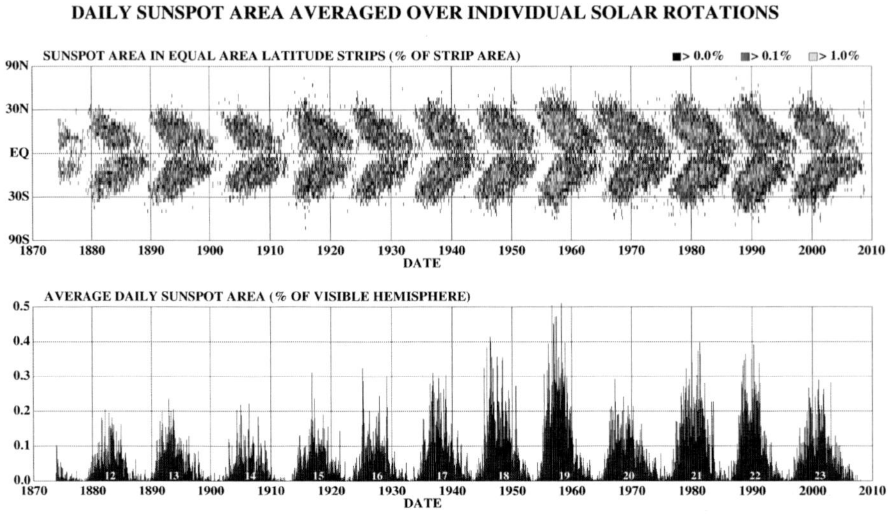

Fig. 6 The *lower panel* depicts daily sunspot area, annual averages of which correspond to the last 130 years or so of Fig. 5. The *upper panel* marks the latitudes of the spots (compiled by David Hathaway)

about the equator. Spots first appear at the beginning of a cycle in the vicinity of latitudes $\pm 30^{\circ}$; as the cycle progresses, the belts migrate equator-wards and eventually merge and disappear as new belts of reverse magnetic polarity emerge at "high" latitudes at the start of the next cycle. That plot is now called the butterfly diagram, a name which hardly needs explanation.

Associated with the sunspot coverage is a variation of solar irradiance, the total radiative flux from the Sun in the plane of the ecliptic, normalized to a distance of one astronomical unit. Irradiance is thus an indicator of the flux of radiation from the Sun in the direction of the Earth. The irradiance has been measured accurately only since detectors could be raised above (most of) the Earth's atmosphere. Figure 7 shows measurements by a variety of instruments. It should be appreciated that it is very difficult to make an absolute measurement, as is evident in the upper panel of the figure, but if the zero points of the fluxes are shifted appropriately, the measurements can be made to lie on top of each other. The lower panel is a (weighted) composite of the shifted curves; the thick line is a running average, which shows quite clearly a roughly 11-year cycle, as one might expect. Interestingly, comparison with Fig. 6 reveals that at sunspot maximum, when one would expect the greatest reduction of light output by the dark spots, the irradiance too is a maximum, as are the magnitudes of the fluctuations in the absolute sunspot number. That demands explanation. One comforting property of the plot is that at sunspot maximum the fluctuations in the irradiance are also at their greatest. We now know that these fluctuations are caused principally by sunspots and their immediate surroundings moving into and out of view as the Sun rotates.

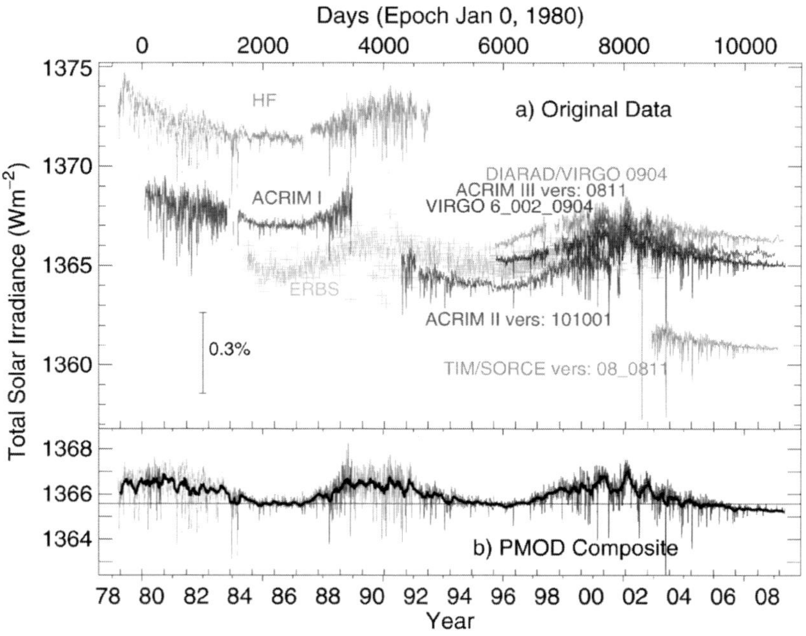

Fig. 7 Measurements of solar irradiance by several different instruments. In the panel below is a combination of those measurements obtained by shifting the zero points to make the results lie on top of each other. The *thick superposed line* is a running mean (Physikalisch-Meteorologisches Observatorium, Davos)

Another property evident in Figs. 5 and 6 is that there is a variation in the value of the sunspot number from one maximum to another, and that the variation has a long-term trend with a characteristic timescale of the order of a century. Included in this variation is the Maunder Minimum, dating from about 1645 to 1715 the last was from, and indeed there is proxy evidence, such as from tree-ring analysis, that there were earlier similar minima, now called grand minima: the last was from about the last took place from 1450 to 1550, and was Spörer Minimum, before which was the Wolf Minimum from 1280 to 1350, the Oort Minimum from 1010 to 1050, and presumably many others earlier. The mean duration of those minima was about 70 years, with standard deviation of 25 years. They have occurred roughly every two and a half centuries, with standard deviation one century. It seems, therefore, that we are now due for another.

What determines the sunspot-cycle period? Or perhaps one should ask more appropriately: what determines the period of the 22-year magnetic cycle? Perhaps the first idea to be put forward was by C. Walén, who suggested that the cycle is essentially a manifestation of a magnetic oscillation of the entire Sun. One can easily estimate the intensity of a global magnetic field required to produce an oscillation with a 22-year period; its precise value depends on the geometry of the field, but all plausible geometries yield fields of the order of 3,000 G, the very value observed to be present in sunspot umbrae. More modern ideas suppose the cycle to be determined by what has been called dynamo action, the complicated process of field augmentation and decay caused by magnetohydrodynamical stretching and twisting moderated by Ohmic diffusion in and immediately beneath the turbulent convection zone. The 22-year cycle period does not emerge from this scenario in so natural a manner as it does from the global-oscillation postulate. But it can be rationalized. However, I shall not attempt to describe in this lecture the panoply of theories that have been invented to explain it, but instead refer to the excellent recently published book on Sunspots and Starspots by Jack Thomas and Nigel Weiss, which also points the reader to more detailed literature.

There has been much discussion about the extent to which the sunspot cycle can be predicted. It seems that most investigators believe that there is a degree of predictability, the interval between, say, one maximum and the next, being influenced by – in the extreme view completely determined by – what transpired before. This notion was advanced some three decades ago by Bob Dicke, who noticed that the unusually early arrivals of the 1778 and the 1788 maxima were followed immediately by some compensating long inter-maximum intervals, apparently trying to restore the cycle to a regular oscillation. Others later have purveyed more complicated relations. They all imply that the mechanism of sunspot production has memory.

An interesting (at least to me) exercise triggered by Dicke's remark was simply to try to answer the question: is the Sun a clock? One can invent two extreme, admittedly highly simplified, models. The first is to presume that the Sun is a clock, whose timing is controlled by a Walén-like oscillation but whose manifestation at the surface through sunspots has a random time lag, random because the information about the interior must travel through the turbulent convection zone, which occupies the outer 30%, by radius, of the Sun (see Fig. 8), yet accounts for but 2% of the mass. At the other extreme one can posit that, as dynamo theorists believe, the

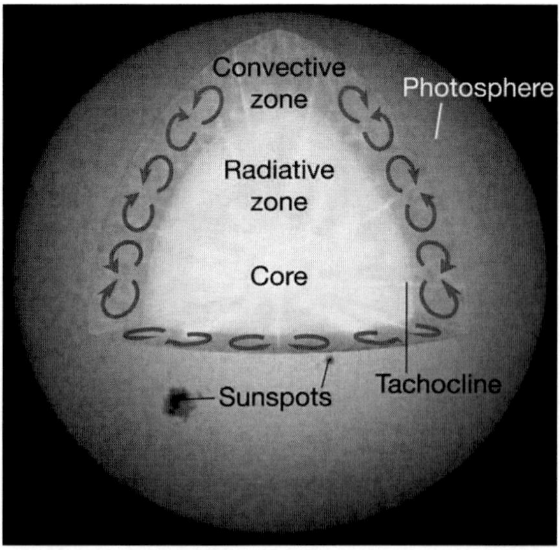

Fig. 8 Simple representation of the Sun, showing in a cut-out the major zones. The curved arrows represent convective overturning

cycle is controlled entirely in or immediately beneath the convection zone where the dynamics is turbulent, and thereby, on a timescale of 22 years, it has no memory at all. Then the cycle period itself is a random function. I hasten to add that this model is actually more extreme than most dynamo theorists accept. The apparent phase maintenance predicted by these two models has been compared with sunspot data by both Dicke and myself, with similar results where our analyses overlap; however, we did not draw similar overall conclusions. I think it is fair to say that the solar data lie between the two extremes, suggesting that the Sun has a modicum of memory, as many dynamo theorists would maintain.

Sunspot-cycle predictability, and with it actual prediction, has come into vogue in recent times. But before remarking on current happenings, I shall relate a pertinent story, which exposes an important variance of opinion concerning scientific inference. Nearly four decades ago I met Charlie Barnes, the chief keeper of time at what was then called the National Bureau of Standards, in Boulder, Colorado, USA. In a digression from his usual activities, he had addressed sunspot-cycle variability from the viewpoint of his modeling the random fluctuations in precision timing by caesium clocks. He had a simple mathematical model, basically a filter which in effect accepted only a part of a time series, concentrated mainly in a given frequency band. Thus, if one sent a random signal through the filter, one received as output a quasi-periodic response which, after rectification, could be compared with sunspot numbers. The only pertinent parameter he could adjust is the ratio of the width of the filter to its central frequency. Barnes calibrated that ratio first by requiring that the variance of the cycle period was the same as that of the sunspot number, and then by requiring that the variance of the heights of the maxima agreed with the variance of the sunspot numbers at maximum. The two calibrations gave the same result. Barnes then pointed out that if one ran the model backwards the original random signal (save for a component that does not influence the output) was recovered, because the whole (linear) process was determinate in both directions. So one could run the machinery backwards feeding it with the actual sunspot data, obtaining an apparently random result, and then run it forwards to recover the original data. What Barnes knew is that if one ran it forwards and, at some moment, stopped the input, the output is the most likely outcome of the process. He therefore had a predicting machine, which he had tested by truncating the apparently random input early, and seeing how well his mathematical machinery "predicted" what should follow. It performed rather well. I was so excited by this result that I went straight up the hill to the High Altitude Observatory, which in those days was situated on a mesa above the National Bureau of Standards at the National Center for Atmospheric Research. There I encountered Peter Gilman, and enthusiastically described to him this fascinating result. "It has no interest whatever," retorted Gilman, "because it contains no physics." But I disagreed strongly, for it is indeed extremely interesting, and the reason for it being so interesting is because it apparently contains no physics; if one wishes to demonstrate the validity of the physics that has been put into a theory by comparing its consequences (I refrain from calling them predictions because so often these consequences are post hoc) with observation, one must surely demonstrate that one has done significantly better than a physics-free procedure.[2]

I now come to real prediction. Or shall I call it sociology? Currently there are (at least) two identical games being played – competitions in waiting whereby scientists have deposited with adjudicators their estimates of the sunspot number at the next maximum. It is supposed to be a bit of harmless fun. I should stress that fun is scientifically useful, a view with which I am sure Vainu Bappu would agree, for it provides rejuvenating relief from the serious pursuit of discovery that occupies most of our lives. But what will the reaction be when the results of the competitions are known? Will the winners claim that the theories they have used are vindicated? Although the entries have been kept confidential by the adjudicators, I do know from talking to some of the competitors that there is substantial diversity amongst the procedures that have been adopted for determining them, procedures which at some level are presumably being tested. One can imagine, for example, that Gilman and his colleague Matsumi Dikpati, who have made much of their ability to predict the solar cycle, will have entered hoping, perhaps, to vindicate their theory. Their model requires several parameters to be calibrated, and so one should heed Pauli's warning. There are also purely mathematical, less deterministic, algorithms, which in a less-easily-appreciated manner incorporate history into a statistical foretelling. At the other extreme, Weiss and David Hughes, for example, believe that the cycle is inherently chaotic, albeit with an underlying control which, turbulent convection aside, is deterministic. Therefore, any prediction must be very uncertain. What might either of them have submitted, if indeed they have entered the fray? There is a diversity too amongst the reasons for entering the competition. I have entered one of the competitions myself, but I shall keep quiet about my motives until the matter is settled. One thing we do know is that there are many competitors, with entries that must surely range from near zero, submitted by those who believe that we are plunging into the next grand minimum (at the time writing there are many fewer sunspots than most spectators have expected) to values comparable with the highest ever recorded. Therefore, the range of possibilities is bound to be densely sampled, as would have been the case had everyone submitted random numbers. So the winners are therefore bound to be very close to the actual result.

5 What Causes Sunspot Darkening?

It is the magnetic field. That field can roughly be thought of as an ensemble of elastic bands imbedded in the fluid, such as the flux tubes illustrated in Fig. 9.

Before embarking on a discussion of the physics of sunspots, I must point out what is actually meant by the term "sunspot." As was evident in my introduction,

[2] Or one must demonstrate that the physics-free procedure happens, by chance, to model the physics of the process under investigation.

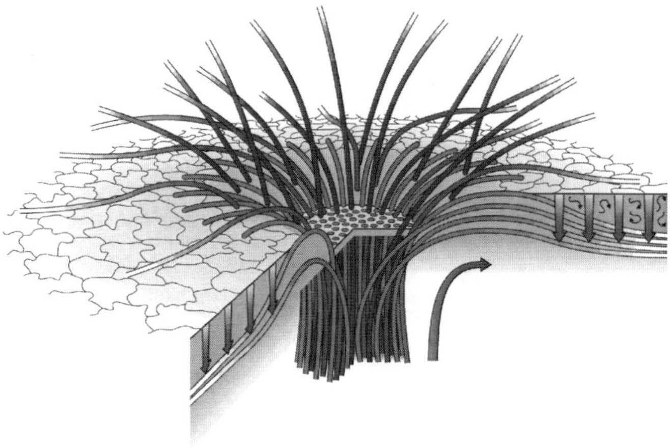

Fig. 9 Sketch by Weiss, Thomas, Brummell, and Tobias of the upper portion of a sunspot. The magnetic field is held together, presumably by converging fluid beneath the region illustrated, in a vertical umbral column, and then splays out through and above the penumbra where the fluid is unable to confine it, alternating between flux tubes rising almost freely into the upper atmosphere and tubes forced back beneath the photosphere by descending convective flow (indicated by the broad vertical *arrows*)

initially the term was considered to denote simply a dark patch on the Sun's surface like those illustrated by Figs. 4 and 10. But now it is considered also to be the entire three-dimensional edifice, extending upwards from the dynamically controlling layers beneath the photosphere into the consequent magnetically active region above it in the atmosphere. I shall use the term in both senses, I hope without ambiguity.

Magnetic field resists being stretched, and therefore opposes any shear in the fluid that would induce stretching – in other words, it reacts against a fluid velocity with a transverse component that varies in the direction of the field. The energy generated by nuclear reactions in the core of the Sun is transported outwards through the majority of the surrounding envelope by photon diffusion, but in the outer 30% (by radius) the fluid is buoyantly unstable, and the energy is carried almost entirely by convection, which consists of overturning eddies (illustrated by curved arrows in Fig. 8). The magnetic field hinders the overturning, and in the umbra is strong enough to stop the normal convection entirely, at least in the very upper layers of the convection zone where the fluid density (inertia) is relatively low and is incapable of overcoming the stresses imposed by the magnetic field. Some motion can occur, however; it provides a weak vehicle to transport energy, and is responsible for the umbral structure observed in the photosphere, but that is of secondary importance to the broad overall picture I am painting here.

The geometry of the field is illustrated by the tube-like structures, sometimes called ropes, in Fig. 9, drawn by Weiss, Thomas, Nic Brummell, and Steve Tobias. The tubes are concentrated in a vertical column underneath the umbra. Some care should be exercised in interpreting the illustration, which should not be taken too

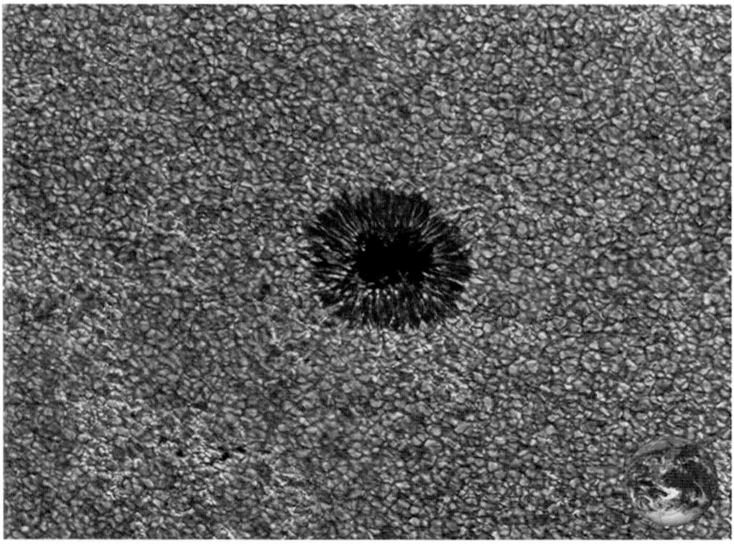

Fig. 10 G-band image of a portion of the surface of the Sun containing a medium-sized sunspot. Superposed in the *bottom right-hand corner* is a image of the Earth to provide a graphic comparison of scale. The mean intensity of the surrounding convective flow, the solar granulation, appears to vary on a scale much larger than the granules, but in patches that are apparently random, with no obvious bright ring around the spot

literally. It gives the impression that the umbral field is contained in the tubes with little or no field between them. That is almost certainly not the case; instead the field is bound to be much smoother on the transverse scale of the tubes. It should be appreciated also that the orientation of the field is the same in all the ropes.

I must now point out another property of magnetic fields: not only do they exert a tension along the tubes, endowing the fluid with a degree of elasticity, but they also exert a transverse pressure – neighboring field tubes of the same polarity (magnetic fields parallel) repel one another; conversely, tubes of opposite polarity (magnetic fields antiparallel) apparently attract, and annihilate each other, dissipating much of their energy into heat and converting the rest into kinetic energy of the fluid. This process, generically called reconnection, is very complicated, and is an arena of very active research. It is of particular interest in the atmosphere above and near sunspots, where the activity is visible. It is no doubt just as important, if not more so, beneath the photosphere where it cannot be seen, and where either the fluid is less of a slave to the field or the field is a slave to the fluid. But I digress. Returning to the concentrated umbral field, it is evident that there must be some force holding the field together. The only possibility is an inward (towards the "axis" of the spot) momentum flux carried by fluid converging at depths where its inertia is great enough to dominate the dynamics. Near the surface of the Sun the fluid can no longer contain the field; the field splays out, becoming weaker and more nearly horizontal. It can no longer prevent the convection (mainly because of the changed orientation), which tends to obviate field stretching by forming elongated eddies, aligned with the field, whose motion is predominantly transverse to the field, producing the penumbral filaments. Moreover, the surrounding fluid no longer converges on the spot, but diverges, at least in places, as was observed by Evershed a 100 years ago.

In the picture provided by Weiss and his colleagues, which is based on prior superficial observation, the field does not splay out smoothly into the penumbra; instead there is an alternation of gradually splaying flux tubes that extend high into the atmosphere and more nearly horizontal tubes that tip back below the photosphere near the edge of the penumbra, pushed down, it is believed, by granular convective motion that is not seriously impeded by magnetic field and which has an up–down asymmetry of such a nature that descending fluid has the greater influence on the magnetic field. That process is called magnetic pumping, and is represented by the downward arrows in the figure. It holds the field down against both the natural tendency of the field to want to be straight (because of its tension) and against buoyancy: magnetic field exerts transverse pressure, which equilibrates with the pressure in the surrounding fluid, the fluid requiring density (inertia, and therefore gravitational mass) to exert pressure, whereas the field has none; regions of concentrated field are less dense than their immediate surroundings and are therefore buoyant. In the inner penumbra where the inclinations of the alternating magnetic flux tubes do not differ greatly, the elongated rolls raise the field where the hot bright fluid ascends and depress it where the cool darker fluid descends. Further out where the inclinations differ substantially, the interaction between the motion in the bright filaments and that in the dark horizontal filaments is probably weaker. It is along the near-horizontal darker tubes that the Evershed motion is driven by a pressure gradient that is insufficient to push fluid high into the atmosphere along the more inclined (from the horizontal) field. What produces that pressure gradient appears not to be well understood. I should point out that other scenarios have been suggested in the literature; once again, I refer the reader to Thomas and Weiss's book for details.

I come back now to the question posed by the title of this section. Except in a very thin superadiabatic boundary layer at the top of the convection zone, almost all the heat from the nuclear reactions in the core is transported through the convection zone by material motion. As I have already indicated, that transport is inhibited in a sunspot by the magnetic field. Therefore, less heat gets through, one might naturally think, and the spot must obviously be dark. That conclusion is basically correct, although with a little more thought one must realize that it is actually not entirely obvious. It depends on certain conditions being satisfied, namely that the spot is a small superficial blemish on a deep convection zone – and by small I mean having both a lateral lengthscale and a depth that are much less than the depth of the convection zone.

A spot is normally considered to have ceased to exist once a depth is reached beyond which significant convective inhibition is no longer in operation. How that comes about depends on the field configuration, which we do not know. But we could consider two extremes. If the field were to extend downwards as a uniform monolithic tube, the stress it would exert would be essentially independent of depth; gas pressure increases monotonically downwards, however, and there must be a level beneath which it overwhelms the magnetic stress, rendering the field incapable of preventing convection. In the opposite extreme, if the field stress were to remain, say, a constant proportion of the gas pressure – I should point out that stress is proportional to the square of the field strength B, and that the magnetic flux, which is the product of B and the cross-sectional area σ of a magnetic flux tube, is invariant along the tube – then the area σ of the region in which the field is contained (whether it remains a monolith or splits into spaghetti, as some investigators have maintained), and in which there is no convection, becomes so tiny at great depths that its presence is irrelevant to the overall picture.

The spot dams up heat beneath it, which nevertheless can readily be transported sideways and upwards around the spot by the highly efficacious convection without substantial modification to the stratification in the surrounds. There is now less heat demanding to be carried through the spot. The flux radiated from the surface of the spot is less than that elsewhere, and therefore the spot is darker; moreover, the surface temperature is lower than that of the normal photosphere, because total radiant flux is proportional to a positive (actually the fourth) power of temperature. With the reduction in temperature in the spot is a consequent reduction in pressure, which causes the material in the spot to sink under gravity (recall that the magnetic field is essentially vertical and the field exerts no longitudinal pressure); that is basically the reason for the Wilson depression. The reduction in pressure is compensated by a lateral pressure-like stress in the horizontal from the magnetic field, enabling the spot to be in pressure equilibrium with the surrounding hotter, more distended, material. Given this apparently straightforward description, one might expect spots not to be a phenomenon associated with only the Sun. Indeed, the presence of dark spots has been inferred from observations of other cool stars having deep convection zones.

The situation is not the same in hot stars. There is overwhelming evidence for spots on Ap stars, for example. Indeed, both magnetic field concentrations and coincident patches of anomalous chemical abundance have been mapped by Doppler imaging. But there is no evident variation in total brightness. (I hasten to add that some such stars exhibit brightness variation in limited optical wavelength bands, but that is due mainly to optical spectrum changes caused by the abundance anomalies, and is not necessarily indicative of total flux variation.) The reason is that these stars have very thin convection zones, and convection is suppressed by the magnetic field in the spot all the way from the top to the bottom of the zone; also the spots are very much larger than those in the Sun, having areas that are a substantial fraction of the total area of the stellar surface, therefore having a linear lateral dimension which is very much greater than the depth of the convection zone. Heat cannot easily escape around the edges of the spot by flowing laterally great distances though the ill-conducting radiative zone beneath. Instead, the stratification in the spot is forced to adjust to accommodate the heat flux demanded by the radiative interior. That adjustment is one in which the spot region becomes more distended, noticeably so if one measures the distension in units of the convection-zone depth, but by only a very small amount relative to the total radius of the star: there is what one might call a Wilson elevation.

I should point out that these two descriptions of spots do not encompass all possibilities: there are also stars whose structure is intermediate between that of the Sun and those of what I have called hot stars; they also support spots, and those spots produce some genuine local diminution of the total radiative flux. Why have I digressed so far from the Sun to describe a situation which is hardly relevant to sunspots? The reason is simply to stress that the physics of sunspots is more subtle than one might have first suspected, and that suppression of the *mechanism* of heat transport in a star does not necessarily result in substantial suppression of the *amount* of heat that is transported.

The process of diverting the heat around a sunspot was first considered seriously by Henk Spruit. The motivation for his study was that others had speculated earlier that the missing heat flux should be radiated from a necessarily bright annulus around the spot of thickness comparable with the spot's radius, but that the brightening had not been observed (see Fig. 10). In his study, Spruit assumed the convective motion to be everywhere on a scale much smaller than the scale of variation of the heat flow, and he ignored the presence of any large-scale flow induced by the disturbance to the temperature variation produced by the suppression of the convective heat transport in the spot. He also ignored the effect of the large-scale temperature disturbance on the convection, so that the heat transport could be described as simply a classical diffusive process with a temporally unvarying diffusion coefficient, the value of which Spruit obtained from mixing-length theory. Spruit considered the evolution of the temperature distribution after suddenly imposing a heat plug in the outer layers of the convection zone to represent the creation of a sunspot. He confirmed a view that was already held by some, although perhaps it had not been well substantiated, that because the turbulent diffusion coefficient and the heat capacity of the convection zone are both so high, transport around the spot is facile and extensive: most of the heat blocked by the spot is distributed throughout the convection zone, almost all of which could easily be retained over the lifetime of a spot (the cooling time of the convection zone is 10^5 years), and that which is radiated around the spot is distributed so widely that its influence on the photosphere is undetectable, in agreement with observation. It should perhaps be commented that the calculation is highly idealized, even in the context of mixing-length theory. The speed of propagation of the greater part of the thermal disturbance produced by the introduction of the plug is comparable with the convective velocities, which invalidates the diffusion equation that was used: purely thermal disturbances cannot travel faster than the convective motion that advects them (admittedly the associated "hydrostatic" readjustment is transmitted at the speed of sound, but the magnitude of the large-scale adjustment is tiny), which is contrary to the formally infinite speed permitted by a classical diffusion equation. Instead, the transport equation should have a wave-like component, somewhat analogous to the telegraph equation. Moreover, temperature fluctuations are not passive, but influence the buoyancy force that drives the very convection that transports them. That back reaction modifies the wave-like term in the transport equation. Nevertheless, because the convection zone is so close to being adiabatically stratified (except in a thin boundary layer), these niceties play little role in the overall structure of the Sun, and Spruit's basic conclusions must surely be right.

6 The Rotation of the Sun

I have already remarked that in the early days Galileo, Fabricius, Scheiner, and others had inferred from the motion of sunspots across the disc that the Sun rotates. Subsequent observations have mapped the angular velocity in greater detail, and in modern times those results have been broadly confirmed by direct Doppler observations of the photospheric layers; the different measures are not precisely the same, but that is because Doppler observations see only the surface of the Sun, while sunspots extend below the surface and presumably rotate with some average over their depth, which we now know is not quite the same. Nevertheless, the basic picture is one of a smooth decline in rotation rate from equator to pole, the rotation period (viewed from an inertial frame of reference, not rotating with the Earth) increasing from about 25.4 days at the equator to something like 36 days at the poles; the latter value is only approximate because it is difficult to view the poles (recall that the axis of solar rotation is inclined by only 7° from the normal to the plane of the ecliptic), and, of course, sunspot motion itself cannot be measured because sunspots are found only equator-ward of latitudes $\pm 30°$ or so, and so other indicators have had to be followed.

Rotation well beneath the surface has only recently been measured, by seismology with acoustic waves. I shall describe briefly how that is done. Acoustic waves are generated essentially as noise by the turbulence in the convection zone and reverberate around the Sun. Any given wave propagates around the Sun, confined (approximately) to a plane, as illustrated in Fig. 11. They are reflected near the

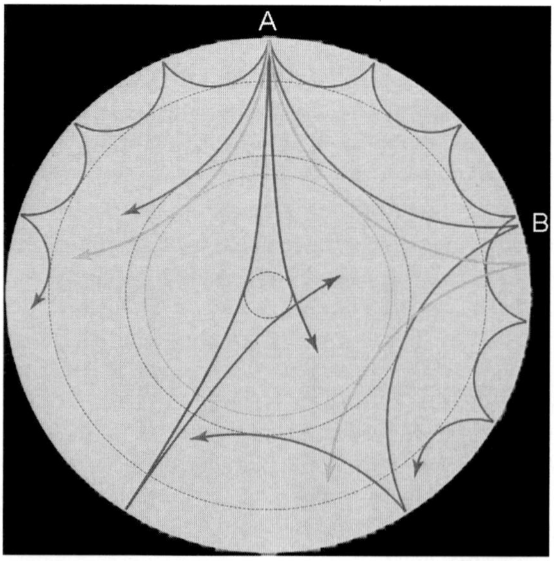

Fig. 11 Segments of ray paths followed by acoustic waves in the Sun. The *dotted circles* represent the envelopes of the lower turning points (lowest points of the ray paths) of the waves

surface of the Sun, typically somewhat below the upper superadiabatic boundary layer of the convection zone where the scale of variation of the density and pressure is comparable with or less than the inverse wavenumber of the waves, thereby preventing those waves from propagating upwards into the atmosphere – the condition for propagation of an acoustic wave to be possible is that, roughly speaking, the scale height of the background state must exceed $1/4\pi$ of the wavelength of the wave. Downwardly propagating waves are refracted back towards the surface by the rising sound speed caused mainly by the increase of temperature with depth. Therefore, waves of a given inclination are trapped in an annulus, whose inner boundary is represented by the dotted circles in the figure (I am assuming for the purposes of the introduction to this discussion that the Sun is basically spherically symmetric), and their properties are determined by conditions in that shell: the relation between the wave frequency and the observable wavenumber at the surface is an indicator of average conditions in the shell, the average being weighted by a function proportional to the time spent by the wave in any particular region. Segments of four sample ray paths (essentially the paths followed by the waves) of differently directed waves are illustrated in Fig. 11; there are other paths, similar to those illustrated, lying in planes through the center of the Sun but inclined to the one illustrated – for example, out of the page towards us at the top and away from us at the bottom, or vice versa.

The essence of the procedure for mapping the solar interior is as follows: Suppose we were to know the wave speed in the Sun down to the bottom of the shell containing, say, the second most deeply penetrating wave illustrated in the figure. Then we can actually calculate the properties of that wave, and also that of the first, shallowest wave and, indeed, of all other waves that are shallower than our selected second wave. Consider now the third wave, which penetrates only slightly more deeply than the second. Evidently we could calculate its progress throughout most of its passage; what is missing is the almost horizontal passage through the very thin annulus occupying the space between its deepest penetration level and that of the second wave: the space between the second and third dotted circles in Fig. 11. We can therefore represent the observable properties of that wave – in particular the relation between its frequency and its horizontal wavenumber at the surface of the Sun – in terms of the average wave speed, I call it $\bar{c}$, in that thin annulus. Measurement of the surface wavenumber and frequency then provides the essential datum to determine $\bar{c}$. We have thereby extended our knowledge of the wave speed down to a lower level. By considering successively more and more deeply penetrating waves we can, provided we have observations of a sufficient range of waves, build up a somewhat blurred view of the wave speed throughout the entire Sun, the blurring being because we are actually measuring averages over the annuli between adjacent lower boundaries of different regions of wave propagation, not point values. One can then combine with that information corresponding results from similar sets of waves propagating in planes inclined to the first, and thereby in principle build up a three-dimensional picture of the wave speed throughout the Sun.

An obvious apparent flaw in my argument is that if all the waves are reflected beneath, rather than at, the surface of the Sun, one cannot know the structure of the Sun all the way to the surface. So how can one proceed? And how can the trapped waves even be observed at the surface? The answer to the second question is that even though the motion at the upper reflecting boundary of the region of propagation cannot formally propagate to the surface, the surface layers do respond as a whole to that motion, being simply lifted up and down in approximate synchronism with the wave below. (I admit to speaking rather loosely here, but as a first approximation it is safe to regard that statement as being true.) Therefore, the wave motion below is observable. Its influence on the motion of the photosphere is portrayed by the Doppler images in Fig. 12. One can now address the first question by simply representing the surface layers by some average impedance, much as we represented the wave speed between the lower boundaries of the regions of propagation of the second and third waves by an appropriate average $\bar{c}$. Fortunately, the upper boundaries of the regions of propagation of all the waves are roughly in the same place, so the impedance for all waves does not vary a great deal. (The range of observable frequencies, roughly 2–4 mHz, which also influence – fortunately only weakly – the impedance somewhat, is not great.) This represents a fundamental uncertainty in the inferences, but that uncertainty becomes smaller and smaller the deeper in the star one's inferences are drawn.

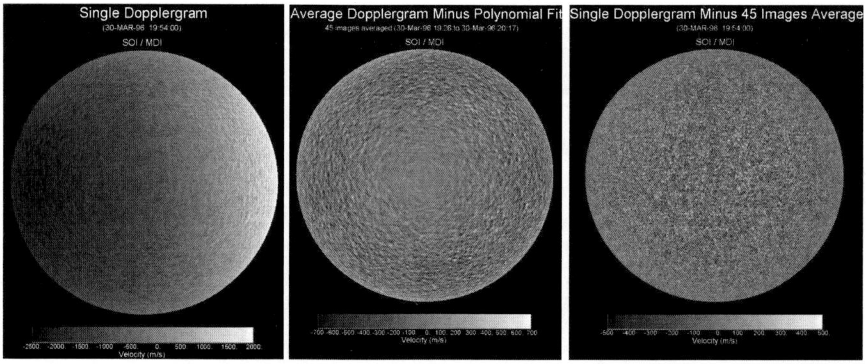

Fig. 12 Doppler images of the Sun obtained by the solar oscillations investigation using the Michelson Doppler imager on the spacecraft SoHO. *Dark shading* represents line-of-sight velocity towards the observer, *light shading* represents velocity away. The values of the velocities represented by the greyscales are indicated at the bottom of each panel. The first panel is a raw Dopplergram; it is dominated by the Sun's rotation, although superposed smaller-scale motion is evident. The second panel is an average of 45 images (which suppresses the oscillations and granular convective motion, although the resolution is inadequate to resolve granules) from which the contribution from rotation has been subtracted; what is left are the tops of the supergranular convective cells, whose velocities are more-or-less horizontal, and therefore is most visible towards the limb (although not too close where foreshortening is severe), and invisible at disc center. The third panel is a single Dopplergram from which the 45-image average has been subtracted, thereby removing rotation and supergranulation, leaving principally the acoustic oscillations, whose velocity in the photosphere is almost vertical; the amplitude observed is therefore greatest at disc center. Notice that the magnitudes of the oscillation velocities are comparable with the convective velocities, approximately $0.5\,\mathrm{km\,s^{-1}}$. For comparison, the sound speed in the photosphere is about $7\,\mathrm{km\,s^{-1}}$. The sound speed at a level near the base of the sunspot (say, 7 Mm) is about $30\,\mathrm{km\,s^{-1}}$

Let me now address what we can deduce from the wave-speed inferences. In the absence of a significant magnetic field, the wave speed relative to the fluid is essentially a local property of the fluid; it is dominated by what we normally call the sound speed, which depends just on pressure and density (and somewhat on chemical composition), but is modified a little by stratification. In addition, the wave is "carried" by the fluid motion, the latter being mainly a consequence of the rotation of the Sun. So one can measure the wave-speed averages in the manner I have just described, first from a set of waves all of which have an eastward component of propagation, and then from a similar set of waves with a westward component. Their average is then the intrinsic wave speed, relative to the fluid, and their difference is twice the rotation velocity of the Sun. Much physics has been learned from the intrinsic wave speed, because it is directly related to the properties of the material of which the Sun is composed, at least in regions where magnetic stresses are negligible. But that is not the subject of this lecture. Instead I shall comment briefly just on the rotation.

The rotation rate in a quadrant of the Sun is depicted in Fig. 13. Plotted are contours of constant rotation rate. Adjacent contours are separated by 10 nHz. The method used to construct this diagram produces only an average of the rotation in the northern and southern hemispheres, which is why only a quadrant is displayed. It is evident that, broadly speaking, the latitudinal variation of the rotation that had been observed at the surface persists with only minor change right through the convection zone. But the radiative zone rotates uniformly. There is a thin shearing layer at the base of the convection zone, called the tachocline, which is too thin to be resolved. It is here that many dynamo theorists believe that magnetic field is augmented and, temporarily, stored, producing the solar cycle. I have already promised

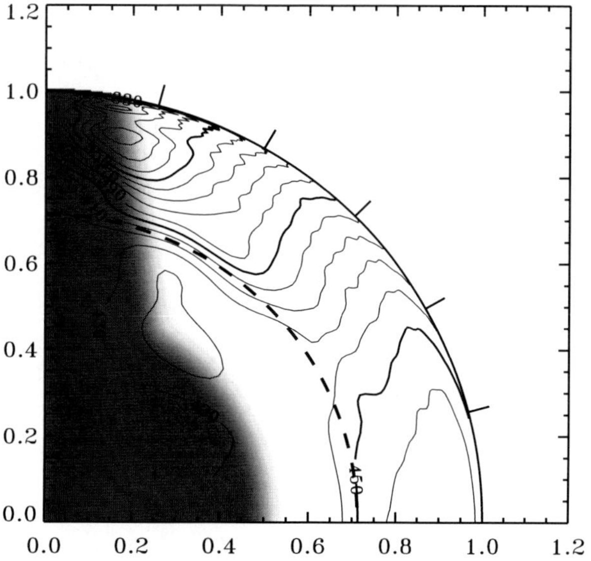

Fig. 13 Contours of constant angular velocity in the Sun. The blacked-out regions mark where it has not been possible to draw reliable inferences (from a study by Jesper Schou and his many participating colleagues)

not to discuss the details. One feature of the plot to which I would like to draw attention, however, is that the shear, and therefore any consequent stretching and winding of the (dynamically weak) magnetic field that might be present reverses direction at a latitude of about 30°. That is just the latitude at which sunspots first form at the beginning of each new solar cycle (Fig. 6). Surely that must provide a clue to the mechanism of the cycle. Or is it mere fortuitous coincidence?

7 The Overall Structure of a Large Sunspot

Only the larger sunspots have a nice well defined structure with surface appearance like those illustrated in Figs. 4 and 10. Small spots contain less magnetic flux and are less able to control the turbulent convective flow in which they are imbedded. They are consequently much less regular. I shall therefore confine my discussion to the relatively clear prototypical case, thereby avoiding having to describe the gamut of smaller magnetic structures that are visible on the surface of the Sun: if I were to do otherwise, this lecture may never end.

The properties of a large sunspot and its immediate surrounds have been mapped by acoustic seismology by Jun Wei Zhao, Sasha Kosovichev, and Tom Duvall. To a large extent they are consistent with the picture I have been building up during this lecture, although one essential ingredient is missing, namely the Evershed flow. In principle, the method of inference that was employed to obtain this picture is much as I described for determining the Sun's rotation; the difference is just in the detail, which is a little more complicated. Consider the three ray-path segments joining observation points A and B in Fig. 14; the point C marks the location of a sunspot. The continuous ray paths are examples from the set considered in Sect. 6, and are drawn simply as a benchmark; they are unperturbed by the shallow sunspot. The dotted ray path passes underneath the sunspot and may feel some influence from it, and the dashed path evidently passes through the spot. By comparing observed propagation times from A to B and from B to A of the dotted and dashed waves with those of similar wave segments in another location where there is no sunspot, the influence of the sunspot can be ascertained. As always, the answer is a new average propagation speed $\bar{c}$ along the ray paths. One must then tackle the complicated geometrical problem of unraveling those averages over a wide variety of rays to obtain genuinely localized averages, of both intrinsic propagation speed and of fluid flow, for such averages are comprehended more easily than the raw ray-path averages. I shall not go into the details of how the unraveling is accomplished; for the purposes of the present discussion, it is adequate to consider the task to be just a technicality, which we know how to handle.

The outcome is illustrated in Fig. 15. What is shown is a section in a rotatable vertical plane of a three-dimensional representation of a measure of the intrinsic wave propagation speed and the large-scale fluid flow – only a single orientation of the plane is illustrated in the figure reproduced here. The shading represents the intrinsic wave speed and the arrows represent the flow, their size denoting the magnitude

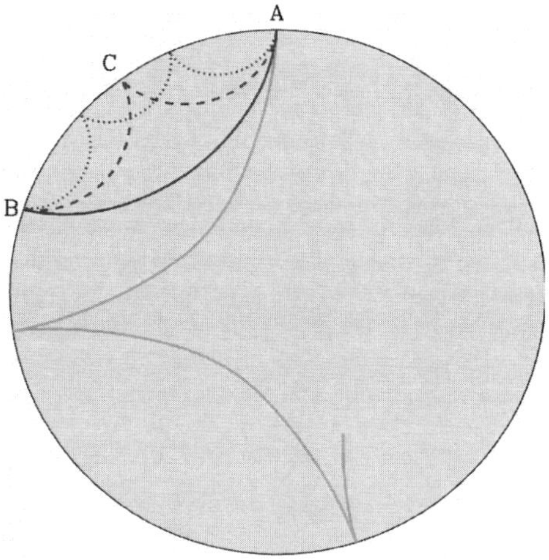

Fig. 14 Sketch of sample ray paths adopted by Zhao, Kosovichev, and Duvall for inferring sunspot structure. Points like A and B mark typical observation points in the quiet Sun; point D is the location of a sunspot. The distances between A, B and C, and hence the depths of the ray paths joining them, have been exaggerated for clarity. The continuous ray paths are typical spot-free paths, like those depicted in Fig. 11, from which the background (spot-free) structure is inferred

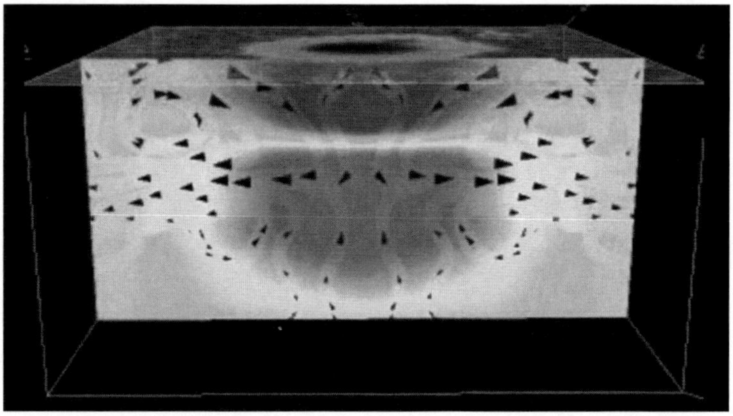

Fig. 15 Seismic image of a sunspot by Zhao, Kosovichev, and Duvall. The *shading* represents the deviation of the wave propagation speed from that in a corresponding spot-free region of the Sun, *dark* (in this black-and-white picture) denoting both positive and negatives values. The *arrowheads* indicate the direction and magnitude (denoted by their size) of the flow

of the velocity. The intrinsic wave speed is difficult to interpret: it is influenced by both the temperature of the fluid and the magnetic field, which the current measurements cannot disentangle; even more uncertainty is added by the fact that the effect of the magnetic field is anisotropic, being a more-or-less increasing function of the inclination of the direction of wave propagation to the direction of the field – what is illustrated by shading in the figure is only a scalar, presumably an average over the particular waves that have been used for the inference, weighted by the relative importance that the localization procedure adopted by the analysis has given to those waves. Interpretation must therefore entail some guesswork. It is likely that the wave speed illustrated in the figure is due predominantly to temperature, because immediately beneath the photosphere both field and acoustic wave propagation are both very nearly vertical, and consequently parallel to each other, and therefore hardly interact. Moreover, as I have already described, at depth the influence of the field declines dramatically either because, unlike the gas pressure, the intensity of the field does not increase significantly with depth, or because the proportion of the volume occupied by the field diminishes greatly. (It is worth pointing out that because the lateral field stress under the umbra balances the gas pressure deficit produced by the lowering of the temperature, a putative horizontally propagating acoustic wave would be influenced by comparable amounts, although oppositely, by field and negative temperature change. Those influences would not exactly cancel, however, because the effective adiabatic compressibilities of field and gas, which control the wave speed, are different.) Therefore, I may lapse into "hotter" and "colder" as a convenient device to describe wave-propagation-speed differences succinctly.

The dark shading in Fig. 15 immediately beneath the upper surface of the spot is to be expected: the surface of the spot is cool, and, as I have already explained, so should be the underlying fluid where convection is suppressed by the magnetic field. There is a second relatively dark region lower down in this black-and-white image, this time representing hotter fluid, presumably beneath the region in which convection is suppressed – in other words, beneath the spot. This is where heat from below is dammed up, being unable to pass easily through the spot. In a broad sense, the fluid flow associated with these temperature (actually wave-speed) anomalies is easy to understand – at least it seems superficially to be that way. The cool plug beneath the surface cools the surrounding fluid, causing it to sink in a negatively buoyant cold collar around the spot, drawing in fluid from the near-surface regions to replace it. The hot fluid beneath the spot is positively buoyant; it is inhibited from rising directly upwards by the magnetic field in the spot, and must therefore first move axially outwards before it can rise around the spot. It collides with the upper descending cold collar, and the two are deflected outwards away from the spot. Some of the diverging fluid then rises and some of that then reconverges, producing a toroidal eddy around the spot; the remainder of the ascending fluid is deflected outwards, flowing away from the spot in the near-surface layers. That motion is quite difficult to perceive in Fig. 15, which is but a single frame of a movie, for there are just two small inclined arrows near each outer edge of the figure, suggesting the outward deflection. But it is quite obvious when the movie is played. However, that outward motion is not the Evershed flow. It is too far from the spot. The structure of the visible spot is shown on the representation of the upper horizontal boundary of the region being depicted, and it is evident that immediately beneath the penumbra, and somewhat beyond, the near-surface flow is axially inwards, towards the spot. This failure to miss the Evershed flow has spread considerable doubt amongst solar physicists, particularly theorists and modelers, on the reliability of the seismological inferences. Perhaps that doubt is justified. After all, Eddington said that one should never trust an observation until it is confirmed by theory. So I shall address theoretical simulations in a moment. But perhaps the doubt was due as much to the reluctance of observers of only the superficial layers of a star to accept more profound methods. Ray Lyttleton once said that if a modern observer were to meet a chimney sweep,[3] he would deduce that the sweep were composed of pure carbon.

It is important to remain aware that, as I described when discussing seismological inference of rotation, we cannot (readily) come to reliable conclusions about conditions very near the solar surface from the seismology of acoustic waves. The top of Fig. 15 is about 2 Mm beneath the photosphere. Therefore, if the situation presented by that figure is correct, one must conclude that the Evershed flow is shallow.

There is yet more seismological inference, which I have not yet described. In addition to acoustic waves there are surface gravity waves, called f waves, whose physics is identical to that of the waves on the surface of the ocean. These waves do not propagate through the interior of the Sun, but remain near the surface, their amplitudes declining exponentially with depth at the same rate as they oscillate horizontally (in other words, the e-folding depth is $(2\pi)^{-1}$ oscillation wavelengths). They too are advected by flow. Surface gravity waves confined essentially to a layer extending to about 2 Mm beneath the photosphere have been analyzed by Laurent Gizon, Duvall, and Tim Larsen, who did indeed find outflow from the spot. The depth-averaged velocity is much less than that observed directly in the photosphere, which is to be expected if the flow is a countercell of the subsurface flow around the spot depicted in Fig. 15, whose center must lie less than 2 Mm beneath the photosphere. It seems that these two complementary seismological analyses essentially complete the basic picture. I hasten to add, however, that the picture is not accepted by a substantial number of theorists; Thomas and Weiss, for example, consider such a shallow countercell to be unlikely.

It is evident from Fig. 15 that the subsurface inflow occurs in an annulus that extends well beyond the penumbra. So does the outflow observed at the surface of the Sun, although the obvious penumbral striations cease once the flow has passed the point at which it is strongly influenced by magnetic field. Therefore, its superficial appearance is different, and solar astronomers of late have given it a different name: moat flow. However, there appears to be no convincing evidence that it is no more than simply the outer extent of the Evershed flow.

Triggered by the doubt cast by solar physicists, helioseismologists have reconsidered the approximations that were used in the construction of Fig. 15: for example, the manner in which the velocities observed at the ends of a ray-path segments (such

[3] It was commonplace in northern Europe up to half a century or so ago for houses to be heated by burning coal, often bituminous, the soft brown lignite coal that burns incompletely, encrusting the insides of chimneys with unwanted soot, which subsequently might fall back into the room being heated or, more seriously, catch fire. What escaped at the top of the chimney polluted the atmosphere, producing, under inclement conditions, dense unhealthy yellow-brown fog. For safety, the soot had to be swept periodically from the insides of the chimneys, and a profession of chimney sweeps was established to perform that task. It was dirty work, and often a sweep's clothes and his exposed skin became covered with soot. By contrast, a modern Danish chimney sweep prides himself of his cleanliness: he is well dressed, in tailcoat, top hat, and white gloves.

as points A and B in Fig. 14) are cross-correlated for inferring travel times, the effect of ignoring the apparent time difference between the reflection of an acoustic wave at its upper turning point and its manifestation in the photosphere, the scattering by inhomogeneities out of and into the ray path, diffraction, and the effect of stratification on acoustic wave propagation. All have some quantitative impact on the inference, but at the moment it appears unlikely that any is severe enough to make a qualitative change to the picture.

There have been several attempts at direct numerical simulation of sunspots. Neal Hurlburt and Alastair Rucklidge have considered the effect of a monolithic axisymmetric concentration of nearly vertical magnetic field on convection in a layer of ideal gas. In all cases, they found the fluid to converge on the field and sink in a cool collar around the field, just as in Fig. 15. They pointed out that they had not modeled the solar atmosphere: they regarded the top of their idealized model to be well below the solar photosphere, just as are the current acoustic seismological inferences, and they too embraced the idea that in the Sun there is a toroidal countercell above the converging fluid, which is manifest as the Evershed flow. They also found an outer toroidal countercell surrounding the main cell, which is diverging from the spot in its upper half, as is (barely) seen in Fig. 15 (but is quite evident in the movie). Hurlburt and Rucklidge suggested that the flow (without a countercell above it) might be the moat flow. The outflow evident at the upper boundary of Fig. 15 (without a countercell above it) is so far from the umbra that it could only be the outer extent of the moat.

The converging subsurface flow offers a natural explanation of how the magnetic field is held together: it is continually advected inwards against diffusion and its natural tendency to expand. The superficial layers that support the reverse Evershed flow have too little inertia to offer significant opposition to that process. In the deep layers, below about 7 Mm or so, the magnetic field has negligible influence on the flow. It surely seems most likely that the field is tangled by the (three-dimensional) turbulent convection into thin flux tubes by a process combining advection and diffusion akin to the pioneering (two-dimensional) numerical studies carried out by Weiss in the 1960s.

8 On the Birth, Death, and Lifespan of Sunspots

Sunspots tend to form in groups in regions in which there is a lot of magnetic activity. These regions are called, naturally enough, active regions. Active regions form, it is believed, from large magnetic flux tubes that had been formed from field intensification possibly in the tachocline beneath the convection zone, and have then risen buoyantly to the surface. The outcome is a pair of regions in which the photosphere is crossed by magnetic field of opposite polarity, moving away from each other and connected in an arch in the atmosphere above, as in the cartoon depicted in Fig. 16. This picture was first adduced after studying the evolution of these regions from observations of the photosphere and the overlying atmosphere; more direct

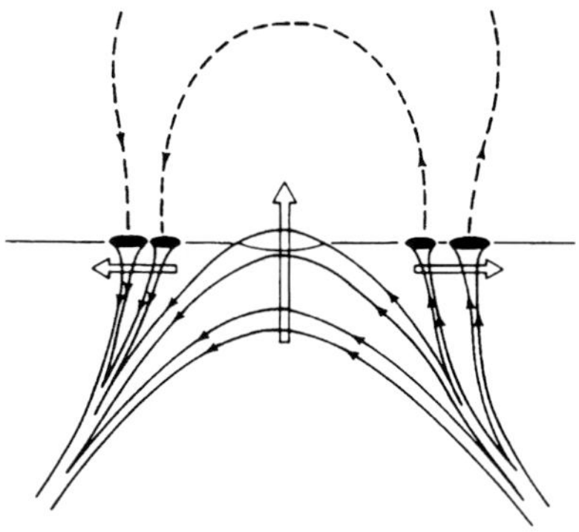

Fig. 16 Cartoon of a rising flux tube producing bipolar pairs of sunspots joined above by a magnetic arch (provided by Carolus Schrijver and Cornelis Zwaan). The *single arrows* indicate the direction of the field and the *double arrows* the motion of the flux tubes

evidence for the rising of flux tubes before their appearance at the surface has since been provided by seismology. Active regions can be up to 100 Mm across. They are temporary phenomena, with lifetimes up to several months. After an active region has disappeared, it is not uncommon for a new one to erupt in about the same place, and on the longer timescale of several years there are so-called active longitudes in the vicinity of which active regions persistently form. Understanding the long-term pattern of the coming and going of these regions, which broadly indicate the locations of the major sunspots in the butterfly diagram depicted in Fig. 6, is the realm of solar dynamo theory.

The magnetic field that emerges in active regions is inhomogeneous, initially being concentrated into flux tubes with cross-sectional diameters of about 200 km, containing field with intensity about 400 G. These tubes are quickly (on a timescale of less than an hour) compressed by the convective flow into tubes 100 km across with field intensity about 1,500 G. The tubes are advected by the supergranular convection in such a manner as to cause them to meet, despite their natural repulsive character, and coalesce into bigger tubes, called pores, which sometimes, on a timescale of days, coalesce into yet bigger tubes that then become fully fledged sunspots with penumbra. The larger sunspots often form in recognizable pairs of opposite polarity, joined by the magnetic arch in the atmosphere, although more complicated groups, and individual sunspots, are common. The image of the solar atmosphere reproduced as Fig. 17 obtained from the space mission TRACE tracing the magnetic field in an active region near the solar limb is dominated by the field joining a large sunspot pair. But there is also more complicated magnetic structure that undergoes reconnection, ejecting charged particles from the Sun and creating what has been called space weather, which is a danger to spacecraft and can upset the Earth's ionosphere, interfering with radio communication and, in extreme circumstances, damaging power lines. Understanding the whole realm of these phenomena is now sometimes referred to as heliophysics, although the word was originally coined to encompass studies of only the (entire) Sun, from the energy-generating core to the corona.

Sunspots of a pair are located very roughly east-west of each other, which is consistent with them having risen from a field that has been stretched into a toroidal coil in the tachocline, but inclined somewhat such that the leading spot is closer

Fig. 17 Image of an active region containing a large sunspot pair, taken by the spaceborne camera on TRACE. The observation was made in extreme-ultraviolet line, which highlights the magnetic field (courtesy Alan Title)

to the equator. The inclination is a result of Coriolis torque (from a point of view in the rotating Sun) as the field and its accompanying fluid moved upwards and away from the axis of rotation – that is simply the tendency of the spot-pair to try to conserve its angular momentum, thereby finding itself rotating more slowly than its surroundings. Moreover, the relative polarities of the spots are opposite in the northern and southern hemispheres, which is consistent with the idea of tachocline winding of a basic large-scale internal dipole magnetic field whose axis is aligned more-or-less with the axis of rotation.

As soon as a sunspot is created, it starts to decay. The decay appears to be consistent with the idea of lateral-surface abrasion by the small-scale granular convection. That is essentially a diffusive process, and occurs much more slowly than sunspot formation – large sunspots are created in the course of days, but it then takes a month or so for them to decline and die. The timescale of diffusion scales with the square of the linear dimension (it takes four times as long to roast a turkey than it does to roast at the same temperature a chicken of half the linear size: the roasting time of birds, or any other food that scales in a homologous fashion, is proportional to the two-thirds power of the weight, contrary to the advice given in many cookery books), and inversely with the magnitude of the diffusion coefficient. If the diffusion coefficient of convective abrasion were constant, the spot lifetime would be proportional to its area, and indeed there is observational evidence corroborating that. Not all spots are as regular as those illustrated in Figs. 4 and 10, however; the scatter in their properties is large, and the result of inferring any age–size relation must be only approximate. From some studies of the observations, it has been concluded that the effective diffusion coefficient is proportional instead to the spot diameter. When the spot becomes small enough, it is essentially a pore.

According to this discussion, it is the convection that controls the sunspot dynamics. The same agent is responsible for both the birth and the death of a spot. How can that be? Admittedly it is the large-scale convection that appears to be responsible for a sunspot's birth, and small-scale convection for its death. But I have seen no cogent explanation of why the large scales dominate in the early stages of life and small scales in the decline – so far as I can see that has in most cases merely been implicitly assumed; otherwise, the matter appears to have been ignored. Perhaps it is simply a stochastic result moderated by the broad evolving conditions in the active region. There can be no sunspot decay in a spot-free region; and a sunspot of any given size is more likely to be decaying than be in a state of being created. Perhaps that is simply because the process of creation dominates the decay, but is only rarely operational. One is reminded of Boltzmann's H theorem. Maybe it is simply the very existence or not of a sunspot that biases future evolution, just as statistical fluctuations in a stable thermodynamic system are at any moment more likely to be decaying than growing, causing entropy, on the whole, to increase.

9 Solar-Cycle Irradiance Variation

I conclude by returning to the question of why it is that on a solar-cycle timescale the solar irradiance at sunspot maximum, when there is more direct darkening of the photosphere, is greater than it is at sunspot minimum. A partial answer to the apparent contradiction has emerged from detailed studies by Peter Foukal, Judith Lean, Judit Pap, Sami Solanki, and their colleagues, who addressed particularly the causes of shorter-term (daily-to-monthly) irradiance variations evident especially at sunspot maximum. They have found that those fluctuations can be very well reproduced as a combination of the reduced radiation from sunspots with enhanced radiation from surrounding regions called faculae. Faculae are structures in active regions that are somewhat hotter than the normal atmosphere, being hotter by about 100 K in the photosphere and by substantially more than that higher in the atmosphere. They are closely associated with sunspots, their total area following the solar cycle, roughly preserving a facular-to-sunspot area ratio. Being only slightly hotter than the normal photosphere, they are difficult to see near disc center, but they stand proud of the normal surface and are therefore relatively more visible near the limb. The radiation they emit exceeds the sunspot deficit, which immediately explains why the irradiance is greatest at sunspot maximum. The extra energy that heats them is presumably transported through the photospheric regions directly by the magnetic field, rather than by convection and radiative transfer, although some time ago I suggested, not without (admittedly incomplete) theoretical justification, that a degree of magnetic enhancement of convective transport under the photosphere of the so-called quiet Sun (away from active regions) might also contribute

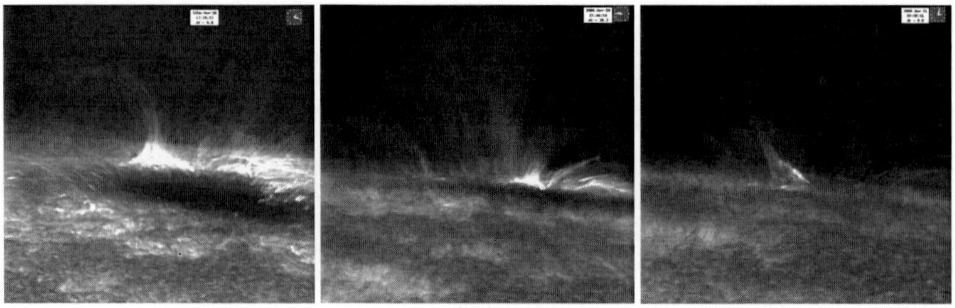

Fig. 18 A sequence of three images of a sunspot approaching the solar limb on 20 and 21 November 2006, taken by the Hinode space mission

to enhanced irradiance, at least on solar-cycle timescales; Gene Parker subsequently embraced this idea, at least for a while, but on the ground that at the time no other plausible explanation could be found.

An extreme example of what I am now talking about is beautifully illustrated in a stunning movie taken recently from the Japanese spacecraft Hinode (meaning Sunrise). Figure 18 is three images selected from that movie. In the first image, the dark sunspot is clearly visible; it is dominated by the umbra, which is evidently lower than the surrounding photosphere. Around it are bright faculae, which extend up into the atmosphere. This image is in contrast to Figs. 4 and 10 depicting sunspots far from the limb, in which little or no facula brightening is readily visible. The faculae have become relatively more prominent in the second image of Fig. 18, partly because the depressed umbral region is substantially obscured by the photosphere in the foreground, while the faculae remain fully visible. In the final image, the spot, which is now quite close to the limb, cannot be seen at all; but faculae are still apparent. This example not only reveals the facular brightening, but also demonstrates that the angular distribution of the brightening is different from that of both the normal photosphere and the sunspots. Sunspot darkening is more evident when viewed from above; facular brightening is more visible, relative to the photosphere, when viewed from the side. This causes the energy from the Sun to be radiated anisotropically. When viewed from the Earth, the sunspots are, on average, the most visible, because they lie in a band near the equator (mainly between latitudes $\pm 30^{\circ}$ or so), which is close to the plane of the ecliptic in which the Earth orbits. If it were to be viewed from the poles, however, the Sun would appear more luminous, because the sunspots would be hardly visible, yet a complete ring of faculae would be seen shining in the vicinity of the limb. Moreover, after a little thought it is evident that when viewed from any latitude away from the equator there is luminal enhancement, although it is less than that when viewed from the poles. The amplitude of solar-cycle luminosity variation, which is an integral over all directions, is therefore greater than that of irradiance variation by about 30%. Of course it is only the irradiance variation that concerns us directly on Earth, for it is that which controls the overall energy budget of the Earth's atmosphere and has an influence on climate. But for those heliophysicists interested in the overall energy budget of the Sun, it is the luminosity that counts. In the past it has been assumed, often without apparent caveat, that the luminosity variation is the same as the irradiance variation. But it is now evident that it is not so.

Finally, I should point out that it is only the temporal variation, not the mean value, of the luminosity of the Sun that is significantly affected by the sunspots and the faculae. As in my discussion of Ap-star spots, on a timescale exceeding the thermal relaxation time of the convection zone – about 10^5 years in the case of the Sun – granted no other agent to produce temporal variation (I ignore the main-sequence evolution of the Sun, which takes place on a 10^{10}-year timescale), the mean luminosity is determined by the rate at which energy is supplied to the convection zone by the radiative envelope, which is itself determined by the energy generated by nuclear reactions in the core, and the outer layers of the Sun just have to adjust to cope with the amount of energy flow required. The energy generation rate depends on the physical conditions in the core, of course, which depends in turn on the weight of the envelope bearing down on it, and on the value of the thermal conductivity of the poorly conducting region beneath the convection zone. Because the convection zone has so little mass, any variation in its structure can have only a very small influence on its weight, and therefore can cause only an almost imperceptible change to the core, leaving the luminosity essentially unchanged.

I have now traveled, rather hastily, from the surface of the Sun to its central core, where the energy that powers the multitude of magnetohydrodynamical processes in the directly observable surface layers is produced. Sunspots are but a single manifestation of these processes, but one which has a long history, and which remains incompletely understood. There is still much research to be carried out, even to acquire a firm understanding of flow discovered by John Evershed 100 years ago.

Acknowledgement I thank Paula Younger for typing the manuscript, Guenter Houdek for converting the figures into a format acceptable to LaTeX, and for assembling an early version of the powerpoint presentation used in the delivery of the lecture, and Sacha Brun for powerpoint advice in its editing. I also thank Rob Rutten for converting this written version of the lecture into the book format.

Part II
Oral Presentations

Angular-Momentum Coupling Through the Tachocline

D.O. Gough

Abstract Astronomical observation of stellar rotation suggests that at least the surface layers of the Sun have lost a substantial amount of the angular momentum that they possessed at the beginning of the main-sequence phase of evolution, and solar-wind observations indicate that magnetic coupling is still draining angular momentum from the Sun today. In addition, helioseismological analysis has shown that the specific angular momentum at the top of the almost uniformly rotating radiative interior is approximately (although not exactly) the same as the spherically averaged value at the base of the (differentially rotating) convection zone, suggesting that angular momentum is being transported through the tachocline. The mechanism by which that transport is taking place is not understood. Nor is there a consensus of opinion. I review some of the suggestions that have been put forward, biassing my discussion, no doubt, according to my own opinions.

1 Introduction

First, I recall that the Sun must have started life with much more angular momentum than it has today. There is considerable circumstantial evidence to support that view, the strongest, in my opinion being that stars are formed from interstellar gas clouds whose typical intrinsic angular momentum per solar mass is some 10^7 times that of the Sun today.

Almost all of the initial angular momenta of almost all stars was lost before the gas was even able to condense to stellar proportions. It is likely that during the later stages of gravitational collapse a circumstellar disc is formed, which removes angular momentum from the star by magnetic coupling. The process by which that occurs is a very important arena of research, and a great deal of very interesting work is being done to understand it. But that is beyond the scope of this discussion.

D.O. Gough (✉)
Institute of Astronomy, University of Cambridge, UK
and
Department of Applied Mathematics and Theoretical Physics, University of Cambridge, UK

S.S. Hasan and R.J. Rutten (eds.), *Magnetic Coupling between the Interior and Atmosphere of the Sun*, Astrophysics and Space Science Proceedings,
DOI 10.1007/978-3-642-02859-5_5,

What I might point out, however, is that angular-momentum loss appears to continue after the star has been established and most of the disc has departed, as is evinced by the observed decline with age in the rotation rates of the surfaces of stars in young clusters. It should also be acknowledged that even today the Sun is losing angular momentum, via magnetic coupling to the solar wind, although at a rate that is rather less than that suggested by the rotation rates obtained by a simple extrapolation to the solar age of apparent angular-momentum loss rates from young-cluster stars.

The "simple" extrapolation is based on the assumption that young stars rotate almost uniformly, as does the Sun today, which may not be realistic. It may be that the mechanical coupling between the surface layers and the deep interior is insufficient to transfer angular momentum fast enough to render the surface rotation rates of young stars, direct indicators of the rotation of their interiors. That would explain the initial rapid slow-down, but it would also imply that the "initial" angular momentum of the main-sequence Sun was lower than average.

Was there a tachocline in those early main-sequence days? And if so, was its structure similar to that of today? I suggest that there was a tachocline, and that its structure was similar, although in the early days the shear was probably stronger.

A second, very important, matter that must be considered is whether the early Sun, and, of course, other similar young stars, harbored substantial large-scale magnetic fields throughout their radiative interiors, and by "substantial" I mean one that is intense enough to transport angular momentum on a timescale not significantly greater than the age of the star. I invite you to entertain the possibility, partly because I regard it as being most unlikely that the Sun had lost all the field that had pervaded the interstellar gas from which it had condensed. Indeed, I believe that there remains a dynamically important residual even today, simply because there appears to be no cogent argument for maintaining the observed uniform rotation of the radiative interior by any nonmagnetic process. I must hastily point out that this opinion is not universally accepted, so I shall return to it later.

Before proceeding, it is useful to state the principal properties of the Sun that must not be ignored: The Sun is basically in hydrostatic balance; approximately the outer 30% by radius (65% by volume, 2% by mass), supporting 15% of the moment of inertia, is in a state of turbulent convection. The rest is relatively quiescent. There, energy that has been generated in the nuclear-reacting core is transported by radiative diffusion. The stratification of both the radiative interior and most of the convection zone has been well established by helioseismology, as also has the angular velocity, Ω, which I now describe.

In the bulk of the convection zone, Ω increases with co-latitude, from about $2.3 \times 10^{-6}\,\mathrm{s}^{-1}$ (corresponding to a period of about 31 days) at the poles (where it cannot be measured accurately) to about $2.86 \times 10^{-6}\,\mathrm{s}^{-1}$ (corresponding to a period of about 25.4 days) at the equator; it is roughly independent of radius, although, as noted by Gilman and Howe (2003), the Ω contours are more nearly inclined by a constant 27° from the axis. The radiative interior (at least outside the energy-generating core in which reliable measurements have not yet been made) rotates approximately uniformly, with angular velocity $\Omega_0 \simeq 2.67\times10^{-6}\,\mathrm{s}^{-1}$, which is 0.93

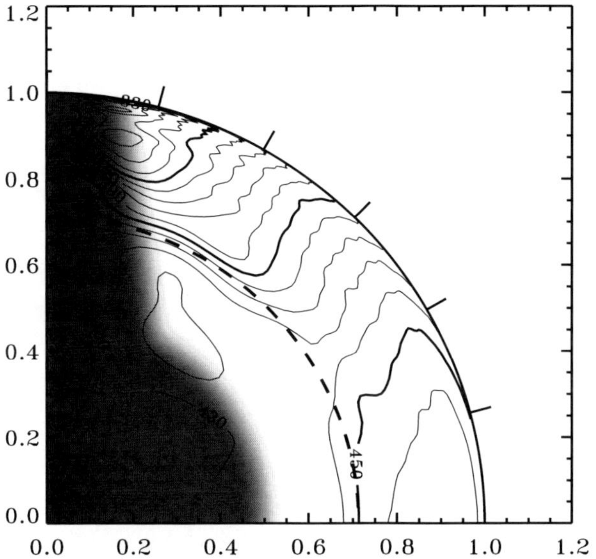

Fig. 1 Contours of constant rotation rate ($\Omega/2\pi$) in a quadrant of the Sun as inferred by Schou et al. (1998). The contours, two of which are labeled, are separated by 10 nHz. The *continuous quarter-circle* represents the photosphere, the *dashed quarter-circle* the base of the adiabatically stratified convection zone. The tachocline is clearly visible, the shear layer extending outwards from the radiative interior into the convection zone, the angular velocity increasing outwards near the equator and decreasing near the poles; the vertical shear vanishes at a latitude of about 30°. The darkened region is where reliable helioseismological inferences could not be drawn

of the equatorial angular velocity in the convection zone (see Fig. 1). The radiative and convective regions are separated by a thin shear layer, known as the tachocline, which is too thin to be reliably resolved directly by seismology (or any other means); indirect estimates range between 2% and about 8% of the solar radius, with some indication that the tachocline is thicker near the poles than near the equator. I should point out that the variation in the estimates is due partly to differences in the way the tachocline is perceived, a matter on which I shall expand later. The base of the convection zone is likely on theoretical grounds to be fairly accurately spherical.

2 Some Initial Remarks on Tachocline Dynamics

My brief for this presentation is to discuss how angular momentum is transported from the radiative interior to the convection zone through the tachocline. It is not unreasonable to suppose that the transport has always been outwards, at least since the Sun arrived on the main sequence. I say this partly because the Sun is, and presumable always has been, losing angular momentum through the photosphere to the solar wind via magnetic stresses, and partly because most of the Sun has been expanding slightly, and preferentially in the convection zone, during its main-sequence evolution (the core has contracted). It is, of course, of interest to understand how the radiative interior has slowed down through this process to its current, uniformly rotating, state, which raises the question of how angular momentum is redistributed to maintain the uniformity of rotation. Here I shall maintain the stance that the rotation is uniform because on timescales of relevance the interior is rigid, held so by a large-scale primordial magnetic field. I shall mention points of view to the contrary at the end of the presentation.

I now discuss a fundamental dynamical phenomenon that occurs universally in rotating shearing fluids, namely that associated with the differential rotation is a meridional circulation. This phenomenon was discussed by Einstein (1926) to explain the meanders of gently flowing rivers, and it is responsible for the rapid spin-down of a stirred cup of tea. Consider the fluid in the cup illustrated in Fig. 2. The fluid is rotating, slowly, about a vertical axis, and the centrifugal force in the body of the fluid, denoted by the double arrows, is resisted by the impermeable wall of the cup, augmenting the pressure to produce an outwardly directed component to the pressure gradient. The higher pressure farther from the axis supports a greater head of fluid (which is essentially in vertical hydrostatic balance under gravity), whose upper surface is therefore concave. Very near the bottom of the cup, however, the rotation of the fluid is impeded by viscous stresses against the rigid base; the centrifugal force is unable to balance the outward component of the pressure gradient (which evidently remains unchanged in view of the vertical hydrostatic

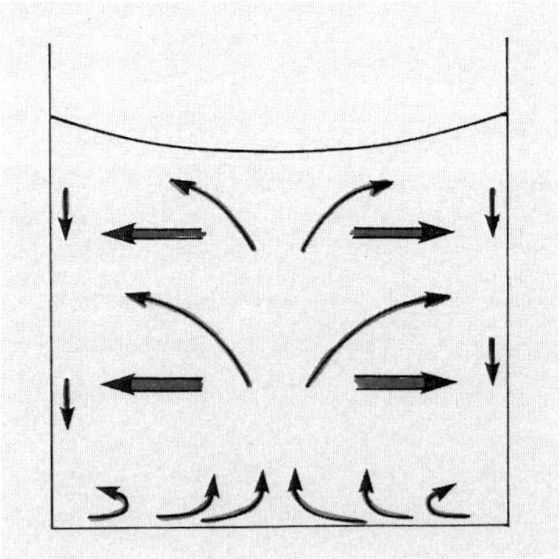

Fig. 2 Cartoon of a stirred cup of tea (with no handle). The *double arrows* represent the centrifugal force excerted on the fluid, which is resisted by the sides of the cup inducing a radial (i.e., pointing away from the axis) pressure gradient, which causes the surface of the tea to be concave. The centrifugal force is much weaker near the bottom of the teacup where viscous stresses inhibit the rotation, leaving a ("statically") unbalanced pressure gradient to drive an inward flow in a thin viscous (Ekman) boundary layer, denoted by the *upwardly curving single arrows*. The return flow completing the circulation is also indicated by *single arrows*

balance), the residual of which drives an inward flow (denoted by the single, curved arrows), drawing in fluid from the side of the cup to replace it. The inward flow is deflected upwards into the body of the fluid, where it expands away from the axis, and subsequently descends near the wall of the cup to complete the circulation. If one were to view the fluid from above, concentrating on a portion of an outer annulus, one can liken it to a bend in a river: the locally rotating stream produced by the bend causes an inward flow near the bottom of the river, which erodes mud, sand, and stones, transporting them from the outer bank to the inner bank and thereby accentuating the bend, resulting finally in a meander.

The inward flow near the base of the teacup occurs in a thin boundary layer, of thickness δ, say, whose basic structure had been analyzed by Vagn Ekman (and has subsequently been named after him) in his doctoral thesis in 1902 for the purpose of explaining why Arctic ice does not drift in the same direction as the prevailing wind. It explains also why a stirred cup of tea slows down faster than the characteristic viscous diffusion time $\tau_{\rm d} = R^2/\nu$, where R is the radius of the teacup and ν is the viscous diffusion coefficient: the angular momentum in the tea is transferred to the cup in the boundary layer on the shorter timescale δ^2/ν, and the angular velocity in the essentially inviscid flow elsewhere declines because the angular-momentum-conserving fluid is drifting away from the axis of rotation (it can be shown that the flow near the outer wall occurs in a layer considerably thicker than δ, so angular-momentum transfer there can be ignored). When all this is put together, the final outcome, as in so many other cases of fluid flow with thin diffusive (boundary) layers, is that global evolution occurs on a timescale, which is the geometric mean of the characteristic dynamical and the large-scale diffusive times. In the case of a cup of tea, $\tau_{\rm d} \simeq 1/2\,\mathrm{h}$, and if one has stirred at a rate of two per second, the dynamical time is 1/2 s; therefore, the tea slows down in about 1/2 min. Note that the spin-down time is also the characteristic circulation timescale.

The solar tachocline is broadly similar. The rigid radiative interior, which is very strongly stably stratified (the characteristic buoyancy restoring timescale is about an hour – the characteristic period of grave g modes – which is very much shorter than even the rotation period), acts like the rigid impermeable base of the teacup, above which is fluid which is caused to rotate differentially by the anisotropic Reynolds stresses, and associated Maxwell stresses, in the convection zone. Even the simplest dynamical description of the boundary layer (tachocline) that separates the two regions is considerably more complicated than an Ekman boundary layer (Gough and McIntyre 1998); the basic dynamics of the meridional circulation in the convection zone is also somewhat different, being maintained by "gyroscopic pumping" by the anisotropic turbulent stresses rather than a spinning down like a cup of tea that is no longer being stirred. Near the equator, where the fluid in the convection zone is rotating faster than the "rigid" interior, there is a (poleward) essentially horizontal flow in the tachocline towards the axis of rotation; contrarily, near the poles, where the convection zone rotates more slowly than the interior, the tachocline flow is equatorward. In both regions the motion draws fluid in from the convection zone (or, if one prefers, is pushed aside by the fluid descending from the convection zone from whence it has been pumped by the rotating anisotropic turbulent stresses). The

Fig. 3 Quadrant of the Sun indicating the gyroscopically driven meridional flow (*lines with arrows*) in the tachocline (from Gough and McIntyre 1998). The width of the tachocline has been artificially exaggerated to render the flow visible. *Circular arcs* represent convective eddies, and the *broad lines* beneath the tachocline represent lines of magnetic field – they are dashed near the upwelling region, where their geometry is even more uncertain than elsewhere and where they are probably drawn into the convection zone, providing a source of Maxwell stress that may locally quench the shear

opposing flows in the tachocline meet somewhere at mid latitudes, and, being unable to penetrate significantly into the radiative zone below, they are forced to rise back into the zone above, as in a cup of tea. The flow is illustrated in Fig. 3, where lines of interior magnetic field are also depicted.

3 The Early History of Dynamical Tachocline Models

The existence of the tachocline was first recognized theoretically by Spiegel (1972) in his study of the spin-down of the Sun. It attracted little attention at the time, and was hardly discussed until the existence of the tachocline was established seismologically (Brown and Morrow 1987; Christensen-Dalsgaard and Schou 1988; Brown et al. 1989; Dziembowski et al. 1989). A prime motivation for the early consequent studies was to explain how the radiative interior of the Sun could rotate uniformly in the face of the nonuniform stresses imposed by the differentially rotating convection zone, although many authors make a different emphasis, addressing why the transition between the differentially rotating convection zone and a putative uniformly rotating interior is so thin. The conclusion is that a thin tachocline is indeed plausible, perhaps even inevitable.

The first dynamical discussion of the matter was provided by Spiegel and Zahn (1992), who proposed that turbulence in the stably stratified tachocline was essentially layerwise two-dimensional, having negligible vertical motion and being horizontally isotropic so that it acted like a powerful (constant) viscosity on horizontal spheres. The turbulence appeared to be generated by the latitudinal rotational shear itself, and was presumed to be sufficiently intense actually to quench at depth the shear that produces it. Subject to those assumptions a tachocline was elegantly constructed so as to reproduce the helioseismological findings.

One of the obvious deficiencies of the Spiegel–Zahn model is that it predicts the wrong angular velocity Ω_0 for the uniformly rotating core: $0.90\,\Omega_{\rm eq}$ rather than the observed value of $0.93\,\Omega_{\rm eq}$, where $\Omega_{\rm eq}$ is the equatorial angular velocity in the convection zone. This quantity had already been discussed by Gough (1985), who pointed out that its value would have been $0.96\,\Omega_{\rm eq}$ had the local shear stress–rate-of-strain relation at the base of the convection been independent of latitude, as would be the case with a thin tachocline exerting a shear stress produced by a uniform isotropic viscosity and (artificially) supporting no meridional flow. It is an important diagnostic of tachocline models, partly because it is an overall measure of the differential angular-momentum transport and partly because its latitudinal average can be measured quite precisely by seismology. Indeed, in those days it appeared that the spherically averaged angular velocity in the radiative interior might drop below even $0.90\,\Omega_{\rm eq}$ (Duvall et al. 1984; Gough 1984), which might actually be the case in the core (Elsworth et al. 1995; Basu et al. 1997). The reduction of Ω_0 in the Spiegel–Zahn model appears to result principally from the latitudinal transport of angular momentum by the shear-turbulent viscous stresses in the tachocline acting upon the meridional circulation, which behave differently from the Reynolds stresses in the convection zone resulting from three-dimensional turbulence driven by buoyancy.

Spiegel and Zahn's assumptions were challenged some time later (Gough and McIntyre 1998), principally because rotating layerwise two-dimensional turbulence in nature does not act in such a way as to quench shear, nor, in general, does any other known purely fluid-dynamical (non-magnetic) motion, such as the dissipation of gravity waves generated near the base of the convection zone, which Kumar and Quataert (1997) and Zahn et al. (1997) had erroneously (e.g. Gough 1997) argued. It was therefore concluded that only if the radiative interior is held rigid by a sufficiently intense magnetic field could its angular velocity be uniform. The flow descending into the tachocline at low and high latitudes (see Fig. 3) advects the outwardly diffusing magnetic field to maintain a field-confining balance, which plays a role in determining the thickness of the tachocline. At mid latitudes, where the rotational shear is small, the ascending return flow is likely to drag field into the convection zone. There the dynamics is difficult to analyze, which is why the field lines in Fig. 3 have been drawn dashed. In their admittedly grossly oversimplified model, Gough and McIntyre assumed the field to be basically dipolar (a reasonable assumption, because the dipole component of a complicated primordial field decays the most slowly), purely poloidal (not so reasonable because such fields are not stable (Markey and Tayler 1973); a numerical simulation of field decay in an idealized star by Braithwaite and Spruit (2004) resulted in a dipole field with a toroidal component), and with, for simplicity, its axis aligned with the axis of rotation.

Although those assumptions are bound not to be satisfied exactly, it is unlikely that any plausible deviation from them would invalidate the basic underlying arguments. In both the Spiegel–Zahn and the Gough–McIntyre models, angular momentum is advected through the bulk of the tachocline by the meridional circulation. It appears to be the case that the overall strength of that flow is insensitive to the details of the tachocline boundary-layer structure, as is perhaps the case also of the cup of tea (recall that the time to spin down a cup of tea is the geometric mean of two global times that take no explicit account of the Ekman boundary-layer structure); the ventilation timescale (quoted by Gough and McIntyre but not explicitly by Spiegel and Zahn) of the meridional circulation in the solar tachocline is of order 10^6 years.

It has often been mentioned that Rüdiger and Kitchatinov (1997) had already pointed out that the radiative interior could be held rigid by a horizontal magnetic field of sufficient intensity, and that the shearing interface could be thin if the vertical variation of the field were somehow maintained. However, no cogent dynamical argument addressing how the field could be confined was offered. Instead, it was merely stated that the field would not diffuse into the convection zone because the (eddy) diffusivity there is so high; no explicit account was taken of the dynamical implications of the necessarily consequent high diffusion rate.

In the model by Gough and McIntyre, an additional angular-momentum conduit is provided by the magnetic field that is presumed to penetrate the tachocline at mid latitudes. In the absence of instability (an unrealistic circumstance, but one to consider as a benchmark), any (imposed) vertical rotational shear would stretch the field into a gyre, creating an increasing opposing torque until diffusion quenches its growth. Perhaps that torque quenches the shear instead, leading to a tachocline with essentially no shear at all (with apologies for the oxymoron) in the region (in each hemisphere) of penetrating field. T. Sekii and I are currently seeking seismological evidence for such a shear-free region, which, according to current seismological inference (see Fig. 1), must be located near latitudes $\pm 30°$.

The 10^6-year tachocline ventilation time is very much shorter than the 10^{10}-year evolution timescale of the Sun. Therefore, on that timescale the net transport of angular momentum across the tachocline essentially vanishes, unless the Sun is undergoing a deeply seated torsional oscillation (for which there is actually some helioseismological evidence, provided by Howe et al. (2000), who report a 1.3-year oscillation in Ω near the equator immediately above and below the tachocline), in which case it is the time-averaged angular momentum transport that vanishes. That constraint, coupled with the (model-dependent) latitudinal variation of the angular-momentum flux, determines the equilibrium (average) angular velocity Ω_0 of the radiative interior.

Note that I have in mind a state of rotation which is essentially steady (with perhaps a gentle superposed oscillation) over the ventilation timescale; on the much longer timescale of the main-sequence evolution of the Sun, a minute shear would transport angular momentum enough to maintain the angular-velocity balance between the braking convection zone and the radiative interior.

I should point out that the confinement of the magnetic field by the generally downwelling meridional flow in the radiative interior, except in the presumably small upwelling zone at mid latitudes, is an essential feature of the dynamics proposed by Gough and McIntyre. Without it, magnetic field would thread the tachocline into the convection zone where it would, in the mean, adopt (much of) the differential rotation of that zone and thereby exert a torque on the radiative region below, causing it to rotate differentially too. Some numerical simulations have provided examples of this general process (e.g., MacGregor and Charbonneau 1999; Garaud 2002; Brun and Zahn 2006).

An essential property of the dynamics, therefore, is that the downwelling velocity is strong enough to dominate the outward diffusion of the field, except of course at the very bottom of the tachocline where there is magnetic boundary layer – the magnetopause – of thickness δ, which is much thinner than the thickness Δ of the essentially diffusion-free body of the tachocline. It goes without saying that in order to model the downwelling from the convection zone, the fluid must be permitted to flow from the convection zone into the tachocline; any theoretical model separating the two regions by an artificial impermeable boundary is bound not to provide a faithful representation of the true dynamics.

4 Some Consequences of Tachocline Circulation

It can hardly pass unnoticed that the latitude at which magnetic field might penetrate the tachocline is the same as that at which sunspots first appear at the start of a new cycle (see Fig. 4). Perhaps it is the radiative interior of the Sun that supplies the

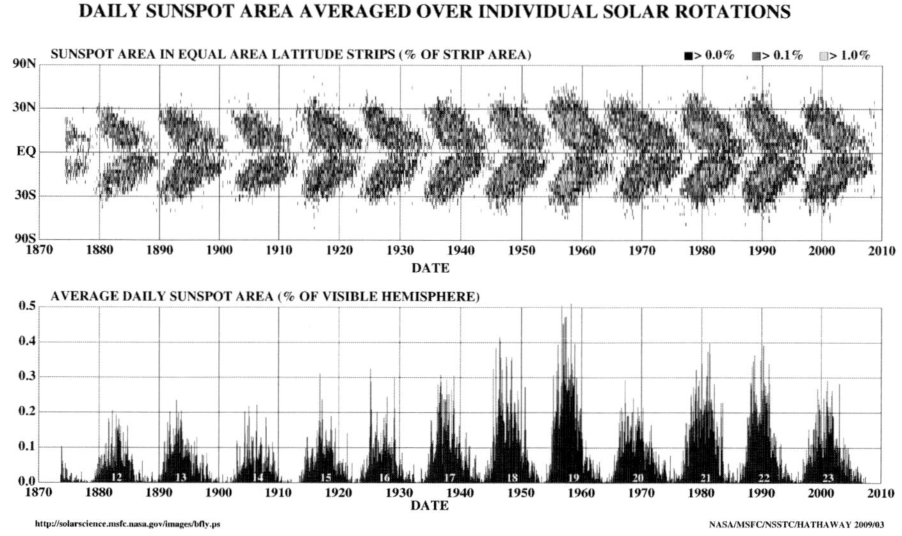

Fig. 4 Sunspot area (averaged over a solar rotation) plotted against time. The *lower panel* is a histogram of the total area; the *upper panel* indicates the latitudes of the spots (courtesy D. Hathaway, NASA Marshall Space Flight Center)

convection zone with a seed field with which to generate the magnetic cycle. I shall leave that remark undeveloped, one on which to ponder. Evidently, magnetohydrodynamical processes, commonly called dynamo action in this context, process the field to produce its cyclic behavior, but those processes may not sustain the field as in a true dynamo, instead declining with the decaying primordial field.

One process of which we can be sure is that the meridional circulation, driven by gyroscopic pumping in the convection zone, ventilates the tachocline and tends to mix any heavy elements that may have settled under gravity back into the convection zone. As the ventilation time is very much less than the age of the Sun, that mixing is essentially complete in the Gough–McIntyre scenario; in the Spiegel–Zahn scenario, the rapid horizontal diffusion opposes the vertical transport, augmenting the mixing time to a substantial fraction of the solar age (Chaboyer and Zahn 1992) and rendering it plausible that mixing is incomplete. In either case, the outcome is a local reduction of the mean molecular mass of the tachocline material below that predicted by standard solar models, and a consequent augmentation of the sound speed, producing what has been called the tachocline anomaly (see Fig. 5). Calibration of this anomaly seismologically is the most precise way to determine the thickness of the tachocline, because it uses the multiplet frequencies (mean over azimuthal order at given principal order and degree) of the seismic modes, rather than

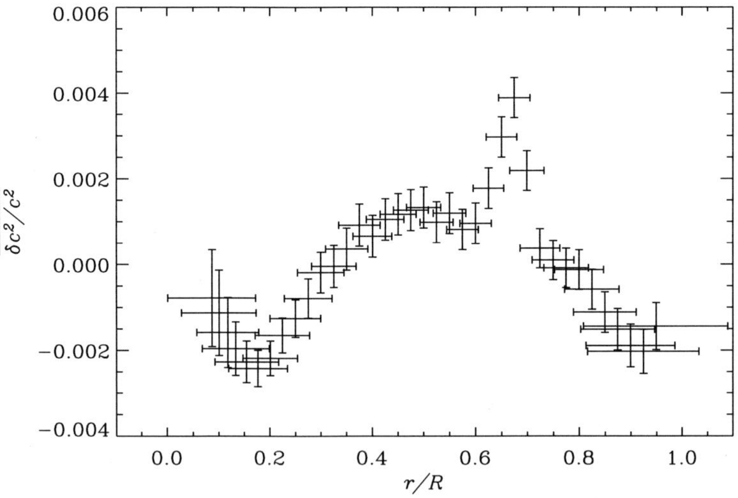

Fig. 5 Localized averages $\overline{\delta c^2/c^2}$ of the relative differences of the (spherically averaged) squares of the sound speeds in the Sun and in a standard theoretical solar model (Model S of Christensen-Dalsgaard et al. 1996), plotted against the location of the centers of the overaging kernels. The *horizontal bars* indicate the extent of the averaging (approximately full width at half maximum of the almost Gaussian averaging kernels), the *vertical bars* the standard errors (which are correlated with one another) arising from the formal uncertainties in the seismic frequency data, which are presumed to be independent of one another. The prominent sharp anomaly is in the region occupied by the tachocline beneath the convection zone, the base of which is situated at radius $r = 0.713\,R$, where here R is the radius of the Sun (Christensen-Dalsgaard et al. 1991)

the less-well determined rotational splitting. However, the outcome depends on the premises upon which the theoretical reference solar model was built, and on the presumed evolution of the thickness of the tachocline. It may therefore not provide the most accurate determination.

It should be pointed out that the molecular-mass gradient (maybe a near discontinuity) created at the base of the tachocline is hydrostatically very stable: that is to say that it cannot be significantly upset by the 10^6-year meridional flow or the magnetic field that the flow confines (a basic point that was missed by Gough and McIntyre when they suggested the possibility of a "polar pit" in the tachocline, which might permit mixing of lithium to depths at which nuclear transmutations could occur today). Therefore, the base of the tachocline is likely to be quite precisely spherical (actually, almost coincident with an equipotential surface).

It is appropriate here to point out that the tachocline calibration yields the thickness of only the layer in the radiative region that is homogenized with the convection zone. That is precisely the layer which Spiegel and Zahn named the tachocline. Spiegel and Zahn, and Gough and McIntyre after them, assumed for simplicity that the natural stresses in the convection zone are so strong that the tachocline does not react back on the angular velocity in the convection zone by an appreciable amount, so that essentially all the rotational shear resides in the radiative layer. Helioseismological evidence suggests that it is not exactly right (see Fig. 1), and that the shear has spread somewhat into the convection zone above, especially at high latitudes where vortex stretching is the greatest. There the dynamics is quite different; so it is perhaps prudent to reserve the term "tachocline" for only the radiative boundary layer, respecting Spiegel and Zahn's precedent notwithstanding its etymological origins. A straightforward calibration of the sound-speed anomaly can then yield a tachocline thickness Δ of about $0.02\,\mathrm{R}_{\odot}$, assuming complete mixing (Elliott and Gough 1999). Incomplete mixing would yield a larger value. (At this point it is worth mentioning that complete mixing down to some level is not consistent with a recent seismological analysis by Christensen-Dalsgaard and Gough (unpublished), but nor is a simple incompletely mixed layer residing entirely beneath the adiabatically stratified convection zone.) Calibrations by fitting parametrized functions to the seismologically inferred shear layer near the base of the convection zone (e.g. Kosovichev 1996; Charbonneau et al. 1999; Corbard et al. 1999), however, depend, of course, on the fitting function adopted. They have yielded values of Δ ranging from about $0.03\,\mathrm{R}_{\odot}$, (near the equator) to $0.08\,\mathrm{R}_{\odot}$, (near the poles). Basu and Antia (2003), for example, have reported that the surface defined by the mid-point of the transition is prolate; one can infer from their results, accepting the quoted uncertainties literally, that the base of the tachocline, defined by the shear, could perhaps be prolate too (with a pole-equator radius difference of about $0.01\,\mathrm{R}_{\odot}$), but in the light of my earlier dynamical remarks on the matter I advise taking such a conclusion with a pinch of salt. There is some seismic evidence for solar-cycle variation of tachocline structure (e.g., Dziembowski and Goode 1989; Baldner and Basu 2008), but its nature is undetermined.

5 The Quest for a Tachocline Simulation

There have been a series of attempts to address whether a magnetic-field-confining tachocline can actually be formed and maintained, using numerical simulations of different degrees of sophistication. Perhaps the most ambitious is a dynamical anelastic computation of Brun and Zahn (2006), who modeled the outer half (by radius) of the radiative zone of a rotating Sun-like star harboring a large-scale magnetic field. The inner and outer boundaries, at radii $0.35\,R_\odot$ and $0.72\,R_\odot$, were presumed to be impermeable but otherwise stress-free, maintaining a constant radial component of the entropy gradient; the magnetic field was matched smoothly onto potential fields outside the domain of computation. Of course, realistic diffusion coefficients could not be adopted; they had to be taken many orders of magnitude greater, but Brun and Zahn wisely maintained the correct ordering of their magnitudes, which they believed captured the essence of the real problem. They chose constant values, formally augmenting the thermal diffusivity κ (in the tachocline) by a factor 6×10^5, the magnetic diffusivity η by 2×10^8 and the viscosity by 3×10^8 (taking the diffusion coefficients in the tachocline to be those quoted by Gough 2007).

Brun and Zahn carried out their computations from a variety of initial states, the most promising (for anyone interested in seeing a positive outcome, one might think) being one with a field well confined within the radiative interior. In all cases, the field diffused through any semblance of a tachocline into the convection zone and took up its differential rotation, imparting it to the radiative interior beneath. In most of the simulations, a potential magnetopause would have occupied the entire tachocline ($\delta \simeq \Delta$, rather than $\delta \ll \Delta$ as found by Gough and McIntyre). A simulation was run with $\delta \simeq \Delta/3$; the differences in the result were barely noticeable. Brun and Zahn concluded that other processes must therefore be invoked to explain the tachocline structure, such as gravity-wave transport of the kind discussed by Charbonnel and Talon (2005) (see also Charbonnel and Talon 2007) or a rigidly generated oscillating dynamo field of the kind considered by Forgács-Dajka and Petrovay (2001, 2002); I shall return to those matters later.

A different approach has been taken by Garaud and her colleagues – the discussion by Garaud (2007) puts it into context – who carried out steady-state axisymmetric computations of a model of the radiative zone, the temporal and geometrical simplifications being made with the intention of rendering it possible to reduce the diffusion coefficients well below those truly attainable in resolved three-dimensional time-dependent simulations, and, it is hoped, to values even low enough for the interior magnetic field to be contained by the tachocline flow. She presumed the convection zone to be magnetically infinitely diffusive, and in her first paper (Garaud 2002) adopted an impermeable but otherwise stress-free boundary between the convection zone and the radiative interior, as did Brun and Zahn (2006) after her. She too always found the magnetic field to penetrate the tachocline, particularly in the polar regions, and so impart a differential rotation on the radiative interior. But with her realization of the importance of the downwelling from the convection zone at low and high latitudes driven by the gyroscopic pumping (Garaud and Brummell 2008), and the development of a new numerical procedure, she succeeded in collaboration with her brother (Garaud and Garaud 2008) to compute models of the outer part of the radiative zone ($0.35\,\mathrm{R}_\odot \leqslant r \leqslant r_c$ with $r_c = 0.7\,\mathrm{R}_\odot$ being the presumed base of the convection zone) in which fluid with yet lesser diffusion is made to descend from and ascend into the convection zone in a prescribed way.

Garaud and Garaud (2008) demonstrated that if the upper boundary were presumed to be impermeable, the vertical flow could never be strong enough to attain a magnetic Reynolds number high enough to contain the field, no matter how low the diffusion coefficients might be, thereby explaining why all previous attempts to obtain a plausible numerical model of the tachocline had failed. When motion through the upper boundary was introduced, however, a dynamical balance between advection and diffusion of magnetic field qualitatively similar to that described by Gough and McIntyre (1998), with angular momentum transported principally by advection in the upper part of the tachocline and by Maxwell stresses in the lower part, could in some situations be achieved, although the solutions also revealed a number of previously unrecognized subtleties. Moreover, the configuration adopted by the flow and the field is not unlike that found in the recent study by Wood and McIntyre (2007), which extended the analysis of the Gough–McIntyre model to the vicinity of the poles.

As in all numerical studies to date, it was necessary for Garaud and Garaud to augment the diffusivities by large factors, which they did in such a way as to maintain the hierarchy of characteristic boundary-layer thicknesses that might be expected to be encountered: the magnetic diffusivity η was augmented by a factor $\tilde{f}$, and the kinematic viscosity ν and radiative thermal diffusivity κ were adjusted accordingly to yield a Prandtl number ν/κ that is 10 times solar and a magnetic Prandtl number ν/η that is 0.1 solar. Figure 6 shows the solution for the least dif-

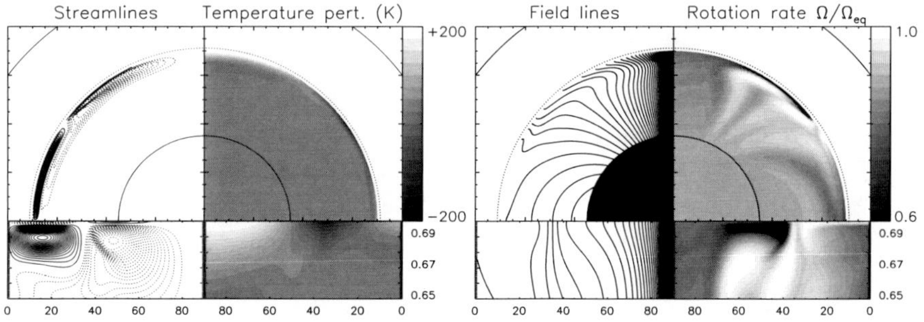

Fig. 6 Tachocline model by Garaud and Garaud (2008), with imposed vertical velocity through the upper boundary as in Fig. 7. In the quadrants are indicated (on the *left*) streamlines – *solid* for clockwise circulation, *dashed* for anticlockwise – and magnetic field lines, and (on the *right*) *greyscale* representations of temperature perturbations from the spherically symmetric background state and the angular velocity Ω measured in units of the equatorial angular velocity Ω_{eq} of the convection zone. Beneath each quadrant is an enlargement of the tachocline region, plotted against latitude and radius

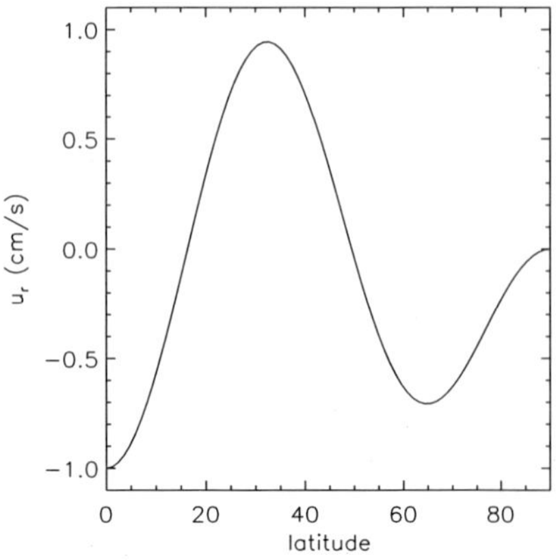

Fig. 7 Radial velocity component imposed by Garaud and Garaud (2008) at the base of the putative convection zone, plotted as a function of latitude

fusive case that Garaud and Garaud were able to compute with an upper permeable boundary, having $\tilde{f} = 8 \times 10^9$. (It should be pointed out that this value is actually more than 10 times higher than what was formally employed by Brun and Zahn (2006), but on the other hand, the spatial resolution of the numerical scheme that was employed was more than 10 times finer, sufficient to resolve the boundary layers that were anticipated. Garaud and Garaud did reproduce the Brun–Zahn coefficients with an impermeable boundary.) Illustrated are streamlines, magnetic field lines, and greyscale representations of temperature perturbations and the angular velocity in quadrants of the Sun, underneath which are enlargements of the tachocline region ($r > 0.65\,R_\odot$) plotted as functions of latitude and radius, r. It is evident that a tachocline has been established, and that, for the first time in a global simulation, the large-scale behavior of the differential rotation of the convection zone has not been imprinted at great depths in the radiative interior. There is some spatial variation in the interior angular velocity, however, on a relatively small scale, which is not evident in Fig. 1. Garaud and Garaud suggest that the averaging kernels for Fig. 1 are broad enough not to have resolved the structure, and that in any case the features are probably much stronger than is likely to exist in the Sun, because it had been necessary to augment the magnitudes of both the imposed flow velocities and the magnetic field to produce a magnetic Reynolds number that exceeds unity with the enormously increased diffusivities that it had been necessary to adopt.

In the face of these encouraging beginnings, there remain some outstanding issues. The most blatant is the angular velocity Ω_0 of the inert core, which is in torque-free equilibrium with the outer part of the radiative zone. Its value is $0.875\,\Omega_{\rm eq}$, whereas seismological analysis of the Sun, as I have already pointed out, yields $0.93\,\Omega_{\rm eq}$, $\Omega_{\rm eq}$ being the equatorial angular velocity of the convection zone. The discrepancy is unexplained, although Garaud and Guervilly (2008) are making some progress towards rectifying that. The discrepancy may be merely a product of not having achieved a low-enough value of the diffusivity augmentation factor $\tilde{f}$.

Another obvious deficiency of the model is that the vertical flow through the base of the convection zone has been imposed artificially. A significant advance would be to couple the tachocline to a simple yet not too unrealistic model of the convection zone, with, say, a representation of the effect of the anisotropic Reynolds stresses on the large-scale flow to generate the gyroscopic pumping. I eagerly await the outcome of such a calculation.

6 Closing Remarks

I conclude with three brief remarks concerning alternative views of the overall dynamics. I trust it is evident, even if I have not given it as much emphasis as others might, that the general picture that I have advocated is far from being universally accepted. As I pointed out in the previous section, Brun and Zahn (2006), for example, believe that the principal model I have discussed lacks some further essential ingredient. I shall do little more than simply mention them.

My first point concerns the angular-momentum transport by the layerwise two-dimensional turbulence envisaged by Spiegel and Zahn (1992) in their tachocline model. As Gough and McIntyre (1998) pointed out, such motion is not expected to lead to diffusive mixing of angular velocity – indeed it is not even everywhere diffusive of anything – but in places transports angular momentum in a wave-like manner – it is what some people have called antidiffusive, with a tendency to mix potential vorticity (Garaud 2001; see also Gough and Lynden-Bell 1968). McIntyre (2003) provides a more extensive discussion, citing a critical analysis by Haynes et al. (1991) showing that at least in the rectilinear case straightforward local momentum mixing by turbulence is not possible without accompanying wave transport. However, there is a more recent, and rather different, numerical simulation by Miesch (2003), which appears to lead Miesch to conclude the contrary, and which therefore demands comment. Miesch considered a spherical shell of fluid between impenetrable, isothermal, horizontal, stress-free boundaries, initially rotating uniformly with angular velocity Ω_0. He supplied the vertical component of the momentum equation with an axisymmetric torque whose magnitude declined with depth, to mimic forcing from above by differentially rotating overshooting convection. He also added a distribution of random sources of either vertical vorticity, to generate Rossby waves, or horizontal velocity divergence, to generate gravity waves, the locations of those sources rotating rigidly (not moving with the fluid), also with angular velocity Ω_0. Not surprisingly, wave drag from dissipation tended to move the differentially forced flow towards the state of uniform rotation assumed by the externally imposed grid of sources, which Miesch interpreted as a tendency for internally generated turbulence to oppose rotational shear. Evidently, a similar calculation in which the wave sources are contained by the fluid, the fluid receiving no externally imposed torque other than the differential forcing from above, would be much more informative.

Propagation and subsequent dissipation of downwardly propagating gravity waves generated in the convection zone is also a way of transporting angular momentum through the tachocline, but when such transport was first mooted as a way of modifying the angular velocity of the Sun (Gough 1977), it was considered not to lead to uniform rotation. I have in mind the mechanism now believed to drive the quasibiennial oscillation of the Earth's atmosphere, a mechanism which was dramatically demonstrated by Plumb and McEwan (1978), and which arises from the phenomenon that prograde gravity waves, transporting positive angular momentum, dissipate more readily than retrograde waves, at a rate which increases with increasing angular velocity of the fluid, thereby enhancing shear. However, this mechanism has not been generally accepted by astrophysicists as being the dominant determinant of gravity-wave angular-momentum transport through the Sun. There are two principal issues: the first is whether or not the waves are generated to an amplitude great enough for the transport to be significant, the second is whether they enhance the shear or suppress it. These issues have been reviewed by Gough (2002) and, from a rather different perspective, by Charbonnel and Talon (2007). I think it is true to say that they have not yet been resolved to everyone's satisfaction. So the debate will continue.

Finally, I comment on a claim by Forgács-Dajka and Petrovay (2001, 2002; also Petrovay 2003; Forgács-Dajka 2004) that an oscillating horizontal magnetic field, presumed to have been generated by a dynamo in the convection zone and diffusing into the radiative interior of the Sun from above, applies a stress on the fluid, which opposes other agents acting on the fluid that tend to transfer latitudinal differential rotation. It is argued that it causes the radiative zone to rotate rigidly, the transition occurring on a lengthscale which, with a suitable choice of conditions, could correspond to the solar tachocline. Even though the early work was criticized at the Solar Tachocline meeting held at the Isaac Newton Institute of Mathematical Sciences, Cambridge in 2004 (Gough 2007), for presuming the field presented to the tachocline to be latitudinally unsheared, and thereby imparting an unsheared Maxwell stress on the radiative zone which would naturally tend to oppose any other tendency to shear, the idea in some circles appears to have been given almost equal credance to the more sophisticated dynamical discussions (e.g. Zahn 2007; Brun and Zahn 2006). Therefore, it behoves anyone wanting to establish the true dynamical balance to look at this model more critically. To this end, Jean-Paul Zahn and I have initiated a simple calculation to assess the effect of the convection-zone shear on the Maxwell stresses and to estimate the outcome for a Forgács-Dajka and Petrovay model. The idea is to imagine the dynamo field to be generated by a suitable imposed current distribution in the convection zone, and to calculate the resulting magnetic field assuming the convection zone to be rotating as is observed, endowing the fluid in the convection zone with a uniform scalar turbulent magnetic diffusivity. We have not yet reached an agreed conclusion concerning the dynamical response of the nonturbulent fluid beneath. But what we have established is that if the radiative interior were initially rotating uniformly, the stress transported would have been such as to induce latitudinal differential rotation in the same sense as that of the convection zone, as one would expect. Whether the subsequent evolution would induce a latitudinal shear that is intense enough to be tested by seismology remains to be seen.

Acknowledgement I am very grateful to Paula Younger for typing the manuscript, to Günter Houdek and Jesper Schou for helping with the figures, and to Rob Rutten for conversion to the publisher's format.

References

Baldner, C. S., Basu, S. 2008, ApJ, 686, 1349
Basu, S., Antia, H. M. 2003, ApJ, 585, 553
Basu, S., Christensen-Dalsgaard, J., Chaplin, W. J., et al. 1997, MNRAS, 292, 243
Braithwaite, J., Spruit, H. C. 2004, Nat, 431, 819
Brown, T. M., Christensen-Dalsgaard, J., Dziembowski, W. A., et al. 1989, ApJ, 343, 526
Brown, T. M., Morrow, C. A. 1987, ApJ, 314, L21
Brun, A. S., Zahn, J.-P. 2006, A&A, 457, 665
Chaboyer, B., Zahn, J.-P. 1992, A&A, 253, 173
Charbonneau, P., Dikpati, M., Gilman, P. A. 1999, ApJ, 526, 523
Charbonnel, C., Talon, S. 2005, Science, 309, 2189
Charbonnel, C., Talon, S. 2007, In: Unsolved Problems in Stellar Physics: A Conference in Honor of Douglas Gough, R. J. Stancliffe, J. Dewi, G. Houdek, R. G. Martin and C. A. Tout (eds.), Am. Inst. Phys. Conf. Series, vol. 948, p. 15
Christensen-Dalsgaard, J., Dappen, W., Ajukov, S. V., et al. 1996, Science, 272, 1286
Christensen-Dalsgaard, J., Gough, D. O., Thompson, M. J. 1991, ApJ, 378, 413
Christensen-Dalsgaard, J., Schou, J. 1988, In: Seismology of the Sun and Sun-Like Stars, E. J. Rolfe (ed.), ESA-SP, vol. 286, p. 149
Corbard, T., Blanc-Féraud, L., Berthomieu, G., Provost, J. 1999, A&A, 344, 696
Duvall, Jr., T. L., Dziembowski, W. A., Goode, P. R., et al. 1984, Nat, 310, 22
Dziembowski, W. A., Goode, P. R. 1989, ApJ, 347, 540
Dziembowski, W. A., Goode, P. R., Libbrecht, K. G. 1989, ApJ, 337, L53
Einstein, A. 1926, Naturwissenschaften, 14, 223
Elliott, J. R., Gough, D. O. 1999, ApJ, 516, 475
Elsworth, Y., Howe, R., Isaak, G. R., et al. 1995, Nat, 376, 669
Forgács-Dajka, E. 2004, A&A, 413, 1143
Forgács-Dajka, E., Petrovay, K. 2001, Solar Phys., 203, 195
Forgács-Dajka, E., Petrovay, K. 2002, A&A, 389, 629
Garaud, P. 2001, MNRAS, 324, 68
Garaud, P. 2002, MNRAS, 329, 1
Garaud, P. 2007, In: The Solar Tachocline, D. W. Hughes, R. Rosner, N. O. Weiss (eds.), p. 147
Garaud, P., Brummell, N. H. 2008, ApJ, 674, 498
Garaud, P., Garaud, J.-D. 2008, MNRAS, 391, 1239
Garaud, P., Guervilly, C. 2008, ArXiv e-print
Gilman, P. A., Howe, R. 2003, In: GONG+ 2002. Local and Global Helioseismology: The Present and Future, H. Sawaya-Lacoste (ed.), ESA-SP, vol. 517, p. 283
Gough, D. O. 1977, In The Energy Balance and Hydrodynamics of the Solar Chromosphere and Corona, R. M. Bonnet, Ph. Delache (eds.), IAU Colloq. vol. 36, p. 3
Gough, D. O. 1984, Adv. Space Res., 4, 85
Gough, D. O. 1985, In: Future Missions in Solar, Heliospheric and Space Plasma Physics, E. J. Rolfe, B. Battrick (eds.), ESA-SP, vol. 235, p. 183

Gough, D. O. 1997, Nat, 388, 324
Gough, D. O. 2002, In: From Solar Min to Max: Half a Solar Cycle with SOHO, A. Wilson (ed.), ESA-SP, vol. 508, p. 577
Gough, D. O. 2007, In The Solar Tachocline, D. W. Hughes, R. Rosner, N. O. Weiss (eds.), p. 3
Gough, D. O., Lynden-Bell, D. 1968, J. Fluid Mech., 32, 437
Gough, D. O., McIntyre, M. E. 1998, Nat, 394, 755
Haynes, P. H., McIntyre, M. E., Shepherd, T. G., Marks, C. J., Shine, K. P. 1991, J. Atmos. Sciences, 48, 651
Howe, R., Christensen-Dalsgaard, J., Hill, F., et al. 2000, Science, 287, 2456
Kosovichev, A. G. 1996, ApJ, 469, L61
Kumar, P., Quataert, E. J. 1997, ApJ, 475, L143
MacGregor, K. B., Charbonneau, P. 1999, ApJ, 519, 911
Markey, P., Tayler, R. J. 1973, MNRAS, 163, 77
McIntyre, M. E. 2003, In: Stellar Astrophysical Fluid Dynamics, M. J. Thompson, J. Christensen-Dalsgaard (eds.), Cambridge University Press, p. 111
Miesch, M. S. 2003, ApJ, 586, 663
Petrovay, K. 2003, Solar Phys., 215, 17
Plumb, R. A., McEwan, A. D. 1978, J. Atmos. Sciences, 35, 1827
Rüdiger, G., Kitchatinov, L. L. 1997, Astron. Nachrichten, 318, 273
Schou, J., Antia, H. M., Basu, S., et al. 1998, ApJ, 505, 390
Spiegel, E. A. 1972, NASA Special Publication, 300, 61
Spiegel, E. A., Zahn, J.-P. 1992, A&A, 265, 106
Wood, T. S., McIntyre, M. E. 2007, In: Unsolved Problems in Stellar Physics: A Conference in Honor of R. J. Stancliffe, J. Dewi, G. Houdek, R. G. Martin and C. A. Tout (eds.), Am. Inst. Phys. Conf. Series, vol. 948, p. 303
Zahn, J.-P. 2007, In: The Solar Tachocline, D. W. Hughes, R. Rosner, N. O. Weiss (eds.), p. 89
Zahn, J.-P., Talon, S., Matias, J. 1997, A&A, 322, 320

Outstanding Issues in Solar Dynamo Theory

D. Nandy

Abstract The magnetic activity of the Sun, as manifested in the sunspot cycle, originates deep within its convection zone through a dynamo mechanism, which involves nontrivial interactions between the plasma and the magnetic field in the solar interior. Recent advances in magnetohydrodynamic dynamo theory have led us closer towards a better understanding of the physics of the solar magnetic cycle. In conjunction, helioseismic observations of large-scale flows in the solar interior has now made it possible to constrain some of the parameters used in models of the solar cycle. In the first part of this review, I briefly describe this current state of understanding of the solar cycle. In the second part, I highlight some of the outstanding issues in solar dynamo theory related to the nature of the dynamo α-effect, magnetic buoyancy, and the origin of Maunder-like minima in activity. I also discuss how poor constraints on key physical processes such as turbulent diffusion, meridional circulation, and turbulent flux pumping confuse the relative roles of these vis-a-vis magnetic flux transport. I argue that unless some of these issues are addressed, no model of the solar cycle can claim to be "the standard model," nor can any predictions from such models be trusted; in other words, we are still not there yet.

1 Introduction

Sunspots have been telescopically observed for centuries, starting with the pioneering observations of many, including Galileo Galilei in the early seventeenth century. Much later in the twentieth century, Schwabe discovered that the number of sunspots on the solar surface vary cyclically, and Carrington discovered that the sunspots appear at lower and lower solar latitudes with the progress of the cycle. With Hale's discovery of magnetic fields within sunspots in 1908, it became clear that the sunspot cycle is in fact a magnetic cycle (see Fig. 1 for an overview of the magnetic butterfly diagram). Efforts to theoretically explain the origin of the solar

D. Nandy (✉)
Indian Institute of Science Education and Research, Kolkata, India

S.S. Hasan and R.J. Rutten (eds.), *Magnetic Coupling between the Interior and Atmosphere of the Sun*, Astrophysics and Space Science Proceedings,
DOI 10.1007/978-3-642-02859-5_6,

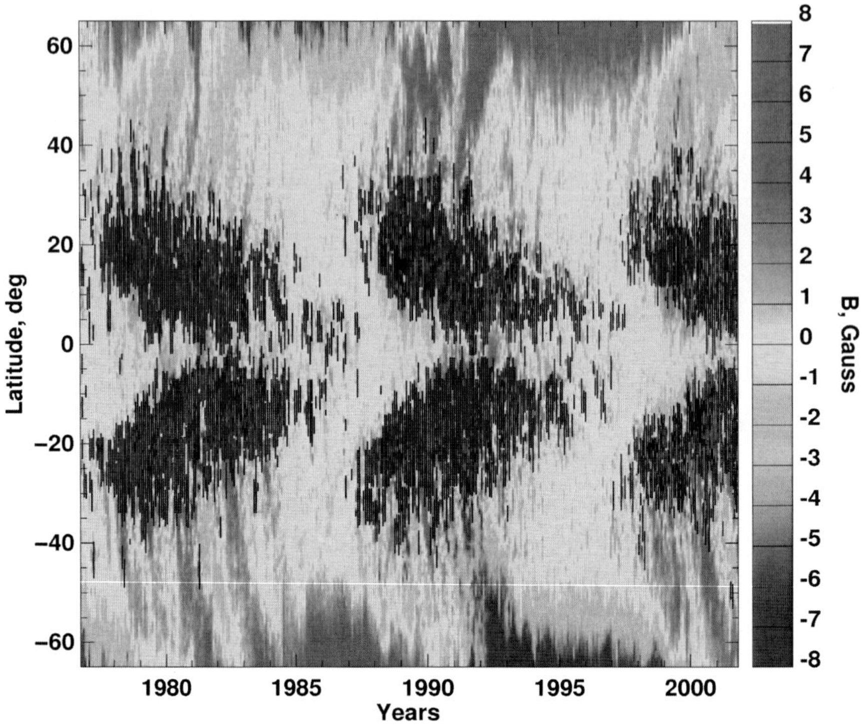

Fig. 1 The solar butterfly diagram depicting the latitude of sunspot appearance (*think dark lines*) with time. The background shows the weak and diffuse field outside of sunspots. Note that while the sunspot formation belt migrates equatorward, the weak field outside of it migrates poleward with the progress of the cycle, reversing the older polar field at the time of sunspot maximum

cycle continued from then on and took a giant leap in 1955 when Parker outlined his theory of the solar cycle based on a magnetohydrodynamic (MHD) dynamo mechanism.

In what follows, I briefly summarize the important concepts underlying the solar cycle that have been developed in the last half of the twentieth century (Sect. 2). I then describe the current state of our understanding, concentrating on those ideas that are widely accepted as important for dynamo action (Sect. 3). Following this, I highlight the outstanding issues that need to be addressed towards developing a "standard model" of the solar cycle (Sect. 4). Finally, I end with some concluding remarks (Sect. 5).

Before we proceed, it is important here to state the scope of this review; this is neither meant to be a comprehensive review of all complementary ideas in solar dynamo theory and modeling, nor is it a reference source for important works in this field. Interested readers who desire these are referred to the recent and comprehensive review on the solar dynamo by Charbonneau (2005). This is a personalized account of the field as I perceive it to be.

2 Basic Concepts

The interior of the Sun consists of highly ionized gas, that is, plasma. The fundamental equation that governs the behavior (and generation) of magnetic fields in such a plasma system is the induction equation

$$\frac{\partial \mathbf{B}}{\partial t} = \nabla \times (\mathbf{v} \times \mathbf{B} - \eta \nabla \times \mathbf{B}), \tag{1}$$

where $\mathbf{B}$ is the magnetic field, $\mathbf{v}$ the velocity field, and η the effective magnetic diffusivity of the system. In astrophysical systems such as the Sun, the plasma has a very high characteristic magnetic Reynolds number (the ratio of the first to the second term on the R.H.S. of the above equation). In such a plasma, the magnetic fields are frozen in the fluid and therefore the field and plasma movement are coupled. This allows the energy of convective flows in the solar convection zone (SCZ) to be drawn into producing and amplifying magnetic fields, which is the essence of the dynamo mechanism.

Under the approximation of spherical symmetry, applicable to a star such as the Sun, the magnetic and velocity fields can be expressed as

$$\mathbf{B} = B_\phi \hat{\mathbf{e}}_\phi + \nabla \times (A \hat{\mathbf{e}}_\phi), \tag{2}$$

$$\mathbf{v} = r \sin(\theta) \Omega \hat{\mathbf{e}}_\phi + \mathbf{v}_p. \tag{3}$$

The first term on the R.H.S. of (2) is the toroidal component (i.e, in the ϕ-direction) and the second term is the poloidal component (i.e., in the r-θ plane) of the magnetic field. In the case of the velocity field (3), these two terms correspond to the differential rotation Ω and meridional circulation v_{p}, respectively. The field of helioseismology has now constrained the profile of the solar differential rotation throughout the solar convection zone and it is therefore no longer a free parameter in models of the solar dynamo. The meridional circulation is observed in the surface and helioseismic inversions constrain it somewhat in the upper 10% of the Sun; however, the deeper counter-flow is not yet observed and is theoretically constructed by invoking mass conservation in conjunction with the solar density stratification.

As the Sun rotates differentially, any preexisting poloidal field would get stretched in the direction of rotation, creating a toroidal component. Such horizontal toroidal flux tubes in the solar interior are subject to magnetic buoyancy (Parker 1955a) and therefore erupt out through the surface creating bipolar sunspot pairs. These bipolar sunspot pairs acquire a tilt due to the action of Coriolis force during their rise through the solar convection zone (SCZ), generating what is commonly known as the Joy's law distribution of solar active region tilt angles. To complete the dynamo chain of events, the toroidal component of the magnetic field has to be converted back into the poloidal component. This necessitates the action of a non-axisymmetric mechanism, that is, with nonzero vorticity. The first such proposed mechanism was due to Parker (1955b) who proposed that small-scale helical turbulence can twist rising toroidal flux tubes back into the r-θ plane, thereby recreating the poloidal field – a process that traditionally came to be known as the dynamo α-effect; this became an essential ingredient in models of the solar dynamo.

3 Current State of Our Understanding

In the last two decades, simulations of the dynamics of buoyantly rising (thin) toroidal flux tubes showed that the initial strength of these flux tubes at the base of the SCZ had to be on the order of 10^5 G to match the morphological properties of active regions observed at the solar surface (D'Silva and Choudhuri 1993; Fan et al. 1993). However, the equipartition field strength of the magnetic field in the SCZ (i.e., the field strength at which magnetic and convective flow energies are in equipartition) is of the order of 10^4 G. If the strength of the sunspot forming toroidal flux tubes are an order of magnitude greater than the equipartition field strength, then the helical convective flows would be unable to twist them as envisaged in the traditional dynamo α-effect formalism.

This realization has now led the dynamo community to explore alternative mechanisms for the regeneration of the poloidal field. Amongst the various contenders, the so-called Babcock–Leighton (BL) mechanism is perceived to be the front-runner. In this mechanism, originally proposed by Babcock (1961) and Leighton (1969), the decay of tilted bipolar sunspot pairs and the subsequent (net) poleward dispersal of their flux by surface processes such as diffusion, differential rotation, and meridional circulation regenerates and reverses the solar poloidal field (Dikpati and Charbonneau 1999; Nandy and Choudhuri 2001). Although this process is now commonly referred to as the BL α-effect, it may be noted that in spirit, this process is very different from the traditional α-effect; in the latter formalism, averaging over small scale turbulence and the first-order-smoothing-approximation is required, while in the former, it is not. The BL mechanism for poloidal field generation is actually observed on the surface and has been substantiated with numerical surface-flux-transport simulations.

As this poloidal field generation mechanism is primarily located in the near-surface layers, the generated poloidal field has to be transported back into the solar interior, where the toroidal field amplification and storage takes place (in the overshoot layer at the base of the SCZ). In most BL models of the solar cycle, this flux transport is achieved by meridional circulation, although turbulent diffusion is also expected to play a significant role. In such models, it is found that the meridional circulation governs the spatio-temporal distribution of sunspots on the solar surface (Nandy and Choudhuri 2002) and its speed determines the period of the solar cycle, even in regimes where the SCZ is diffusion dominated (Yeates et al. 2008).

The amplitude of the sunspot cycle is found to be weakly correlated with the speed of the meridional circulation and the coefficient of turbulent diffusion. However, the threshold of magnetic buoyancy, that is, the field strength at which stored toroidal flux tubes become magnetically buoyant and escape out of the overshoot layer, is a limiting factor on the amplitude of the solar cycle (Nandy 2002).

4 Outstanding Issues

Although it seems that we have made much progress in the last decade or so in understanding many aspects of the solar cycle, this progress has also uncovered multiple aspects that pose a challenge to dynamo theory. I take this opportunity to discuss some of these outstanding issues.

4.1 *Nature of the Dynamo α-Effect*

While the BL mechanism for poloidal field regeneration is observed on the solar surface, a few other α-effect mechanisms have been proposed in recent times, which do not have any direct observational confirmation, but nevertheless may be functional in the solar interior. These α-effects are driven by magnetic field instabilities or differential rotation instabilities and are spatially located around the base of the SCZ (for an overview of the various proposed dynamo α-effects, see Charbonneau 2005).

What is unclear is the extent to which these proposed α-effects may contribute to poloidal field regeneration in the Sun and the relative efficacy of these compared to the BL mechanism; this is connected to the following question (I believe first articulated by Manfred Schüssler): Is the observed BL mechanism a by-product of a dynamo mechanism that completely resides in the solar interior, or is it actually an integral part of the dynamo mechanism? Although the success of dynamo models based on the BL mechanism argues for the latter scenario, in my view we still cannot rule out the possibility that other α-effect mechanisms may contribute at least in parts to poloidal field generation. If this were to be the case, it creates an interesting dilemma, which is explained below.

For some time now, many of us are content with the perception that we can observe one complete half of the solar dynamo mechanism, namely the poloidal field regeneration at the solar surface through active regions decay and dispersal. These surface observations have been used widely to constrain and fine tune dynamo models. Moreover, because the poloidal field of a given cycle feeds directly into producing the toroidal field of future cycle(s), these surface observations provide an useful tool to predict future cycle amplitudes. However, if some other observationally unconstrained mechanism for poloidal field regeneration is actually more dominant than the BL mechanism, this would pose a serious challenge to our current perceptions of the solar cycle and would negatively impact attempts to predict future solar activity. Therefore, any evidence related to mechanisms of solar poloidal field generation has to be seriously evaluated to illuminate whether there are multiple mechanisms, and if yes, what are their relative contributions to the overall dynamo.

4.2 *Treatment of Magnetic Buoyancy*

An issue coupled to the nature of the poloidal field regeneration mechanism is the treatment magnetic buoyancy and bipolar sunspot creation in models of the solar cycle. It is believed that those strong toroidal flux tubes stored in the stably stratified region beneath the SCZ, which exceed a certain threshold (on the order of 100 kG), become magnetically buoyant when they emerge out in the SCZ and subsequently produce poloidal field. This whole process of buoyant eruption and poloidal field generation is treated in dynamo models through diverse implementations – almost all of which are not fully consistent with the philosophy of the BL mechanism.

The most popular approach, in the context of the BL dynamo, has been to approximate this process with a poloidal field source term that is located at near-surface layers (constrained by a prescribed spatially dependent function) and which is proportional to the toroidal field strength at the base of the SCZ. Although this approach typically has a upper quenching threshold, which stops poloidal field creation when the toroidal field exceeds a certain threshold (in accordance with flux tube rise simulations which show that very strong toroidal flux tubes come out without any tilt), most modelers do not use a lower operational threshold. This causes even weak, sub-kG toroidal field to contribute to poloidal field creation through the BL mechanism, which goes against the spirit of the BL idea. An alternative approach, which has been used by some modelers, is to employ an explicit algorithm for magnetic buoyancy, which searches for strong toroidal fields exceeding the buoyancy threshold and transporting this field to the surface layers, conserving flux in the process. However, this process too over-simplifies the surface-flux-dispersal process as it still uses a source term at the surface. Basically, the usage of this source-term preempts the surface-flux-transport process, which results in dynamo simulations giving results that are not consistent with surface-flux-transport simulations when a variable meridional flow is used (Schrijver and Liu 2008).

A more realistic but rarely used approach is to buoyantly erupt spaced double rings of opposite radial field (akin to bipolar sunspot pairs) to the solar surface using a explicit buoyancy algorithm (for a comparative study of various buoyancy algorithms, including the double-ring approach, see Nandy and Choudhuri 2001). The source-term is discarded with in this approach. Surface differential rotation, meridional circulation, and diffusion subsequently acts on these erupted double rings to generate the poloidal field in a truer representation of the BL philosophy. However, this approach is computationally intensive as it demands a very high grid-resolution, which is sufficiently close to solar AR spatial-scale. This being a impractical task, the over-simplified and somewhat questionable buoyancy prescriptions continue to be used in solar dynamo models. If an alternative, physically correct algorithm cannot be devised, it seems the brute-force solution to this problem is to implement more computationally efficient numerical algorithms for the solar dynamo that can handle very high grid-resolutions.

4.3 Origin of Grand Minima

Small, but significant variations in solar cycle amplitude is commonly observed from one cycle to another, and models based on either stochastic fluctuations, or nonlinear feedback, or time-delay dynamics exist to explain such variability in cycle amplitude (for overviews, see Charbonneau 2005; Wilmot-Smith et al. 2006). However, most models find it difficult to switch off the sunspot cycle completely for an extended period of time – such as that observed during the Maunder minimum – and subsequently recover back to normal activity.

Two important and unresolved questions in this context are what physical mechanism stops active region creation completely and how does the dynamo recover from this quiescent state. The first question is the more vexing one and still eludes a coherent and widely accepted explanation. The second question is less challenging in my opinion; the answer possibly lies in the continuing presence of another α-effect (could be the traditional dynamo α-effect suggested by Parker), which can work on weaker, sub-equipartition toroidal fields – to slowly build up the dynamo amplitude to eventually recover the sunspot cycle from a Maunder-like grand minima.

These are speculative ideas and one thing that can be said with confidence at this writing is that we are just scratching the surface as far as the physics of grand minima like episodes is concerned.

4.4 Parametrization of Turbulent Diffusivity

Typically, in many dynamo models published in the literature, the coefficient of turbulent diffusivity employed is much lower than that suggested by mixing-length theory (about $10^{13}\,\mathrm{cm}^2\,\mathrm{s}^{-1}$; Christensen-Dalsgaard et al. 1996). This is done to ensure that the flux transport in the SCZ in advection dominated (i.e., meridional circulation is the primary flux transport process). There are many disadvantages to using a higher diffusivity value in these dynamo models. Usage of higher diffusivity values makes the flux transport process diffusion dominated, reducing the dynamo period to values somewhat lower than the observed solar cycle period. It also makes flux storage and amplification difficult and shortens cycle memory; the latter is the basis for solar cycle predictions. Nevertheless, this inconsistency between mixing-length theory and parametrization of turbulent diffusivity in dynamo models is, in my opinion, a vexing problem.

In the absence of any observational constraints on the depth-dependence of the diffusivity profile in the solar interior, this problem can be addressed only theoretically. One possible solution to resolving this inconsistency is by invoking magnetic quenching of the mixing-length theory suggested diffusivity profile. The idea is simple enough; as magnetic fields have an inhibiting effect on turbulent convection, strong magnetic fields should quench and thereby be subject to less diffusive mixing. The magnetic quenching of turbulent diffusivity is challenging to implement numerically, but seems to me to be the best bet towards reconciling this inconsistency within the framework of the current modeling approach.

4.5 Role of Downward Flux Pumping

An important physical mechanism for magnetic flux transport has been identified recently from full MHD simulations of the solar interior. This mechanism, often referred to as turbulent flux pumping, pumps magnetic field preferentially downwards, in the presence of rotating, stratified convection such as that in the SCZ (see, e.g., Tobias et al. 2001). Typical estimates yield a downward pumping speed, which can be as high as $10\,\mathrm{m\,s^{-1}}$; this would make flux pumping the dominant downward flux transport mechanism in the SCZ, short-circuiting the transport by meridional circulation and turbulent diffusion. However, turbulent flux pumping is usually ignored in kinematic dynamo models of the solar cycle.

If indeed the downward pumping speed is as high as indicated, then turbulent flux pumping may influence the solar cycle period, crucially impact flux storage and amplification, and also affect solar cycle memory. Therefore, turbulent flux pumping must be properly accounted for in kinematic dynamo models and its effects completely explored; this remains an issue to be addressed adequately.

5 Concluding Remarks

Now let us elaborate on and examine some of the consequences of the outstanding issues highlighted in the earlier section.

5.1 A Story of Communication Timescales

To put a broader perspective on some of these issues facing dynamo theory, specifically in the context of the interplay between various flux-transport processes, it will be instructive here to consider the various timescales involved within the dynamo mechanism. Let us, for the sake of argument, consider that the BL mechanism is the predominant mechanism for poloidal field regeneration. Because this poloidal field generation happens at surface layers, but toroidal field is stored and amplified deeper down near the base of the SCZ, for the dynamo to work, these two spatially segregated layers must communicate with each other. In this context, magnetic buoyancy plays an important role in transporting toroidal field from the base of the SCZ to the surface layers – where the poloidal field is produced. The timescale of buoyant transport is quite short, on the order of 0.1 year and this process dominates the upward transport of toroidal field.

Now, to complete the dynamo chain, the poloidal field must be brought back down to deeper layers of the SCZ where the toroidal field is produced and stored. There are multiple processes that compete for this downward transport, namely meridional circulation, diffusion, and turbulent flux pumping.

Considering the typical meridional flow loop from mid-latitudes at the surface to mid-latitudes at the base of the SCZ, and a peak flow speed of $20\,\mathrm{m\,s^{-1}}$, one gets a typical circulation timescale $\tau_v = 10$ years. Most modelers use low values of diffusivity on the order of $10^{11}\,\mathrm{cm^2\,s^{-1}}$, which makes the diffusivity timescale (L^2_{SCZ}/η, assuming vertical transport over the depth of the SCZ) $\tau_\eta = 140$ years; that is, much more than τ_v, therefore making the circulation dominate the flux transport. However, if one assumes diffusivity values close to that suggested by mixing length theory (say, $5 \times 10^{12}\,\mathrm{cm^2\,s^{-1}}$), then the diffusivity timescale becomes $\tau_\eta = 2.8$ years; that is, shorter than the circulation timescale – making diffusive dispersal dominate the flux transport process.

If we now consider the usually ignored process of turbulent pumping, the situation changes again. Assuming a typical turbulent pumping speed on the order of $10\,\mathrm{m\,s^{-1}}$ over the depth of the SCZ gives a timescale $\tau_{\mathrm{pumping}} = 0.67$ years, shorter than both the diffusion and meridional flow timescales. This would make turbulent pumping the most dominant flux transport mechanism for downward transport of poloidal field into the layers where the toroidal field is produced and stored.

5.2 *Solar Cycle Predictions*

As outlined in Yeates et al. (2008), the length of solar cycle memory (defined as over how many cycles the poloidal field of a given cycle would contribute to toroidal field generation) determines the input for predicting the strength of future solar cycles. The relative timescales of different flux transport mechanisms within the dynamo chain of events and their interplay, based on which process (or processes) dominate, determine this memory. For example, if the dynamo is advection (circulation)-dominated, then the memory tends to be long, lasting over multiple cycles. However, if the dynamo is diffusion (or turbulent pumping) dominated, then this memory would be much shorter.

Now, within the scope of the current framework of dynamo models, I have argued that significant confusion exists regarding the role of various flux transport processes. So much so that we do not yet have a consensus on which of these processes dominate; therefore, we do not have a so-called *standard-model* of the solar cycle yet. Should solar cycle predictions be trusted then?

Taking into account this uncertainty in the current state of our understanding of the solar dynamo mechanism, I believe that any solar cycle predictions – that does not adequately address these outstanding issues – should be carefully evaluated. In fact, under the circumstances, it is fair to say that if any solar cycle predictions match reality, it would be more fortuitous than a vindication of the model used for the prediction. This is not to say that modelers should not explore the physical processes that contribute to solar cycle predictability; indeed that is where most of our efforts should be. My concern is that we do not yet understand all the physical processes that constitute the dynamo mechanism and their interplay well enough to begin making predictions. Prediction is the ultimate test of any model, but there are many issues that need to be sorted out before the current day dynamo models are ready for that ultimate test.

Acknowledgement This work has been supported by the Ramanujan Fellowship of the Department of Science and Technology, Government of India and a NASA Living with a Star Grant NNX08AW53G to the Smithsonian Astrophysical Observatory at Harvard University. I gratefully acknowledge many useful interactions with colleagues at the solar physics groups at Montana State University (Bozeman) and the Harvard Smithsonian Center for Astrophysics (Boston). I am indebted to my friends at Bozeman, Montana, from where I recently moved back to India, for contributing to a very enriching experience during the 7 years I spent there.

References

Babcock, H. W. 1961, ApJ, 133, 572
Charbonneau, P. 2005, Living Reviews in Solar Physics, 2, 2
Christensen-Dalsgaard, J., et al. 1996, Science, 272, 1286
Dikpati, M., Charbonneau, P. 1999, ApJ, 518, 508
D'Silva, S., Choudhuri, A. R. 1993, A&A, 272, 621
Fan, Y., Fisher, G. H., Deluca, E. E. 1993, ApJ, 405, 390
Leighton, R. B. 1969, ApJ, 156, 1
Nandy, D. 2002, Ap&SS, 282, 209
Nandy, D., Choudhuri, A. R. 2001, ApJ, 551, 576
Nandy, D., Choudhuri, A. R. 2002, Science, 296, 1671
Parker, E. N. 1955a, ApJ, 121, 491
Parker, E. N. 1955b, ApJ, 122, 293
Schrijver, C. J., Liu, Y. 2008, Solar Phys., 252, 19
Tobias, S. M., Brunnell, N. H., Clune, T. L., Toomre, J. 2001, ApJ, 549, 1183
Wilmot-Smith, A. L., Nandy, D., Hornig, G., Martens, P. C. H. 2006, ApJ, 652, 696
Yeates, A. R., Nandy, D., Mackay, D. H. 2008, ApJ, 673, 544

Status of 3D MHD Models of Solar Global Internal Dynamics

A.S. Brun

Abstract This is a brief report on the decade-long effort by our group to model the Sun's internal magnetohydrodynamics in 3D with the ASH code.

1 Introduction: Solar Global MHD

The Sun is a complex magnetohydrodynamic object that requires state-of-the-art observations and numerical simulations in order to pin down the physical processes at the origin of such diverse activity and dynamics. We here give a brief summary of recent advances made with the Anelastic Spherical Harmonic (ASH) code (Clune et al. 1999; Brun et al. 2004) in modeling global solar magnetohydrodynamics.

2 Global Convection

A series of papers has been published on this important topic (Miesch et al. 2000; Elliott et al. 2000; Brun and Toomre 2002), most recently by Miesch et al. (2008). In this paper, for the first time, a global model of solar convection with a density contrast of 150 from top to bottom and a resolution equivalent to $1{,}500^3$ has been achieved. This has lead to significant results regarding the turbulent convection spectra from large-scale (like giant cells) down to supergranular-like convection patterns and their correlation with the temperature fluctuations, leading to large (150% $L_{\odot}$) convective luminosity.

A.S. Brun (✉)
CEA/CNRS/Université Paris 7, France

S.S. Hasan and R.J. Rutten (eds.), *Magnetic Coupling between the Interior and Atmosphere of the Sun*, Astrophysics and Space Science Proceedings,
DOI 10.1007/978-3-642-02859-5_7, © Springer-Verlag Berlin Heidelberg 2010

3 Differential Rotation and Meridional Circulation

A recent review by Brun and Rempel (2008) discussed the respective role of Reynolds stresses, latitudinal heat transport, and baroclinic effect in setting the peculiar conical differential rotation profile observed in the Sun. Indeed, basic rotating fluid dynamic considerations imply that the differential rotation should be invariant along the rotation axis, yielding a cylindrical rotation profile. As this is not observed, it is necessary to find the source of the breaking of the so-called Taylor–Proudman constraint. In particular, in a recent paper by Miesch et al. (2006), we have been able to show that baroclinic effects are associated with latitudinal variation of the temperature and that convection by transporting heat poleward contributes a significant part of that variation, but not all. A temperature contrast of about 10 K is compatible with helioseismic inferences for the inner solar angular velocity profile. Meridional flows in most cases are found to be multicellular, and fluctuate significantly over a solar rotation. These flows contribute little to the heat transport and to the kinetic energy budget (accounting for only 0.5% of the total kinetic energy). However, it plays a pivotal role in the angular momentum redistribution by opposing and balancing the equatorward transport by Reynolds stresses (Brun and Toomre 2002; Brun and Rempel 2008).

4 Global Dynamo

In continuation of Gilman (1983) and Glatzmaier (1985), we have studied, at much higher resolution, dynamo action in turbulent convective shells (Brun 2004; Brun et al. 2004). We have found that dynamo action is reached above a critical magnetic Reynolds number and that the magnetic field is mostly intermittent and small-scale (Fig. 1), with the large-scale axisymmetric field only contributing for about 3% of the total magnetic energy. Reversals of the field occur on a time scale of about 1.5 year, as opposed to the observed 11 year cycle of solar activity. This is partly due to the absence of a tachocline at the base of the convective envelope. In an attempt to resolve this issue, we have in Browning et al. (2006) computed the first 3D MHD model of a convection zone with an imposed stable tachocline. In that layer, the field that has been transported or pumped down from the turbulent convection zone above it, is found to be organized in strong axisymmetric toroidal ribbons with dominant antisymmetry with respect to the equator. The poloidal field in the convection zone is stabilized by the presence of that layer with much less frequent, if any, reversals. The magnetic energy reaches in both cases about 10% of the total kinetic energy. We also find that the differential rotation is reduced in amplitude due to the nonlinear feedback of the field on the flow via the Lorentz force. In a recent study by Jouve and Brun (2007, 2009), we have also studied flux emergence in isentropic and turbulent rotating convection zone. We confirmed that a certain amount of field concentration and twist is required for the structure to emerge at the

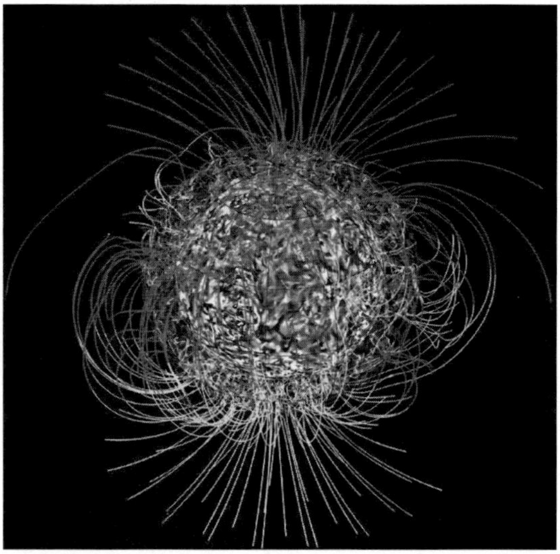

Fig. 1 Three-dimensional magnetic field rendering (*yellow* corresponds to positive B_r, i.e., field lines directed toward the observer) inside the convection zone (below the transparent sphere) and in a current-free corona (potential-field reconstruction) for the turbulent dynamo case M3 of Brun et al. (2004). We note the intricacy of the field lines in the convection zone, often located in strong downflows, and the connectivity of the field over large distances at the surface

surface and that convection may pin down and recycle weak field. At the surface in the convective case, horizontal converging motions modify the orientation of the emerging structure.

5 Solar Radiation Zone and Its Coupling to the Convection Zone

The thinness of the solar tachocline ($h < 0.05\,\mathrm{R}_\odot$) implies the presence of processes that transport angular momentum horizontally to prevent its viscous and thermal spread along the solar evolution. Horizontal turbulence in stratified layers, the presence of a fossil magnetic field or internal waves, may provide such a transport (Spiegel and Zahn 1992; Gough and McIntyre 1998). Brun and Zahn 2006 have studied for the first time in 3D the nonlinear evolution of an horizontal shear (similar to the solar angular profile) in a stable radiative zone and the role of a fossil field. We found that such field, by connecting to the surface shear, will ease rather than oppose the spread of the tachocline. It is thus difficult to invoke only the magnetic field to explain the solid body rotation of the radiative zone and the thin tachocline. Another important finding in Brun and Zahn (2006) is that simple magnetic configurations undergo non-axisymmetric MHD instabilities first discovered by Tayler in the 1970s. In particular, a purely poloidal (dipolar) fossil field would undergo high azimuthal wavenumber instabilities. To obtain a stable configuration for a magnetic field burried in the solar radiative interior, it is necessary to have a coupled poloidal/toroidal topology. Such non-axisymmetric instabilities may lead to dynamo action in radiative interiors as has also been advocated by Spruit (2002) and discussed in detail by Zahn et al. (2007).

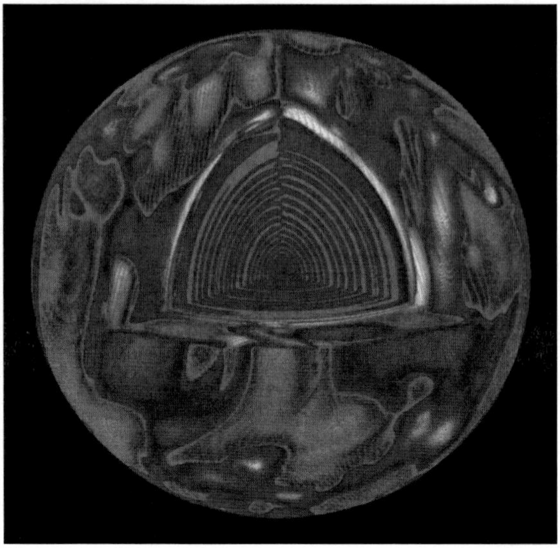

Fig. 2 First 3D integrated solar model coupling nonlinearly the convective envelope to the radiative interior. Shown is a 3D rendering of the density perturbations, with *red* corresponding to positive fluctuations. We have omitted an octant in order to be able to see the equatorial and meridional planes within the domain. We note the clear presence of internal waves in the radiative zone

6 Towards a 3D Integrated Model of the Sun

Coupling nonlinearly the convection zone with the radiative interior is the key to understand the solar global dynamo and inner dynamics. Brun (in preparation) has developed the first 3D solar integrated model from $r = 0.07\,R_\odot$ up to $0.97\,R_\odot$. We show in Fig. 2 a 3D rendering of the density fluctuations over the whole computational domain. The presence of internal waves is obvious in the radiative interior. The penetrative convection is at the origin of these gravito-inertial waves. We are currently studying in detail the source function at every depth in the model and the resulting power spectrum at different locations in the radiative interior and find that a large spectrum near the base of the convection zone is excited. The tachocline is kept thin in this model by using a step function at the base of the convection zone for the various diffusion parameters, making the thermal and viscous spread of the latitudinal shear imposed by the convective envelope slow with respect to the convective overturning time. We intend in the near future to redo the study of Brun and Zahn (2006) by introducing in the integrated model a fossil field, taking advantage of the more realistic boundary conditions realized in this new class of models.

Acknowledgement I am thankful to my friends and colleagues J. Toomre, J.-P. Zahn, M. Miesch, M. Derosa, M. Browning, and L. Jouve without whom the results reported in this paper would not have been obtained. I also thank the IFAN network for partial funding during my visit to India. Finally, I am grateful to Profs. S. Hasan, K. Chitre, and H.M. Antia for the wonderful time I spent in Bangalore and Mumbai.

References

Browning, M., Miesch, M. S., Brun, A.S., Toomre, J. 2006, ApJ, 648, L157
Brun, A. S. 2004, Solar Phys., 220, 333
Brun, A. S., Rempel, M. 2008, Space Science Reviews, Volume 144, Issue 1–4, pp. 151
Brun, A. S., Toomre, J. 2002, ApJ, 570, 865
Brun, A. S., Zahn, J.-P. 2006, A&A, 457, 665
Brun, A. S., Miesch, M. S., Toomre, J. 2004, ApJ, 614, 1073
Clune, T. L., et al. 1999, Parallel Comput., 25, 361
Elliott, J. R., Miesch, M. S., Toomre, J. 2000, ApJ, 533, 546
Gilman, P. A. 1983, ApJS, 53, 243
Glatzmaier, G. A. 1985, ApJ, 291, 300
Gough, D. O., McIntyre, M. 1998, Nature, 394, 755
Jouve, L., Brun, A. S. 2007, Astron. Nachr., 328, 1104
Jouve, L., Brun, A. S. 2009, ApJ, 701, 1300
Miesch, M. S., et al. 2000, ApJ, 532, 593
Miesch, M.S., Brun, A.S., Toomre, J. 2006, ApJ, 641, 618
Miesch, M.S., Brun, A.S., Derosa, M., Toomre, J. 2008, ApJ, 673, 557
Spiegel, E. A., Zahn, J.-P. 1992, A&A, 265, 106
Spruit, H. C. 2002, A&A, 381, 923
Zahn, J.-P., Brun, A.S., Mathis, S. 2007, A&A, 474, 145

Measuring the Hidden Aspects of Solar Magnetism

J.O. Stenflo

Abstract 2008 marks the 100th anniversary of the discovery of astrophysical magnetic fields, when George Ellery Hale recorded the Zeeman splitting of spectral lines in sunspots. With the introduction of Babcock's photoelectric magnetograph, it soon became clear that the Sun's magnetic field outside sunspots is extremely structured. The field strengths that were measured were found to get larger when the spatial resolution was improved. It was therefore necessary to come up with methods to go beyond the spatial resolution limit and diagnose the intrinsic magnetic-field properties without dependence on the quality of the telescope used. The line-ratio technique that was developed in the early 1970s revealed a picture where most flux that we see in magnetograms originates in highly bundled, kG fields with a tiny volume filling factor. This led to interpretations in terms of discrete, strong-field magnetic flux tubes embedded in a rather field-free medium, and a whole industry of flux tube models at increasing levels of sophistication. This magnetic-field paradigm has now been shattered with the advent of high-precision imaging polarimeters that allow us to apply the so-called "Second Solar Spectrum" to diagnose aspects of solar magnetism that have been hidden to Zeeman diagnostics. It is found that the bulk of the photospheric volume is seething with intermediately strong, tangled fields. In the new paradigm, the field behaves like a fractal with a high degree of self-similarity, spanning about 8 orders of magnitude in scale size, down to scales of order 10 m.

1 The Zeeman Effect as a Window to Cosmic Magnetism

2008 marks the 100th anniversary of the discovery of magnetic fields outside the Earth (cf. Fig. 1). George Ellery Hale had suspected that the Sun might be a magnetized sphere from the appearance of the solar corona seen at total solar eclipses, and from the structure of Hα fibrils around sunspots, which was reminiscent of iron files in a magnetic field. The proof came when Hale placed the spectrograph slit in

J.O. Stenflo (✉)
Institute of Astronomy, ETH Zurich, Zurich, Switzerland

S.S. Hasan and R.J. Rutten (eds.), *Magnetic Coupling between the Interior and Atmosphere of the Sun*, Astrophysics and Space Science Proceedings,
DOI 10.1007/978-3-642-02859-5_8, © Springer-Verlag Berlin Heidelberg 2010

Fig. 1 George Ellery Hale in 1905 in his office at the Mount Wilson observatory (*left*) and his discovery of the Zeeman splitting in sunspots (*right*)

the solar tower of his newly constructed Mt. Wilson Observatory across a sunspot and the splitting of the spectral lines was revealed (Hale 1908). The Dutch physicist Pieter Zeeman had discovered such splitting in the laboratory the decade before, in 1896, when an external magnetic field was applied to the gas in which the spectral lines were formed. The frequency of Larmor precession of the atomic dipoles around the field mixes with the atomic resonance frequencies, resulting in the splitting of these frequencies. As the magnetic field breaks the spatial symmetry, the split line components get polarized in ways that depend on the orientation of the magnetic field vector relative to the line of sight.

Often the splitting is too small and subtle to measure by itself except in sunspots. Instead, it is the polarization effects that are the telltale signature of the Zeeman effect and the magnetic field. While the line-of-sight component of the magnetic field gives rise to circular polarization (longitudinal Zeeman effect), the perpendicular component causes linear polarization (transverse Zeeman effect). Thus one can in principle measure the full magnetic field vector (strength and orientation) by recording the full state of polarization in a spectral line, the full Stokes vector, with the four Stokes parameters I, Q, U, V, representing intensity, two states of linear polarization (differing in orientation by 45°), and circular polarization, respectively. Detailed treatments can be found in several monographs (Stenflo 1994; del Toro Iniesta 2003; Landi Degl'Innocenti and Landolfi 2004).

The Zeeman effect, which outside terrestrial laboratories was first seen in sunspots, opens a window to the exploration of cosmic magnetism. As usual, the Sun provides us with a laboratory that serves as a test bed for the exploration of various new physical processes before they can be applied to the investigation of stars and galaxies elsewhere in the universe. Our increasing empirical knowledge about the Sun's magnetism has helped guide the development and understanding of various theoretical tools, like plasma physics and magnetohydrodynamics. The experimental tool is spectro-polarimetry, which needs the Zeeman effect (and more recently also the Hanle effect, see below) as an interpretational tool to connect theory and observation.

Outside sunspots, the polarization signals of the transverse Zeeman effect are much smaller than those of the longitudinal Zeeman effect. For weak fields, the linear polarization from the transverse Zeeman effect is approximately proportional to the square of the transverse field strength rather than in linear proportion, and it is limited by a 180° ambiguity. In contrast, the circular polarization is easy to measure, and to first order it is proportional to the line-of-sight component of the field, with sign. Therefore, magnetic-field measurements have been dominated by recordings of the circular polarization due to the longitudinal Zeeman effect. The breakthrough in these measurements came with the introduction by Babcock of the photoelectric magnetograph (Babcock 1953). Soon afterwards, full-disk magnetograms (maps of the circular polarization) were being produced on a regular basis, forming a unique data base for the understanding of stellar magnetism and dynamos.

2 Emergence of the Flux Tube Paradigm

When directly resolved magnetic-field observations are not available, like for magnetic Ap-type stars, one usually makes models assuming that the star has a dipole or low-degree multipolar field. The solar magnetograms, however, showed the Sun's field to be highly structured. It was found that the measured field strength increases with the angular resolution of the instrument used (Stenflo 1966). As the measured field strength also depended on the spectral line used, many believed that this was a calibration problem that could be solved by a coordinated campaign, organized by an IAU committee, to record the same regions on the Sun with different instruments.

It was only with the introduction of the line-ratio technique (Stenflo 1973) that the cause for this apparent "calibration problem" could be found. The magnetic flux is highly intermittent, with most of the flux concentrated in elements that were far smaller than the available spatial resolution. The magnetograph calibration (conversion of measured polarization to field strength) was based on the shape of the spatially averaged line profile and the assumption of weak fields (linear relation between polarization and field strength). The average line profile is, however, not representative of the line formation conditions within the flux concentrations, and also the weak-field approximation is not valid there (we have "Zeeman saturation"), as the concentrated fields are intrinsically strong. Inside the strong-field regions, the thermodynamic conditions are very different from the rest of the atmosphere, which leads to temperature-induced line weakenings.

The magnitude of the line-weakening and Zeeman saturation effects vary from line to line, which leads to the noticed dependence of the field-strength values on the spectral line used. This effect cannot be calibrated away, as the line-formation properties in the flux concentrations are not accessible to direct observations when they are not resolved. A further effect is that different lines are formed at different atmospheric heights, and the field expands and weakens with height. All these effects contribute jointly in an entangled way to the "calibration error." The line-ratio technique was introduced to untangle them, and it is described in Fig. 2.

The trick is to use a combination of lines, for which all the various entangled factors are identical, except one. Thus it was possible to isolate the Zeeman saturation (nonlinearity) effect from all the thermodynamic and line formation effects by choosing the line pair Fe I 5250.22 and 5247.06 Å. Both these lines belong to multiplet No. 1 of iron, have the same line strength and excitation potential, and therefore have identical thermodynamic response and line-formation properties. The only significant difference between them is their Landé factors, which are 3.0 and 2.0, respectively. No other line combination has since been found, which can so cleanly isolate the Zeeman saturation effect from the other effects.

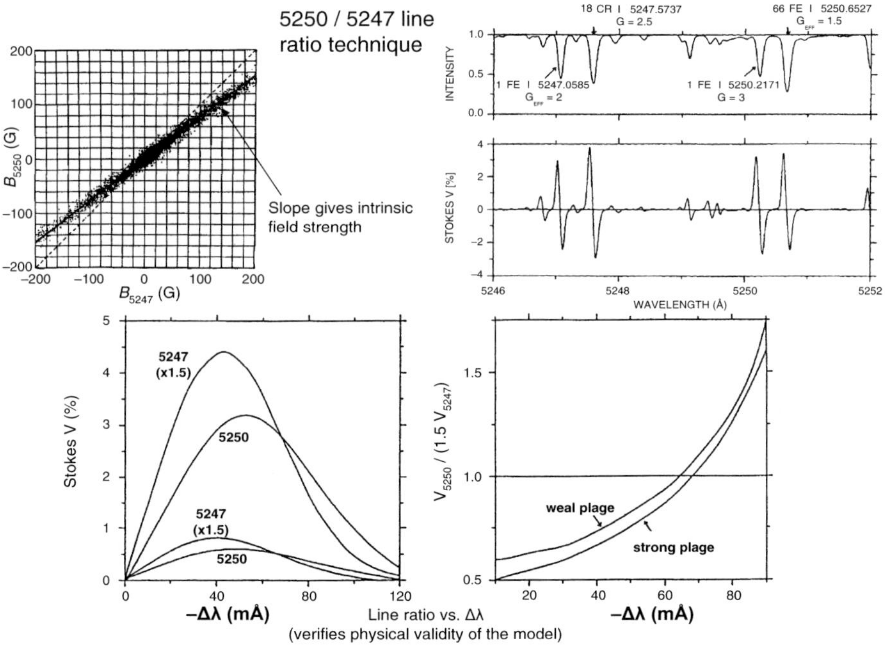

Fig. 2 Illustration of the various aspects of the 5,250/5,247 line ratio technique (Stenflo 1973). The linear slope in the diagram to *upper left* (from Frazier and Stenflo 1978) determines the differential Zeeman saturation, from which the intrinsic field strength can be found. The portion of the FTS Stokes V spectrum to *upper right*, from Stenflo et al. (1984), shows that the amplitudes of the 5,250 and 5,247 iron lines are not in proportion to their Landé factors, but are closer to 1:1. In the *bottom* diagram, from Stenflo and Harvey (1985), the Stokes V profiles and line ratios are plotted as functions of wavelength distance from line center. This profile behavior verifies that the line difference is really due to differential Zeeman saturation

If we were in the linear, weak-field regime, the circular polarization measured in the two lines should scale in proportion to their Landé factors, but as the field strength increases, the deviation from this ratio increases (*differential Zeeman saturation*). Thus the circular-polarization line ratio is a direct measure of the intrinsic field strength. The observed ratio showed that the intrinsic field strength was 1–2 kG at the quiet-sun disk center, although the apparent magnetograph field strengths there were only a few Gauss, a discrepancy of 2–3 orders of magnitude (Stenflo 1973)!

A further surprising result was that there seemed to be no dependence of the intrinsic field strength on the apparent field strength (which in a first approximation represents magnetic flux divided by the spatial resolution element). This property is seen in the scatter-plot diagram to the upper left in Fig. 2 (from Frazier and Stenflo 1978). The line ratio or differential Zeeman saturation is represented by the *slope* in the diagram (in comparison with the 45° slope that represents the case without Zeeman saturation). There is no indication that the slope changes as we go from smaller to larger apparent field strengths. A statistical analysis led to the conclusion that more than 90% of the photospheric flux (that is "seen" by the magnetographs with the resolution of a few arcsec that was used then) is in strong-field form (Howard and Stenflo 1972; Frazier and Stenflo 1972; Stenflo 1994), and that strong-field flux elements have "unique" internal properties, meaning that the statistical spread in their field strengths and thermodynamic properties was small and not dependent on the amount of flux in the region. Thus active-region plages and the quiet-sun network gave very similar intrinsic field strengths.

These findings lay the foundation for the validity of the two-component model that was used as the interpretational tool: one "magnetic" component with a certain filling factor (fractional area of the resolution element covered), which was the source of all the circular-polarization signals seen in magnetograms, and another component, which was called "nonmagnetic," as it did not contribute anything to the magnetograms. The line-ratio method showed that the field strength of the magnetic component was nearly independent of the magnetic filling factor, which could vary by orders of magnitude (but had typical values of order 1% on the quiet Sun).

The empirical foundation for the two-component model was further strengthened by the powerful Stokes V multiline profile constraints provided by FTS (Fourier transform spectrometer) polarimetry (Stenflo et al. 1984), and by the use of the larger Zeeman splitting in the near infrared (cf. Rüedi et al. 1992).

This empirical scenario found its theoretical counterpart in the concept of strong-field magnetic flux tubes embedded in field-free surroundings (Spruit 1976). Semi-empirical flux tube models of increasing sophistication could be built, in particular thanks to the powerful observational constraints provided by the FTS Stokes V spectra (cf. Solanki 1993). In these models, the observational constraints were combined with the MHD constraints that included the self-consistent expansion of the flux tubes, with height in a numerically specified atmosphere with pressure balance.

With these successes, the unphysical nature of the two-component model tended to be forgotten, according to which something like 99% of the photosphere was "nonmagnetic." In the electrically highly conducting solar plasma, the concept of such a field-free volume is non-sensical. When the two-component model was introduced nearly four decades ago, the introduction of a "nonmagnetic" component was done for the sake of mathematical simplicity, with the purpose of isolating the properties of the magnetic component, but not with the intention of making a statement about the intrinsic nature of the "nonmagnetic" component. As the longitudinal Zeeman effect was "blind" to this component (as it did not contribute to anything in the magnetograms), the quest began to find another diagnostic tool to access its hidden magnetic properties, to find a diagnostic window to the aspect of solar magnetism that represents 99% of the photosphere. This window was found through the Hanle effect.

3 The Hanle Effect as a Window to the Hidden Fields

The circular polarization from the longitudinal Zeeman effect is to first order proportional to the net magnetic flux through the angular resolution element. If the magnetic field has mixed-polarity fields inside the resolution element with equal total amounts of positive and negative polarity flux, the net flux and therefore also the net circular polarization is zero. Although the strength and magnetic energy density of such a tangled field can be arbitrarily high, it is invisible to the longitudinal Zeeman effect as long as the individual flux elements are not resolved.

If this were merely a matter of insufficient angular resolution, one might hope that this tangled field could be mapped by magnetograms in some future. However, even if we would have infinite angular resolution, the cancelation problem of the opposite polarities would not go away, as the spatial resolution *along the line of sight* is ultimately limited by the thickness of the line-forming layer, which is of order 100 km in the photosphere (the photon mean free path). For optically thin magnetic elements with opposite polarities along the line of sight, the cancelation effect remains, regardless of the angular resolution.

The task therefore becomes to find a physical mechanism that is not subject to these cancelation effects. Magnetic line broadening from the Zeeman effect is one such mechanism, as it scales with the square of the field strength, the magnetic energy, and therefore is of one "sign," in contrast to the circular polarization. As, however, these effects are tiny, and many other factors affect the width of spectral lines, only a 1-σ upper limit of about 100 G could be set for the tangled field from a statistical study of 400 unblended Fe I lines (Stenflo and Lindegren 1977). In contrast, the Hanle effect is sensitive to much weaker tangled fields.

In contrast to the Zeeman effect, the Hanle effect is a coherence phenomenon that occurs only when coherent scattering contributes to the formation of the spectral line. It was discovered in Göttingen in 1923 by Wilhelm Hanle and played a significant role in the conceptual development of quantum mechanics, as it demonstrated explicitly the fundamental concept of the coherent superposition of quantum states (later sometimes called "Schrödinger cats").

Fig. 3 Wilhelm Hanle (*right*) visits ETH Zurich in 1983 on the occasion of the 60th anniversary of his effect

Coherent scattering polarizes the light. The term Hanle effect covers all the magnetic-field modifications of this scattering polarization. In the absence of magnetic fields, the magnetic m substates are degenerate (coherently superposed). A magnetic field breaks the spatial symmetry and lifts the degeneracy, thereby causing partial decoherence. One can also speak of quantum interferences between the m states. For details, see Moruzzi and Strumia (1991); Stenflo (1994); Landi Degl'Innocenti and Landolfi (2004).

A good intuitive understanding of the Hanle effect can be obtained with the help of the classical oscillator model. The incident radiation induces dipole oscillations in the transverse plane (perpendicular to the incident beam). For a 90° scattering angle, the plane in which the oscillations take place is viewed from the side and due to this projection appear as 1D oscillations. The scattered radiation therefore gets 100% linearly polarized perpendicular to the scattering plane.

For scattering polarization to occur one needs anisotropic radiative excitation. For a spherically symmetric Sun (when we neglect local inhomogeneities), the anisotropy is a consequence of the limb darkening, which implies that the illumination of a scattering particle inside the atmosphere occurs more in the vertical direction from below than from the sides. In the hypothetical case of extreme limb darkening, when all illumination is in the vertical direction, we would have 90° scattering at the extreme limb. The scattering angle decreases towards zero when we move towards disk center, where for symmetry reasons the scattering polarization (in the nonmagnetic case) is zero. As the scattering polarization gets larger as we approach the limb, most scattering and Hanle-effect observations are performed on the disk relatively close to the limb, with the spectrograph slit parallel to the nearest limb. The nonmagnetic scattering polarization is then expected to be oriented along the slit direction, which we in our Stokes vector representations define as the positive Stokes Q direction. Stokes U then represents polarization oriented at 45° to the slit.

Let us now introduce a magnetic field along the scattering direction. The damped oscillator is then subject to Larmor precession around the magnetic field vector, which results in the Rosette patterns illustrated in Fig. 4. The pattern gets tilted and more randomized as the field strength increases (from the left to the right Rosette diagram in the figure). The line profile and polarization properties are obtained from Fourier transformations of the Rosette patterns.

The magnetic field has two main effects on the polarization of the scattered radiation: (1) Depolarization, as the precession randomizes the orientations of the oscillating dipoles. In terms of the Stokes parameters, this corresponds to a reduction of the Q/I amplitudes. (2) Rotation of the plane of linear polarization, as the net effect of the precession is a skewed or tilted oscillation pattern. This corresponds

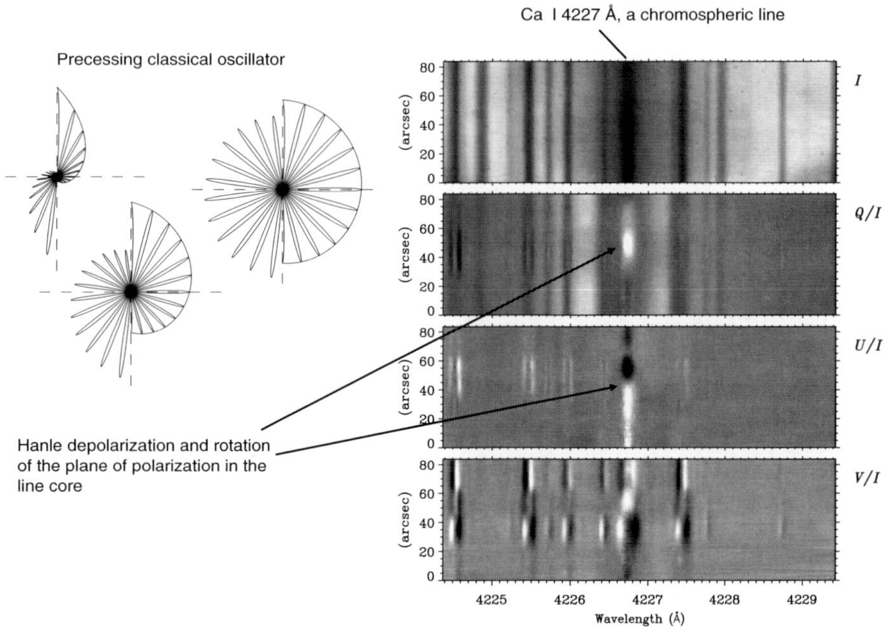

Fig. 4 *Left*: Rosette patterns of a classical oscillator in a magnetic field oriented along the line of sight, illustrating the Hanle depolarization and rotation effects. *Right*: Spectral image of the Stokes vector (the four Stokes parameters in terms of intensity I and the fractional polarizations Q/I, U/I, and V/I) recorded with the spectrograph slit across a moderately magnetic region 5 arcsec inside and parallel to the solar limb. The Hanle signatures appear in stokes Q and U in the core of the Ca I 4,227 Å line, while the surrounding lines exhibit the characteristic signatures of the transverse Zeeman effect. In stokes V all the lines show the antisymmetric signatures of the longitudinal Zeeman effect

to the creation of signatures in Stokes U/I, which can be of either sign, depending on the sense of rotation (orientation of the field vector). The magnitudes of these two effects depend on the competition between the Larmor precession rate and the damping rate, or, equivalently, the ratio between the Zeeman splitting and the damping width of the line. In contrast, the polarization caused by the ordinary Zeeman effect depends on the ratio between the Zeeman splitting and the Doppler width of the line. As the damping width is smaller by typically a factor of 30 than the Doppler width, the Hanle effect is sensitive to much weaker fields than the Zeeman effect. Equally important, the two effects have different symmetry properties and therefore respond to magnetic fields in highly complementary ways.

Assume for instance that we are observing a magnetic field that is tangled on subresolution scales, such that there is no net magnetic flux when one averages over the spatial resolution element due to cancelation of the contributions of opposite signs. Such a magnetic field gives no observable signatures in the circular polarization (longitudinal Zeeman effect, on which solar magnetograms are based) or in the Hanle rotation (Stokes U/I) due to cancelations of the opposite signs. In contrast, the Hanle depolarization is not subject to such cancelations, as it has only one "sign" (depolarization), regardless of the field direction. The Hanle depolarization therefore opens a diagnostic window to such a subresolution, tangled field (Stenflo 1982).

4 The "Standard Model" and Its Shortcomings

The "standard model" that has emerged from Zeeman and Hanle observations of the quiet Sun, and which is illustrated in Fig. 5, refers to the magnetic-field structuring in the spatially unresolved domain. Only recently, with advances in angular resolution, are we beginning to resolve individual flux tubes, but in general, their existence and properties have only been inferred from indirect techniques (line-ratio method, FTS Stokes V spectra, Stokes V line profiles in the near infrared). As the fields are not resolved, all such indirect techniques must be based on interpretative models.

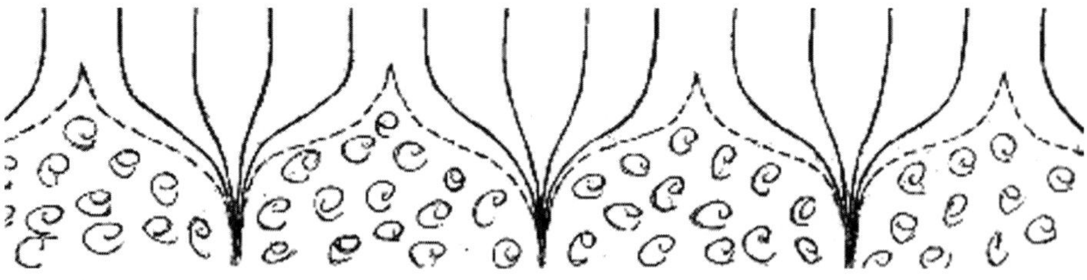

Fig. 5 Standard model of quiet-sun solar magnetism (here illustrated for a region where the different flux tubes have the same polarity). The atmosphere is described in terms of two components, one representing the flux tubes, which contribute to the Zeeman effect, the other component representing the tangled field in between, which contributes to the Hanle effect

The dominating interpretative model in the past has been a two-component model, consisting of (1) the flux tube component, which is responsible for practically all the magnetic flux that is seen in solar magnetograms, and (2) the "turbulent" component in between the flux tubes, with tangled fields of mixed polarities on sub-resolution scales, which are invisible to the Zeeman effect. The filling factor of the flux tube component is of order 1% in the quiet solar photosphere, which implies that the turbulent component represents 99% of the photospheric volume. Because of the exponential pressure drop with height, the flux tubes expand to reach a filling factor of 100% in the corona.

The question about the strength of the volume-filling "turbulent" field representing 99% of the photosphere could be given an answer from observations of the Hanle depolarization of the scattering polarization, in particular with the Sr I 4607 Å line. As with one such line we only have one observable (the amount of Hanle depolarization), the interpretative model could not have more than one free parameter. The natural choice of one-parameter model that was adopted in the initial interpretations of the Hanle data was in terms of a tangled field consisting of optically thin elements with a random, isotropic distribution of the magnetic field vectors and a single-valued field strength (Stenflo 1982). Detailed radiative-transfer modeling of the Sr I 4607 Å observations (Faurobert-Scholl 1993; Faurobert-Scholl et al. 1995) gave values of typically 30 G, but more recent applications of 3D polarized radiative transfer for much more realistic model atmospheres generated by hydrodynamic simulations of granular convection give field strengths of about 60 G, twice as large (Trujillo Bueno et al. 2004).

The dualistic nature of the world that is represented by this "standard model" is, however, much an artefact of having two mutually almost exclusive diagnostic tools at our disposal. The Zeeman effect is blind to the turbulent fields due to flux cancelation. The Hanle effect is blind to the flux tube fields for several reasons: (1) With filling factors of order 1% only, the flux tube contribution to the Hanle depolarization is insignificant. (2) The Hanle effect is insensitive to vertical fields (for symmetry reasons, when the illumination is axially symmetric around the field vector), and the flux tubes tend to be vertical because of buoyancy. (3) The Hanle effect saturates for the strong fields in the flux tubes.

We always see a filtered version of the real world, filtered by our diagnostic tools in combination with the interpretational models (analytical tools) used. Thus, when we put on our "Zeeman goggles," we see a magnetic world governed by flux tubes, while when we put on our "Hanle goggles," we see a world of tangled or turbulent fields. We should, however, not forget that these are merely idealized aspects of the real world, shaped by our models. Instead of having the dichotomy of two discrete components, the real world should rather be described in terms of continuous probability density functions (PDFs), as indicated by the theory of magnetoconvection and by numerical simulations (Cattaneo 1999; Nordlund and Stein 1990). Moreover, exploration of the magnetic pattern on the spatially resolved scales indicates a high degree of self-similarity that is characteristic of a fractal (Stenflo and Holzreuter 2002; Janßen et al. 2003).

When Trujillo Bueno et al. (2004) used an interpretational model based on a realistic PDF rather than a single-valued field strength, their 3D modeling of the Sr I 4,607 Å observations gave substantially higher average field strengths (in excess of 100 G) as compared with the single-valued model. This suggests that the hidden, turbulent field contains a magnetic energy density that may be of significance for the overall energy balance of the solar atmosphere. The question whether or not the magnetic energy dominates the energy balance remains unanswered due to the current model dependence of these interpretations.

5 The Second Solar Spectrum and Solar Magnetism

The term Hanle effect stands for the magnetic-field modifications of the scattering polarization. The Sun's spectrum is linearly polarized as coherent scattering contributes to the formation of the spectrum (like the polarization of the blue sky by Rayleigh scattering at terrestrial molecules). Because of the small anisotropy of the radiation field in the solar atmosphere and the competing nonpolarizing opacity sources, the amplitudes of the scattering polarization signals are small, of order 0.01–1% near the limb, varying from line to line. Although a number of the polarized line profiles could be revealed in early surveys of the linear polarization (Stenflo et al. 1983a,b), it was only with the advent of highly sensitive imaging polarimeters that the rich spectral world of scattering polarization became fully accessible to observation. The breakthrough came with the implementation in 1994 of the ZIMPOL (Zurich Imaging Polarimeter) technology, which allowed imaging spectro-polarimetry with a precision of 10^{-5} in the degree of polarization (Povel 1995, 2001; Gandorfer et al. 2004). At this level of sensitivity everything is polarized, even without magnetic fields. It came as a big surprise, however, that the polarized spectrum was as richly structured as the ordinary intensity spectrum but without resembling it, as if a new spectral face of the Sun had been unveiled, and we had to start over again to identify the various spectral structures and their physical origins. It was therefore natural to call this new and unfamiliar spectrum the "Second Solar Spectrum" (Ivanov 1991; Stenflo and Keller 1997). A spectral atlas has been produced, which in three volumes covers the Second Solar Spectrum from 3,160 to 6,995 Å (Gandorfer 2000, 2002, 2005).

The Second Solar Spectrum exists as a fundamentally nonmagnetic phenomenon, but it is modified by magnetic fields, and it is the playground for the Hanle effect. Because of the rich structuring of the Second Solar Spectrum and the diverse behavior of the different spectral lines, it contains a variety of novel opportunities to diagnose solar magnetism in ways not possible with the Zeeman effect. Here we will only illustrate a few examples of this. Further details can be found in the proceedings of the series of Solar Polarization Workshops (Stenflo and Nagendra 1996; Nagendra and Stenflo 1999; Trujillo-Bueno and Sanchez Almeida 2003; Casini and Lites 2006).

The structuring in the Second Solar Spectrum is governed by previously unfamiliar physical processes, like quantum interference between atomic levels, hyperfine structure and isotope effects, optical pumping, molecular scattering, and enigmatic, as yet unexplained phenomena that appear to defy quantum mechanics as we know it (cf. Stenflo 2004). The identification and interpretation of the various polarized structures have presented us with fascinating theoretical challenges, and we have now reached a good qualitative understanding of the underlying physics in most but not all of the cases. Here we will limit ourselves to illustrate the case of molecular scattering.

The spectral Stokes vector images (intensity I, linear polarizations Q/I and U/I, circular polarization V/I) in Fig. 6 illustrate the behavior of scattering polarization in the CN molecular lines in the wavelength range 3,771–3,775 Å, in solar regions of different degrees of magnetic activity. The CN lines have the appearance of emission lines in Q/I with little if any spatial variations along the

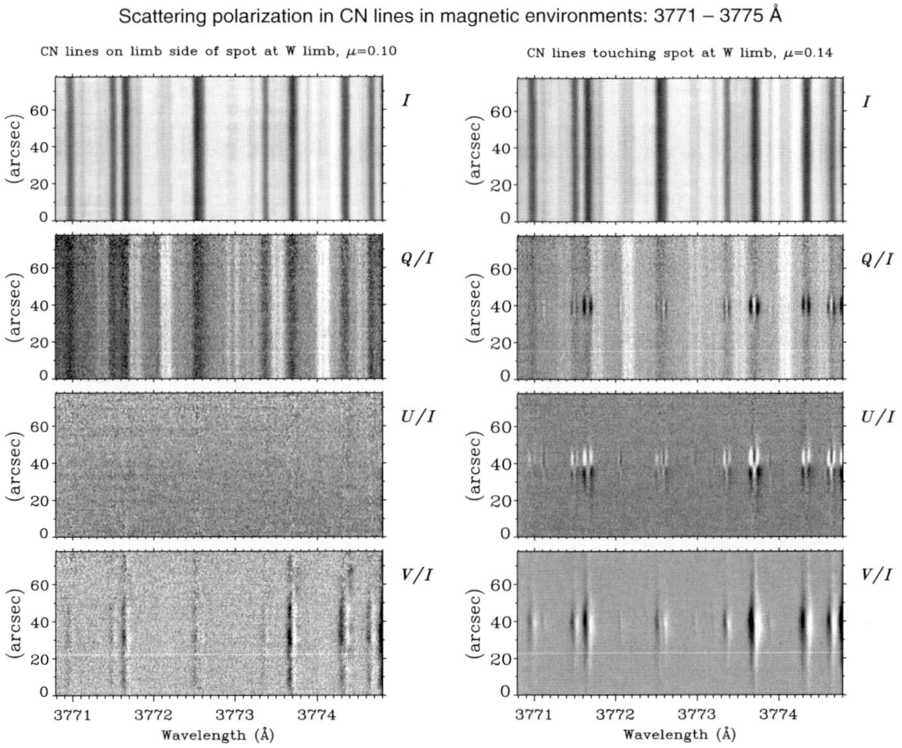

Fig. 6 Molecular CN lines in the second solar spectrum (the bright bands in Stokes Q/I). Note the absence of scattering polarization in U/I and significant variation of Q/I along the slit, in contrast to the surrounding atomic lines, which show the familiar signatures of the transverse and longitudinal Zeeman effects. The recording was made with ZIMPOL at Kitt Peak at $\mu = 0.1$ inside the west solar limb (Stenflo 2007)

spectrograph slit, in contrast to the surrounding atomic lines, which exhibit the characteristic signatures of the transverse Zeeman effect. This would seem to imply that the molecular lines are not affected by magnetic fields, as we see no spatial structuring due to the Hanle effect, in contrast to the chromospheric Ca I 4227 Å line in Fig. 4, where we see dramatic Q/I and U/I variations along the slit due to the Hanle effect. A careful analysis of the observed Q/I amplitudes in the molecular lines reveal, however, that they are indeed affected (depolarized) by the Hanle effect, and by a magnetic field that is tangled and structured on subresolution scales, and therefore does not show resolved variations along the slit or any U/I signatures (Hanle rotation).

The model dependence in the translation of polarization amplitudes to field strengths can be suppressed by using combinations of spectral lines that behave similarly in all respects except for their sensitivity to the Hanle effect. This *differential Hanle effect* (Stenflo et al. 1998) is similar to the line-ratio technique for the Zeeman effect that we discussed in Sect. 2. Its effectiveness depends on our ability to find optimum line combinations that allow us to isolate the Hanle effect from all the other effects. It turns out to be much easier to find optimum line pairs among the molecular lines than among the atomic lines. This technique has been successfully used by Berdyugina and Fluri (2004) with a pair of C_2 molecular lines to determine the strength (15 G) of the tangled or turbulent field. The molecular lines are found to give systematically lower field strengths than the atomic lines, which can be explained in terms of spatial structuring of the turbulent field on the granulation scale (Trujillo Bueno et al. 2004). Three-dimensional radiative transfer modeling shows that the molecular abundance is highest inside the granules, which implies that the turbulent field is preferentially located in the intergranular lanes while containing structuring that continues far below the granulation scales. In the next section, we will consider how far down this structuring is expected to continue.

6 Scale Spectrum of the Magnetic Structures

Magnetic fields permeate the Sun with its convection zone. The turbulent convection, which penetrates into the photosphere, tangles the frozen-in magnetic field lines and thereby structures the field on a vast range of scales. The structuring continues to ever smaller scales, until we reach the scales where the frozen-in condition ceases to be valid and the field decouples from the turbulent plasma. This happens when the time scale of magnetic diffusion becomes shorter than the time scale of convective transport. The ratio between these two time scales is represented by the magnetic Reynolds number

$$R_{\rm m} = \mu_0 \, \sigma \, \ell_{\rm c} \, v_{\rm c} \tag{1}$$

in SI units. σ is the electrical conductivity, $\ell_{\rm c}$ the characteristic length scale, $v_{\rm c}$ the characteristic velocity. $\mu_0 = 4\pi \times 10^{-7}$. For large scales, when $R_{\rm m} \gg 1$, the field lines are effectively frozen in and carried around by the convective motions. For sufficiently small scales $R_{\rm m} \ll 1$, the field decouples and diffuses through the plasma. The end of the scale spectrum is where the decoupling occurs, namely where $R_m \approx 1$.

To calculate the decoupling scale we need to know how the characteristic turbulent velocity v_c scales with ℓ_c. Such a scaling law is given in the Kolmogorov theory of isotropic turbulence. In the for us relevant inertial range it is

$$v_c = k\ \ell_c^{1/3}\,, \tag{2}$$

where k is a constant. An estimate of $k \approx 25$ can be obtained from the observed properties of solar granulation (Åke Nordlund, private communication).

Note that this type of scaling should apply to the photosphere in spite of its stratification, as the inertial range that we are considering occurs at scales much smaller than the photospheric scale height. This small-scale turbulence does not "feel" the stratification and is therefore nearly isotropic, in contrast to the larger scales.

For $R_m = 1$, these two equations give us the diffusion scale

$$\ell_{\rm diff} = 1/(\mu_0\,\sigma\,k)^{3/4}. \tag{3}$$

Inserting the Spitzer conductivity in SI units,

$$\sigma = 10^{-3}\ T^{3/2}, \tag{4}$$

we obtain

$$\ell_{\rm diff} = 5\times 10^5\,/\,T^{9/8}. \tag{5}$$

For $T = 10^4$ K (a rounded value that is representative of the lowest part of the photosphere or upper boundary of the convection zone), $\ell_{\rm diff} \approx 15$ m.

Note that the ordinary, nonmagnetic Reynolds number is still very high at these 10 m scales. Thus the turbulent spectrum continues to much smaller scales down to the viscous diffusion limit, but without contributing to magnetic structuring at these scales.

The present-day spatial resolution limit in solar observations lies around 100 km. This is four orders of magnitude larger than the smallest magnetic structures that we can expect. Therefore, in spite of conspicuous advances in high-resolution imaging, much of the structuring will remain unresolved in any foreseeable future.

7 Beyond the Standard Model: Scaling Laws and PDFs for a Fractal-Like Field

Time has come to replace the previous dualistic magnetic-field paradigm or two-component "standard model" with a scenario characterized by PDFs. While the strong-field tail of such a distribution corresponds to the "flux tubes" of the standard model, the bulk of the PDF corresponds to the "turbulent field" component. Instead of using two different interpretational models for the Zeeman and Hanle effects when diagnosing the spatially unresolved domain, it is more logical to apply a single, unified interpretational model based on PDFs for both these effects. The diagnostic tools for this unified and much more realistic approach are currently being developed (cf. Sampoorna et al. 2008, Sampoorna 2010).

This task is complicated because there are PDFs for both field strength and field orientation, and they appear to vary spatially on the granulation scale, as suggested by the different Hanle behavior of atomic and molecular lines. To clarify this we need to resolve the solar granulation in Hanle effect observations. Furthermore, we know much less about the PDF for the angular field distribution than we know about the PDF for the vertical field strengths. For theoretical reasons we expect the angular and strength distributions to be coupled to each other. Strong fields are more affected by buoyancy forces, which make the angular distribution more peaked around the vertical direction. Small-scale, weak fields, on the other hand, are passively tangled by the turbulent motions and are therefore expected to have a more isotropic distribution. The issue is confused by the recent Hinode finding that there appears to be substantially more horizontal than vertical magnetic flux on the quiet Sun (Lites et al. 2008), which finds support in some numerical simulations (Schüssler and Vögler 2008). The implications of these findings for the angular PDFs have not yet been clarified.

Another fundamental issue is the dependence of these various PDFs on scale size. To wisely select the interpretational models to be used to diagnose the unresolved domain we need to understand the relevant scaling laws. Explorations of the magnetic-field pattern in magnetograms (the spatially resolved domain) and in numerical simulations indicate a high degree of self-similarity and fractal-like behavior. This would justify the use of PDF shapes that are found from the resolved domain to be applied to diagnostics of the unresolved domain. On the other hand, there are reasons to expect possible deviations from such scale invariance. We have already seen indications for a difference between the PDFs in granules and in intergranular lanes. The current spatial resolution limit (about 100 km) also marks the boundary between optically thick and thin elements, as well as between elements governed by the atmospheric stratification effects (scale height) and elements that are too small to "feel" this stratification. The 100 km scale is therefore expected to be of physical significance and may influence the behavior of the scaling laws.

The fractal nature of the field is illustrated in Fig. 7 as we zoom in on the quiet-sun magnetic pattern at the center of the solar disk. There is a coexistence of weak and strong fields over a wide dynamic range. The PDF for the vertical field-strength component is nearly scale invariant and can be well represented by a Voigt function with a narrow Gaussian core and "damping wings" extending to kG values (Stenflo and Holzreuter 2002; Stenflo and Holzreuter 2003). A fractal dimension of 1.4 has been found from both observations and numerical simulations (Janßen et al. 2003). The simulations indicate that this fractal behavior extends well into the spatially unresolved domain.

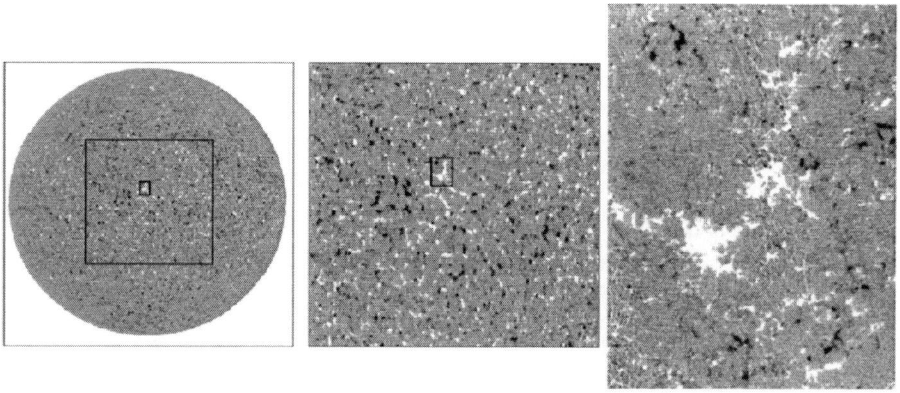

Fig. 7 The fractal-like nature of quiet-sun magnetic fields becomes apparent as we zoom in on ever smaller scales. The two *left* maps are from a Kitt Peak full-disk magnetogram of 9 February 1996, while the *right*, high-resolution magnetogram was recorded on the same day at the Swedish La Palma Observatory (courtesy Göran Scharmer). The La Palma magnetogram covers an area that is only 0.35% of the map next to it

The Second Solar Spectrum opens a new window to explorations of previously inaccessible aspects of solar magnetism. With the vast amounts of hidden magnetic energy in the spatially unresolved magnetoconvective spectrum, the determination of the properties of the hidden field is a central task for contemporary solar physics. The quality of the determination depends on the interpretational models that we choose. For an optimum choice we need to understand the scaling laws of the fractal-like magnetic field and the role of various physical scales that may cause deviations from scale invariance of the pattern. Insight into the scaling behavior can be advanced with improved spatial resolution of the observations, combined with guidance from numerical simulations.

References

Babcock, H. W. 1953, ApJ, 118, 387
Berdyugina, S. V., Fluri, D. M. 2004, A&A, 417, 775
Casini, R., Lites, B. W., eds. 2006, Solar Polarization 4, ASP Conf. Ser., 358
Cattaneo, F. 1999, ApJ, 515, L39
del Toro Iniesta, J. C. 2003, Introduction to Spectropolarimetry, Cambridge University Press, Cambridge
Faurobert-Scholl, M. 1993, A&A, 268, 765
Faurobert-Scholl, M., Feautrier, N., Machefert, F., Petrovay, K., Spielfiedel, A. 1995, A&A, 298, 289
Frazier, E. N., Stenflo, J. O. 1972, Solar Phys., 27, 330
Frazier, E. N., Stenflo, J. O. 1978, A&A, 70, 789

Gandorfer, A. 2000, The Second Solar Spectrum: A High Spectral Resolution Polarimetric Survey of Scattering Polarization at the Solar Limb in Graphical Representation. Vol. I: 4625 Å to 6995 Å, VdF, Zurich
Gandorfer, A. 2002, The Second Solar Spectrum: A High Spectral Resolution Polarimetric Survey of Scattering Polarization at the Solar Limb in Graphical Representation. Vol. II: 3910 Å to 4630 Å, VdF, Zurich
Gandorfer, A. 2005, The Second Solar Spectrum: A High Spectral Resolution Polarimetric Survey of Scattering Polarization at the Solar Limb in Graphical Representation. Vol. III: 3160 Å to 3915 Å, VdF, Zurich
Gandorfer, A. M., Steiner, P., Povel, H. P., et al. 2004, A&A, 422, 703
Howard, R., Stenflo, J. O. 1972, Solar Phys., 22, 402
Ivanov, V. V. 1991, in Stellar Atmospheres – Beyond Classical Models, NATO ASIC Proc. 341, 81
Janßen, K., Vögler, A., Kneer, F. 2003, A&A, 409, 1127
Landi Degl'Innocenti, E., Landolfi, M. 2004, Astrophysics and Space Science Library, Vol. 307, Polarization in Spectral Lines, Kluwer, Dordrecht
Lites, B. W., Kubo, M., Socas-Navarro, H., et al. 2008, ApJ, 672, 1237
Moruzzi, G., Strumia, F., eds. 1991, The Hanle Effect and Level-Crossing Spectroscopy, Plenum Press
Nagendra, K. N., Stenflo, J. O., eds. 1999, Solar polarization, Astrophys. Space Sci. Library, 243
Nordlund, Å., Stein, R. F. 1990, In: Solar Photosphere: Structure, Convection, and Magnetic Fields, J. O. Stenflo (ed.), IAU Symp. vol. 138, p. 191
Povel, H. 1995, Optical Engineering, 34, 1870
Povel, H. P. 2001, In: Magnetic Fields Across the Hertzsprung-Russell Diagram, G. Mathys, S. K. Solanki, D. T. Wickramasinghe (eds.), ASP Conf. Ser., vol. 248, p. 543
Rüedi, I., Solanki, S. K., Livingston, W., Stenflo, J. O. 1992, A&A, 263, 323
Sampoorna, M., Nagendra, K. N., Frisch, H., Stenflo, J. O. 2008, A&A, 485, 275
Sampoorna, M., 2010, In: Magnetic Coupling between the Interior and the Atmosphere of the Sun, S. S. Hasan, R. J. Rutten (eds.), Astrophys. Space Sci. Procs., Springer, Heidelberg, these proceedings
Schüssler, M., Vögler, A. 2008, A&A, 481, L5
Solanki, S. K. 1993, Sp. Sci. Reviews, 63, 1
Spruit, H. C. 1976, Solar Phys., 50, 269
Stenflo, J. O. 1966, Arkiv Astron., 4, 173
Stenflo, J. O. 1973, Solar Phys., 32, 41
Stenflo, J. O. 1982, Solar Phys., 80, 209
Stenflo, J. O. 1994, Solar Magnetic Fields – Polarized Radiation Diagnostics, Kluwer, Dordrecht
Stenflo, J. O. 2004, In: Reviews in Modern Astronomy, R. E. Schielicke (ed.), vol. 17, p. 269
Stenflo, J. O. 2007, Mem. della Soc. Astron. Italiana, 78, 181
Stenflo, J. O., Harvey, J. W. 1985, Solar Phys., 95, 99
Stenflo, J. O., Holzreuter, R. 2002, In: Magnetic Coupling of the Solar Atmosphere, H. Sawaya-Lacoste (ed.), ESA-SP, vol. 505, p. 101
Stenflo, J. O., Holzreuter, R. 2003, In: Current Theoretical Models and Future High Resolution Solar Observations: Preparing for ATST, A. A. Pevtsov, H. Uitenbroek (eds.), ASP Conf. Ser., vol. 286, p. 169
Stenflo, J. O., Keller, C. U. 1997, A&A, 321, 927
Stenflo, J. O., Keller, C. U., Gandorfer, A. 1998, A&A, 329, 319
Stenflo, J. O., Lindegren, L. 1977, A&A, 59, 367
Stenflo, J. O., Nagendra, K. N. 1996, Solar Phys., 164
Stenflo, J. O., Solanki, S., Harvey, J. W., Brault, J. W. 1984, A&A, 131, 333
Stenflo, J. O., Twerenbold, D., Harvey, J. W. 1983a, A&AS, 52, 161
Stenflo, J. O., Twerenbold, D., Harvey, J. W., Brault, J. W. 1983b, A&AS, 54, 505
Trujillo-Bueno, J., Sanchez Almeida, J., 2003, Solar Polarization, ASP Conf. Ser., 307
Trujillo Bueno, J., Shchukina, N., Asensio Ramos, A. 2004, Nat, 430, 326

Recent Advances in Chromospheric and Coronal Polarization Diagnostics

J. Trujillo Bueno

Abstract I review some recent advances in methods to diagnose polarized radiation with which we may hope to explore the magnetism of the solar chromosphere and corona. These methods are based on the remarkable signatures that the radiatively induced quantum coherences produce in the emergent spectral line polarization and on the joint action of the Hanle and Zeeman effects. Some applications to spicules, prominences, active region filaments, emerging flux regions, and the quiet chromosphere are discussed.

1 Introduction

The fact that the anisotropic illumination of the atoms in the chromosphere and corona induces population imbalances and quantum coherences between the magnetic sublevels, even among those pertaining to different levels, is often considered as a hurdle for the development of practical diagnostic tools of "measuring" the magnetic field in such outer regions of the solar atmosphere. However, as we shall see throughout this paper, it is precisely this fact that gives us the hope of reaching such an important scientific goal. The price to be paid is that we need to develop high-sensitivity spectropolarimeters for ground-based and space telescopes and to interpret the observations within the framework of the quantum theory of spectral line formation. As J. W. Harvey put it, "this is a hard research area that is not for the timid" (Harvey 2006).

Rather than attempting to survey all of the literature on the subject, I have opted for beginning with a very brief introduction to the physics of spectral line polarization, pointing out the advantages and disadvantages of the Hanle and Zeeman effects as diagnostic tools, and continuing with a more detailed discussion of selected

J. Trujillo Bueno (✉)
Instituto de Astrofísica de Canarias, La Laguna, Tenerife, Spain
and
Consejo Superior de Investigaciones Científicas, Spain

S.S. Hasan and R.J. Rutten (eds.), *Magnetic Coupling between the Interior and Atmosphere of the Sun*, Astrophysics and Space Science Proceedings,
DOI 10.1007/978-3-642-02859-5_9,

developments. Recent reviews where the reader finds complementary information are Harvey (2006); Stenflo (2006); Lagg (2007); López Ariste and Aulanier (2007); Casini and Landi Degl'Innocenti (2007); and Trujillo Bueno (2009a).

2 The Physical Origin of the Spectral Line Polarization

Solar magnetic fields leave their fingerprints in the polarization signatures of the emergent spectral line radiation. This occurs through a variety of rather unfamiliar physical mechanisms, not only via the Zeeman effect. In particular, in stellar atmospheres there is a more fundamental mechanism producing polarization in spectral lines. There, where the emitted radiation can escape through the stellar surface, the atoms are illuminated by an anisotropic radiation field. The ensuing radiation pumping produces population imbalances among the magnetic substates of the energy levels (i.e., atomic level polarization) in such a way that the populations of substates with different values of $|M|$ are different (M being the magnetic quantum number). This is termed *atomic level alignment*. As a result, the emission process can generate linear polarization in spectral lines without the need for a magnetic field. This is known as scattering line polarization (e.g., Stenflo 1994; Landi Degl'Innocenti and Landolfi 2004). Moreover, radiation is also selectively absorbed when the lower level of the transition is polarized (Trujillo Bueno and Landi Degl'Innocenti 1997; Trujillo Bueno 1999; Trujillo Bueno et al. 2002b; Manso Sainz and Trujillo Bueno 2003b). Thus, the medium becomes dichroic simply because the light itself has the chance of escaping from it.

Upper-level polarization produces *selective emission* of polarization components, while lower-level polarization produces *selective absorption* of polarization components. A useful expression to estimate the amplitude of the emergent fractional linear polarization is the following generalization of the Eddington–Barbier formula (Trujillo Bueno 2003a), which establishes that the emergent Q/I at the center of a sufficiently strong spectral line when observing along a line of sight (LOS) specified by $\mu = \cos\theta$ (with θ being the angle between the local solar vertical and the LOS) is approximately given by

$$\frac{Q}{I} \approx \frac{3}{2\sqrt{2}}(1-\mu^2)\,[\mathcal{W}\,\sigma_0^2(J_u) - \mathcal{Z}\,\sigma_0^2(J_l)] = \frac{3}{2\sqrt{2}}(1-\mu^2)\,\mathcal{F}, \qquad (1)$$

where $\mathcal{W}$ and $\mathcal{Z}$ are numerical factors that depend on the angular momentum values (J) of the lower (l) and upper (u) levels of the transition (e.g., $\mathcal{W} = \mathcal{Z} = -1/2$ for a line with $J_l = J_u = 1$), while $\sigma_0^2 = \rho_0^2/\rho_0^0$ quantifies the fractional atomic alignment of the upper or lower level of the spectral line under consideration, calculated in a reference system whose Z-axis (i.e., the quantization axis of total angular momentum) is along the local solar vertical.[1] The $\sigma_0^2(J)$ values

[1] For example, $\rho_0^0(J=1) = (N_1 + N_0 + N_{-1})/\sqrt{3}$ and $\rho_0^2(J=1) = (N_1 - 2N_0 + N_{-1})/\sqrt{6}$, where N_1, N_0, and N_{-1} are the populations of the magnetic sublevels.

quantify the degree of population imbalances among the sublevels of level J with different $|M|$-values. They have to be calculated by solving the statistical equilibrium equations for the multipolar components of the atomic density matrix (see Chap. 7 of Landi Degl'Innocenti and Landolfi 2004). In a weakly anisotropic medium like the solar atmosphere, the $\sigma_0^2(J_l)$ and $\sigma_0^2(J_u)$ values of a resonance line transition are proportional to the so-called anisotropy factor $w = \sqrt{2}J_0^2/J_0^0$ (e.g., Sect. 3 in Trujillo Bueno 2001), where J_0^0 is the familiar mean intensity and $J_0^2 \approx \oint \mathrm{d}\boldsymbol{\Omega}/(4\pi)\, 1/(2\sqrt{2})\,(3\mu^2-1)\, I_{\nu,\boldsymbol{\Omega}}$ quantifies whether the illumination of the atomic system is preferentially vertical ($w > 0$) or horizontal ($w < 0$). Note that in (1) the σ_0^2 values are those corresponding to the atmospheric height where the line-center optical depth is unity along the LOS.

The most practical aspect is that a magnetic field inclined with respect to the symmetry axis of the pumping radiation field modifies the atomic level polarization via the Hanle effect (e.g., the reviews by Trujillo Bueno 2001, Trujillo Bueno 2005; see also Landi Degl'Innocenti and Landolfi 2004). Approximately, the amplitude of the emergent spectral line polarization is sensitive to magnetic strengths between $0.1\, B_{\mathrm{H}}$ and $10\, B_{\mathrm{H}}$, where the critical Hanle field intensity (B_{H}, in gauss) is that for which the Zeeman splitting of the J-level under consideration is equal to its natural width:

$$B_{\mathrm{H}} = 1.137{\times}10^{-7}/(t_{\mathrm{life}}\, g_J), \tag{2}$$

with t_{life} the lifetime, in seconds, of the J-level under consideration and g_J its Landé factor. As the lifetimes of the upper levels (J_u) of the transitions of interest are usually much smaller than those of the lower levels (J_l), clearly diagnostic techniques based on the lower-level Hanle effect are sensitive to much weaker fields than those based on the upper-level Hanle effect.

The Hanle effect gives rise to a rather complex magnetic-field dependence of the linear polarization of the emergent spectral line radiation. In the saturation regime of the upper-level Hanle effect (i.e., when the magnetic strength $B > B_{\mathrm{satur}} \approx 10\, B_{\mathrm{H}}(J_u)$, with $B_{\mathrm{H}}(J_u)$ the critical Hanle field of the line's upper level), it is possible to obtain manageable formulae for the line-center amplitudes of the emergent linear polarization profiles, which show that in such a regime the Q/I and U/I signals depend only on the orientation of the magnetic field vector. Assume, for simplicity, a deterministic magnetic field with $B > B_{\mathrm{satur}}$ inclined by an angle θ_B with respect to the local solar vertical (i.e., the Z-axis) and contained in the Z–Y plane. Consider any LOS contained in the Z–X plane, characterized by $\mu = \cos\theta$. Choose the Y-axis direction as the reference direction for Stokes Q. It can be shown that the following approximate expressions hold for the emergent linear polarization amplitudes in an electric-dipole transition:[2]

[2] For magnetic dipole transitions it is only necessary to change the sign of the Q/I and U/I expressions given in this paper. To understand the reason for this, see Sect. 6.8 of Landi Degl'Innocenti and Landolfi (2004).

$$\frac{Q}{I} \approx \frac{3}{8\sqrt{2}} \Big[(1-\mu^2)(3\cos^2\theta_B - 1) + (1+\mu^2)(\cos^2\theta_B - 1)\Big] (3\cos^2\theta_B - 1)\,\mathcal{F}, \tag{3}$$

$$\frac{U}{I} \approx -\frac{3}{2\sqrt{2}} \sqrt{1-\mu^2}\,\sin\theta_B \cos\theta_B\,(3\cos^2\theta_B - 1)\,\mathcal{F}, \tag{4}$$

where $\mathcal{F} = \mathcal{W}\,\sigma_0^2\,(J_u) - \mathcal{Z}\,\sigma_0^2(J_l)$ is identical to that of (1), which depends on the σ_0^2 values for the *unmagnetized* reference case.

It is of interest to consider the following particular cases, ignoring for the moment that in a stellar atmosphere the $\mathcal{F}$ value tends to be the larger the smaller μ. First, the $B = 0$ case of (1) can be easily recovered by choosing $\theta_B = 0°$ in (3) and (4), because there is no Hanle effect if the magnetic field is parallel to the symmetry axis of the incident radiation field. Second, note that for $\theta_B = 90°$ (horizontal magnetic field) $U/I = 0$ and that for this case we find exactly the same Q/I amplitude for all LOSs contained in the Z–X plane, including that with $\mu = 1$, which corresponds to forward-scattering geometry. Note also that (3) implies that the amplitude of the forward-scattering Q/I signal created by the Hanle effect of a horizontal magnetic field with a strength in the saturation regime is only a factor two smaller than the Q/I signal of the unmagnetized reference case in 90° scattering geometry (i.e., the case of a LOS with $\mu = 0$). Some interesting examples of detailed numerical calculations of the emergent Q/I and U/I amplitudes for a variety of θ_B and μ values can be seen in Fig. 9 of Asensio Ramos et al. (2008), which the reader will find useful to inspect. Such results for the lines of the He I 10830 Å multiplet can be easily understood via (3) and (4).

It is easy to generalize (3) and (4) for any magnetic field azimuth χ_B. Such general equations show clearly that there are two particular scattering geometries (i.e., those with $\mu = 0$ and $\mu = 1$) for which the Stokes profiles corresponding to $\theta_B^* = 180° - \theta_B$ and $\chi_B^* = -\chi_B$ are identical to those for which the magnetic field vector has θ_B and χ_B (i.e., the familiar ambiguity of the Hanle effect). If the observed plasma structure is not located in the plane of the sky, or if it is outside the solar disk center, one then has a quasi-degeneracy, which can disappear when μ is considerably different from 1 or from 0. This fact can be exploited for removing the 180° azimuth ambiguity present in vector magnetograms (Landi Degl'Innocenti and Bommier 1993; see also Fig. 2 below).

For the case of a magnetic field with a fixed inclination θ_B and a random azimuth below the spatial scale of the mean free path of the line photons we have $U/I = 0$, while

$$\frac{Q}{I} \approx \frac{3}{8\sqrt{2}} (1-\mu^2)\,[3\cos^2\theta_B - 1]^2\,\mathcal{F}. \tag{5}$$

This expression shows that under such circumstances there is no forward scattering polarization. It shows also that the Q/I amplitude of a scattering signal produced in the presence of a horizontal magnetic field with a random azimuth and $B > B_{\rm satur}$ is a factor 4 smaller than $[Q/I]_0$ (i.e., than the Q/I amplitude corresponding to the unmagnetized reference case). It is also possible to show that $Q/I \approx (1/5)\,[Q/I]_0$ for the case of a microturbulent magnetic field with an isotropic distribution of field directions.

Finally, it is important to emphasize that a rigorous modeling of the polarization produced by the joint action of the Hanle and Zeeman effects in many spectral lines of diagnostic interest requires calculating the wavelength positions and the strengths of the π and σ components within the framework of the Paschen–Back effect theory. This theory allows us to model the important level-crossing regime in which the energy eigenvectors are gradually evolving from the form $|LSJM\rangle$ (with M the projection of the total angular momentum $J = L + S$ along the quantization axis) to the form $|LSM_LM_S\rangle$ as the magnetic field increases. This range between the limiting cases of "weak" fields (Zeeman effect regime) and "strong" fields (complete Paschen–Back regime) is called the incomplete Paschen–Back effect regime. The reason why it is so important for a correct modeling of the spectral line polarization in fine-structured and in hyperfine-structured multiplets is because the level crossings and repulsions that take place in this regime give rise to subtle modifications of the atomic level polarization and, therefore, to a number of remarkable effects on the emergent spectral line polarization (e.g., Bommier 1980; Landi Degl'Innocenti 1982; Trujillo Bueno et al. 2002a; Belluzzi et al. 2007). Of particular interest is the so-called alignment-to-orientation transfer mechanism studied by Landi Degl'Innocenti (1982) for the He I D_3 multiplet, by means of which a fraction of the atomic level alignment produced by anisotropic pumping processes can lead to *atomic level orientation* (i.e., an atomic excitation situation such that the populations of substates with magnetic quantum numbers M and $-M$ are different). Obviously, the observational signature of the presence of a significant amount of atomic level orientation is a Stokes $V(\lambda)$ profile dominated by one of its lobes.

3 The Zeeman Effect vs. the Hanle Effect

Good news is that the mere detection of Zeeman polarization signature(s) implies the presence of a magnetic field. One disadvantage of the polarization of the Zeeman effect as a diagnostic tool is that it is blind to magnetic fields that are tangled on scales too small to be resolved. Another drawback is that it is of limited practical interest for the determination of magnetic fields in hot (chromospheric and coronal) plasmas because the circular polarization induced by the longitudinal Zeeman effect scales with the ratio, $\mathcal{R}$, between the Zeeman splitting and the Doppler width (which is much larger than the natural width of the atomic levels!). Likewise, given that for not too strong fields the Stokes Q and U signals produced by the transverse Zeeman effect scale as $\mathcal{R}^2$, their amplitudes are normally below the noise level of present observational possibilities for intrinsically weak fields (typically, $B \lesssim 100$ gauss in solar spectropolarimetry).

The Hanle effect is especially sensitive to magnetic fields for which the Zeeman splitting is comparable to the natural width of the upper (or lower) level of the spectral line used, regardless of how large the line width due to Doppler broadening is.

Therefore, it is sensitive to weaker magnetic fields than the Zeeman effect: from at least 1 mG to a few hundred gauss (2). Moreover, it is sensitive to magnetic fields that are tangled on scales too small to be resolved (e.g., Stenflo 1994; Trujillo Bueno et al. 2004). Finally, note that the diagnostic use of the Hanle effect is *not* limited to a narrow solar limb zone. In particular, in forward scattering at the solar disk center, the Hanle effect can create linear polarization in the presence of inclined magnetic fields (Trujillo Bueno et al. 2002b). The disadvantage is that the Hanle effect signal saturates for magnetic strengths $B > 10\, B_{\rm H}(J_u)$, a regime where the linear polarization signals are sensitive only to the orientation of the magnetic field vector.

Fortunately, both effects can be suitably complemented for exploring magnetic fields in solar and stellar physics.

4 Diagnostic Tools Based on Spectral Line Polarization

The determination of the magnetic, dynamic, and thermal properties of solar plasma structures via the interpretation of the observed Stokes profiles requires the development of suitable diagnostic tools. The aim is to find the physical properties of the adopted model such that the difference between the synthetic and the observed Stokes profiles is the smallest possible one. Depending on the observed plasma structure, the model chosen to represent it can be rather simple (e.g., a constant-property slab) or more sophisticated (e.g., a stratified, one-dimensional atmosphere model). The first step is to develop an efficient way to compute the emergent Stokes profiles for any plausible realization of the model's physical properties. Such spectral synthesis tools can be used for doing forward modeling calculations (e.g., in snapshot models taken from MHD simulations) or for developing inversion codes of Stokes profiles induced by various physical mechanisms. At present there are two Stokes inversion approaches. One employs searching in databases of theoretical Stokes profiles computed with the spectral synthesis tool, ideally for all possible configurations of the model's physical properties. The other employs iterative algorithms aiming to minimize the merit function used to quantify the goodness of the fit of the model properties. This is done by combining the spectral synthesis tool with suitable minimization algorithms, such as the Levenberg–Marquardt method.

4.1 Methods for the Chromosphere and Transition Region

The intensity and polarization of the spectral lines that originate in the bulk of the solar chromosphere (e.g., the IR triplet and the K-line of Ca II) and in the transition region (e.g., Ly α and Mg II k) contain precious information on these atmospheric regions. In general, their linear polarization is due to the joint action of the atomic level polarization and the Hanle and transverse Zeeman effects, while their circular polarization is dominated by the longitudinal Zeeman effect.

In regions with high concentrations of magnetic flux, such as in sunspots, the polarization signals are dominated by the Zeeman effect. Therefore, diagnostic techniques based on this effect are quite useful. For example, the non-LTE inversion code of Stokes profiles induced by the Zeeman effect developed by Socas-Navarro et al. (2000b) has led to several interesting applications (e.g., Socas-Navarro et al. 2000a; Socas-Navarro 2005; Pietarila et al. 2007).

To model spectropolarimetric observations of chromospheric and transition region lines outside sunspots, it is necessary to take into account the atomic polarization that anisotropic radiation pumping processes induce in the atomic levels. This requires solving a significantly more complicated radiative transfer problem, known as the non-LTE problem of the second kind (see Landi Degl'Innocenti and Landolfi 2004). It consists in calculating, at each spatial point of any given atmospheric model and for each J-level of the chosen atomic model, the density matrix elements that are consistent with the intensity and polarization of the radiation field generated within the (generally magnetized) medium under consideration. Once such density matrix elements are known, it is possible to solve the Stokes vector transfer equation for any desired LOS with an accurate and efficient formal solution method, such as the DELOPAR technique discussed by Trujillo Bueno (2003b).

To that end, Manso Sainz and Trujillo Bueno (2003a) developed MULTIPOL, a general radiative transfer computer program for solving multilevel scattering polarization problems, including the Hanle and Zeeman effects of a weak magnetic field (see also Manso Sainz 2002). MULTIPOL is based on the multilevel atom model of the quantum theory of spectral line formation (see Sect. 7.2 in Landi Degl'Innocenti and Landolfi 2004), which allows us to take into account that the mean intensity and anisotropy of the various line transitions pertaining to any given multiplet can be different. A similar spectral synthesis code based also on the DELOPAR technique and on the iterative scheme proposed by Trujillo Bueno (1999) has been recently developed by Štěpán (2008). Solving the ensuing Stokes inversion problem for the magnetic field vector is possible, but requires to adopt a model (or a few plausible models) for the thermal and density stratifications.

The quantum theory of spectral line polarization on which the above-mentioned computer programs are based treats the scattering line polarization phenomenon as the temporal succession of first-order absorption and re-emission processes, interpreted as statistically independent events (complete redistribution in frequency). This theory is very suitable for modeling the polarization observed in many diagnostically important spectral lines, such as the IR triplet of Ca II and Hα. It can also be used for estimating the line-center polarization amplitude in lines for which frequency correlations between the incoming and outgoing photons are significant (e.g., Ca II K, Lyα, Mg II k), especially in forward-scattering geometry at the solar disk center. However, for interpreting particular spectral features that are observed in the wings of some strong lines, it is necessary to apply a theory not based on the Markov approximation. At present, such a formulation is available only for the particular case of a two-level atom without lower-level polarization (e.g., Sampoorna et al. 2007 and references therein). Fortunately, the lower level of several resonance line transitions cannot be aligned, so that modeling efforts in this direction are of interest (e.g., Holzreuter and Stenflo 2007).

4.2 *Methods for Chromospheric and Coronal Structures*

There are several possibilities for determining the magnetic field vector that confines and/or channels the plasma of structures embedded in the optically thin outer layers of the solar atmosphere, such as prominences, spicules, active region filaments, etc. At present, the best available option is to choose spectral lines entirely produced by the plasma structures themselves, such as those of the He I 10,830 Å and 5,876 Å (D_3) multiplets, and to interpret observations of their intensity and polarization within the framework of the multiterm atom model of the quantum theory of spectral line formation (see Sect. 7.6 in Landi Degl'Innocenti and Landolfi 2004). The spectral lines of the He I 10,830 Å and 5,876 Å multiplets result from transitions between terms of the triplet system of helium (ortho-helium), whose respective J-levels (with J the level's total angular momentum) are far less populated than the ground level of helium (the singlet level $^1S_0$). The lower term ($2s^3S_1$) of the He I 10,830 Å multiplet is the ground level of ortho-helium, while its upper term ($2p^3P_{2,1,0}$) is the lower one of 5,876 Å (whose upper term is $3d^3D_{3,2,1}$).

The Stokes profiles of the He I 10,830 Å and 5,876 Å multiplets depend on the strengths and wavelength positions of the π and σ transitions, which can be calculated correctly only within the framework of the Paschen–Back theory. In fact, the sublevels of the $J = 2$ and $J = 1$ upper levels of the He I 10,830 Å triplet cross between 400 and 1,600 G, approximately, while the sublevels of the $J = 3$ and $J = 2$ upper levels of the He I 5,876 Å multiplet show several crossings for field strengths of the order of 10 G (e.g., Fig. 3 of Asensio Ramos et al. 2008). Moreover, the emergent Stokes profiles can be seriously affected by the presence of atomic level polarization produced by anisotropic radiative pumping processes, which can be very significant even in the metastable (long-lived) lower level of the He I 10,830 Å multiplet (Trujillo Bueno et al. 2002b). Elastic collisions with the neutral hydrogen atoms of the solar chromospheric and coronal structures are unable to destroy the atomic polarization of the He I levels.

It is important to put reliable codes for the synthesis and inversion of Stokes profiles at the disposal of the astrophysical community. To this end, Asensio Ramos et al. (2008) developed a user-friendly computer program called HAZEL (from HAnle and ZEeman Light), which takes into account all the relevant physical mechanisms and ingredients (optical pumping, atomic level polarization, level crossings and repulsions, Zeeman, Paschen–Back, and Hanle effects). The user can either calculate the emergent intensity and polarization for any given magnetic field vector, or can infer the dynamical and magnetic properties from the observed Stokes profiles via an efficient inversion algorithm based on global optimization methods. The influence of radiative transfer on the emergent spectral line radiation is taken into account through a suitable constant-property slab model, in which the radiatively induced atomic level polarization is assumed to be dominated by the photospheric continuum radiation. At each point of the observed field of view, the slab's optical thickness is chosen to fit the observed Stokes I profile, a strategy that accounts implicitly for the true physical mechanisms that populate the He I triplet levels (e.g., the photoionization-recombination mechanism discussed by Avrett et al. 1994; Centeno et al. 2008, and others). The observed Stokes Q, U, and V profiles are then used to infer the magnetic field vector.

It is important to clarify that the assumption of a constant line source function within the slab is reasonable for the He I 10,830 Å and D_3 multiplets, as can be deduced from non-LTE calculations of the populations of the He I levels that take into account the influence of the EUV radiation that penetrates the chromosphere from the overlying corona (e.g., Fig. 6 of Centeno et al. 2008). This can be seen more clearly in Fig. 1, which shows the height variation of the line source function in any of the He I 10,830 Å transitions for increasing values of the EUV irradiance. Avrett et al. (1994) also find that the line source function of the He I 10,830 Å triplet is essentially constant, and equal to about $0.4 \times I_c$ (with I_c the continuum intensity at 10,830 Å).

Figure 2 shows an example of a model calculation carried out with the synthesis option of HAZEL. For the case of a plasma structure levitating at a height of 2,200 km above the visible solar surface and permeated by a magnetic field of 1,200 G, the figure shows two types of calculations of the emergent Stokes profiles

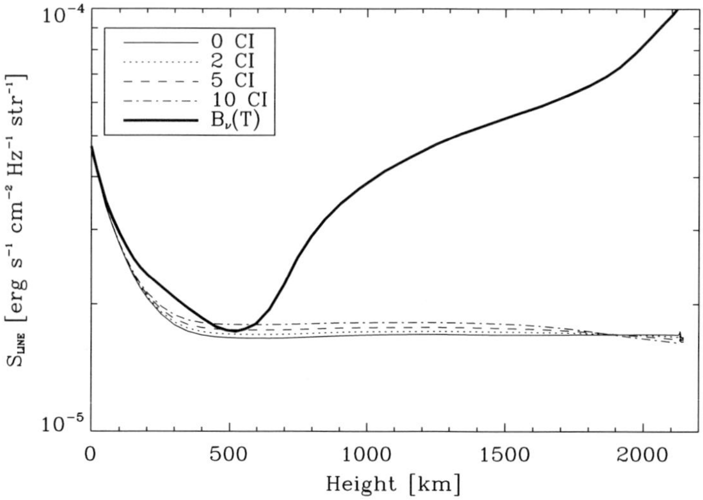

Fig. 1 The variation with height in a semi-empirical solar-atmosphere model of the line source function of the He I 10,830 Å transitions, calculated from the non-LTE population values given in Fig. 6 of Centeno et al. (2008), which indicates that significant absorption of the photospheric radiation at 10,830 Å is to be expected mainly around 2,000 km in the atmosphere model. Each *curve* corresponds to the indicated value of the EUV coronal irradiance (CI), in units of the nominal value

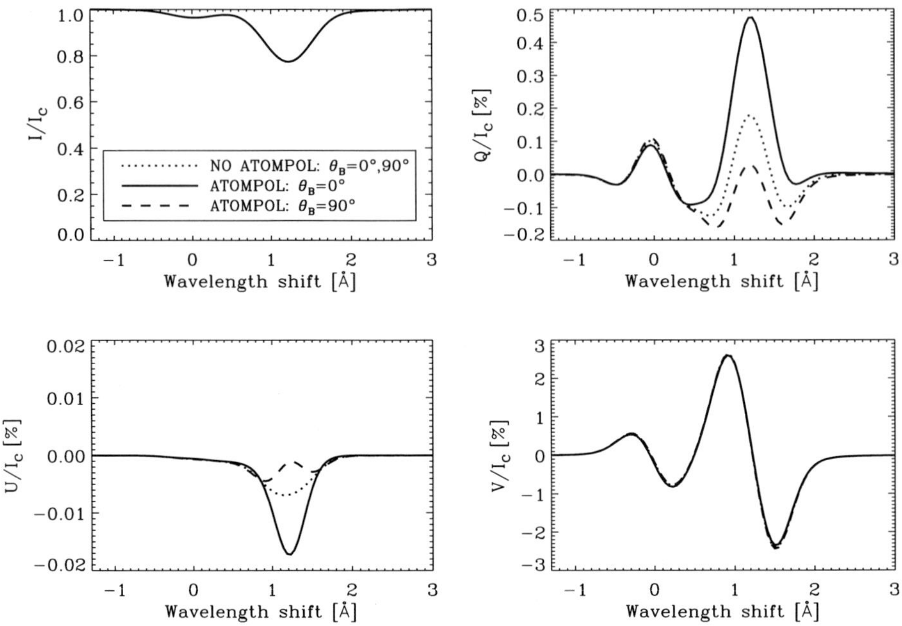

Fig. 2 The emergent stokes profiles in the He I 10,830 Å triplet calculated with the spectral synthesis option of HAZEL for a LOS with $\theta = 45°$ ($\mu = 0.707$) contained in the Z–X plane, with the Z-axis along the local solar vertical. The positive reference direction for stokes Q is along the Y-axis. The calculations assume a magnetized ($B = 1,200$ G!) slab of helium atoms located at a height of about 2,200 km above the visible solar surface. Two magnetic field orientations are considered: (**a**) vertical field along the Z-axis ($\theta_B = 0°$ and $\chi_B = 0°$) and (**b**) horizontal field along the X-axis ($\theta_B = 90°$ and $\chi_B = 0°$). The slab's optical thickness at the wavelength of the red blended component is $\Delta\tau_{\rm red} = 0.5$. The *dotted curves* show the emergent stokes profiles for cases a and b when only the Zeeman effect is taken into account. The other curves show the emergent Stokes profiles when taking into account also the influence of atomic level polarization in case a (*solid*) and in case b (*dashed*). Note that the influence of atomic level polarization on the linear polarization profiles is very significant and that it removes the 180° azimuth ambiguity present in the Zeeman-effect profiles. Interestingly, the nonzero stokes U signals are caused by the anomalous dispersion terms

in the He I 10,830 Å multiplet for a LOS with $\theta = 45°$. The calculations have been carried out for the two magnetic field orientations indicated in the figure legend: (a) vertical field and (b) horizontal field. When the atomic level polarization is neglected, the Zeeman effect caused by both magnetic field vectors produce exactly the same Stokes profiles (see the dotted lines). This is because the circular polarization of the Zeeman effect depends on the inclination of the magnetic field vector with respect to the LOS (which is identical for cases (a) and (b)) and the linear polarization is invariant when the component of the field in the plane perpendicular to the LOS is rotated by 180° (i.e., the well-known azimuth ambiguity of the Zeeman effect). However, when the influence of atomic polarization is taken into account, then the linear polarization profiles corresponding to such magnetic field orientations are very different (see the solid and dashed curves), simply because in the presence of atomic level polarization, the emergent Stokes profiles not only depend on the orientation of the magnetic field with respect to the LOS, but also on its inclination with respect to the local solar vertical (e.g., Landi Degl'Innocenti and Bommier 1993). In fact, the information provided by (3) and (4) is contained in the following single formula (cf., Trujillo Bueno 2003a)

$$\frac{Q}{I} \approx -\frac{3}{4\sqrt{2}} \sin^2 \Theta \, (3 \cos^2 \theta_B - 1) \, \mathcal{F}, \tag{6}$$

where Θ is the angle between the magnetic field vector and the LOS, θ_B the angle between the magnetic field vector and the local solar vertical, and the reference direction for Stokes Q is that for which $U/I = 0$ (i.e., the parallel to the projection of the magnetic field onto the plane perpendicular to the LOS). The sign of $\mathcal{F}$ is established by the solution of the statistical equilibrium equations for the elements of the atomic density matrix. This formula shows clearly why the so-called Van-Vleck angle, $\theta_{\mathrm{V}} = 54.74°$, is *magic*. As $\cos^2 \theta_{\mathrm{V}} = 1/3$, it is clear that Q/I has the sign of $\mathcal{F}$ for $\theta_{\mathrm{V}} < \theta_B < \pi - \theta_{\mathrm{V}}$, but the opposite sign for $0 < \theta_B < \theta_{\mathrm{V}}$ or $\pi - \theta_{\mathrm{V}} < \theta_B < \pi$. Equation (6) can also be used to understand what the Van-Vleck ambiguity of the Hanle effect is (e.g., Casini and Judge 1999; López Ariste and Casini 2005; Merenda et al. 2006).

Interestingly, as shown also in Fig. 2, for some spectral lines, such as those of the He I 10,830 Å and D_3 multiplets, the influence of atomic level alignment on the emergent linear polarization can be very important, even in the presence of magnetic fields as strong as 1,200 G (Trujillo Bueno and Asensio Ramos 2007). Therefore, inversion codes that neglect the influence of atomic level polarization, such as the Milne–Eddington codes of Lagg et al. (2004) and Socas-Navarro et al. (2004), should ideally be used only for the inversion of Stokes profiles emerging from strongly magnetized regions (e.g., with $B > 2{,}000$ G for the case of the He I 10,830 Å triplet) or when the observed Stokes Q and U profiles turn out to be dominated by the transverse Zeeman effect, as happens with some active region filaments (see Sect. 6.2).[3]

The inversion option of HAZEL is based on the Levenberg–Marquardt (LM) method for locating the minimum of the merit function that quantifies the goodness of the fit between the observed and synthetic Stokes profiles. To improve the convergence properties of the LM method, HAZEL uses a novel initialization technique based on the DIRECT algorithm, a deterministic global optimization technique that is significantly more efficient than the stochastic method PIKAIA considered by Charbonneau (1995) and used by Lagg et al. (2004) in their inversion code of Stokes profiles induced by the Zeeman effect. This code, called HELIX, has been recently improved by combining its inversion approach with the

[3] It is, however, important to note that the positions and strengths of the π- and σ-components must be calculated within the framework of the Paschen–Back effect theory, even in the Zeeman-dominated case (see Socas-Navarro et al. 2004).

(Hanle+Zeeman) spectral synthesis calculation core of HAZEL (see Lagg et al. 2009). An alternative inversion procedure is the Principal Component Analysis (PCA) technique described by López Ariste and Casini (2002), which necessitates first creating a suitable database of emergent Stokes profiles for a comprehensive set of illumination, thermodynamic, and magnetic conditions in the plasma structure under consideration. Another inversion strategy based on databases was applied by Trujillo Bueno et al. (2005b) and Merenda et al. (2006) to spectropolarimetric observations of solar spicules and prominences in the He I 10,830 Å multiplet, which is of particular interest for the determination of the magnetic field vector in plasma structures with $10\,\mathrm{G} \lesssim B \lesssim 100\,\mathrm{G}$.

5 The Quiet Chromosphere

A very suitable diagnostic window for mapping the magnetic fields of the "quiet" regions of the solar chromosphere is that provided by the polarization signals of the Ca II IR triplet (Manso Sainz and Trujillo Bueno 2007). In such regions, the circular polarization of the Ca II IR lines is caused by the longitudinal Zeeman effect, while the Stokes Q and U profiles are dominated by atomic level polarization and the Hanle effect. Interestingly, while the linear polarization in the 8,498 Å line shows a strong sensitivity to inclined magnetic fields with strengths between 1 mG and 10 G, the emergent linear polarization in the 8,542 and 8,662 Å lines is very sensitive to magnetic fields in the milligauss range. The reason for this very interesting behavior is that the scattering polarization in the 8498 Å line gets a significant contribution from the selective emission processes that result from the atomic polarization of the short-lived upper level, while that in the 8,542 and 8,662 Å lines is dominated by the selective absorption processes that result from the atomic polarization of the metastable (long-lived) lower levels (Manso Sainz and Trujillo Bueno 2003b). Therefore, in the quiet chromosphere the magnetic sensitivity of the linear polarization of the 8,542 and 8,662 Å lines is mainly controlled by the lower-level Hanle effect, which implies that in regions with $1\,\mathrm{G} \lesssim B \lesssim 50\,\mathrm{G}$ the Stokes Q and U profiles are only sensitive to the orientation of the magnetic field vector. The 8,498 Å line is, however, sensitive to both the orientation and the strength of the magnetic field through the upper-level Hanle effect.

Figure 3 shows a high-sensitivity spectropolarimetric observation of the quiet solar chromosphere in the strongest (8,542 Å) line of the Ca II IR triplet. It was obtained by R. Ramelli (IRSOL), R. Manso Sainz (IAC) and me using the Zürich Imaging Polarimeter (ZIMPOL) attached to THEMIS. The observed Stokes V/I profiles are clearly caused by the longitudinal Zeeman effect, while the Stokes Q/I and U/I signals are produced mainly by the influence of atomic level polarization. As seen in Fig. 3, although the spatio-temporal resolution of this spectropolarimetric observation is rather low (i.e., no better than 3″ and 20 min), the fractional polarization amplitudes fluctuate between 0.01 and 0.1% along the spatial direction of the spectrograph's slit, with a typical spatial scale of 5″. Interestingly enough, while

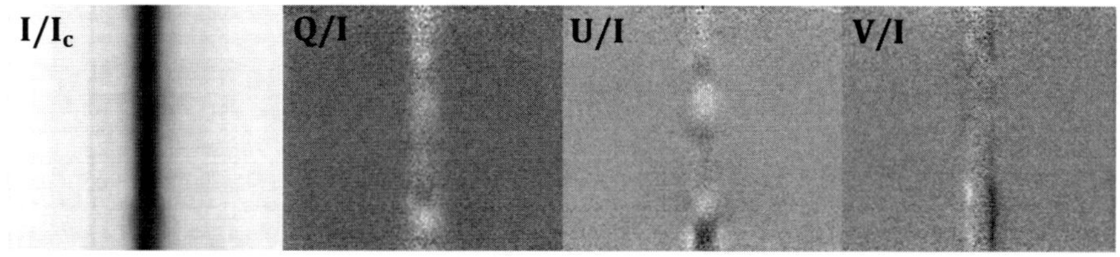

Fig. 3 An example of our recent spectropolarimetric observations of the Ca II 8542 Å line in a very quiet region close to the solar limb, using ZIMPOL at the French–Italian telescope THEMIS. The reference direction for Stokes Q is the tangent to the closest limb. From Trujillo Bueno (2009a)

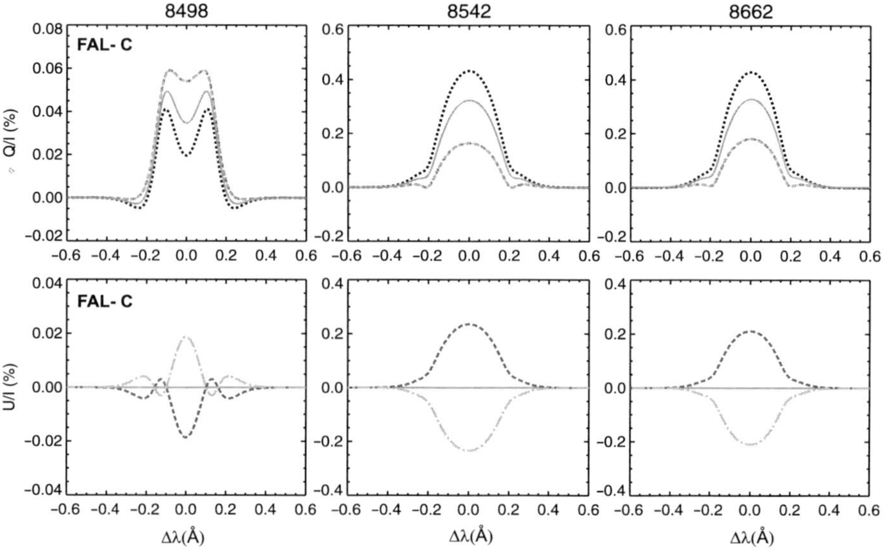

Fig. 4 The emergent fractional linear polarization of the Ca II IR triplet calculated for a LOS with $\mu = 0.1$ in a semi-empirical model of the solar atmosphere. For the $B = 0$ case (*dotted curves*) only Q/I is nonzero. The other curves refer to a horizontal 5 mG magnetic field pointing towards the observer (*red dashed curves*), perpendicular to the observer (*green solid curves*), and away from the observer (*blue dash-dotted curves*). The reference direction for Stokes Q is the parallel to the closest limb. Note that while the Q/I amplitude fluctuates but remains always positive, the sign of U/I depends on the azimuth of the magnetic field. From Manso Sainz and Trujillo Bueno (2009b) (color illustration are available in the on-line version)

the Stokes Q/I signal changes its amplitude but remains always positive along that spatial direction, the sign of the Stokes U/I signal fluctuates.

The physical interpretation of this type of spectropolarimetric observations requires solving the non-LTE problem of the second kind for the Ca II IR triplet. Figure 4 shows examples of the emergent fractional linear polarization calculated with MULTIPOL in a semi-empirical model of the solar atmosphere. The top panels show the emergent Q/I profiles for a LOS with $\mu = 0.1$, for the unmagnetized reference case and for three possible orientations of a 5 mG horizontal magnetic field. The bottom panels show the corresponding U/I signals, which are of course zero for the unmagnetized case. Note that while the amplitudes of the theoretical Q/I profiles change with the strength and orientation of the magnetic field and are always positive, the sign of U/I is sensitive to the azimuth of the magnetic field vector. Therefore, as expected, the spatial variations in the observed fractional linear polarization (see Fig. 3) are mainly due to changes in the orientation of the chromospheric magnetic field.

These types of polarization signal resulting from atomic level polarization and the Hanle and Zeeman effects can be exploited to explore the thermal and magnetic structure of the solar chromosphere. They can also be used to evaluate the degree of realism of magneto-hydrodynamic simulations of the photosphere-chromosphere system via careful comparisons of the observed Stokes profiles with those obtained through forward-modeling calculations.

6 Plasma Structures in the Chromosphere and Corona

As mentioned earlier, a suitable diagnostic window for inferring the magnetic field vector of plasma structures embedded in the solar atmosphere is that provided by the polarization produced by the joint action of atomic level polarization and the Hanle and Zeeman effects in the He I 10,830 and 5876 Å multiplets.

The resolved components of both helium multiplets stand out in emission when observing off-limb structures at a given height above the visible limb. As their respective spectral lines have different sensitivities to the Hanle effect, one would benefit from observing them simultaneously in spicules and prominences. At present, such simultaneous spectropolarimetric observations can be carried out with THEMIS and with the polarimeter SPINOR attached to the Dunn Solar Telescope. The main uncertainty with off-limb observations is that we do not know whether the observed plasma structure was really in the plane of the sky during the observing period (i.e., it is not known whether the observed Stokes profiles were produced in 90° scattering geometry or not).

Concerning on-disk observations it is clear that the He I 10,830 Å triplet is the most suitable one, given that it shows significantly more absorption than the He I D_3 multiplet when observing a variety of plasma structures against the bright background of the solar disk. The additional fact that the Hanle effect in forward scattering creates measurable linear polarization signals in the lines of the He I 10,830 Å multiplet when the magnetic field is inclined with respect to the local solar vertical direction (Trujillo Bueno et al. 2002b), and that there is a nearby photospheric line of Si I, makes the 10,830 Å spectral region very suitable for investigating the coupling between the photosphere and the corona. The main uncertainty with on-disk observations is that we do not know the exact height above the solar visible surface where the observed plasma structure is located.

6.1 Off-Limb Diagnostics of Prominences and Spicules

The off-limb observational strategy consists in doing spectropolarimetric observations with the image of the spectrograph's slit at given, ideally consecutive distances from the visible solar limb. The Stokes profiles measured at each pixel (or after downgrading the original spatial resolution to increase the signal-to-noise ratio) are then used to obtain information on the strength, inclination, and azimuth of the magnetic field vector via the application of Stokes inversion techniques like those discussed in Sect. 4.2. All published applications to the He I D_3 multiplet observations are based on the optically thin plasma assumption (e.g., Landi Degl'Innocenti 1982; Casini et al. 2003; López Ariste and Casini 2005), while some of the reported He I 10,830 Å observations were interpreted taking into account radiative transfer effects through the constant-property slab model discussed in Sect. 4.2 (see Trujillo Bueno et al. 2002b; Trujillo Bueno et al. 2005b; Asensio Ramos et al. 2008).

For example, Casini et al. (2003) interpreted new spectropolarimetric observations of quiescent solar prominences in the He I D_3 multiplet via the application of a PCA inversion code (López Ariste and Casini 2002) based on an extensive database of theoretical Stokes profiles. These profiles were calculated using the optically thin approximation, covering all the magnetic configurations and scattering geometries of interest. For one of the observed prominences, Casini et al. (2003) provided two-dimensional maps of the inferred magnetic field vector. Such maps showed the surprising feature of areas reaching up to 80 G within a background of prominence plasma with predominantly horizontal fields having 10–20 G (see also Casini et al. 2005). Paletou et al. (2001) and Trujillo Bueno et al. (2002b) also inferred prominence fields significantly stronger than those found by Leroy and collaborators (see Leroy 1989). For a detailed He I 10,830 Å investigation of the magnetic field vector in a polar-crown prominence, see Merenda et al. (2006).

Concerning spicules, it is important to emphasize that the observation and theoretical modeling of the Hanle and Zeeman effects in such spike-like jet features provides a suitable tool for investigating the magnetism of the solar chromosphere. The paper by Centeno et al. (2010) discusses briefly the results that several researchers have obtained through the interpretation of off-limb observations and reports on a recent investigation based on the application of the inversion code HAZEL to He I 10,830 Å spectropolarimetric observations of spicules in the quiet solar chromosphere. They find magnetic fields with $B \approx 40$ G in a localized area of the slit-jaw image, which could represent a possible lower value for the field strength of organized network spicules.

Very interesting He I D_3 observations of spicules are shown in Fig. 5. The Stokes profiles of the left panel correspond to a quiet region, while those of the right panel were observed close to an active region. Obviously, for the magnetic strengths of spicules (i.e., $B \lesssim 100$ G), the observed linear polarization is fully dominated by the selective emission processes that result from the atomic alignment of the upper levels of the $3d^3D$ term, without any significant influence of the transverse Zeeman effect. Note that in both regions Stokes U is nonzero, which is the observational signature of the Hanle effect of an inclined magnetic field. The change of sign in

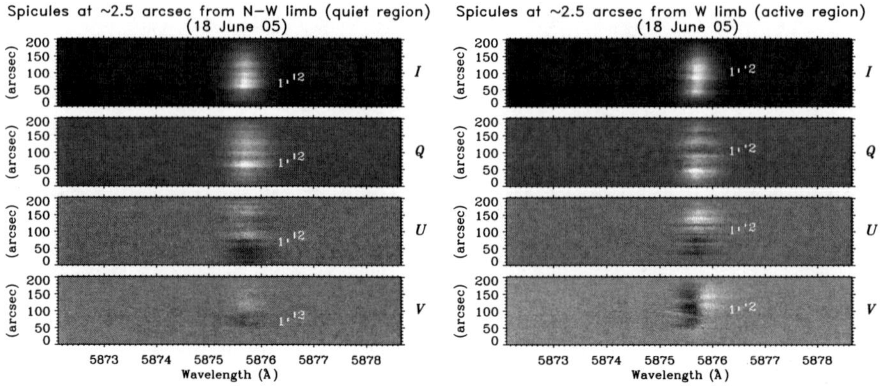

Fig. 5 Illustrative examples of the off-limb stokes profiles of the He I 5876 Å multiplet observed in a quiet region (*left panels*) and close to an active region (*right panels*). Interestingly, the inferred magnetic field vector has $B \approx 10$ G in the quiet region, while a best fit to the profiles observed close to the active region requires $B \approx 50$ G. The reference direction for stokes Q is the parallel to the solar limb. From Ramelli et al. (2006)

Stokes U along the spatial direction of the spectrograph's slit can be easily explained by variations in the azimuth of the magnetic field vector. In contrast, the circular polarization profiles of the D_3 multiplet are the result of the joint action of the longitudinal Zeeman effect and of atomic level orientation (see Sect. 2). Interestingly, the Stokes V profiles corresponding to the observed quiet region are dominated by atomic level orientation, while those observed in the spicules close to the active region are caused mainly by the longitudinal Zeeman effect.

6.2 *On-Disk Diagnostics of Filaments*

A significant difference between quiet region (QR) filaments and active region (AR) filaments is that the former are weakly magnetized (i.e., with $B < 100$ G) and embedded in the 10^6 K solar corona, while the latter are strongly magnetized (i.e., with $100\,\mathrm{G} \lesssim B \lesssim 1{,}000\,\mathrm{G}$) and located at much lower heights above the solar visible surface.

For magnetic strengths $B \lesssim 100$ G, the linear polarization of the He I 10,830 Å triplet is fully dominated by the atomic level polarization that is produced by anisotropic radiation pumping (Trujillo Bueno et al. 2002b). For stronger magnetic fields, the contribution of the transverse Zeeman effect cannot be neglected. However, in principle, the emergent linear polarization should still show an important contribution caused by the presence of atomic level polarization, even for the unfavorable case of low-lying plasma structures having magnetic field strengths as large as 1,000 G (see Fig. 2 in Trujillo Bueno and Asensio Ramos 2007; see also Fig. 2 of Sect. 4.2). Surprisingly, at the Fourth International Workshop on Solar Polarization,

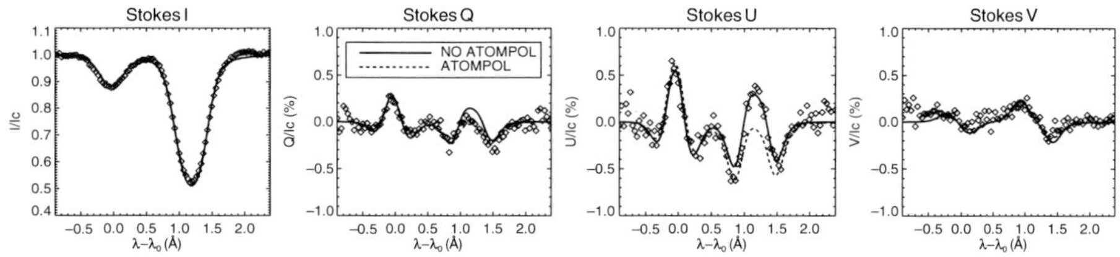

Fig. 6 Example of stokes profiles dominated by the Zeeman effect observed in an AR filament. Both best fits obtained with HAZEL give $B \approx 750\,\mathrm{G}$ and $\theta_B = 97°$. This figure results from an ongoing collaboration with Y. Katsukawa (NAOJ)

V. Martínez Pillet and collaborators reported that the Stokes Q and U profiles of the He I 10,830 Å multiplet from an active-region filament had the typical shape of polarization profiles produced by the transverse Zeeman effect. Such observations of an AR filament on top of a dense plage region at the polarity inversion line (i.e., an "abutted" plage region) have been recently analyzed in detail by Kuckein et al. (2009), showing that the filament magnetic fields were mainly horizontal and with strengths between 600 and 700 G. Figure 6 shows another example of Stokes profiles dominated by the Zeeman effect, observed in a (low-lying) filament in an AR with large sunspots.

What is the explanation of this enigmatic finding? According to Trujillo Bueno and Asensio Ramos (2007), the AR filament analyzed by Kuckein et al. (2009) had significant optical thickness so that the radiation field generated by the structure itself reduced the positive contribution to the anisotropy factor caused by the radiation from the underlying solar photosphere (see Fig. 4 of Trujillo Bueno and Asensio Ramos 2007). At meaningful optical thickness in the "horizontal" direction (e.g., along the filament axis), the amount of atomic level alignment in the filament can be significantly reduced to the extent that the transverse Zeeman effect of its *strong* horizontal field dominates the emergent linear polarization (e.g., a vertical light beam of intensity I_c and two oppositely directed horizontal beams of intensity $I_{\mathrm{filament}} = S = I_c/4$ produce *zero anisotropy* in the horizontal reference system). An alternative possibility, suggested by Casini et al. (2008), requires the presence of a randomly oriented field entangled with the main filament field and of similar magnitude (700 G!).

Another important question is whether all AR filaments permeated by a significantly strong (e.g., $B \approx 500\,\mathrm{G}$) and predominantly horizontal magnetic field show linear polarization profiles dominated by the transverse Zeeman effect. The fact that the answer to this question is *negative* can be seen in Fig. 7, which shows an example of the Stokes profiles that A. Asensio Ramos (IAC), C. Beck (IAC), and I observed on 9 June 2007 in an AR filament several hours before its eruption. The solid curves in the upper panels show the best theoretical fits to the observed Stokes profiles that the inversion code HAZEL gives when only the Zeeman effect is considered, while the lower panels demonstrate that the fit to the observed linear polarization is dramatically improved when the influence of atomic level polarization is also taken into

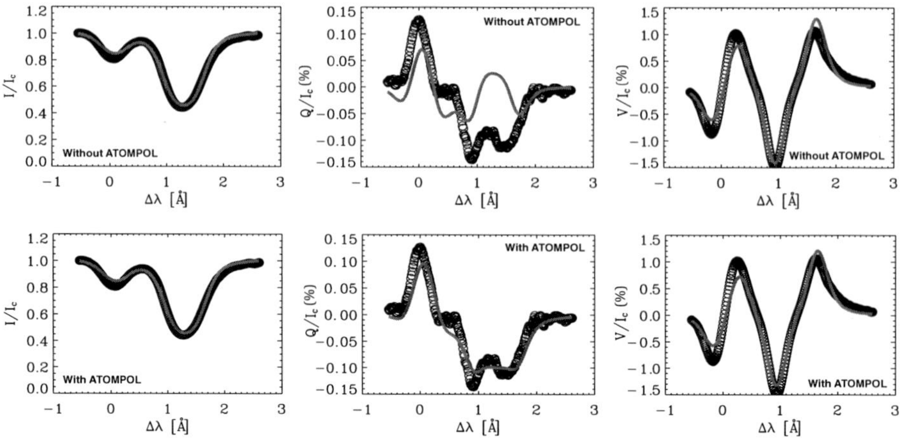

Fig. 7 Example of the stokes profiles observed in an AR filament a few hours before its eruption, using the Tenerife infrared polarimeter attached to the German VTT. *Upper panels*: the best theoretical fit obtained with HAZEL when neglecting atomic level polarization corresponds to $B = 460$ G. *Lower panels*: the best theoretical fit when atomic level polarization is taken into account corresponds to $B = 390$ G. In both cases the inferred inclination is $\theta_B = 115°$. The Stokes U profile (not shown here) is more or less similar to the observed Stokes Q, but of opposite sign

account. It is interesting to note that, while the inferred inclination of the magnetic field with respect to the local vertical is $\theta_B \approx 115°$ in both cases, the magnetic field strength turns out to be 70 G stronger when Stokes inversion is performed neglecting atomic level polarization.

As seen in Fig. 7, the intensity profiles observed in this AR filament show significant absorption, which implies that the slab's optical thickness along the LOS needed to fit them with HAZEL is significant. Therefore, following Trujillo Bueno and Asensio Ramos (2007), one may argue that the anisotropy of the radiation field within the filament was small and that, accordingly, the emergent linear polarization should be dominated by the transverse Zeeman effect. However, this was not the case (see the Stokes Q panels of Fig. 7). I believe that the solution to this enigma has to do with the degree of compactness of the multitude of individual magnetic fibrils, stacked one upon another, that together determine the structure of a filament. Very likely, the plasma of the AR filament analyzed by Kuckein et al. (2009) had a high degree of compactness, such that a single constant-property slab or tube model with a significant optical thickness in the horizontal direction provides a suitable representation. As a result, the radiation field generated by the plasma structure itself may have produced a negative contribution to the anisotropy factor, so that the anisotropy of the true radiation field that illuminated the helium atoms in the filament body was significantly smaller than that corresponding to the optically thin case (see Fig. 4 of Trujillo Bueno and Asensio Ramos 2007). On the contrary, in my opinion, a more suitable model for the AR filament we observed on 9 June 2007 a few hours before its eruption is that of a multitude of *optically thin* threads of magnetized plasma, each of them illuminated by the anisotropic radiation coming from the underlying photosphere, and such that the total optical thickness along the LOS is that needed by HAZEL to fit the observed Stokes $I(\lambda)$ profiles. The observational and theoretical support in favor of prominence thread structure is overwhelming (e.g., the review by Heinzel 2007), but it is interesting to note that the signature of atomic level polarization in the linear polarization profiles observed in some AR filaments may provide information on the degree of compactness of the structure that results from the agglomeration of multitude of individual magnetized fibrils.

Finally, it is of interest to mention that on 11 September 2003, Merenda et al. (2007) carried out He I spectropolarimetric observations of a filament in a moderately active region that was located relatively close to the solar disk center, and found that the Stokes Q and U profiles were dominated by the influence of atomic level polarization in most of the filament body, except in a small region of the filament apparently close to one of its footpoints (where the observed linear polarization clearly resulted from the joint action of atomic level polarization and the transverse Zeeman effect). A detailed interpretation of these observations via the Stokes inversion strategy described in Merenda et al. (2006) allowed us to infer a full magnetic map of the filament, with the azimuth, inclination, and strength of the magnetic field vector at each point within the filament (see Merenda et al. 2007 and Merenda 2008). Interestingly, while the inferred magnetic field vector was predominantly horizontal, practically aligned with the local filament axis and with $B \approx 100\,\mathrm{G}$ in most points of the filament body, it was found to be significantly stronger (a few hundred gauss) in the part of the filament apparently close to one of its footpoints.

6.3 *Magnetic Field "Reconstruction" in Emerging Flux Regions*

Figure 8 shows results from spectropolarimetric measurements of a region with small pores observed in the 10,830 Å spectral region. The left panel is a continuum image of the observed field of view, while the bright features in the right

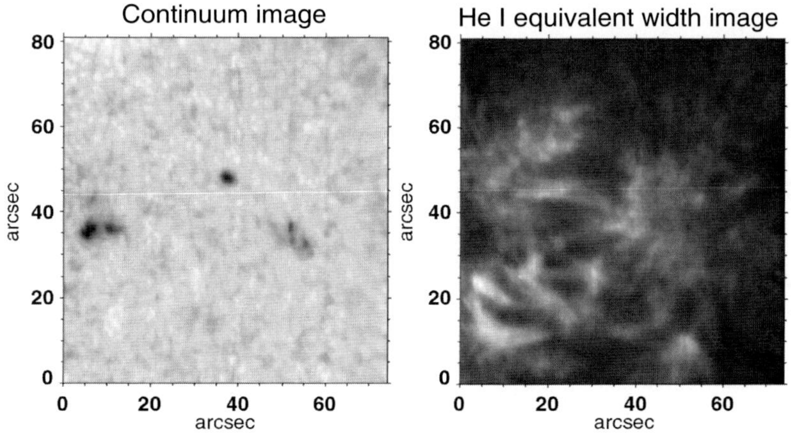

Fig. 8 Observation of a region with small pores obtained in collaboration with M. Collados (IAC) using the Tenerife infrared polarimeter attached to the VTT

panel show the pixels where the equivalent width of the Stokes $I(\lambda)$ profile of the He I 10,830 Å red blended component was significant. This is one of the most spectacular He I 10,830 Å equivalent width images I have ever seen. It shows loop-like structures in an emerging flux region with the largest equivalent width values localized at the apex of the loops.

It is not obvious to identify loops in the equivalent width image of the region observed by Solanki et al. (2003). In a recently published research note, Judge (2009) has pointed out that if there were really loops in the region analyzed by Solanki et al. (2003), they must have been near the plane containing the loop footpoints and the LOS, otherwise their loop images would have appeared significantly curved. The conclusion by Solanki et al. (2003) that their data demonstrate the presence of a current sheet *in the upper solar atmosphere* is based on the assumption that the observed absorption in the He I 10,830 Å triplet forms along magnetic field lines, which allowed them to associate a formation height to the magnetic field vector inferred at each pixel via the application of their Milne–Eddington inversion code. Judge (2009) has criticized this assumption on which the magnetic field "reconstruction" technique of Solanki et al. (2003) is based. He argues that He I 10,830 Å formation along a horizontal slab is a more reasonable assumption that along the skin of the emerging flux region and that the magnetic field reconstructions by Solanki et al. (2003) lead to spurious results. I think, however, that the He I equivalent width image of Fig. 8 supports Solanki et al.'s assumption that in regions of emerging flux the He I triplet is formed within loops, but it is not clear to me whether "freshly emerged loops" were really present in the region they observed. Fortunately, it seems clear that loops were really present in the region corresponding to Fig. 8. Hence, we will soon try to invert the ensuing data with HAZEL in order to check carefully if the application of the reconstruction technique of Solanki et al. (2003) leads to loops coinciding with those seen in Fig. 8. This is important because, as far as I know, the magnetic field reconstruction technique proposed by Solanki et al. (2003) is the only available to deliver three-dimensional information about the magnetic field vector from He I 10,830 Å spectropolarimetry.

7 How to Map the Magnetic Fields of Coronal Loops?

The primary emission of the 10^6 K solar coronal plasma lies in the EUV and soft X-rays, two spectral regions that can only be observed from space. At such short wavelengths, the coronal magnetic fields are unable to produce any significant Zeeman polarization signal. Therefore, a key question is: can we expect scattering polarization in permitted lines at EUV wavelengths for which the underlying quiet solar disk is seen completely dark? Manso Sainz and Trujillo Bueno (2009a) think that the answer is yes for the following reasons. For some EUV lines, their lower level is the ground level of a forbidden line at visible or near-IR wavelengths, which is polarized due to the anisotropic illumination of the atoms at the forbidden line wavelength. This lower-level atomic alignment is transferred to the upper level of the EUV line by collisional excitation. Therefore, as the upper level of the EUV line can be polarized, we may have measurable scattering polarization signals caused by the ensuing selective emission processes in the allowed EUV line. The linear polarization thus generated would be sensitive to the electronic density (N_e) and to the orientation of the magnetic field vector, although not to its strength, because for permitted EUV lines $B \gg B_H(J_l)$ and $B \ll B_H(J_u)$ (with B the magnetic strength in the coronal plasma). Interestingly enough, contrary to the case of forbidden line polarimetry (e.g., Casini and Judge 1999; Tomczyk et al. 2008), such linear polarization in allowed EUV transitions would be observable also in forward scattering at the solar disk center. This is extremely important because it provides a way for mapping the magnetic field of the extended solar atmosphere all the way up from the photosphere to the corona.

There are several interesting EUV lines satisfying these requirements. For example, the theoretical prediction of Manso Sainz and Trujillo Bueno (2009a) for the Fe X line at 174.5 Å is that the ensuing Q/I amplitude in 90° scattering geometry at a height of 0.1 solar radii above the solar surface varies between about 0.1% for $N_e = 10^9\,\mathrm{cm}^{-3}$ and 5% for $N_e = 10^7\,\mathrm{cm}^{-3}$, being a factor 2 smaller for the case of a horizontal magnetic field observed against the solar disk in forward-scattering geometry.

8 Concluding Comments

We should put high-sensitivity spectropolarimeters on ground-based and space telescopes for simultaneously measuring the polarization in photospheric and chromospheric lines. A very good choice would be the spectral region of the IR triplet of Ca II and/or that of the He I 10,830 Å triplet. If, in addition, we want to do something technologically challenging, we could then put a EUV imaging polarimeter on a space telescope (i.e., a TRACE-like instrument, but capable of obtaining also linear polarization images of coronal loops).

It would also be of great scientific interest to put a high-sensitivity spectropolarimeter in space to simply *discover* what the linearly polarized UV spectrum of the Sun looks like. For example, lines like Mg II k at 2,795 Å and the Ly α and Ly β lines of hydrogen are expected to show measurable linear polarization signals, also when pointing at the solar disk center where we observe the forward scattering case (Trujillo Bueno et al. 2005a). Such observations would provide precious information on the magnetic field structuring of the solar transition region from the chromosphere to the 10^6 K solar coronal plasma.

Concerning improvements in the diagnostic tools of chromospheric and coronal fields, our next step will be to acknowledge that "the Sun is a wolf in sheep's clothing" and to generalize the methods reported here to consider more sophisticated radiative transfer models and to account for variations of the magnetic fields and flows at sub-resolution scales. In fact, although many of the Stokes profiles observed in chromospheric lines can be fitted with one-component models, some of the observations discussed in this review and many others (e.g., Socas-Navarro et al. 2000a; Centeno et al. 2005; Lagg et al. 2007; Sasso et al. 2007) indicate the need to consider at least two magnetic field components with different flows and/or orientations.

Acknowledgement It is a pleasure to thank my colleagues and friends from Tenerife, Locarno, Firenze, Boulder, and Tokyo for fruitful discussions and collaborations on the topic of this review. The author is also grateful to the editors and the other SOC members of the Evershed centenary meeting for their invitation to participate in a very good conference. Financial support by the Spanish Ministry of Science through project AYA2007-63881 and by the European Commission via the SOLAIRE network (MTRN-CT-2006-035484) are gratefully acknowledged.

References

Asensio Ramos, A., Trujillo Bueno, J., Landi Degl'Innocenti, E. 2008, ApJ, 683, 542
Avrett, E. H., Fontenla, J. M., Loeser, R. 1994, in Infrared Solar Physics, eds. D. M. Rabin, J. T. Jefferies, and C. Lindsey, IAU Symp. 154, 35
Belluzzi, L., Trujillo Bueno, J., Landi Degl'Innocenti, E. 2007, ApJ, 666, 588
Bommier, V. 1980, A&A, 87, 109
Casini, R., Judge, P. G. 1999, ApJ, 522, 524
Casini, R., Landi Degl'Innocenti, E. 2007, in Plasma Polarization Spectroscopy, eds. T. Fujimoto and A. B. S. Iwamae, 247
Casini, R., López Ariste, A., Tomczyk, S., Lites, B. W. 2003, ApJ, 598, L67
Casini, R., Bevilacqua, R., López Ariste, A. 2005, ApJ, 622, 1265
Centeno, R., Socas-Navarro, H., Collados, M., Trujillo Bueno, J. 2005, ApJ, 635, 670
Casini, R., Manso Sainz, R., Low, B. C. 2008, ArXiv e-prints
Centeno, R., Trujillo Bueno, J., Uitenbroek, H., Collados, M. 2008, ApJ, 677, 742
Centeno, R., Trujillo Bueno, J., Asensio Ramos, A. 2010, in Magnetic Coupling between the Interior and the Atmosphere of the Sun, eds. S. S. Hasan and R. J. Rutten, Astrophys. Space Sci. Procs., Springer, Heidelberg, these proceedings
Charbonneau, P. 1995, ApJS, 101, 309
Harvey, J. W. 2006, in Solar Polarization 4, eds. R. Casini and B. W. Lites, ASP Conf. Ser., 358, 419
Heinzel, P. 2007, in The Physics of Chromospheric Plasmas, eds. P. Heinzel, I. Dorotovič, and R. J. Rutten, ASP Conf. Ser., 368, 271
Holzreuter, R., Stenflo, J. O. 2007, A&A, 467, 695
Judge, P. G. 2009, A&A, 493, 1121
Kuckein, C., Centeno, R., Martinez Pillet, V., et al. 2009, A&A, 501, 1113
Lagg, A. 2007, Advances in Space Research, 39, 1734
Lagg, A., Woch, J., Krupp, N., Solanki, S. K. 2004, A&A, 414, 1109
Lagg, A., Woch, J., Solanki, S. K., Krupp, N. 2007, A&A, 462, 1147
Lagg, A., Ishikawa, R., Merenda, L., et al. 2009, in The Second Hinode Science Meeting, eds. M. Cheung, B. W. Lites, T. Magara, J. Mariska and K. Reeves, ASP Conf. Ser. (in press)
Landi Degl'Innocenti, E. 1982, Solar Phys., 79, 291
Landi Degl'Innocenti, E., Bommier, V. 1993, ApJ, 411, L49
Landi Degl'Innocenti, E., Landolfi, M. 2004, Polarization in Spectral Lines, Kluwer, Dordrecht
Leroy, J. L. 1989, in Dynamics and Structure of Quiescent Solar Prominences, ed. E. R. Priest, Astrophys. Space Sci. Lib., 150, 77
López Ariste, A., Aulanier, G. 2007, in The Physics of Chromospheric Plasmas, eds. P. Heinzel, I. Dorotovič, and R. J. Rutten, ASP Conf. Ser., 368, 291
López Ariste, A., Casini, R. 2002, ApJ, 575, 529

López Ariste, A., Casini, R. 2005, A&A, 436, 325
Manso Sainz, R. 2002, PhD thesis, University of La Laguna
Manso Sainz, R., Trujillo Bueno, J. 2003a, in Solar Polarization 3, eds. J. Trujillo-Bueno and J. Sánchez Almeida, ASP Conf. Ser., 307, 251
Manso Sainz, R., Trujillo Bueno, J. 2003b, Phys. Rev. Letters, 91, 111102
Manso Sainz, R., Trujillo Bueno, J. 2007, in The Physics of Chromospheric Plasmas, eds. P. Heinzel, I. Dorotovič, and R. J. Rutten, ASP Conf. Ser., 368, 155
Manso Sainz, R., Trujillo Bueno, J. 2009a, in Solar Polarization 5, eds. S. Berdyugina, K. N. Nagendra, and R. Ramelli, ASP Conf. Ser. Vol. 405, 423
Manso Sainz, R., Trujillo Bueno, J. 2009b (in preparation)
Merenda, L. 2008, PhD thesis, University of La Laguna
Merenda, L., Trujillo Bueno, J., Landi Degl'Innocenti, E., Collados, M. 2006, ApJ, 642, 554
Merenda, L., Trujillo Bueno, J., Collados, M. 2007, in The Physics of Chromospheric Plasmas, eds. P. Heinzel, I. Dorotovič, and R. J. Rutten, ASP Conf. Ser., 368, 347
Paletou, F., López Ariste, A., Bommier, V., Semel, M. 2001, A&A, 375, L39
Pietarila, A., Socas-Navarro, H., Bogdan, T. 2007, ApJ, 670, 885
Ramelli, R., Bianda, M., Merenda, L., Trujillo Bueno, T. 2006, in Solar Polarization 4, eds. R. Casini and B. W. Lites, ASP Conf. Ser., 358, 448
Sampoorna, M., Nagendra, K. N., Stenflo, J. O. 2007, ApJ, 663, 625
Sasso, C., Lagg, A., Solanki, S. K., Aznar Cuadrado, R., Collados, M. 2007, in The Physics of Chromospheric Plasmas, eds. P. Heinzel, I. Dorotovič, and R. J. Rutten, ASP Conf. Ser., 368, 467
Socas-Navarro, H. 2005, ApJ, 631, L167
Socas-Navarro, H., Trujillo Bueno, J., Ruiz Cobo, B. 2000a, Science, 288, 1398
Socas-Navarro, H., Trujillo Bueno, J., Ruiz Cobo, B. 2000b, ApJ, 530, 977
Socas-Navarro, H., Trujillo Bueno, J., Landi Degl'Innocenti, E. 2004, ApJ, 612, 1175
Solanki, S. K., Lagg, A., Woch, J., Krupp, N., Collados, M. 2003, Nat, 425, 692
Stenflo, J. O. 1994, Solar Magnetic Fields: Polarized Radiation Diagnostics, Kluwer, Dordrecht
Stenflo, J. O. 2006, in Solar Polarization 4, eds. R. Casini and B. W. Lites, ASP Conf. Ser., 358, 215
Štěpán, J. 2008, PhD thesis, Observatoire de Paris
Tomczyk, S., Card, G. L., Darnell, T., et al. 2008, Solar Phys., 247, 411
Trujillo Bueno, J. 1999, in Solar Polarization, eds. K. N. Nagendra and J. O. Stenflo, Astrophys. Space Sci. Lib., 243, 73
Trujillo Bueno, J. 2001, in Advanced Solar Polarimetry – Theory, Observation, and Instrumentation, ed. M. Sigwarth, ASP Conf. Ser., 236, 161
Trujillo Bueno, J. 2003a, in Solar Polarization 3, eds. J. Trujillo Bueno and J. Sánchez Almeida, ASP Conf. Ser., 307, 407
Trujillo Bueno, J. 2003b, in Stellar Atmosphere Modeling, eds. I. Hubeny, D. Mihalas, and K. Werner, ASP Conf. Ser., 288, 551
Trujillo Bueno, J. 2005, in The Dynamic Sun: Challenges for Theory and Observations, ESA-SP, 600
Trujillo Bueno, J., Asensio Ramos, A. 2007, ApJ, 655, 642
Trujillo Bueno, J., Landi Degl'Innocenti, E. 1997, ApJ, 482, L183
Trujillo Bueno, J., Casini, R., Landolfi, M., Landi Degl'Innocenti, E. 2002a, ApJ, 566, L53
Trujillo Bueno, J., Landi Degl'Innocenti, E., Collados, M., Merenda, L., Manso Sainz, R. 2002b, Nat, 415, 403
Trujillo Bueno, J., Shchukina, N., Asensio Ramos, A. 2004, Nat, 430, 326
Trujillo Bueno, J., Landi Degl'Innocenti, E., Casini, R., Martínez Pillet, V. 2005a, in ESA Special Publication, eds. F. Favata, J. Sanz-Forcada, A. Giménez, and B. Battrick, ESA-SP, 588, 203
Trujillo Bueno, J., Merenda, L., Centeno, R., Collados, M., Landi Degl'Innocenti, E. 2005b, ApJ, 619, L191
Trujillo Bueno, J. 2009a, in Solar Polarization 5, eds. S. Berdyugina, K. N. Nagendra, and R. Ramelli, ASP Conf. Ser. Vol. 405, 65

Probability Density Functions to Represent Magnetic Fields at the Solar Surface

M. Sampoorna

Abstract Numerical simulations of magneto-convection and analysis of solar magnetogram data provide empirical probability density functions (PDFs) for the line-of-sight component of the magnetic field. In this paper, we theoretically explore effects of several types of PDFs on polarized Zeeman line formation. We also propose composite PDFs to account for randomness in both field strength and orientation. Such PDFs can possibly mimic random fields at the solar surface.

1 Introduction

Magneto-convection on the Sun has a size spectrum that spans several orders of magnitude and develops eddies with sizes much smaller than the spatial resolution of current spectro-polarimeters (about 0.2 arcsec or 150 km at the photospheric level). Thus the Stokes profiles that we observe are always averages over space, time, and along the line-of-sight (LOS). This suggests that it would be appropriate to characterize the magnetic field responsible for spectral line polarization by a probability density function (PDF). The case of Gaussian PDFs was studied by Dolginov and Pavlov (1972); Domke and Pavlov (1979); Frisch et al. (2005, 2006a,b, 2007); Sampoorna et al. (2008a). In this paper, we study PDFs determined from observations by Stenflo and Holzreuter (2002, 2003a,b) and from magneto-convection simulations by Stein and Nordlund (2006, see also Vögler et al. 2005). Sampoorna et al. (2008b) used these PDFs to compute mean solutions in the macro and micro-turbulent limits. Macro and micro-turbulence are also referred to as optically thick and optically thin limits. In this paper, we discuss in detail the mean solutions computed for a more general regime of meso-turbulence.

It is well known that photospheric photon mean free paths (50–100 km in the optical) correspond approximately to the sizes of meso-turbulent magnetic eddies. Calculation of mean Stokes parameters in this regime was considered by Landi

M. Sampoorna (✉)
Indian Institute of Astrophysics, Koramangala, Bangalore 560 034, India

S.S. Hasan and R.J. Rutten (eds.), *Magnetic Coupling between the Interior and Atmosphere of the Sun*, Astrophysics and Space Science Proceedings,
DOI 10.1007/978-3-642-02859-5_10,

degl'Innocenti (1994); Frisch et al. (2006a,b); Carroll and Staude (2003, 2005); Carroll and Kopf (2007). We use a Kubo–Anderson Process (KAP) with correlation length $1/\nu$ (where ν is number of jumps per unit optical depth) and a PDF to characterize the random magnetic field (Frisch et al. 2006b). Using KAP in a Milne-Eddington model atmosphere, Frisch et al. (2006b) deduce explicit expressions for the mean and rms fluctuations of the emergent Stokes parameters. In this paper, we use those expressions to compute mean solutions for magnetic eddies of arbitrary size.

2 Scalar PDFs

Recently, Stenflo and Holzreuter (2002, 2003a,b) have found from an analysis of high resolution La Palma and MDI solar magnetograms that the PDF for the LOS component is nearly independent of the spatial scale and can be well represented by a Voigt function. This PDF has a Gaussian core (centered around zero field) with a magnetic width Δ_B, and Lorentzian wings with a magnetic damping parameter a_B. The quantity Δ_B is a measure of the rms fluctuations of the LOS component.

A Voigt PDF with $\Delta_B = 6\,\mathrm{G}$ and $a_B = 1.5$ best fits the empirical PDF for LOS field strength derived from La Palma magnetograms. A symmetric Voigt function has zero net flux. In real magnetograms, magnetic flux is locally unbalanced. Nonzero net flux can be generated by a Voigt PDF that has a symmetric core, but different a_B for positive and negative polarities.

Magneto-convection simulation near the solar surface by Stein and Nordlund (2006) show that the magnetic field is intermittent with a stretched exponential distribution. Here we restrict ourselves to asymmetric scalar PDFs as they represent the photospheric conditions better.

2.1 Asymmetric Voigt PDF

In terms of parameters a_B and Δ_B, symmetric Voigt PDF has the form

$$P_{\mathrm{V}}(y, a_B) = \frac{a_B}{\pi^{3/2}} \int_{-\infty}^{+\infty} \frac{\mathrm{e}^{-u^2}}{(y-u)^2 + a_B^2}\, \mathrm{d}u. \tag{1}$$

In the above equation, we have introduced nondimensional parameters

$$y = B/\Delta_B, \quad u = B_1/\Delta_B, \quad \gamma_B = \Delta_B/B_{\mathrm{D}}, \tag{2}$$

where $1/B_{\mathrm{D}} = ge/(4\pi mc\Delta\nu_{\mathrm{D}})$, in standard notation. Here B is the random magnetic field component along a given direction, and $\Delta\nu_{\mathrm{D}}$ is the frequency Doppler

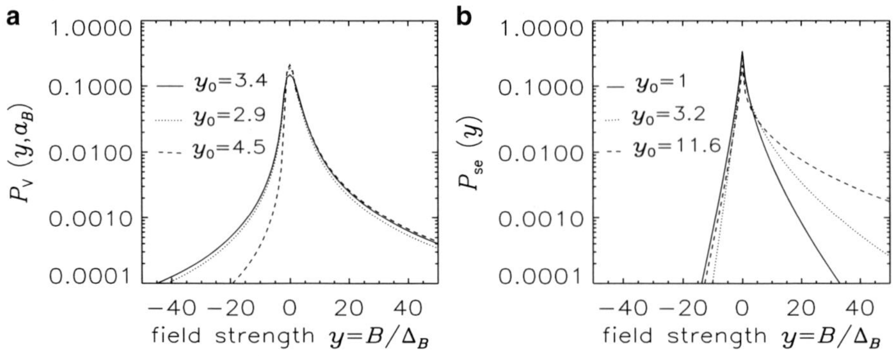

Fig. 1 (**a**) Asymmetric Voigt and (**b**) Stretched exponential PDFs. The y-scale is related to the B-scale through Δ_B (= 6 G according to Stenflo and Holzreuter 2002)

width. Thus, γ_B represents rms fluctuations Δ_B converted to Zeeman shift in Doppler width units.

Asymmetric Voigt PDFs can be constructed by choosing different values of a_B for different parts of the PDF while keeping the Gaussian core symmetrical. Figure 1a shows examples of asymmetric PDFs (see Sampoorna et al. 2008b, for details), which more or less resemble the PDF for the La Palma magnetogram shown in Fig. 2 of Stenflo and Holzreuter (2002). The mean magnetic field y_0 is the average of y over $P_V(y, a_B)$.

Mean emergent residual Stokes parameters $\langle r_{I,Q,U,V} \rangle$ are computed using asymmetric scalar PDFs for magnetic eddies of arbitrary size. The random field has a fixed orientation with respect to the LOS defined by polar angles $(\gamma, \chi) = (60°, 30°)$. Emergent residual Stokes parameters are defined by $r_I = [I_c - I]/C_1$ and $r_X = -X/C_1$, with the symbol X denoting Q, U, or V. The constant C_1 is the slope of continuum source function $S(\tau_c) = C_0 + C_1 \tau_c$, with τ_c the continuum optical depth. The continuum intensity at the surface is $I_c = C_0 + C_1$. This model has damping parameter $a = 0$ and line strength parameter (ratio of line to continuum absorption coefficient) $\beta = k_0/k_c = 10$. We also assume that the spectral line has a wavelength around 5,000 Å, a Landé factor of 2, and a Doppler width of 1.5 km s^{-1}. For this typical line, $B_D = 1.07 \times 10^3$ G. Hence $\gamma_B = 0.0056$ for an rms magnetic field fluctuation $\Delta_B = 6$ G. As a result, the $\langle r_I \rangle$ profiles in all figures in this paper remain insensitive to the PDF parameters.

Figure 2 shows $\langle r_{I,Q,U,V} \rangle$ computed using the three PDFs in Fig. 1a for $\nu = 5$. Note that $\langle r_V \rangle$ is zero if one uses a symmetric PDF with zero mean field. The $\langle r_{Q,U} \rangle$ profiles show very small sensitivity to PDF asymmetry, while the $\langle r_I \rangle$ profiles indeed are insensitive. For all three PDFs, $\langle r_V \rangle$ peaks around $x \approx 1.5$. The peak amplitudes increase with the mean field.

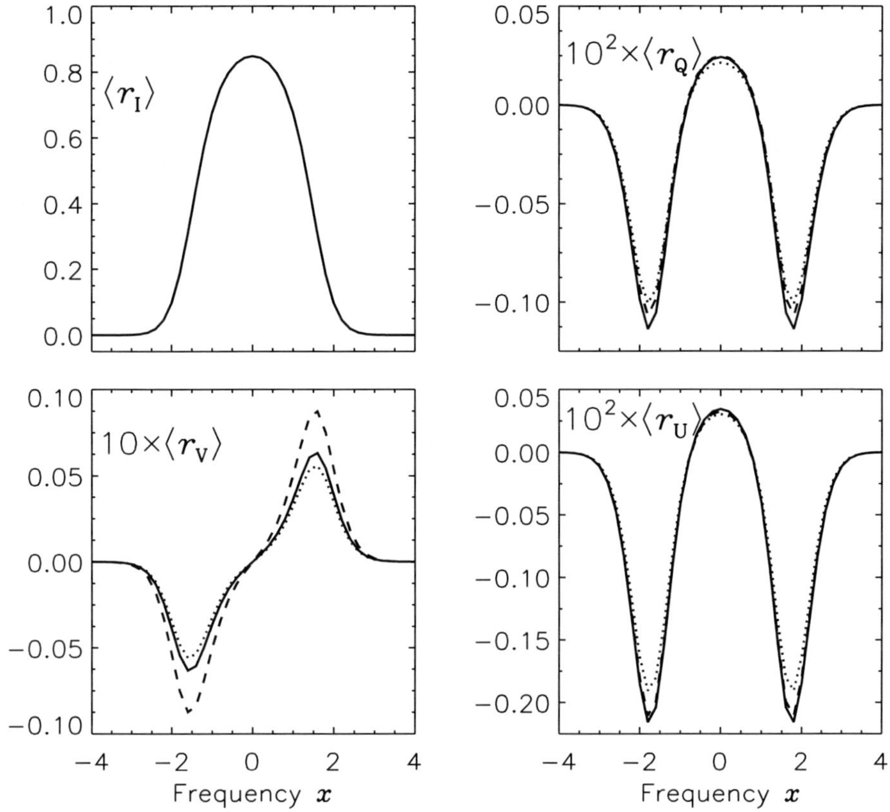

Fig. 2 $\langle r_{I,Q,U,V} \rangle$ computed with asymmetric Voigt PDFs in Fig. 1a for $\nu = 5$. Reference to line types is the same as in Fig. 1a

2.2 Asymmetric Stretched Exponential PDF

A symmetric stretched exponential may be written in functional form as

$$P_{\mathrm{se}}(y)\,\mathrm{d}y = C\,\mathrm{e}^{-|y|^k}\,\mathrm{d}y. \tag{3}$$

The quantity k takes values between 0 and 1 and is referred to as the stretching parameter. C is the normalization constant.

We construct asymmetric stretched exponential PDFs with nonzero mean field by choosing different k values for positive and negative polarities. Figure 1b shows three examples (see Sampoorna et al. 2008b, for details).

Mean residual Stokes profiles computed with these three PDFs are shown in Fig. 3. For $\nu = 5$, we observe a clear increase in the peak amplitudes of $\langle r_{Q,U,V} \rangle$ when the mean field y_0 increases. However, the peak positions are essentially

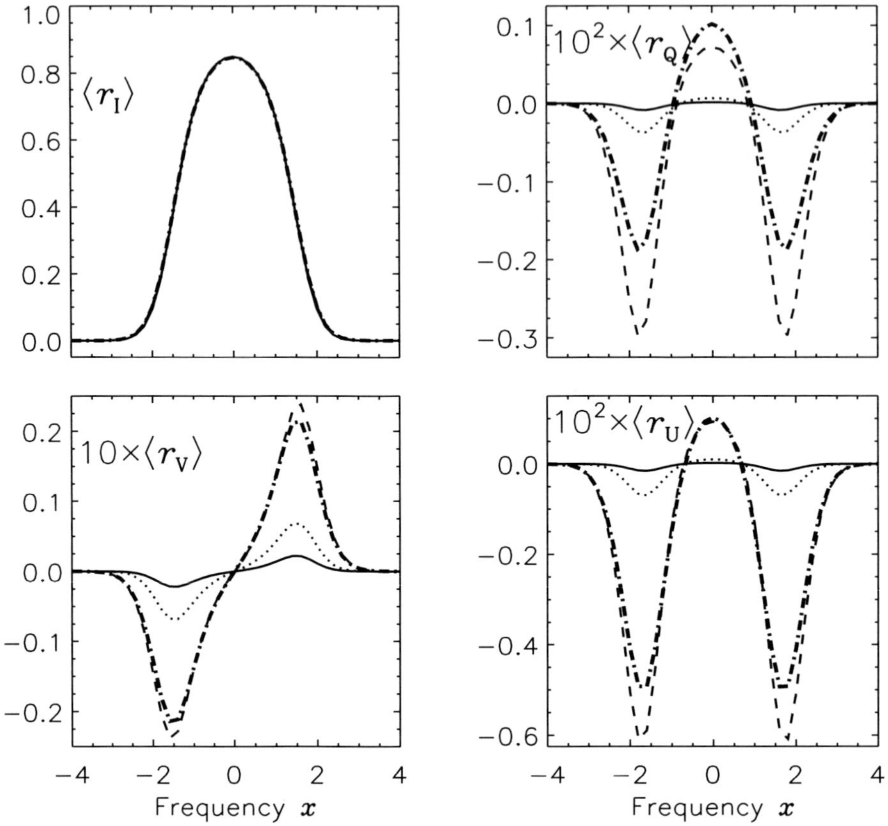

Fig. 3 $\langle r_{I,Q,U,V}\rangle$ for asymmetric stretched exponential PDFs. *Solid, dotted* and *dashed lines* refer to $\nu = 5$ (the same line types as in Fig. 1b). *Heavy dot-dashed line* refers to macro-turbulent limit for $y_0 = 11.6$

insensitive to y_0. The $\langle r_I\rangle$ profiles remain insensitive to PDF asymmetry. Differences between the solutions for $\nu = 5$ and the macro-turbulent limit appear for $\langle r_{Q,U}\rangle$ when $y_0 = 11.6$ due to the extended tail of the PDF for positive polarities. For $\langle r_V\rangle$, the differences remain small even for $y_0 = 11.6$. The relative insensitivity of $\langle r_V\rangle$ to the scale of magnetic field fluctuations has been observed for Voigt type PDFs (see Sampoorna et al. 2008b) and also for symmetric Gaussian PDFs with nonzero mean field (Frisch et al. 2006b). It is due to the fact that the PDF asymmetry is sharply peaked around y_0. In the limit of a Dirac distribution, there would be no difference between solutions for different ν values as the magnetic field would be deterministic. It thus appears that a mean value of Stokes V can be calculated with reasonable confidence using the micro-turbulent limit, a remark made already in Frisch et al. (2006b).

3 Angular PDFs

A large fraction of the solar atmosphere is filled with mixed polarity fields, and the inter-granular lanes contain fields that are preferentially directed upward or downward. To represent this situation, we consider magnetic fields that have a fixed value of the strength B but random orientations. For such a random field, following angular distribution has been suggested by Stenflo (1987):

$$P_{\rm pl}(\mu_B) = \frac{(p+1)}{4\pi}\,|\mu_B|^p, \quad -1 \le \mu_B \le +1. \tag{4}$$

Here, $\mu_B = \cos\theta_B$, with θ_B the field orientation with respect to the atmospheric normal. The power law index p can take any value; the $p = 0$ case corresponds to an isotropic distribution. As p increases the distribution becomes more and more peaked in the vertical direction.

Mean profiles $\langle r_{I,Q}\rangle$ are calculated for $\nu = 5$, magnetic field strength $B/B_{\rm D} = 0.1$, line strength $\beta = 10$, damping parameter $a = 0$, and the power law index p a free parameter. Figure 4 shows $\langle r_{I,Q}\rangle$ at the limb ($\mu = 0.1$). A comparison of this figure with Fig. 12 of Sampoorna et al. (2008b) computed for the micro-turbulent limit shows that there is very little difference between the two. This is because the absolute value of the magnetic field along the LOS is bounded by the condition $B/B_{\rm D} = 0.1$. As the field is weak, $\langle r_Q\rangle \ll \langle r_I\rangle$. Further, $\langle r_Q\rangle$ is zero for $p = 0$. As p gets larger, $\langle r_Q\rangle$ first increases with p and then saturates for $p \simeq 100$ (see Fig. 4).

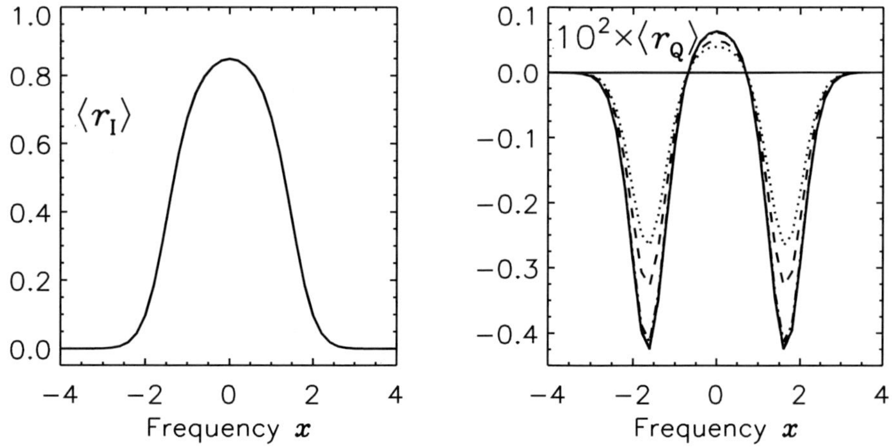

Fig. 4 $\langle r_{I,Q}\rangle$ at $\mu = 0.1$ (limb observation) computed for $\nu = 5$ and angular power law PDF (Eq. (4)). Line types: $p = 0$ (*solid*), 5 (*dotted*), 10 (*dashed*), 100 (*dot-dashed*), 500 (*dash-triple-dotted*), and 1,000 (*long dashed*). In this case $\langle r_{U,V}\rangle = 0$

4 Vector Magnetic Field Distributions

To describe a random vector magnetic field, we need a PDF that simultaneously accounts for strength and angle fluctuations. *From physical considerations one may argue that angular variation should be strongly field-strength dependent* (see de Wijn et al. 2009). For the strongest fields, the distribution should be peaked around the vertical direction, as strong fields tend to have intermittent fluxtube morphology. The weakest fields, on the other hand, would be passively moved and bent by turbulent fluid motions and so get so tangled up that their distribution would be nearly isotropic. The transition from isotropic to peaked distributions would probably be gradual (possibly around 50 G).

Based on this scenario we propose PDFs here which are combinations of angle and field strength distributions. For the angular part, we use the power law distribution introduced in Sect. 3, while for the field strength part, we consider either a Voigt function (see Sect. 2.1) or a stretched exponential (see Sect. 2.2). The functional form of such a composite PDF is

$$P(y, \mu_B, \phi_B)\,\mathrm{d}\mu_B \mathrm{d}\phi_B \mathrm{d}y = \frac{(p+1)}{2\pi} \begin{cases} P_V(y, a_B)\,\mu_B^p\,\mathrm{d}\mu_B \mathrm{d}\phi_B \mathrm{d}y, \\ C\,e^{-|y|^k}\;\mu_B^p\,\mathrm{d}\mu_B \mathrm{d}\phi_B \mathrm{d}y. \end{cases} \tag{5}$$

Here y varies in the range $[-y_{\max}, +y_{\max}]$, μ_B in the range $[0, 1]$, and ϕ_B in the range $[0, 2\pi]$. $P_V(y, a_B)$ is given in (1). Asymmetric composite PDFs can be constructed by choosing different a_B (or k) values for positive and negative polarities of the Voigt (or stretched exponential). The angle and strength distributions are coupled by letting the power law index p depend on y. We have chosen $p = |y|/y_t$ with $y_t = B_t/\Delta_B$, where B_t marks the transition between isotropic and peaked. Note that $y_t = \infty$ corresponds to fully isotropic distribution for all field strengths.

We calculate mean residual Stokes parameters for the composite PDF containing an asymmetric Voigt with mean field $y_0 = 4.5$ (dashed curve in Fig. 1a). The model parameters are $(a, \beta, \gamma_B) = (0, 10, 0.0056)$ and $\mu = 0.1$. Figure 5 shows the solutions for $\nu = 5$. The different curve types correspond to different y_t values. $\langle r_I \rangle$ is insensitive to the PDF asymmetry and to the variation of y_t due to the very weak value of γ_B. When $y_t \to \infty$, the PDF becomes fully isotropic, and hence $\langle r_Q \rangle \to 0$. As y_t decreases, the PDF becomes more and more anisotropic and hence $\langle r_Q \rangle$ as well as $\langle r_V \rangle$ increase in magnitude. In this case $\langle r_U \rangle$ is generated through the magneto-optical effects. Therefore, $\langle r_U \rangle$ is very small, with a behavior similar to $\langle r_Q \rangle$.

5 Conclusions

We have presented mean Stokes profiles formed in media having spatially unresolved magnetic structures with sizes that are comparable to photon mean free paths, using PDFs that describe fluctuations of the ambient field. A Gaussian PDF with

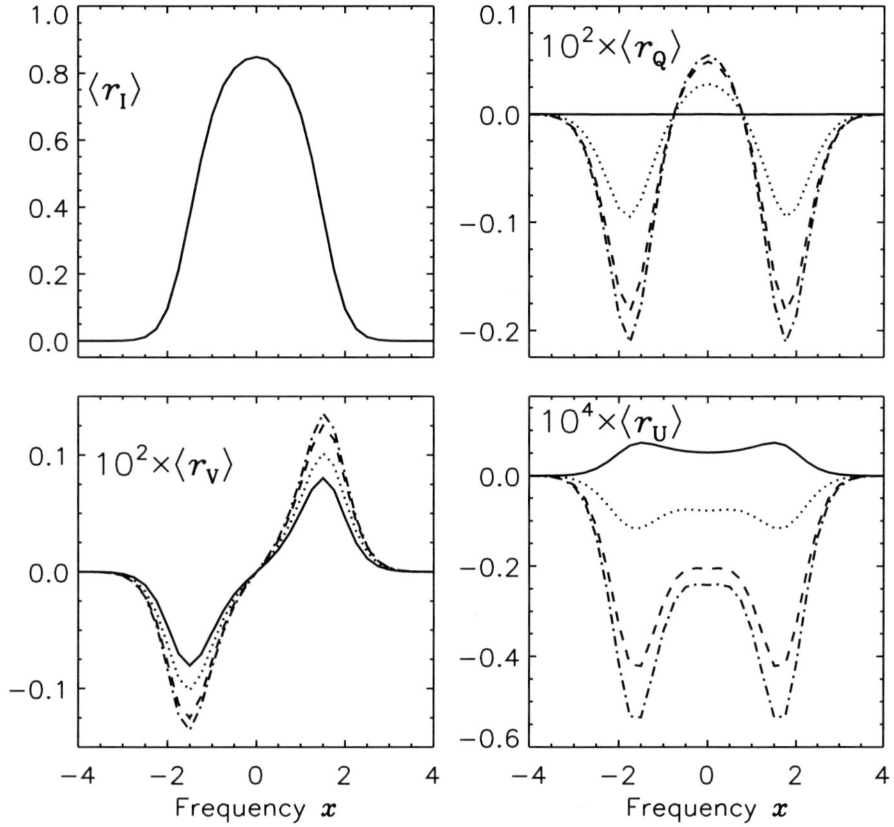

Fig. 5 $\langle r_{I,Q,U,V} \rangle$ for $\nu = 5$ and $\mu = 0.1$. Composite PDF containing asymmetric Voigt function with $y_0 = 4.5$ (*dashed line* of Fig. 1a) is used. Line types are : $y_t = \infty$ (*solid*), $y_t = 50$ (*dotted*), $y_t = 10$ (*dashed*), and $y_t = 5$ (*dash-dotted*)

isotropic or anisotropic fluctuations was considered in Frisch et al. (2005); here, we experimented with other types of PDFs with restriction to asymmetric PDFs, which can generate nonzero net flux as diagnosed by the shape of $\langle r_V \rangle$ profile. We consider very weak fluctuations of the magnetic field ($\gamma_B = 0.0056$). Thus $\langle r_I \rangle$ profiles remain insensitive to the shape of PDF. In contrast, other mean Stokes profiles are quite sensitive to the choice of PDF.

For a complete description of the random vector magnetic field, we need PDFs that describe both the angular and strength fluctuations. We constructed such empirical PDFs by combining a power law (for angular distribution) with a Voigt function or a stretched exponential (for field strength). At the solar surface, weak fields are observed to be nearly isotropic and strong fields more vertical. Composite PDFs can simulate such a situation.

Acknowledgement I am grateful to Dr. K. N. Nagendra for very useful suggestions and comments.

References

Carroll, T. A., Kopf, M. 2007, A&A, 468, 323
Carroll, T. A., Staude, J. 2003, Astronomische Nachrichten, 324, 392
Carroll, T. A., Staude, J. 2005, Astronomische Nachrichten, 326, 296
de Wijn, A. G., Stenflo, J. O., Solanki, S. K., Tsuneta, S. 2009, Space Sci. Rev., 144, 275
Dolginov, A. Z., Pavlov, G. G. 1972, Soviet Astronomy, 16, 450
Domke, H., Pavlov, G. G. 1979, Ap&SS, 66, 47
Frisch, H., Sampoorna, M., Nagendra, K. N. 2005, A&A, 442, 11
Frisch, H., Sampoorna, M., Nagendra, K. N. 2006a, In: Solar Polarization 4, R. Casini, B. W. Lites (eds.), ASP Conf. Ser., vol. 358, p. 126
Frisch, H., Sampoorna, M., Nagendra, K. N. 2006b, A&A, 453, 1095
Frisch, H., Sampoorna, M., Nagendra, K. N. 2007, Memorie della Societa Astronomica Italiana, 78, 142
Landi degl'Innocenti, E. 1994, In: Solar Surface Magnetism, R. J. Rutten, and C. J. Schrijver (eds.), p. 29
Sampoorna, M., Frisch, H., Nagendra, K. N. 2008a, New Astronomy, 13, 233
Sampoorna, M., Nagendra, K. N., Frisch, H., Stenflo, J. O. 2008b, A&A, 485, 275
Stein, R. F., Nordlund, Å. 2006, ApJ, 642, 1246
Stenflo, J. O. 1987, Solar Phys., 114, 1
Stenflo, J. O., Holzreuter, R. 2002, In: SOLMAG 2002. Proceedings of the Magnetic Coupling of the Solar Atmosphere Euroconference, H. Sawaya-Lacoste (ed.), ESA Special Publication, vol. 505, p. 101
Stenflo, J. O., Holzreuter, R. 2003a, In: Current Theoretical Models and Future High Resolution Solar Observations: Preparing for ATST, A. A. Pevtsov, H. Uitenbroek (eds.), ASP Conf. Ser., vol. 286, p. 169
Stenflo, J. O., Holzreuter, R. 2003b, Astronomische Nachrichten, 324, 397
Vögler, A., Shelyag, S., Schüssler, M., et al. 2005, A&A, 429, 335

Spectropolarimetry with CRISP at the Swedish 1-m Solar Telescope

A. Ortiz and L.H.M. Rouppe van der Voort

Abstract CRISP (Crisp Imaging Spectro-polarimeter), the new spectropolarimeter at the Swedish 1-m Solar Telescope, opens a new perspective in solar polarimetry. With better spatial resolution (0.13″) than Hinode in the Fe I 6302 Å lines and similar polarimetric sensitivity reached through postprocessing, CRISP complements the SP spectropolarimeter onboard Hinode. We present some of the data that we obtained in our June 2008 campaign and preliminary results from LTE inversions of a pore containing umbral dots.

1 Introduction

CRISP (CRisp Imaging Spectro-Polarimeter) is a new imaging spectropolarimeter installed at the Swedish 1-m Solar Telescope (SST, Scharmer et al. 2003) in March 2008. The instrument is based on a dual Fabry-Pérot interferometer system similar to that described by Scharmer (2006). It combines a high spectral resolution, high reflectivity etalon with a low resolution, and low reflectivity etalon. It has been designed as compact as possible, that is, with a minimum of optical surfaces, to avoid straylight as well as possible.

For polarimetric studies, nematic liquid crystals are used to modulate the light. These crystals change state in less than 10 ms, which is faster than the CCD readout time. A polarizing beam splitter close to the focal plane splits the beam onto two 1024×1024 synchronized CCDs that measure the two orthogonal polarization states simultaneously. This facilitates a significant reduction of seeing crosstalk in the polarization maps.

A third, synchronized, CCD camera records wide-band images through the prefilter of the Fabry-Pérot system. These images serve as an anchor channel for Multi-Object Multi-Frame Blind Deconvolution (MOMFBD) image restoration (van Noort et al. 2005), which enables near-perfect alignment between the sequentially recorded polarization and line position images. For more details on MOMFBD processing of polarization data see van Noort and Rouppe van der Voort (2008).

A. Ortiz (✉) and L.H.M. Rouppe van der Voort
Institute of Theoretical Astrophysics, University of Oslo, Norway

S.S. Hasan and R.J. Rutten (eds.), *Magnetic Coupling between the Interior and Atmosphere of the Sun*, Astrophysics and Space Science Proceedings,
DOI 10.1007/978-3-642-02859-5_11, © Springer-Verlag Berlin Heidelberg 2010

The etalons can sample spectral lines between 510 and 860 nm. The field of view (FOV) is 70″× 70″; the pixel size is 0.07″/pixel. The instrument has been designed to allow diffraction-limited observation at 0.13″ angular resolution in the Fe I 630 nm lines.

2 The June 2008 Campaign Data and Processing

The data displayed here were recorded on 12 June 2008 as part of a campaign during June 2008. The target was a pore (AR 10998) located at S09 E24 ($\mu = 0.79$). The field of view was 70″ × 70″.

The images recorded correspond to complete Stokes measurements at 15 line positions in steps of 48 mÅ, from −336 mÅ to +336 mÅ, in each of the Fe I lines, 6301.5 and 6302.5 Å. In addition, images were recorded at one continuum wavelength. Each camera operated at 35 Hz frame rate. For each wavelength and

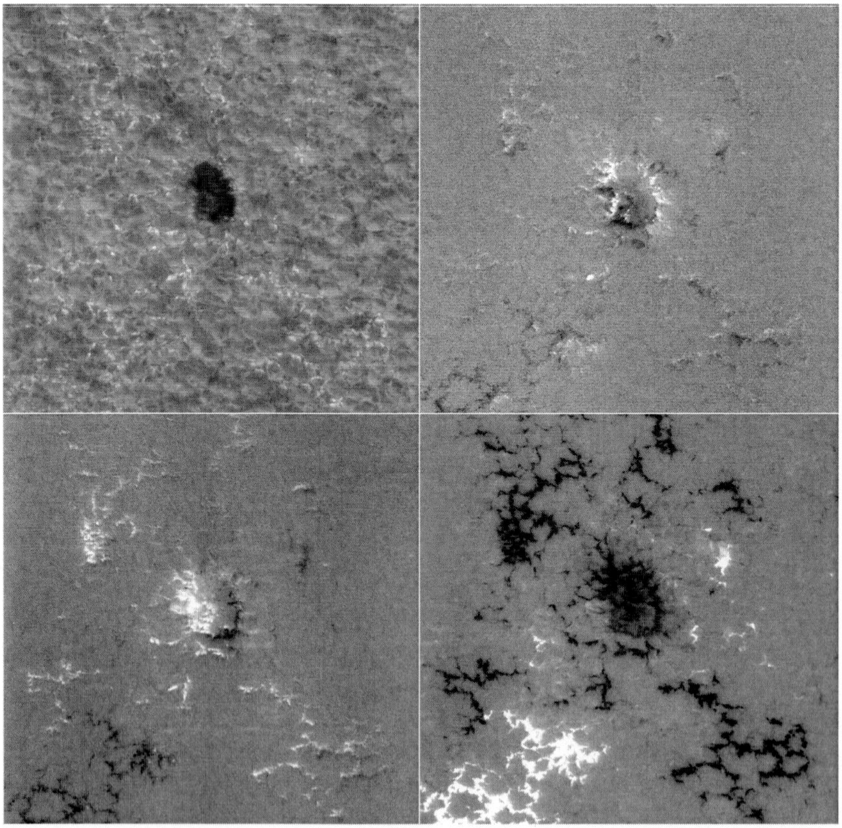

Fig. 1 *Clockwise*: Stokes I, Q, V, and U images taken in the blue wing of Fe I 6302.5 Å at $\Delta\lambda = -48$ mÅ, on 12 June 2008

LC state, seven images were so recorded per camera. Each sequence for subsequent MOMFBD processing consists of about 870 images per CCD (2,600 in total), recorded during 30 s. The images were divided into overlapping 64 × 64 pixel subfields sampling different isoplanaic patches with overlaps. All images from each subfield were then processed as a single MOMFBD set. They were demodulated with respect to the polarimeter and a detailed telescope polarization model. In addition, the resulting Stokes images were corrected for remaining I to Q, U, and V crosstalk by subtraction of the Stokes continuum images. Figure 1 shows an example of the resulting Stokes images.

The theoretical diffraction limit of the SST is $\lambda/D = 0.13''$ at 6,303 Å. We measured the real resolution obtained in our June observations by identifying the smallest intensity feature and fitting a Gaussian to it. Figure 2 shows a cut through a bright point with 80 km FWHM for the Gaussian fit. This value is equivalent to $0.11''$, which is slightly lower than the theoretical resolution $0.13''$ but consistent with it, due to the MOMFBD post-processing performed to the data. We estimated the noise level for the Stokes profiles to be around 2×10^{-3} for Stokes Q/I_c, U/I_c and V/I_c.

3 Inversions and Results

To derive the atmospheric parameters from the observed Stokes images, we use a least-square inversion code, LILIA (Socas-Navarro 2001), based on LTE atmospheres. We assume a one component, laterally homogeneous atmosphere together with stray light contamination. The inversions return nine free parameters as a function of optical depth, including the three components of the magnetic field vector

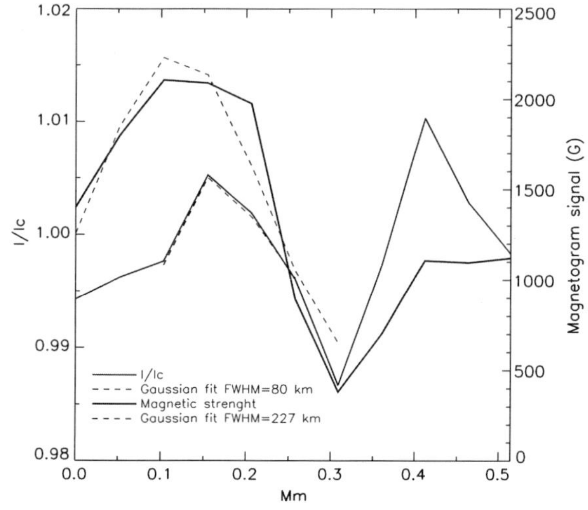

Fig. 2 Cut along brightenings in the stokes I image (*thin line*) and magnetic field obtained from inversions (*thick line*). We have fitted a gaussian to the smallest feature we can observe, both in the intensity image and the resulting magnetic field (*dotted lines*). The fits give us FWHMs of 80 km for I/I_c and 227 km for the magnetic field

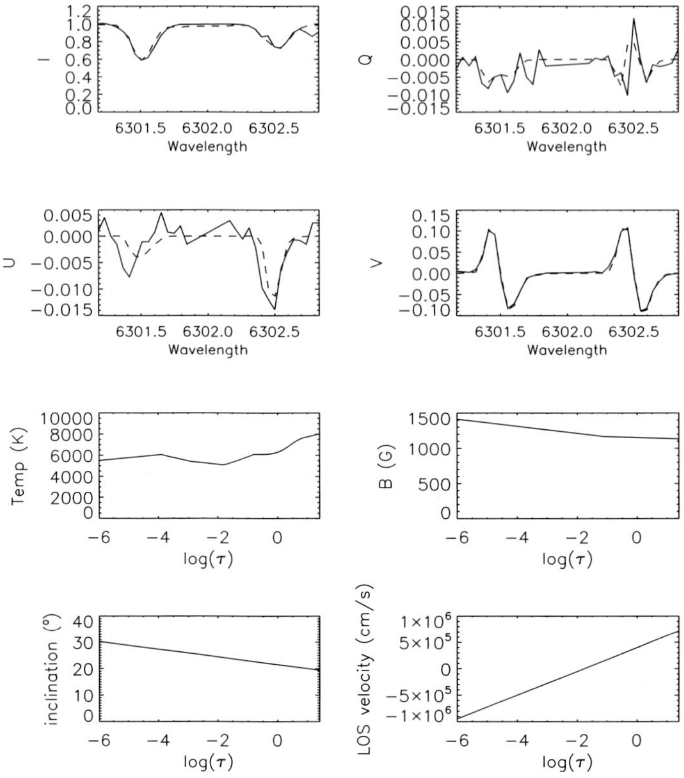

Fig. 3 Results from the LILIA inversion of a bright point observed in an intergranular lane. Observed (*solid*) and fitted (*dashed*) I/I_c, Q/I_c, U/I_c, and V/I_c profiles (*upper panels*), as well as atmospheric parameters (temperature, magnetic field, inclination, and line-of-sight velocity) obtained through the inversion as a function of optical depth (*lower panels*)

(strength, inclination, and azimut), LOS velocity, and temperature among others. We apply the inversion to both the Fe I 6301.5 and 6302.5 Å lines simultaneously.

Figure 3 shows an example of the inversion of an individual pixel belonging to a bright point. In this particular case the inversion code yielded a field strength of 1,100 G, inclination of 25°, and LOS velocity of 0.6 km s^{-1}, (downflow) at $\log(\tau) = -1.5$.

Figures 4 and 5 show maps of the obtained magnetic field strength and line-of-sight (LOS) velocity at different heights. Figure 4 shows a micro-pore as well as brightenings produced by emergent magnetic fields. Ribbons (Berger et al. 2004) can be distinguished. Upflows are correlated with the positions of the center of the granules, while downflows are correlated with the intergranular lanes, except in those areas where the magnetic field is emerging, in which velocities are lower due to the supression of convection. Figure 5 presents a pore with several umbral dots and structures within. These brighter umbral structures show lower magnetic field strengths than the darker parts of the umbra as well as higher temperatures.

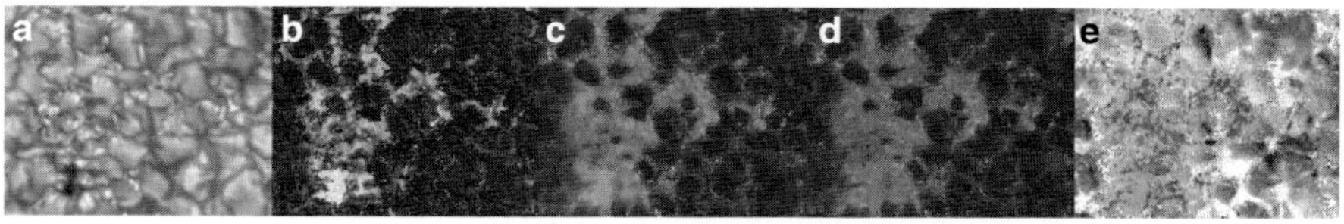

Fig. 4 (**a**) Intensity image of a region containing a micro-pore and intergranular magnetic fields including ribbons and bright points. (**b**)–(**d**): magnetic field strength maps resulting from the inversion at $\log(\tau) = 0, -1.5$, and -2 respectively. (**e**) LOS velocity map at $\log(\tau) = -2$. Downflows are observed in the intergranular lanes, while upflows are observed in the center of granules. The FOV is $14'' \times 10''$

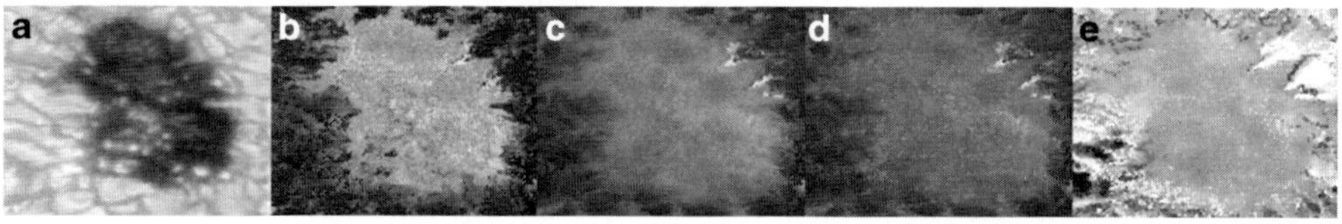

Fig. 5 (**a**) Intensity image of a region containing a sunspot. Umbral dots and other structures can be identified. (**b**)–(**d**): magnetic field strength maps resulting from the inversion at $\log(\tau) = 0$, -1.5, and -2, respectively. (**e**) LOS velocity map at $\log(\tau) = -2$. The FOV is $10'' \times 10''$

Velocities inside the pore tend to be lower than in normal granulation patterns. The velocity field of the umbral structures will be analyzed in a future publication.

4 Discussion

We have presented the first LTE inversions based on the data obtained with the CRISP imaging spectropolarimeter, used with the 1-m SST. The spatial resolution of the Stokes data presented here represents a major improvement compared to other ground-based data and even to the recent Hinode data. We have shown here both the capabilities of the CRISP instrument and of the inversion code applied to this data. We expect to make use of such capabilities for exploring in detail the umbral dots and other structures found in the micro-pore present on 12 June 2008.

Acknowledgement The Swedish 1-m Solar Telescope is operated on the island of La Palma by the Institute for Solar Physics of the Royal Swedish Academy of Sciences in the Observatorio del Roque de los Muchachos of the Instituto de Astrofísica de Canarias.

References

Berger, T. E., Rouppe van der Voort, L. H. M., Löfdahl, M. G., Carlsson, M., Fossum, A., Hansteen, V. H., Marthinussen, E., Title, A., Scharmer, G. 2004, A&A, 428, 613
van Noort, M., Rouppe van der Voort, L., Löfdahl, M. G. 2005, Solar Phys., 228, 191
van Noort, M. J., Rouppe van der Voort, L. H. M. 2008, A&A, 489, 429
Scharmer, G. B. 2006, A&A, 447, 1111
Scharmer, G. B., Bjelksjo, K., Korhonen, T. K., Lindberg, B., Petterson, B. 2003, In: SPIE Conf. Ser., 4853, S. L. Keil, S. V. Avakyan (ed.), p. 341
Socas-Navarro, H. 2001, In: Advanced Solar Polarimetry: Theory, Observation and Instrumentation, M. Sigwarth (ed.), ASP Conf. Ser., vol. 236, p. 487

Spectropolarimetry with the NLST

K. Sankarasubramanian, S.S. Hasan, and K.E. Rangarajan

Abstract India's National Large Solar Telescope (NLST) will provide opportunities to observe the Sun with high spatial, spectral, and polarimetric resolution. The large aperture also enables high-cadence spectropolarimetry with moderate spatial resolution. A multi-slit spectropolarimeter is planned as one of the back-end instruments for this powerful telescope, primarily to measure vector magnetic fields in both active and quiet regions. An integral-field unit added with the multi-slit spectropolarimeter will enable fast-cadence observation. Here we discuss the scientific requirements for such an instrument, along with advantages and limitations of the concept and preliminary design details.

1 Introduction

The National Large Solar Telescope (NLST henceforth) is being planned as a 2 m-class state-of-the-art solar telescope to be installed at a superior site compared to any of the existing solar facilities in India. A state-of-the-art active and adaptive optics system will be incorporated in the telescope design to provide diffraction-limited imaging over the entire wavelength range of interest under favorable seeing conditions. NLST will be one of the best solar observing facilities around the world and be comparable to the next-generation solar facilities elsewhere. It will also provide complementary observations along with current as well as future solar space missions. While space missions can provide uninterrupted coverage of the Sun, the NLST will provide observations with higher spectral, spatial, and polarimetric resolution.

At present, the largest solar telescope for solar research in India is the Kodaikanal Tower Tunnel Telescope, which has been in use for the last 35 years. This telescope, along with its high-dispersion spectrograph, is used primarily for spectral

K. Sankarasubramanian (✉)
Space Astronomy and Instrumentation Division, ISRO Satellite Centre, Bangalore, India

S.S. Hasan and K.E. Rangarajan
Indian Institute of Astrophysics, Bangalore, India

S.S. Hasan and R.J. Rutten (eds.), *Magnetic Coupling between the Interior and Atmosphere of the Sun*, Astrophysics and Space Science Proceedings,
DOI 10.1007/978-3-642-02859-5_12,

and synoptic studies. It has been upgraded with an instrument package for polarimetry at high spectral resolution (Nagaraju et al. 2008). However, this telescope has the disadvantage of large-angle coelostat reflections, which restrict the accuracy of polarization measurements. Also, the site is affected by the regular monsoon seasons and its seeing quality has degraded over the years.

There is also a moderate size (50-cm) solar telescope, named the Multi Application Solar Telescope (MAST), under construction (Venkatakrishnan 2006) at the Udaipur Solar Observatory, which will be the first new solar facility since several decades in India. However, the smaller aperture size of this telescope, the moderate seeing quality of the site, and the limited wavelength coverage will restrict the usage of this telescope to observations with moderate spatial resolution.

In contrast, NLST is planned to study small-scale magneto-hydrodynamical processes, the dynamical evolution of small-scale magnetic structures, active regions, sunspots, magnetoconvection, the thermodynamics of the chromosphere, and turbulent magnetic fields at the highest possible spatial resolution. Table 1 lists the science

Table 1 Science requirements and proposed back-end instruments for the NLST

Physical process or region to be observed	Physical quantity to be measured	Telescope or instrument requirement	Proposed instrument
MHD waves and oscillations	Intensity variation of 1% or less; magnetic fields; velocities of 200 m s^{-1} or less; properties of oscillations	High photon flux; spectral resolution of a few mÅ; polarization accuracy 0.1% or better; high time cadence	High resolution spectrograph; narrow band filters; spectropolarimeter with spatial information; fast cameras
Structuring of the solar atmosphere	Temperature, velocity, and magnetic field with height	Both visible and infrared capabilities	Filters and polarimeters for Ca II K, Hα, G-band He I 10830 Å, 1.6 μm
Active region evolution	Velocity and vector magnetic field	Fast cadence; FOV $\approx 300''$	Spectropolarimeter with spatial information
Hanle– Zeeman effect	Vector magnetic field	Polarization accuracy of 0.01%	Spectropolarimeter
Photospheric small-scale structures	Velocity and vector magnetic field	Fast cadence; FOV $\approx 30''$; polarization accuracy of 0.1%; fast cadence	Spectropolarimeter with spatial information
Off-limb observations	Intensity, velocity, and vector magnetic field	Low scattered light; polarization accuracy of 0.01%	Spectropolarimeter; high-resolution spectrograph

Table 2 Specifications for the NLST

Physical parameter	Value
Aperture of primary	2 m
Focal length of primary	4 m
Optical configuration	Three-mirror on-axis Gregorian
Field of view (FOV)	300″
Final focal fatio	f-40
Image scale at the science focus	2.6″ mm^{-1}
Optical quality	Better than 0.1″ over the FOV
Wavelength coverage	3,800 Å–2.5 μm
Polarization accuracy	Better than 1 part in 10,000
Scattered light within the telescope	Less than 1%
Active and adaptive optics	Integrated in the telescope
Spatial resolution	0.1″ at 500 nm

requirements and the planned back-end instruments that are required to achieve the proposed science objectives. This table is only an indication of the proposed science and is not exhaustive. More details about the NLST and its scientific goals are given in Hasan and NLST Team (2006). A brief technical specifications of the NLST telescope is listed in Table 2.

2 Polarimetry Package

Solar spectropolarimetry is the only way to quantitatively study solar magnetic fields. Unfortunately, detecting polarization is a highly difficult task due to its sensitive nature to any anisotropic reflection. Any large optical imaging system requires reflective optics. To minimize the effects of reflections on the polarization state of the incident light, the polarization analysis (or the polarimetry) for the NLST will be carried out as early as possible in the optical train. The scientific requirements of the NLST also warrant a very accurate polarization modulation and analyzing unit. These science requirements translate into the following instrument requirements:

- Large wavelength coverage, from 3,800 Å to 2.5 μm. Unfortunately, no current technology is available to have a single polarization modulator covering this extended wavelength range. However, it is feasible with two or three modulators covering a broad range of wavelengths each.
- Preferably a fixed package without any moving parts.
- High stability over at least a day in order to reduce the need for polarization calibrations.
- Polarization accuracy of 10^{-5} and a precision of a few times 10^{-4}.
- Good optical quality, as it is close to the focus.
- A calibration unit located in front of the modulator in the optical path and before any large angle reflections.
- Preferably the use of a balanced modulation scheme in order to avoid seeing-induced spurious polarization if off-the-shelf CCD cameras are used.

– The modulator should have good transmission at all wavelengths from 3,800 Å to 2.5 μm as it is located early in the optical path.

Two possible options for the polarization packages are being examined for the NLST, keeping in mind that there should be a possibility to do polarimetry over the entire wavelength range of interest (3,800 Å–2.5 μm). The first option is a rotating waveplate retarder very similar to the modulator used in SPINOR (Socas-Navarro et al. 2006), but optimized for larger wavelength coverage (using the bi-crystalline or Pancharatnam technique). A bi-crystalline modulator is already in operation at the Dunn Solar Telescope (DST) with limited wavelength coverage (from 500 nm to about 1.6 μm). The second option is to use liquid crystals. There are design studies for an achromatic liquid crystal modulator, but these are still at an early stage (Gisler et al. 2003). Liquid-crystal variable retarders are an attractive option, but these also will not cover the full wavelength range, and hence at least two or three modulator packages may be required. A detailed comparison between the two options, including their merits and demerits, will be carried out shortly.

The most preferred location for the modulator is at the second focus of the on-axis three-mirror Gregorian system. Initial studies indicate that the best position for the analyzer (which is the last optical component in any modulation scheme) is next to the modulator. Then the modulation must be restricted to a single beam scheme due to alignment issues through the whole optical path of the telescope as well as the limitation of adaptive optics systems in handling two beams. Hence, a fast chopping mechanism is necessary if the analyzer is kept at the Gregorian focus. This poses severe requirements on the detector. The CCD must be a fast readout camera or a custom-made camera like the ZIMPOL (Povel et al. 1994) or C^3PO (Keller 2005).

The second best option is to keep the analyzer closer to the detector. This introduces cross-talk between the linear polarization (Stokes Q and U) that varies over the day due to the rotation of the image. This can be overcome by having a rotating analyzer as compensator. Furthermore, using a calibration unit at the Gregorian focus and before the modulator will help to reduce any residual cross-talks in the system. However, this calibration should be robust and highly accurate to achieve the required polarimetric accuracy and precision. A detailed polarization model of the telescope and the polarimetric system will be carried out to bring out the subtle differences between different locations of the modulator and analyzer. In essence, the first option, which is the best one in terms of polarization precision and accuracy, will require a ZIMPOL or C^3PO-type CCD detector, while the second option, if calibrated to high accuracy, can work even with off-the-shelf CCD detectors in a balanced two-beam modulation scheme.

3 Spectropolarimetry with NLST

Table 1 illustrates that all the proposed science goals cannot be met with a single spectropolarimetric instrument due to the large range of field-of-view (FOV) requirements (from 30″ to 300″), the time cadence (from 1 min to at most several minutes), and the spectral resolution (from 10 to 100 mÅ). We propose to realize three different instruments (independent or semi-independent) to cover all the science requirements. They are (1) Multi-Slit Imaging Capable (MuSIC) SpectroPolarimeter (hereafter SP) using an integral field unit, (2) Single-slit high spectral resolution SP, and (3) Fabry-Pérot based imaging SP. Given that MAST will realize instruments similar to the second and third type, the priority is given to the first instrument for NLST.

3.1 Multi-slit SP

It is obvious that a single-slit SP has the inherent deficiency of low temporal cadence for an intermediate FOV. This is due to the long scanning time required by the moving spectrograph slit. The left column of Fig. 1 illustrates the single-slit spectral imaging concept. The single slit, marked as a dark vertical line in the continuum intensity image (top-left), is scanned across the FOV in order to produce two-dimensional spectropolarimetric mapping. The bottom-left four images are the four Stokes images at the marked slit position. Almost all of the useful solar polarimetric

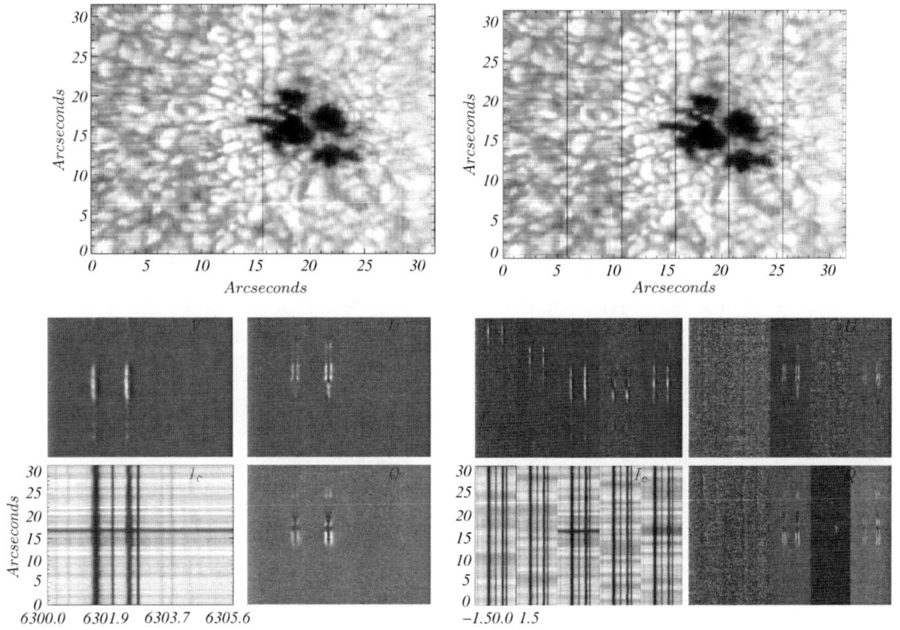

Fig. 1 The top two images show red-continuum intensity image of an active region along with vertical markings for a single-slit (*top-left*) and five slits (*top-right*). The bottom two figures show the corresponding Stokes spectrum (marked as I, Q, U, and V in the respective images). *Bottom-left* is for a single slit and *bottom-right* for five slits. The single-slit data were obtained using the DLSP (Sankarasubramanian et al. 2004). The multi-slit images are simulated data

studies are carried out using observations of a single line or a line pair (in Fig. 1 the two strong and broad lines are the 6,302 Å line pair) with high dispersion along with 2D detector (usually a CCD). The other spectral lines of interest are usually covered using multiple detectors (unless the separation is very close to fit inside a single detector). In most of the cases, a few hundred pixels of a CCD along the wavelength direction is sufficient to cover the single line or the line pair under study. However, to cover a large FOV along the spatial direction, large format CCDs are required. It can be seen in Fig. 1 that to cover the 6,302 Å line pair, a smaller CCD is sufficient but that then has limited FOV.

To overcome this inefficient usage of CCD detector area, the multi-slit spectrograph concept is proposed. As can be seen in the right-hand images in Fig. 1, multi-slit geometry can be used to efficiently utilize the detector area. This figure illustrates this with five slits positioned at equal distances from the central slit. A full Stokes observation will cover five different regions on the solar surface simultaneously. By scanning the five slits synchronously, two-dimensional spectral data are obtained with a time cadence faster than with a single-slit spectrograph.

Multi-slit spectrographs have been a reality in the astronomical community for quite some time (Srivastava and Mathew 1999). However, this concept was not very successful due to the use of conventional filters with Gaussian transmission profiles. In such filters, the transmission changes rapidly around the passband center, and also the wings extend several times further than the FWHM of the filter. Because of these properties, large-FWHM filters are required to keep similar signal-to-noise ratios across spectral-line profiles. Hence the requirement of larger slit separation, due to this large FWHM as well as the extended wings of the filter profiles. Examples of such filter profiles are shown in Fig. 2 for 2, 4, and 6 Å FWHM. Overplotted is the solar spectrum from an atlas in the wavelength region around the 6,302 Å line pair. It is clear from this figure that to cover this line pair with a conventional filter, 6 Å FWHM is a must and that the wings of the profiles can extend as far as 15–20 Å away from the line center. These difficulties practically rule out the multi-slit concept.

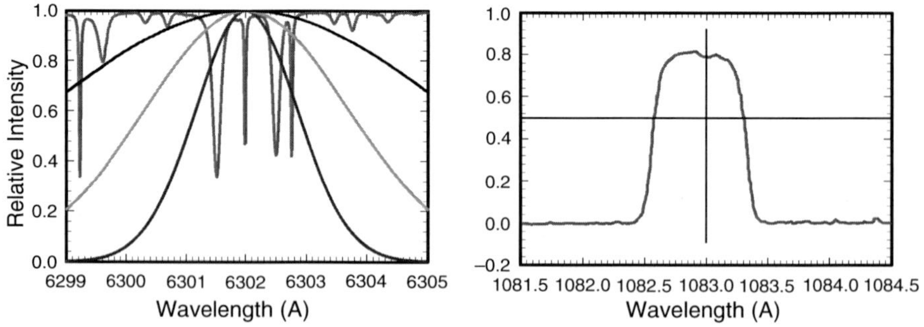

Fig. 2 *Left*: solar intensity from a spectral atlas is plotted as red. Three different Gaussian profiles of 2, 4, and 6 Å FWHM represent filter passbands. Note how the transmission changes at the wings of the line profiles when the filter FWHM becomes narrower. *Right*: A typical square filter profile of 10 Å FWHM in the near infrared (adapted from Koshy et al. 2008)

However, the technological advancement of filter designs over the last several years have resulted in filters with square profile shapes as well as narrow bandwidths. Such filters are being developed regularly for infrared wavelengths for the telecommunications industry. There are efforts to make such filters also in the visible wavelength region (Lin and Versteegh 2006).

3.2 Integral Field Unit

In observing the 6,302 Å line pair, a useful dispersion is about 15 mÅ per pixel. A wavelength coverage of about 3 Å is required to cover the line pair. Given the dispersion and the wavelength coverage, it is easy to find the lowest slit separation to avoid spectral overlap from nearby slits. For this example, it works out to be about 200 pixels. Given the requirements of a two-beam scheme and of separation along the wavelength axis (in order to cover a larger FOV), the minimum slit separation will be 400 pixels. With a slit width comparable to the spatial resolution and with the pixel size equal to the Nyquist sampling interval, the system must still scan 200 steps for obtaining the full two-dimensional FOV. Even though the number of steps is much smaller compared to that required for a single slit to cover this large FOV, the time cadence for such a system remains longer than the typical dynamical time scales on the solar surface. This is one of the serious limitation of the multi-slit configuration.

However, this limitation can be overcome by using an integral field unit (IFU) that redistributes the two-dimensional FOV along the slits. The addition of an IFU provides significant improvement in the temporal cadence but through limitation of the FOV. The IFU can be a fiber bundle or an image slicer, rearranging the two-dimensional image information into multiple one-dimensional slits. The FOV coverage is equal to the total number of slits multiplied by the number of pixel points along each slit. Figure 3 illustrates how a fiber bundle can be used to rearrange a 2D FOV into 1D slits. In summary, multi-slit spectropolarimetry can significantly improve the cadence for vector-field observations for a large FOV without an IFU and for a small FOV with an IFU.

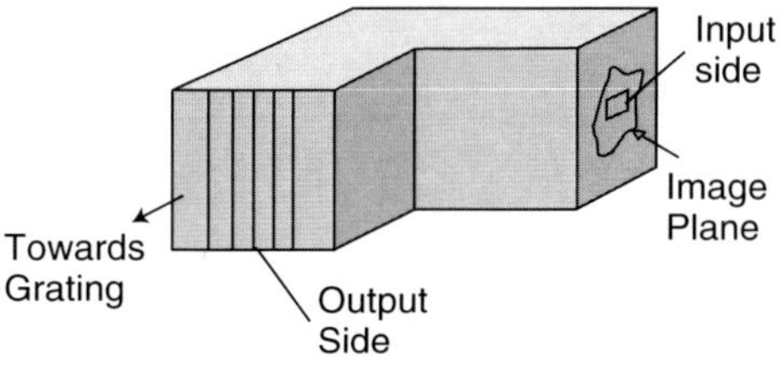

Fig. 3 Conceptual picture of a fiber bundle. A square input field on the image plane is transformed into five different slits at the output plane. These five slits then will act as the multi-slit for the spectrograph

4 Concept Realization

To study, realize, and implement this concept, a prototype instrument is being developed with the aim of using it as one of the back-end instruments at the MAST. The Fe I line pair at 6,302 Å is chosen as the wavelength of interest, and the optical design is optimized for this wavelength. This prototype instrument development is being carried out in two phases. First the multi-slit spectrograph will be designed and developed. At the end of this developmental phase, the instrument can be operated for large FOV observing by scanning over the FOV. In this phase, the square-filter profile developments, its usage, and its limitations will be studied. During the second phase of this prototype instrument, an IFU will be added and studied for usage and limitations. Even though the instrument is being optimized for a defined wavelength, the IFU developed for this prototype will be tested for its wavelength coverage at our laboratory.

Figure 4 shows the optimized optical design of this prototype instrument, which is presently under construction. More details can be found in Koshy et al. (2008). The instrument is being developed jointly by the Space Astronomy and Instrumentation Division of ISAC and the Udaipur Solar Observatory, with the aim of utilizing the experience so gained for the versatile instrument requirement of the NLST. The design parameters and the expected performance of this prototype instrument are consolidated in Table 3.

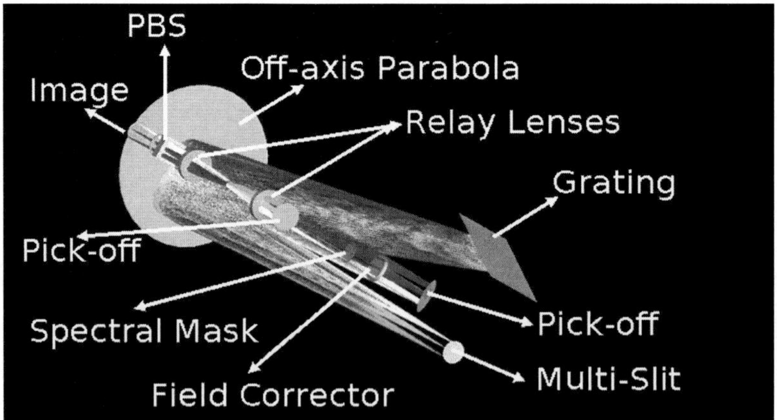

Fig. 4 The optimized design of the prototype instrument, which will be deployed as a back-end instrument for the MAST. It is a Littrow-type spectrograph with an intermediate spectral image mask

Table 3 Design parameters and performance of the prototype instrument

Parameter	Phase I	Phase II
Slit	5 slits	5 linear fiber arrays
	30 mm height	30 mm height
	width 30 μm	width 32 μm
Spectrograph	Dispersion 16 mÅ/pixel	Same
	Spectral resolution $\leq$ 35 mÅ	Same
Spectral mask	28 × 30 mm^2 closed	Same
	28 × 30 mm^2 open	Same
FOV	$300'' \times 300''$	$23'' \times 23''$
FOV scanning	190 steps	Instantaneous
Temporal cadence	25 min/map	Less than 1 min/map

5 Conclusions

Our novel spectropolarimetric concept has the potential to become a dedicated back-end instrument for the National Large Solar Telescope. It requires a narrow-band filter with a square transmission profile, as well as an integral field unit to provide high temporal cadence (less than a minute) for vector magnetic field measurement over a reasonable FOV. This instrument can in principle be operated either as a scanning spectrograph in large-FOV studies or with high time cadence in small-FOV studies. A similar concept was discussed using a fiber bundle but with a single-slit spectrograph by Rutten (1999). To study the feasibility of the proposed concept, a prototype instrument is being developed for use at the focal plane of the MAST. It will be optimized for the 6,302 Å line pair. It is expected that this instrument will be able to provide vector-field information over a large FOV ($300'' \times 300''$) for every half hour without an IFU. Along with an IFU, this instrument can provide high-cadence (less than half-a-minute) small-FOV (about $23'' \times 23''$) vector field information.

The experience gained in this prototype instrument will be translated for NLST spectropolarimetry, which would demand still larger spatial resolution as well as simultaneous multi-line studies. Assuming the realisability of large-format ($4 \times 4K^2$) C^3PO detectors (Keller 2005), the availability of a 3 Å square-profile filter, and with 21.5 mÅ/pixel dispersion, it is feasible to observe vector fields over a $30'' \times 30''$ FOV at the diffraction limit of the NLST. The large format of the CCD allows the accommodation of about 29 slits and would demand a large-format fiber bundle of 172 × 689 elements.

Most ground-based high resolution imaging, also when using Adaptive Optics, requires post-processing to obtain diffraction-limited resolution. The Multi-Object Multi-Frame Blind Deconvolution is becoming the most popular post-processing method (van Noort et al. 2005). The current post-processing algorithms require two-dimensional image sampling and cannot be applied to spectroscopic data. The concept presented here has the advantage of enabling post-processing of spectroscopic data through its use of integral field units.

Acknowledgment The authors thank the NLST team members as well as the solar community in the Indian Institute of Astrophysics for valuable suggestions and inputs. KS thanks R. Bayanna and P. Venkatakrishnan from Udaipur Solar Observatory for their constant encouragement and valuable support during the spectrograph design stage. He also thanks the program organizers for providing the opportunity and the financial support to deliver this contribution. ADS and the editors are acknowledged for facilitating bibtex.

References

Gisler, D., Feller, A., Gandorfer, A. M. 2003, SPIE, 4843, 45
Hasan, S. S., NLST Team. 2006, NLST Proposal, 1
Keller, C. U. 2005, In: Astronomical Polarimetry: Current Status and Future Directions, A. Adamson, C. Aspin, C. Davis, T. Fujiyoshi (eds.), ASP Conf. Ser., vol. 343, p. 33
Koshy, G., Sankarasubramanian, K., Bayanna, R., Lin, H., Venkatakrishnan, P. 2008, In: 2007 ESO Instrument Calibration Workshop, A. Adamson, C. Aspin, C. Davis, T. Fujiyoshi (eds.), p. 515
Lin, H., Versteegh, A. 2006, SPIE, 6269
Nagaraju, K., Sankarasubramanian, K., Rangarajan, K. E., et al. 2008, Bull. Astron. Soc. India, 36, 99
Povel, H. P., Keller, C. U., Yadigaroglu, I. A. 1994, Appl. Optics, 33, 4254
Rutten, R. 1999, In: High Resolution Solar Physics: Theory, Observations, and Techniques, T. R. Rimmele, K. S. Balasubramaniam, R. R. Radick (eds.), ASP Conf. Ser., vol. 183, p. 296
Sankarasubramanian, K., Gullixson, C., Hegwer, S., et al. 2004, SPIE, 5171, 207
Socas-Navarro, H., Elmore, D., Pietarila, A., et al. 2006, Solar Phys., 235, 55
Srivastava, N., Mathew, S. K. 1999, Solar Phys., 185, 61
van Noort, M., Rouppe van der Voort, L., Löfdahl, M. G. 2005, Solar Phys., 228, 191
Venkatakrishnan, P. 2006, In: 2nd UN/NASA Workshop on International Heliophysical Year and Basic Space Science

Magnetic Coupling in the Quiet Solar Atmosphere

O. Steiner

Abstract Three kinds of magnetic couplings in the quiet solar atmosphere are highlighted and discussed, all fundamentally connected to the Lorentz force: first, the coupling of the convecting and overshooting fluid in the surface layers of the Sun with the magnetic field. Here, the plasma motion provides the dominant force, which shapes the magnetic field and drives the surface dynamo. Progress in the understanding of the horizontal magnetic field is summarized and discussed. Second, the coupling between acoustic waves and the magnetic field, in particular the phenomenon of wave conversion and wave refraction. It is described how measurements of wave travel times in the atmosphere can provide information about the topography of the wave conversion zone, that is, the surface of equal Alfvén and sound speed. In quiet regions, this surface separates a highly dynamic magnetic field with fast moving magnetosonic waves and shocks around and above it from the more slowly evolving field of high-beta plasma below it. Third, the magnetic field also couples to the radiation field, which leads to radiative flux channeling and increased anisotropy in the radiation field. It is shown how faculae can be understood in terms of this effect. The article starts with an introduction to the magnetic field of the quiet Sun in the light of new results from the Hinode space observatory and with a brief survey of measurements of the turbulent magnetic field with the help of the Hanle effect.

1 The Magnetic Field of the Quiet Sun

Over the past three and a half years, the Sun stayed in a minimum state of magnetic activity as it has ended cycle 23 and is about to start with cycle 24 (if not pausing for yet a longer period of time). In this period of quiescence, it was possible to observe the Sun with an exceptional instrument, the Solar Optical Telescope SOT onboard the Hinode space observatory (Kosugi et al. 2007). The Japanese Hinode satellite was put in orbit on 22 September 2006. It is not so much the spatial resolution

O. Steiner (✉)
Kiepenheuer-Institut für Sonnenphysik, Freiburg, Germany

S.S. Hasan and R.J. Rutten (eds.), *Magnetic Coupling between the Interior and Atmosphere of the Sun*, Astrophysics and Space Science Proceedings, DOI 10.1007/978-3-642-02859-5_13, © Springer-Verlag Berlin Heidelberg 2010

of 0.3″ that makes this instrument exceptional for quiet-Sun observing, but rather the absence of seeing in combination with high pointing accuracy. This allows for unprecedented "deep" (long-exposure) polarimetry with correspondingly high polarimetric sensitivity at a spatial resolution of 0.3″. A similar polarimetric accuracy at this high spatial resolution has not been achieved from the ground in the past.

Immediately evident in the total or circular polarization maps over a large field of view from Hinode (Lites et al. 2008) is the magnetic network, which persisted existing during the most quiet states of the Sun. The magnetic network consists of an accumulation of magnetic fields in the borders between supergranular cells. The origin of these magnetic fields remains an enigma. Is it the result of advection by the supergranular flow to which it is often described, or, vice versa, is the supergranular flow rather a consequence of the existence of the magnetic network (Stein, private communication)? Is the network field generated locally, near the surface, or is it an integral part of the globally acting dynamo (Stein et al. 2003), or is it just the decay product of sunspots and/or ephemeral active regions?

Also omnipresent in this quiescent state of the Sun are small-scale magnetic field concentrations, visible as delicate, bright objects within and at the vortices of intergranular lanes. The structure made up of ensembles of bright elements is known as the *filigree* (Dunn and Zirker 1973). Mehltretter (1974), while observing in the visible continuum, referred to them as *facular points* because they are the footpoints of magnetic field concentrations that appear as faculae near the solar limb. In more recent times, these objects were mostly observed in the G band (a technique originally introduced by Muller 1985) because the molecular band-head of CH that constitutes the G band acts as a leverage for the intensity contrast (Rutten 1999; Rutten et al. 2001; Sánchez Almeida et al. 2001; Steiner et al. 2001; Shelyag et al. 2004). Being located in the blue part of the visible spectrum, this choice also helps in improving the diffraction-limited spatial resolution and the contrast in the continuum. Recent observational investigations of the dynamics, morphology, and properties of small-scale magnetic field concentrations of the quiet Sun include Berger et al. (2004); Langhans et al. (2004); Lites and Socas-Navarro (2004); Sánchez Almeida et al. (2004); Socas-Navarro and Lites (2004); Wiehr et al. (2004); Rouppe van der Voort et al. (2005); Domínguez Cerdeña et al. (2006a,b); Berger et al. (2007); Bovelet and Wiehr (2007); Centeno et al. (2007); Ishikawa et al. (2007); Langangen et al. (2007); Martínez González et al. (2007); Orozco Suárez et al. (2007); Rezaei et al. (2007a,b); Tritschler et al. (2007); Bello González and Kneer (2008); Bello González et al. (2008); de Wijn et al. (2008); Orozco Suárez et al. (2008); Bello González et al. (2009) and references therein. It would be interesting to learn to which degree the abundance of small-scale magnetic field concentrations persists in a grand minimum, to find out more about their origin.

With the help of the spectropolarimeter of the Solar Optical Telescope (SOT) onboard Hinode, it became for the first time possible to reliably determine the transversal (with respect to the line-of-sight) magnetic field component of the quiet Sun. These measurements indicate that, seen with a spatial resolution of 0.3″, the quiet internetwork regions harbor a photospheric magnetic field whose mean field strength of its horizontal component considerably surpasses that of the vertical component (Lites et al. 2007, 2008; Orozco Suárez et al. 2007). According to these papers, the vertical fields are concentrated in the intergranular lanes, while the horizontal fields occur most commonly at the edges of the bright granules, aside from the vertical fields. Lites et al. (2008) determine for the horizontal field component a mean apparent field strength (averaged over a large field of view including network and internetwork regions) of 55 G, while the corresponding mean absolute vertical field strength was only 11 G. Harvey et al. (2007) find from recordings with GONG and SOLIS at moderate angular resolution a "seething magnetic field" with a line-of-sight component increasing from disk center to limb as expected for a nearly horizontal field orientation. It is reasonable to assume that the horizontal fields of Lites et al. (2008) and those of Harvey et al. (2007) are different manifestations of the same magnetic field. Ishikawa et al. (2008) detected transient horizontal magnetic fields in plage regions as well. Previously, Lites et al. (1996) and Meunier et al. (1998) reported observations of weak and strong horizontal fields in quiet Sun regions.

The anisotropy of the quiet-Sun magnetic field as revealed by the Hinode measurements is not necessarily in contradiction with the observed depolarization of scattered light through the Hanle effect. For the quantitative interpretation of the Hanle effect, it was customarily assumed in the past that the quiet Sun magnetic field was "turbulent" in the sense that it was isotropically distributed on the scales relevant for the analysis. Theories have now been developed to include distribution functions for the field strength and angular directions (see, e.g., Carroll and Kopf 2007; Sampoorna et al. 2008; Sampoorna 2010) for a better description of the turbulent magnetic field in radiation transfer and first steps are taken to use such formulations for the computation of the Hanle effect in scattering media (Frisch 2006; Nagendra et al. 2010).

The isotropy assumption may still be valid on scales smaller than $0.3''$. In fact, simulations suggest constant angular distribution within $\pm 50°$ from the horizontal direction on a scale of $0.05''$. On the other hand, should the field be strongly anisotropic on all scales, it would still produce Hanle depolarization, but its interpretation would be less straightforward. From a theoretical point of view, the anisotropy of the magnetic field comes not as a surprise, and homogeneous turbulence of the magnetic field seems unlikely as the convective flow is far from homogeneously turbulent, given the scale size of 1,000 km of granules vs. the pressure scale height of 100 km.

Stenflo (1982) roughly estimated the strength of the weak turbulent magnetic field based on the Hanle technique between 10 and 100 G. This range was narrowed to 4–40 G by Stenflo et al. (1998). More precise estimates by Faurobert-Scholl (1993) and Faurobert-Scholl et al. (1995) yielded values in the range of 30–60 G in the deep photosphere and 10–20 G in the middle and upper photosphere, where the latter values were increased to 20–30 G by Faurobert et al. (2001). The values reported by Stenflo et al. and by Faurobert et al. are not without controversy, however. Trujillo Bueno et al. (2004) find from three-dimensional radiative transfer modeling of scattering polarization in atomic and molecular lines an ubiquitous tangled magnetic field with an average strength of 130 G around 300 km height in the photosphere, which is much stronger in the intergranular than in the granular regions. They estimate that the energy density of this field would amount to 20% of the kinetic energy density of the convective motion at a height of 200 km. If this high value is correct it also indicates that the Zeeman measurements with Hinode (Lites et al. 2007, 2008; Orozco Suárez et al. 2007) have not captured quite all of the existing quiet-Sun fields, presumably because of polarimetric cancelation which Zeeman measurements are subject to in contrast to Hanle measurements. Adopting the idealized model of a single-valued microturbulent field, Trujillo Bueno et al. (2004) obtained a mean field strength that varies between 50 and 70 G in the height range from 400 to 200 km, respectively, and only when taking an exponential probability density function for the field-strength distribution into account do they obtain the higher value of 130 G. Relatively high values are also reported by Bommier et al. (2005).

More in line with Faurobert et al. are Shapiro et al. (2007), who obtain from differential Hanle measurements with the CN violet system a field strength in the range from 10 to 30 G in the upper solar photosphere, while the analyses of the observed scattering polarization in C2 lines by Faurobert and Arnaud (2003) and Berdyugina and Fluri (2004) imply a field strength of about 10 G. However, measurements of the scattering polarization in molecular lines may be quite sensitive to the thermal structure in the atmosphere. In fact, Trujillo Bueno et al. (2004) obtain from measurements with C_2 a field strength of the order of 10 G too, but they also show that these measurements sample the atmosphere mainly above granules only, where, correspondingly, the turbulent field must be much weaker than in the downflows of the intergranular space (see the review by Trujillo Bueno et al. 2006 for a detailed presentation of their Hanle-effect measurements).

In any case, it seems that "deep" polarimetric measurements with Hinode have discovered a large part of the hitherto "hidden" magnetic field that was known to us only through Hanle measurements. It made this field accessible to Zeeman analysis and therefore to a more reliable determination of its angular distribution, at least down to $0.3''$ spatial resolution. In the next chapter, we review results from recent simulation that aim at explaining the predominance of the mean horizontal over the mean vertical field in the quiet-Sun photosphere.

2 Coupling of Convection with Magnetic Fields

It was mentioned in the previous chapter that a few observational studies prior to Hinode already hinted at a frequent occurrence of horizontally oriented magnetic fields in the quiet Sun. Likewise, the horizontal fields did not come unannounced to theoretical solar physics. Grossmann-Doerth et al. (1998) noted "we find in all simulations also strong horizontal fields above convective upflows," and Schaffenberger et al. (2005, 2006) found frequent horizontal fields in their three-dimensional simulations, which they describe as "small-scale canopies." Also, the three-dimensional simulations of Abbett (2007) display "horizontally directed ribbons of magnetic flux that permeate the model chromosphere," not unlike the figures shown by

Schaffenberger et al. (2006). However, these reports did not receive wide attention because actual measurement of the weak transversal component was not possible or unreliable prior to the advent of Hinode.

More recently and after the discovery of the horizontal field with Hinode, two theoretical works (Schüssler and Vögler 2008; Steiner et al. 2008) specifically aimed at finding out more about its nature and origin. Both papers present results of three-dimensional magnetohydrodynamic numerical simulations of the internetwork magnetic field with regard to the intrinsically produced horizontal magnetic field. In the following, I briefly summarize and compare part of their results.

The two simulation runs presented by Steiner et al. (2008) and the "local dynamo run" of Schüssler and Vögler (2008) differ substantially in their initial and boundary conditions for the magnetic field. Yet, they all show a clear dominance of the horizontal field in parts or the full height range where the spectral lines used for the Hinode observations are formed. Thus, the intrinsic production of a predominantly horizontal magnetic field in the photosphere of three-dimensional magnetohydrodynamic simulations is a rather robust result. Figure 1 shows the horizontal and the vertical magnetic field strengths as a function of height in the atmosphere of the three simulations. Left and right boundaries of the left panel correspond to approximately $z = 400$ km and $z = -1{,}000$ km of the right panel, respectively. Note that the scale of the ordinate is logarithmic and in gauss in the left panel but linear and in mT in the right panel. Also account for the nonlinear relation between the abscissa of the two panels. Interestingly, both simulation runs of Steiner et al. (2008) show

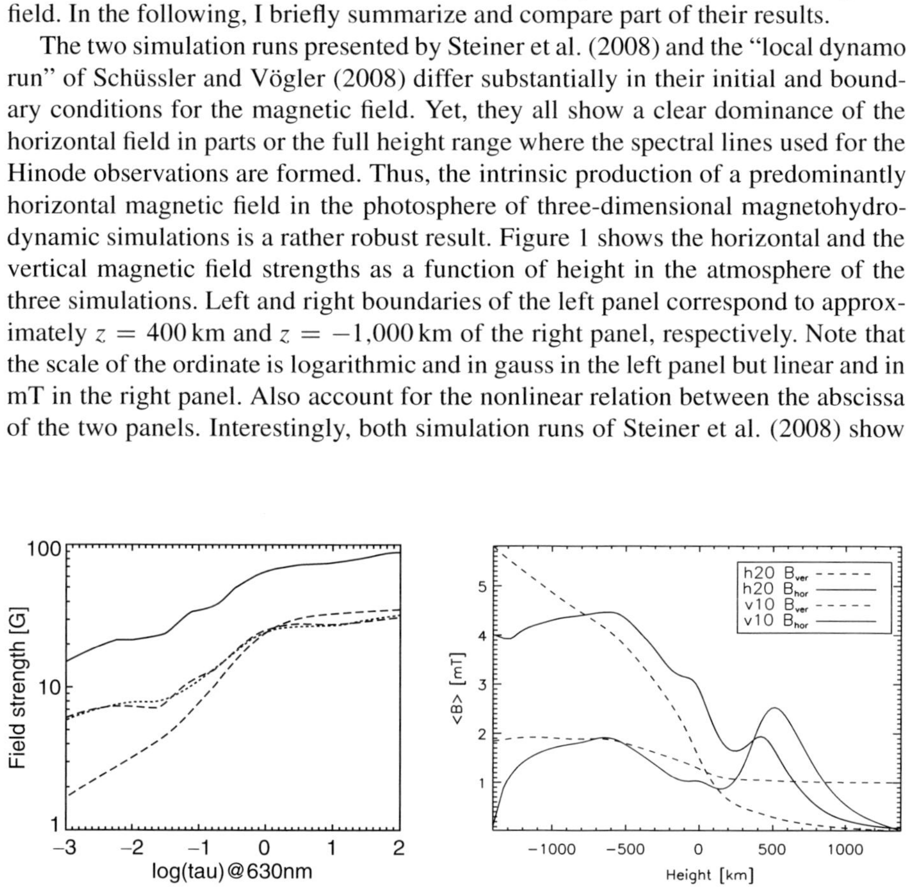

Fig. 1 *Left*: Mean absolute horizontal magnetic field components, $\langle|B_x|\rangle$ (*dashed-dotted curve*), and $\langle|B_y|\rangle$ (*dotted curve*), and absolute vertical field component, $\langle|B_{\rm ver}|\rangle$, (*dashed curve*) as a function of optical depth $\log\tau_{630\,\rm nm}$ of the dynamo run of Schüssler and Vögler (2008). The averaging refers to surfaces of constant $\tau_{630\,\rm nm}$. The *solid curve* is the rms $\langle B_x^2 + B_y^2\rangle^{1/2}$. *Right*: $\langle B_{\rm hor}\rangle = \langle(B_x^2 + B_y^2)^{1/2}\rangle$ (*solid curve*) and $\langle|B_{\rm ver}|\rangle$ (*dashed curve*) as a function of height z from the simulation run h20 (heavy) and run v10 (thin) of Steiner et al. (2008). v10 and h20 substantially differ in their initial and boundary conditions for the magnetic field. Note the different physical meanings of the abscissa and the different units and scales in the ordinates of the two plots

a local maximum of the horizontal field component near 500 km height and this is also the case for a local dynamo run when the top (open) boundary is located at $z = 650$ km (Schüssler, private communication).

How do these results compare with Hinode? For a fair comparison it is indispensable to synthesize the Stokes profiles of the 630 nm Fe I spectral line pair from the simulations and subsequently derive whatever parameters were derived from the actual observations. The analysis of the synthetic data must proceed in the very same manner as done with the observed profiles. Applying the appropriate point spread function (Wedemeyer-Böhm 2008) to the synthetic profiles and subjecting them to the same procedure for conversion to apparent flux densities as done by Lites et al. (2008) for the observed profiles, Steiner et al. (2008) obtain spatial and temporal averages for the transversal and longitudinal apparent magnetic flux densities, $|B^{\mathrm{T}}_{\mathrm{app}}|$ and $|B^{\mathrm{L}}_{\mathrm{app}}|$ of, respectively, 21.5 G and 5.0 G for run h20 and 10.4 G and 6.6 G for run v10. Thus, the ratio $r = \langle|B^{\mathrm{T}}_{\mathrm{app}}|\rangle/\langle|B^{\mathrm{L}}_{\mathrm{app}}|\rangle = 4.3$ for h20 and 1.6 for v10. Lites et al. (2008) obtain from Hinode SP data $\langle|B^{\mathrm{T}}_{\mathrm{app}}|\rangle = 55\,\mathrm{G}$ and $\langle|B^{\mathrm{L}}_{\mathrm{app}}|\rangle = 11\,\mathrm{G}$ resulting in $r = 5.0$. Run v10 was judged to rather reflect network fields because of its preference to produce vertically directed, unipolar magnetic fields, enforced by its initial and boundary conditions. For the internetwork field, h20 is more appropriate. Correspondingly, the r-value of h20 better agrees with the measurements of Lites et al. (2008), which measures mainly internetwork magnetic fields. It should be cautioned that r is quite dependent on spatial resolution in the sense that lower resolution overestimates this value. The reason for this behavior is that horizontal fields have a more patchy, smoother, and less intermittent character than the vertical fields and are therefore less subject to polarimetric cancelation.[1] For a comparison between synthetic and observed center-to-limb data see Steiner et al. (2009).

What kind of physical process produces the horizontal fields? Schüssler and Vögler (2008) and Steiner et al. (2008) offer two different but not necessarily exclusive explanations. Rather they emphasize two different aspects of the coupling of convection with magnetic fields that is at the origin of the horizontal fields. Steiner et al. (2008) emphasize the aspect of the flux expulsion process (Weiss 1966; Galloway and Weiss 1981), which describes the expulsion of magnetic field from the interior of an eddy flow like that of granules. Thus, the fact that the magnetic field tends to be located in the intergranular space and not within granules is considered a consequence of the flux expulsion process. However, it should be noted that the granular flow is not bounded alone by intergranular lanes but also by the overlaying photosphere, which efficiently damps overshooting flow owing to its superadiabatic stratification. Hence, magnetic field tends not only to be expelled in the lateral direction to the intergranular lanes but also in the vertical direction, where it accumulates in the upper photosphere and lower chromosphere. In fact, vertical

[1] Remember that polarimetric cancelation also occurs for independent field components of the horizontal field when they are perpendicular to each other within a single pixel area. Polarimetric cancelation occurs for horizontal fields in the same way as for vertical fields – transversal perpendicular fields lead to polarimetric cancelation just as antiparallel longitudinal fields do.

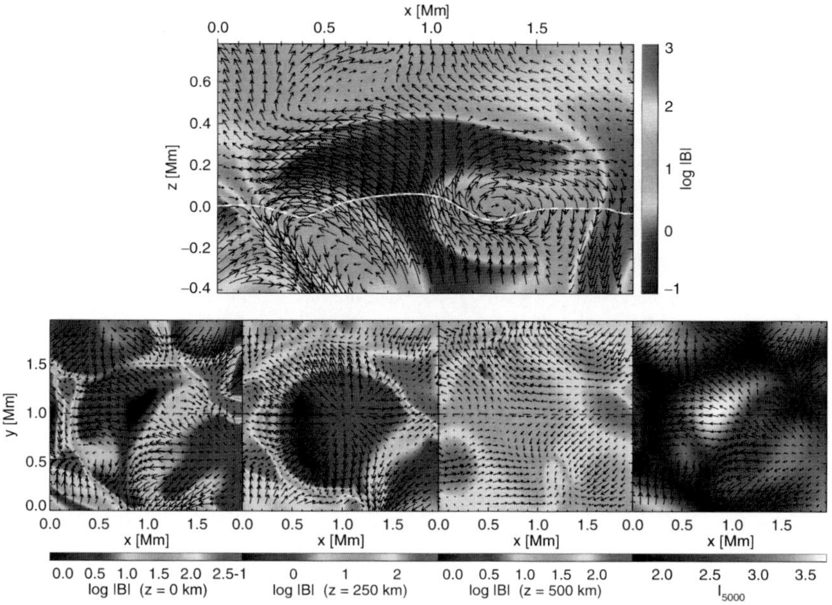

Fig. 2 Flux expulsion in a close-up from a MHD simulation by Schaffenberger et al. (2005): Logarithmic magnetic field strength in a vertical cross-section (*top*) and in three horizontal cross-sections (*bottom*) at heights of 0, 250, and 500 km. The emergent intensity is displayed in the rightmost panel. The *arrows* represent the velocity field in the shown projection planes. The *white curve* in the upper panel marks the height of optical depth unity. From Wedemeyer-Böhm et al. (2008)

sections through the computational domain such as Fig. 1 of Schüssler and Vögler (2008) and Fig. 3 of Schaffenberger et al. (2005) show magnetic voids where the granular flow is most vigorous as a consequence of the flux expulsion process. The voids are arched by horizontally directed magnetic field. This can be nicely seen in the close-up shown in Fig. 2.

Schüssler and Vögler (2008) emphasize the aspect from the local dynamo that operates in the convectively unstable layers beneath the surface of continuum optical depth unity. Near the surface, weak magnetic field gets quickly stretched and thus amplified by the convective flow, in particular also by the small-scale turbulent flow of intergranular downflows.[2] On the other hand, in the convectively stable photosphere above, the flows become weaker and field amplification rapidly drops

[2] There is nothing mystic about this amplification, which is a natural consequence of the field being tied to the plasma in (quasi) ideal MHD, which does work against the Lorentz force on the expense of kinetic energy. However, Vögler and Schüssler (2007) were able to demonstrate that a local dynamo operates in these layers, which means that a magnetic field of constant mean energy density is maintained without the need of continuous supply of a weak (seed) field. It even survives when downflows continuously pump magnetic field out of the simulation box.

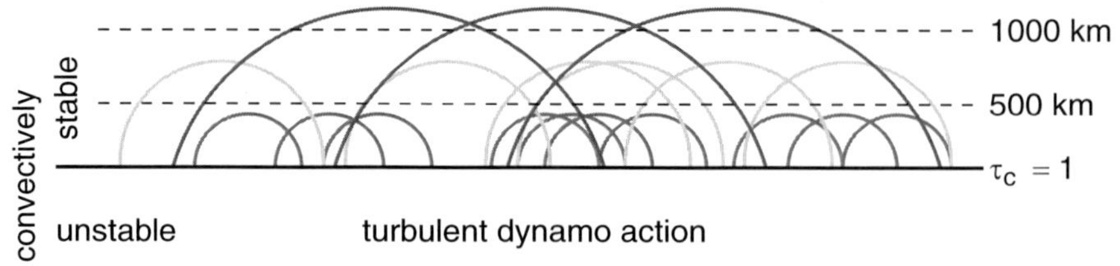

Fig. 3 Dynamo action amplifies and maintains the magnetic field in the convectively unstable layer below $\tau_c = 1$. In the layers above, the field is mainly determined by its distribution at the $\tau_c = 1$ surface. This configuration leads to a steep decline of the absolute vertical flux with height as can be seen by counting the loop foot-points at each indicated level because small loops are more abundant than large loops. On the other hand, this configuration leads to a less steep decline of the mean horizontal field strength

with height. Thus, the magnetic field in the photosphere and its decay with height is mainly determined by the field distribution at the surface $\tau_c = 1$, in particularly by its energy spectrum as a function of horizontal wave number, which in turn is determined by the turbulent dynamo beneath this surface. This results in a steep decline of the absolute vertical magnetic flux with height as can be seen from Fig. 3. While many loops of small scales (where the energy spectrum is maximal) contribute to the vertical flux in the deep photosphere, fewer loops of large scales (where the energy is less) add to it in the higher layers. On the other hand, this configuration leads to a less steep decline of the horizontal field and hence, the mean horizontal field strength starts to dominate the mean absolute vertical field strength as a function of height. In this picture the dominance of the horizontal field component is a natural outcome of the dynamo-generated field.

It was argued in the course of this conference by Stenflo that a predominance of the horizontal field over the vertical one was in contradiction with the solenoidality condition for the magnetic field. Leaving aside that the simulations strictly maintain solenoidality and still show a predominance of the horizontal over the vertical component, Fig. 4 provides another counter example to this conjecture. Assume that within an area A of granular size L, there are two vertical flux concentrations of, say $B_\mathrm{v} = 500\,\mathrm{G}$, that occupy an area of $f_\mathrm{v} A$, where $f_\mathrm{v} \approx 0.02$. The two flux concentrations of opposite polarity are connected by a photospheric, solenoidal arch of thickness $h \approx 0.1L$ as it occurs in simulations (viz. the "small-scale canopies" of Schaffenberger et al. 2005). Then flux conservation demands that in a cross section of the loop (as the one indicated by the dashed line) $\Phi = B_\mathrm{h} L h = B_\mathrm{v} f_\mathrm{v} A/2$, where B_v is the vertical field strength at $\tau_c \approx 1$ and B_h the strength of the horizontal field of the arch. It follows that $B_\mathrm{h} = 5 a f_\mathrm{v} B_\mathrm{v}$. With $a \approx 1$ we obtain $B_\mathrm{h} \approx 50\,\mathrm{G}$. When the horizontal field of the arch fills about $f_h = 0.8$ of the area A, we obtain a mean horizontal field of $\langle |B_\mathrm{h}|\rangle = B_\mathrm{h} f_h = B_\mathrm{v} f_\mathrm{h} f_\mathrm{v} 5a = 40\,\mathrm{G}$ while $\langle |B_\mathrm{v}|\rangle = B_\mathrm{v} f_\nu = 10\,\mathrm{G}$. The ratio $\langle |B_\mathrm{h}|\rangle / \langle |B_\mathrm{v}|\rangle = 5 a f_\mathrm{h}$ can be made arbitrarily large by increasing a, that is, by stretching the arch.

Confusion may arise because of misinterpreting transversal Zeeman measurements as being a measure of magnetic flux. It is true that for longitudinal Zeeman measurements, the measured mean flux density is directly proportional to the

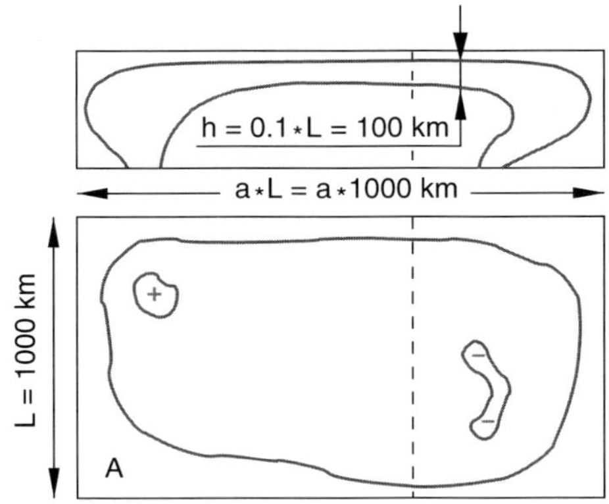

Fig. 4 Two intergranular magnetic flux concentrations of opposite polarity are connected by a small-scale loop (canopy) in the photosphere, spanning a distance of granular scale. Even though this magnetic field configuration is divergence free, it follows that the measured mean absolute horizontal field is much stronger than the mean absolute vertical field: $\langle |B_{\mathrm{h}}|\rangle/\langle |B_{\mathrm{v}}|\rangle \approx 5a$

magnetic flux (disregarding polarimetric cancelation effects for now). Such a generalization is not valid for the transversal Zeeman effect, where the mean flux density is an average value over an area that is not perpendicular but parallel to the measured field component. In this case the apparent flux density is a spatially averaged flux density, not more. Assuming that the horizontal field has the vertical extent of the atmospheric scale height, h, with a filling factor of 1, then the horizontal magnetic flux is $\Phi_{\mathrm{h}} = \langle B_{\mathrm{h}}\rangle hL$, where L is the scale of the field of view. The vertical flux is $\Phi_{\mathrm{v}} = \langle B_{\mathrm{v}}\rangle L^2$ and consequently, using flux conservation $\langle B_{\mathrm{h}}\rangle/\langle B_{\mathrm{v}}\rangle \propto L/h$. On granular scales, $L/h \approx 1{,}000\,\mathrm{km}/100\,\mathrm{km} = 10$, which again suggests that the predominance of the apparent horizontal flux density over the vertical one is a direct consequence of the anisotropic nature of convective turbulence in the highly stratified atmosphere of the Sun, where the granular scale is ten times larger than the pressure scale height. However, remember that this calculation does not apply to subgranular scales at which the field may still be isotropic.

3 Coupling of Waves with Magnetic Fields

Acoustic waves are generated and emitted by the convectively unstable layers beneath the solar surface ($\tau_{500\,\mathrm{nm}} = 1$). They couple to the magnetic field when they enter the atmospheric layers above. Such coupling can be seen to take place in Fig. 5, which shows a time instant of a two-dimensional simulation. Starting from a state where a magnetic flux concentration has formed at approximately $x = 4{,}150\,\mathrm{km}$ (left panel), a plane-parallel, monochromatic acoustic wave is introduced at the bottom of the computational domain (right panel), which propagates within a time span of about 200 s across the nonstationary atmosphere – from the convection-zone,

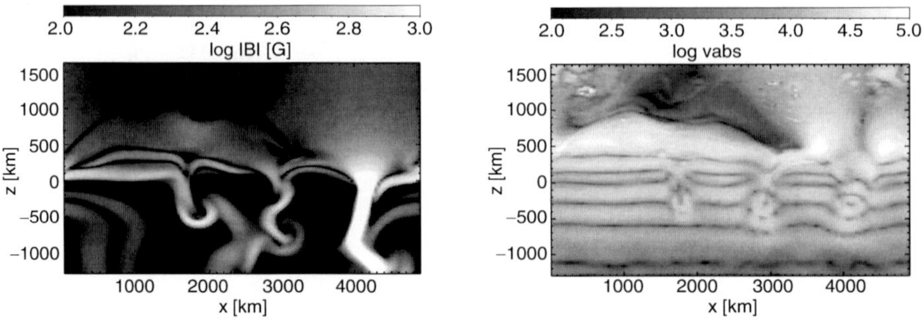

Fig. 5 *Left*: Logarithmic absolute magnetic flux density in a two-dimensional simulation domain. Magnetic flux concentrations form in the downdrafts of convection. A particularly strong one has formed near $x = 4{,}150$ km. *Right*: A plane-parallel wave of frequency 20 mHz travels through the convecting plasma into the magnetically structured photosphere and further into the low-β (magnetically dominated) chromosphere. The panel shows the difference in absolute velocity between the perturbed and the unperturbed solution 168 s after launching the wave. The magnetic field at launch time corresponds to that of the panel to the left. Optical depth $\tau_{500\,\mathrm{nm}} = 1$ is close to $z = 0$. The velocity scaling is logarithmic with dimension [cm s^{-1}]. At the location of the magnetic flux concentration the initially fast (acoustic) wave has converted character to fast magnetic and it underwent refraction to such a degree that the wave front extending from $(x, z) = (2400, 1500)$ to $(3400, 500)$ has already completely turned around and travels back into the atmosphere again. A similar fanning out of the wave front starts to occur around $x = 1{,}100$ km. Adapted from Steiner et al. (2007) courtesy of Ch. Nutto

$z \in [-1200, 0]$ km, through the photosphere, $z \in [0, 500]$ km, into the magnetically dominated chromosphere, $z \in [500, 1600]$ km, where it gets partially refracted and reflected by interaction with the magnetic field.

The perturbation of the wave front in the convection zone in Fig. 5 is not due to the presence of a magnetic field, but rather to the vigorous intergranular downflows and associated temperature deficit. However, as soon as the wave front enters the magnetically dominated atmosphere where $\beta \leq 1$, coupling with the magnetic field kicks in and part of the wave gets primarily magnetically driven. (β is the ratio of magnetic to thermal pressure.) The major effects of this interaction are that (1) the wave front speeds up as the Alfvén velocity becomes the characteristic speed, which sharply increases with height when magnetic pressure drops less quickly than the gas pressure, and (2) the waves refract because of the inhomogeneous magnetic field, defining an inhomogeneous refractive index for the magneto-acoustic wave.

When measuring the wave travel time between two fixed geometrical height levels in the atmosphere (representing the formation height of two spectral lines), we observe a decrease at locations of strong magnetic field concentrations because of (1). Moreover, because of (2) the wave behaves like evanescent because of the strong refraction that effectively leads to a reflection of the wave. Therefore, the wave travel time betrays the presence of the magnetic field concentration and it can be used to map the topography of the magnetic field in the solar atmosphere. In fact, this effect was employed by Finsterle et al. (2004) to obtain the three-dimensional topography of the "magnetic canopy" in and around active regions by determining the travel time of high-frequency acoustic waves in the solar chromosphere. Steiner et al. (2007), from where Fig. 5 is derived from, demonstrated with the help of numerical experiments that wave travel-times can indeed serve this purpose.

The theory and theoretical aspects of magneto-acoustic waves in a gravitationally stratified atmosphere, sometimes called magneto-acoustic-gravity (MAG) waves or magneto-atmospheric waves, have received much attention in recent years. Pioneering works include those of Thomas (1982) and Zhugzhda and Dzhalilov (1982, 1984a,b,c). Shibata (1983) carried out initial numerical computer experiments with magneto-atmospheric waves. In more recent times, Rosenthal et al. (2002) and Bogdan et al. (2003) published two comprehensive papers on the subject. These works include several numerical experiments with nonuniform magnetic field equilibria in a two-dimensional, stratified atmosphere. They recognized and highlighted the role of refraction of fast magnetic waves and the role of the surface of equal Alfvén and sound speed as a wave conversion zone. Aiming at applications in local helioseismology, Cally (2005) derives gravito-magneto-acoustic dispersion relations and then uses these to examine how acoustic rays entering regions of strong field split into fast and slow components and the subsequent fates of each. Cally (2007) presents the theory in a particularly instructive manner. Results from numerical simulations of MAG-wave propagation in three-dimensional space are presented by Cally and Goossens (2008) and Moradi et al. (2009).

Khomenko and Collados (2006) carried out numerical simulations of magneto-acoustic wave propagation in sunspots and found that the fast (magnetic) mode in the region where $c_{\mathrm{s}} < v_{\mathrm{A}}$ does not reach the chromosphere but reflects back to the photosphere due to wave refraction, caused primarily by the vertical and horizontal gradients of the Alfvén speed. For small-scale flux-tubes, Khomenko et al. (2008b) find that deep horizontal motions of the flux tube initially generate a slow (magnetic) mode and a surface mode that are efficiently transformed into a slow (acoustic) mode when the magnetic field starts to dominate. This slow mode propagates vertically along the magnetic field remaining always within the flux tube, where it steepens to a shock. Only a small part of the driver energy is returned to the photosphere by the fast magneto-acoustic mode. Khomenko et al. (2008a) demonstrate that photospheric 5 min oscillations can leak into the chromosphere inside small-scale vertical magnetic flux tubes as a consequence of radiative damping, which leads to a significant reduction of the cutoff frequency and they provide observational evidences of this effect. This effect is not to be confounded with the "ramp effect" (Cally 2007), which lowers the cutoff frequency when the flux tube is inclined with respect to the gravitational acceleration (Suematsu 1990; Jefferies et al. 2006). Both these works of Khomenko et al. suggest that vertical magnetic field concentrations play an essential role in coupling the dynamics of the photosphere to the chromosphere through efficient channeling and conversion of magneto-acoustic waves.

Figure 6 demonstrates the complexity of magneto-acoustic wave propagation in a magnetically structured, stratified atmosphere. A magnetic flux sheet (two-dimensional) of a strength of 1,600 G at its base (where $\beta < 1$) is shifted to the right in the transverse direction with a single impulse of 12 s duration and a maximal

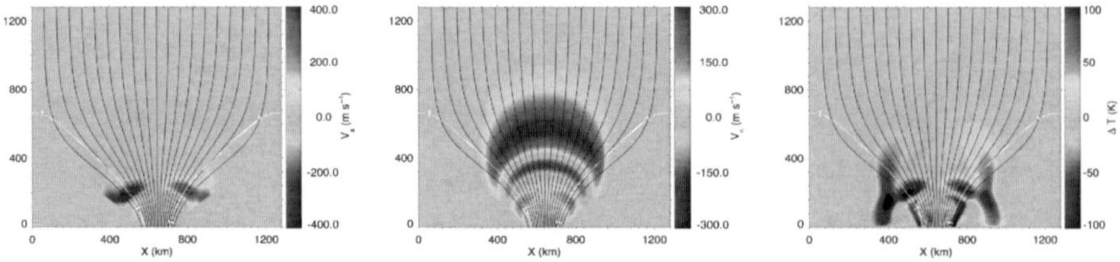

Fig. 6 Snapshot of magneto-acoustic waves generated by an initial impulsive movement of the equilibrium magnetic flux concentration (*black field lines*) to the right. *Left*: Longitudinal velocity of the slow (predominantly acoustic) wave that propagates along the magnetic field. *Middle*: Transversal velocity due to the fast (predominantly magnetic) wave within the flux sheet. It undergoes differential refraction. *Right*: Temperature perturbation due to these two waves and due to the fast acoustic wave in the ambient medium. The *white* contours indicates $\beta = 1$. Velocities are only shown where $|B| > 50\,$G. Courtesy G. Vigeesh

velocity of $0.75\,\mathrm{km\,s^{-1}}$. As a consequence of this sudden movement, four types of waves emanate from the base of the flux sheet: (1) a fast (magnetic) wave in the low-β regime of the flux-sheet interior, visible in the middle panel, (2) a slow (acoustic) wave that propagates along the magnetic field within the flux sheet, visible in the left panel, (3) a fast acoustic wave that propagates spherically into the ambient medium, visible in the right panel, and (4) a slow magnetic wave that propagates in the boundary layer of the flux sheet. The predominantly acoustic waves show an antisymmetric pattern with respect to the sheet axis because of the movement to the right, which causes a compression at the leading edge and a decompression at the trailing edge of the flux sheet. However, as the slow wave propagates it becomes asymmetric (not visible in Fig. 6). On the left side, the wave crest quickly steepens into a shock because of being preceded by a wave trough. On the right side, the inverse sequence causes the wave to spread, which impedes at first the development of a shock.

The shape of the fast (magnetic) wave becomes crescent (middle panel) because of the nonuniform Alfvén speed. As the tips of the crescents move essentially sideways, they enter the low-β periphery of the flux sheet where they convert to predominantly acoustic (fast) and therefore become visible in δT (right panel) as the wing like feature that extends in the vertical direction. From a rough estimate of the acoustic energy flux generated by such impulsive transverse motions, Vigeesh et al. (2009) conclude that this flux would hardly balance the chromospheric energy requirements in the network. Previously, similar numerical experiments with impulsive and periodic driving have been carried out by Hasan et al. (2005) and Hasan and van Ballegooijen (2008).

Until here we have considered the coupling of waves with intense, small-scale magnetic flux concentrations like they occur in network and plage regions. The numerical experiments mentioned above all make use of highly idealized wave drivers: typically a monochromatic transversal or longitudinal periodic motion, or a single impulsive motion is imposed on an initially static equilibrium configuration. Certainly for the internetwork, such quasi static states are unrealistic. Furthermore, magnetic flux concentrations in the intergranular lanes of the internetwork attain typically hectogauss not kilogauss field strength and they rather connect with the omnipresent horizontal field than vertically extending into the chromosphere. In this case, the strength of the internetwork field can be expected to exponentially decrease with height like the gas pressure and the density do, so that the magnetic field would not become dominant in the upper layers of the photosphere and the chromosphere and consequently no substantial coupling between waves and the magnetic field would occur in these layers. On the other hand, if there is a predominance of one magnetic polarity, part of the magnetic flux can be expected to connect to the outer solar atmosphere or to a region of opposite polarity further away, in which case the field strength would decrease less steeply leading to $\beta \leq 1$ in the chromosphere (but not yet in the photosphere). How do waves couple to the magnetic field under these circumstances?

First to the waves. Figure 7 shows the temperature in three horizontal cross sections through the computational domain of 9.6 Mm side length. At 200 km below the $\tau_{500\,\mathrm{nm}} = 1$ surface we see the cool intergranular lanes and hot granules (left panel), and at +200 km the hot intergranular lanes and cool granules of the inverse granulation (seen a bit higher up in spectroscopic quantities). Both these patterns evolve on roughly the granular time scale. In the cross section at +1,000 km, we see a totally different pattern that evolves on a much shorter time scale and is due to shock waves that have formed at this height range and travel in all directions, forming a network of hot material. This shock-wave pattern that emerges from acoustic waves, which are generated by the convective granular motion at the base of the atmosphere, was first shown by Wedemeyer et al. (2004) to exist in three-dimensional simulations without magnetic field. It leads to large fluctuations in the tenuous atmosphere above the classical temperature minimum to a veritable "fluctosphere" (Wedemeyer-Böhm et al. 2008), earlier termed the "clapotisphere" by Rutten and Uitenbroek (1991) and Rutten (1995) with regard to peak fluctuations caused by the interference of (shock-)waves.

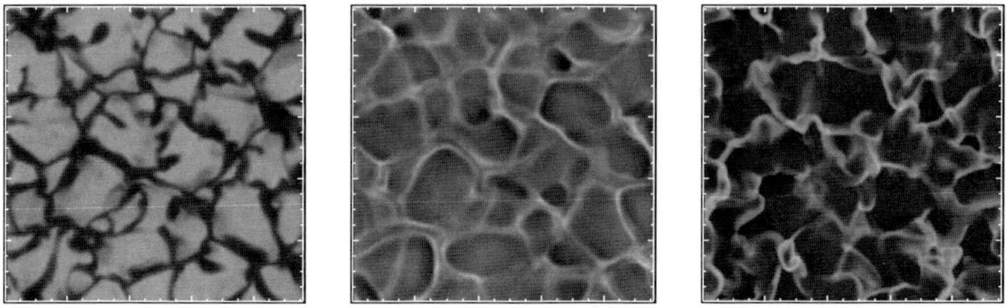

Fig. 7 Three horizontal cross sections through a simulation domain with 9.6 Mm side length displaying the temperature. *Left*: Section at $z = -200$ km showing granules (hot) and intergranular lanes (cool). *Middle*: Section at $z = +200$ km showing inverse granulation. *Right*: Section at $z = +1{,}000$ km showing the "fluctosphere" consisting of a rapidly changing network pattern of hot material compressed in traveling shock waves. Pockets of cool, expanded material reside in between the hot plasma

In combination with magnetic fields, these disturbances give rise to a rich variety of magneto-acoustic wave phenomena. As detailed earlier, the gas pressure in the gravitationally stratified atmosphere may drop more quickly with height than the magnetic energy density does, giving rise to a height range where sound speed and Alfvén speed are of similar magnitude. Within this region, which forms a corrugated surface excursive over a wide height range in the three-dimensional atmosphere, propagating wave modes change nature from acoustic to magnetic and from slow to fast and vice versa. Above this surface there is a predominant tendency for magnetic modes to get refracted and reflected due to the dispersive nature of the inhomogeneous magnetic field.

Schaffenberger et al. (2005) have simulated this case with a field of a predominant polarity of constant mean net vertical flux density 10 G. Their three-dimensional simulation domain encompasses a height range from $-1{,}500$ to $+1{,}500$ km (where zero corresponds to $\tau_{500\,\mathrm{nm}} = 1$). Immediately apparent from a movie that shows the field strength (http://www.kis.uni- freiburg.de/ steiner/ vsec.mov) is that the surface of $\beta = 1$ (where $c_{\mathrm{s}} \approx v_{\mathrm{A}}$) separates a region of highly dynamic magnetic fields with fast moving magnetosonic waves and shocks around and above it from the more slowly evolving field of high-beta plasma below it. This surface is located at roughly 1,000 km in this case. It is corrugated and its local height strongly varies in time over a range of about 1,000 km.

The magnetic field in the chromosphere of this simulation continuously rearranges itself on a time scale of less than 1 min, much shorter than in the photosphere or in the convection-zone layers. The field has a strength between 2 and 40 G. Different from the surface magnetic field, it is more homogeneous and fills practically the entire space so that the magnetic filling factor in the top layer is close to unity. There seems to be no spatial correlation between chromospheric flux accumulations and the small-scale field concentrations in the photosphere. Magnetoacoustic waves that form transient filaments of stronger than average magnetic field are a ubiquitous phenomenon in the chromosphere. They form in the compression zone downstream and along propagating shock fronts. These magnetic filaments that have a field strength rarely exceeding 40 G rapidly move with the shock fronts and quickly form and dissolve with them. Hence, the coupling of waves with the magnetic field leads to a continuous agitation of the magnetic field in the chromosphere by shock waves. It is not yet clear what the significance and the consequences of these perturbations are, especially in view of electro-magnetic dissipation processes.

4 Coupling of Radiation with Magnetic Fields

The radiative flux in the solar atmosphere couples to the magnetic field not only microscopically (changing polarization state) but also macroscopically through modification of the gas pressure and density by the magnetic field. The magnetic field has an energy density, e_{mag}, often called the magnetic pressure, like gas pressure has $p_{\mathrm{g}} = e_{\mathrm{therm}}$. The ratio of the two is $\beta = e_{\mathrm{therm}}/e_{\mathrm{mag}}$. In locations where

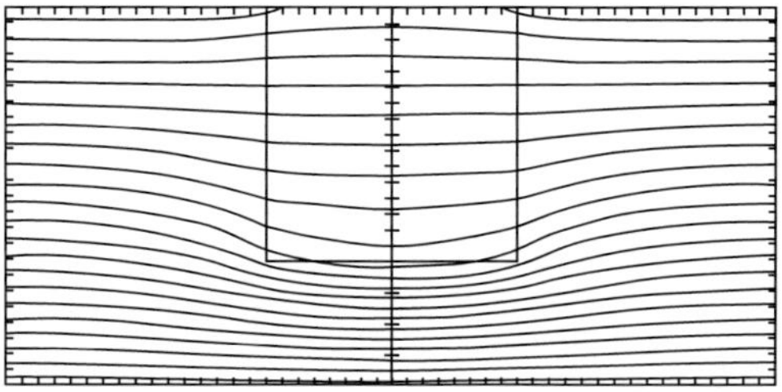

Fig. 8 Temperature contours of a two-dimensional schematic magnetic flux sheet in radiative equilibrium with its surroundings. The opacity in the center rectangle is one fifth of the opacity in the rest of the domain, causing a redirection (channeling) of the radiative flux. In this state of radiative equilibrium, the flux channeling causes a lowering of the temperature in the *bottom* part of the "flux sheet" and a temperature enhancement in the *top* part, where radiation starts escaping to the ambient free space. Adapted from Steiner (1991)

$\beta \leq 1$, magnetic pressure substitutes gas pressure in the transversal direction, which at the same time lowers the material density (given thermodynamic equilibrium) and with it the opacity. Hence, as a rule of thumb, opacity is lower where the magnetic field is strong, which leads to a redirection of the radiative flux. This effect was called "flux channeling" by Cannon (1985), because the net radiative flux, which is strictly vertical and outwardly directed in a plane parallel stellar atmosphere, becomes partially and laterally redirected in the presence of a strong, vertically directed magnetic flux concentration. In radiative equilibrium, this effect leads to a temperature perturbation in the sense that the temperature is slightly enhanced in the surface layers of the magnetic flux concentration but lowered in the deep layers as is shown in Fig. 8.

The coupling of the radiation field to the magnetic field via the channeling effect introduces a substantial anisotropy in the radiation field at the location of a magnetic flux concentration and in its surroundings. This anisotropy gives rise to the facular phenomenon.

Faculae can be explained in terms of the "hot wall model" (Spruit 1976). In this model, the hot wall corresponds to the wall of the depression that is caused by a magnetic flux concentration, like the Wilson depression of a sunspot. The wall separates the magnetic flux concentration from the surrounding, practically field-free plasma. However, the term "hot wall" may be misleading as it is by far not as hot as the plasma in the same height range in the unperturbed ambient atmosphere. In fact, the rectangular wall of the low-opacity region of the radiative equilibrium model shown in Fig. 8 has in most parts a lower temperature. In the case of photospheric magnetic flux concentrations, the temperature of the wall is determined by a delicate balance between radiative losses and the convective energy supply, where the latter is actually reduced in the transverse direction close to the flux concentration because the magnetic field acts like a solid wall to plasma motion in this direction. Equilibrium models and magnetohydrodynamic simulations, however, invariably show a bright edge where the wall of the "Wilson depression" merges with the horizontal surface

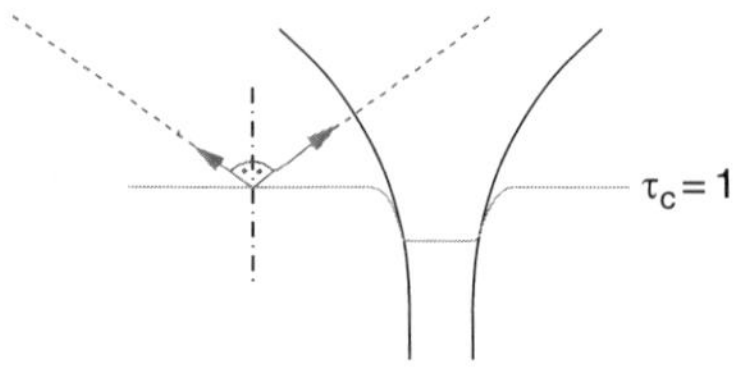

Fig. 9 Photons preferentially escape along the line of sight that traverses the magnetic flux concentration because of its rarified (less opaque) atmosphere. Hence, the radiation field lateral to the flux concentration is asymmetric due to radiative flux channeling

(Carlsson et al. 2004; Keller et al. 2004; Steiner 2005; De Pontieu et al. 2006). Seen at an oblique angle (corresponding to an observation off disk center) this "hot edge" leads to a characteristic center-to-limb variation in contrast.

It was only recently realized when analyzing high resolution filtergrams of faculae from the Swedish Solar Telescope (SST) that their contrast enhancement extends quite a distance in the limbward direction, typically for half a granular size, which is much further than the depth of the depression wall would possibly be (Lites et al. 2004; Hirzberger and Wiehr 2005). Berger et al. (2007) measure an average radial width of faculae of 400 km. This means that the contrast enhancement of faculae extends beyond the depression proper in the limbward direction. The reason for this behavior is explained with the help of Fig. 9 as follows.

A material parcel located in the solar atmosphere lateral to a magnetic flux concentration "sees" a more transparent atmosphere in the direction toward the flux concentration as compared to a direction under equal zenith angle but pointing away from it because of its rarefied atmosphere. Consequently, from a wide area surrounding the magnetic flux concentration, radiation escapes more easily in the direction towards the flux sheet so that a single flux concentration impacts the radiative escape in a cross-sectional area that is much wider than the magnetic field concentration proper. This means that the presence of a magnetic flux concentration introduces an anisotropy in the radiation field such that when observed close to the limb, a granule limbward of (seen across) the flux concentration appears brighter than normal. Hence, a facula is not to be identified with bright plasma that sticks, as the name may insinuate, like a torch out of the solar surface. Rather it is the manifestation of photospheric granulation, seen across a magnetic flux concentration – granulation that appears brighter than normal in the form of so called "facular granules."

Another consequence of the flux channeling effect is an enhancement of radiative loss from the solar surface at the location of the small-scale magnetic flux concentration. One could say that magnetic flux concentrations add a roughness to the solar surface that increases its effective area and thus increases the radiative loss from it. Therefore, the coupling between radiation and magnetic field, and with it faculae, play a key role in the solar irradiance variation over a solar cycle and on shorter time scales (Fligge et al. 2000; Wenzler et al. 2005; Foukal et al. 2006; Krivova and Solanki 2008).

5 Conclusion

Results obtained with the spectropolarimeter of the Solar Optical Telescope onboard the Hinode space observatory have greatly extended our knowledge of the magnetism of the quiet Sun. It has now become apparent that virtually every location on the surface of the Sun harbors magnetic field that can be detected in the 630 nm Fe I spectral line via the Zeeman effect, exceeding a 1σ noise level of 0.6 and 20 G for the longitudinal and transversal magnetic field, respectively (Lites et al. 2008). At the scale of $0.3''$, this magnetic field has a preferential direction, which is parallel to the solar surface, that is, it is anisotropic. It seems now that Hinode has discovered the majority of the hitherto "hidden" magnetic field that was known to us only through Hanle measurements, and it made it accessible to Zeeman analysis.

The predominantly horizontal direction of the weak internetwork magnetic field is also a robust result of magnetohydrodynamic simulations of the surface layers of the Sun. It can be explained in terms of the coupling between magnetic field and convective plasma motion. Two aspects of this interaction are evident. First, the process of flux expulsion, which displaces magnetic field from eddy cells. It leads to a concentration of predominantly vertical fields in the interganular lanes and predominantly horizontal fields in the stable layers above and at the edges immediately adjacent to granules. Second, the turbulent dynamo, which operates in the top surface layers of the convection zone, leads in the adjacent, stably stratified photospheric layer to a multi-scale system of magnetic loops. This loop system naturally leads to a stronger decline with height of the vertical than of the horizontal component of the magnetic field, hence, to a predominance of the horizontal fields in the height range where the spectral lines used for the Hinode observations are formed. It is still a matter of future research to find out to which degree the internetwork magnetic field is due to the turbulent surface dynamo, the remnants of pre-existing magnetic flux of active regions, and the emergence of magnetic flux from the deep convection zone, or due to yet other, additional sources. Also the role of the horizontal field in the heating of the chromosphere, for example, by Ohmic dissipation of associated current sheets, needs yet to be clarified.

The study of the propagation of magnetoacoustic waves in a magnetically structured, stratified atmosphere is a relatively new field of research. It has applications in magnetoatmospheric seismology, for example, when determining the topography of the magnetic canopy by measuring wave travel times in the photosphere and chromosphere. The surface defined by equality between sound speed and Alfvén speed is a zone of mode conversion. It separates the magnetically dominated tenuous region of fast moving magnetic modes and magnetic shock waves from the more slowly evolving atmosphere beneath it. The strong gradients in the Alfvén speed of a magnetically structured atmosphere lead to refraction and reflection of the magnetically driven modes. Future research in this field is directed at improving diagnostics for magnetoatmospheric seismology. Measurement and theory of magnetoatmospheric wave propagation need to be improved and linked for a reliable interpretation and exploitation of observations. The role of wave mode conversion and the channeling of slow modes in magnetic flux concentrations for the heating of the outer atmosphere must yet be quantified.

Magnetic flux concentrations lead to radiative flux channeling and increased anisotropy in the radiation field. Faculae can be understood in terms of this effect, which is apparent from magnetohydrodynamic simulations in two and three spatial dimensions. Future research in this field should include a thorough statistical analysis of (synthetic) faculae of three-dimensional simulations to understand the role of magnetic field strength, size, shape, etc. for the center-to-limb behavior of the facular contrast and for the energy balance of faculae. The latter is crucial for a better understanding of solar radiance variability.

Acknowledgment I am grateful to Ch. Nutto, S. Wedemeyer-Böhm, and G. Vigeesh for their help with the figures; M. Schüssler and J. Trujillo Bueno for clarifications regarding the turbulent dynamo and Hanle depolarization; and Ch. Bethge, J. Bruls, R. Hammer, and Ch. Nutto for proofreading individual sections. This work was supported by German Academic Exchange Service (DAAD) grant D/05/57687.

References

Abbett, W. P. 2007, ApJ, 665, 1469
Bello González, N., Kneer, F. 2008, A&A, 480, 265
Bello González, N., Okunev, O., Kneer, F. 2008, A&A, 490, L23
Bello González, N., Yelles Chaouche, L., Okunev, O., Kneer, F. 2009, A&A, 494, 1091
Berdyugina, S. V., Fluri, D. M. 2004, A&A, 417, 775
Berger, T. E., Rouppe van der Voort, L. H. M., Löfdahl, M. 2007, ApJ, 661, 1272
Berger, T. E., Rouppe van der Voort, L. H. M., Löfdahl, M. G., et al. 2004, A&A, 428, 613
Bogdan, T. J., Carlsson, M., Hansteen, V. H., et al. 2003, ApJ, 599, 626
Bommier, V., Derouich, M., Landi degl'Innocenti, E., Molodij, G., Sahal-Bréchot, S. 2005, A&A, 432, 295
Bovelet, B., Wiehr, E. 2007, Solar Phys., 243, 121
Cally, P. S. 2005, MNRAS, 358, 353
Cally, P. S. 2007, Astron. Nachrichten/AN, 328, 286
Cally, P. S., Goossens, M. 2008, Solar Phys., 251, 251
Cannon, C. J. 1985, The transfer of spectral line radiation (Cambridge University Press, 1985)
Carlsson, M., Stein, R. F., Nordlund, Å., Scharmer, G. B. 2004, ApJ, 610, L137
Carroll, T. A., Kopf, M. 2007, A&A, 468, 323
Centeno, R., Socas-Navarro, H., Lites, B., et al. 2007, ApJ, 666, L137
De Pontieu, B., Carlsson, M., Stein, R., et al. 2006, ApJ, 646, 1405
de Wijn, A. G., Lites, B. W., Berger, T. E., et al. 2008, ApJ, 684, 1469
Domínguez Cerdeña, I., Sánchez Almeida, J., Kneer, F. 2006a, ApJ, 636, 496
Domínguez Cerdeña, I., Sánchez Almeida, J. S., Kneer, F. 2006b, ApJ, 646, 1421
Dunn, R. B., Zirker, J. B. 1973, Solar Phys., 33, 281
Faurobert, M., Arnaud, J. 2003, A&A, 412, 555
Faurobert, M., Arnaud, J., Vigneau, J., Frisch, H. 2001, A&A, 378, 627
Faurobert-Scholl, M. 1993, A&A, 268, 765
Faurobert-Scholl, M., Feautrier, N., Machefert, F., Petrovay, K., Spielfiedel, A. 1995, A&A, 298, 289
Finsterle, W., Jefferies, S., Cacciani, A., Rapex, P., McIntosh, S. 2004, ApJ, 613, L185
Fligge, M., Solanki, S. K., Unruh, Y. C. 2000, A&A, 353, 380
Foukal, P., Fröhlich, C., Spruit, H., Wigley, T. M. L. 2006, Nat, 443, 161
Frisch, H. 2006, A&A, 446, 403

Galloway, D. J., Weiss, N. O. 1981, ApJ, 243, 945
Grossmann-Doerth, U., Schüssler, M., Steiner, O. 1998, A&A, 337, 928
Harvey, J. W., Branston, D., Henney, C. J., Keller, C. U. 2007, ApJ, 659, L177
Hasan, S. S., van Ballegooijen, A. A. 2008, ApJ, 680, 1542
Hasan, S. S., van Ballegooijen, A. A., Kalkofen, W., Steiner, O. 2005, ApJ, 631, 1270
Hirzberger, J., Wiehr, E. 2005, A&A, 438, 1059
Ishikawa, R., Tsuneta, S., Kitakoshi, Y., et al. 2007, A&A, 472, 911
Ishikawa, R., Tsuneta, S., Ichimoto, K., et al. 2008, A&A, 481, L25
Jefferies, S. M., McIntosh, S. W., Armstrong, J. D., et al. 2006, ApJ, 648, L151
Keller, C. U., Schüssler, M., Vögler, A., Zakharov, V. 2004, ApJ, 607, L59
Khomenko, E., Collados, M. 2006, ApJ, 653, 739
Khomenko, E., Centeno, R., Collados, M., Trujillo Bueno, J. 2008a, ApJ, 676, L85
Khomenko, E., Collados, M., Felipe, T. 2008b, Solar Phys., 251, 589
Kosugi, T., Matsuzaki, K., Sakao, T., et al. 2007, Solar Phys., 243, 3
Krivova, N. A., Solanki, S. K. 2008, J. Astrophys. Astron., 29, 151
Langangen, Ø., Carlsson, M., Rouppe van der Voort, L., Stein, R. F. 2007, ApJ, 655, 615
Langhans, K., Schmidt, W., Rimmele, T. 2004, A&A, 423, 1147
Lites, B. W., Socas-Navarro, H. 2004, ApJ, 613, 600
Lites, B. W., Leka, K. D., Skumanich, A., Martinez Pillet, V., Shimizu, T. 1996, ApJ, 460, 1019
Lites, B. W., Scharmer, G. B., Berger, T. E., Title, A. M. 2004, Solar Phys., 221, 65
Lites, B. W., Socas-Navarro, H., Kubo, M., et al. 2007, PASJ, 59, 571
Lites, B. W., Kubo, M., Socas-Navarro, H., et al. 2008, ApJ, 672, 1237
Martínez González, M. J., Collados, M., Ruiz Cobo, B., Solanki, S. K. 2007, A&A, 469, L39
Mehltretter, J. P. 1974, Solar Phys., 38, 43
Meunier, N., Solanki, S., Livingston, W. 1998, A&A, 331, 771
Moradi, H., Hanasoge, S. M., Cally, P. S. 2009, ApJ, 690, L72
Muller, R. 1985, Solar Phys., 100, 237
Nagendra, K., Anusha, L., Sampoorna, M., Frisch, H. 2010, In: Magnetic Coupling Between the Interior and the Atmosphere of the Sun, S. Hasan, R. Rutten (eds.), this volume
Orozco Suárez, D., Bellot Rubio, L. R., del Toro Iniesta, J. C., et al. 2007, ApJ, 670, L61
Orozco Suárez, D., Bellot Rubio, L. R., del Toro Iniesta, J. C., Tsuneta, S. 2008, A&A, 481, L33
Rezaei, R., Schlichenmaier, R., Beck, C. A. R., Bruls, J. H. M. J., Schmidt, W. 2007a, A&A, 466, 1131
Rezaei, R., Steiner, O., Wedemeyer-Böhm, S., et al. 2007b, A&A, 476, L33
Rosenthal, C. S., Bogdan, T. J., Carlsson, M., et al. 2002, ApJ, 564, 508
Rouppe van der Voort, L. H. M., Hansteen, V. H., Carlsson, M., et al. 2005, A&A, 435, 327
Rutten, R. J. 1995, In: Helioseismology, ESA-SP, vol. 376, p. 151
Rutten, R. J. 1999, In: Third Advances in Solar Physics Euroconference: Magnetic Fields and Oscillations, B. Schmieder, A. Hofmann, J. Staude (eds.), ASP Conf. Ser., vol. 184, p. 181
Rutten, R. J., Uitenbroek, H. 1991, Solar Phys., 134, 15
Rutten, R. J., Kiselman, D., Rouppe van der Voort, L., Plez, B. 2001, In: Advanced Solar Polarimetry – Theory, Observation, and Instrumentation, M. Sigwarth (ed.), ASP Conf. Ser., vol. 236, p. 445
Sampoorna, M. 2010, In: Magnetic Coupling Between the Interior and the Atmosphere of the Sun, S. Hasan, R. Rutten (eds.), this volume
Sampoorna, M., Nagendra, K. N., Frisch, H., Stenflo, J. O. 2008, A&A, 485, 275
Sánchez Almeida, J., Asensio Ramos, A., Trujillo Bueno, J., Cernicharo, J. 2001, ApJ, 555, 978
Sánchez Almeida, J., Márquez, I., Bonet, J. A., Domínguez Cerdeña, I., Muller, R. 2004, ApJ, 609, L91
Schaffenberger, W., Wedemeyer-Böhm, S., Steiner, O., Freytag, B. 2005, In: Chromospheric and Coronal Magnetic Fields, D. E. Innes, A. Lagg, S. A. Solanki (eds.), ESA-SP, p. 596
Schaffenberger, W., Wedemeyer-Böhm, S., Steiner, O., Freytag, B. 2006, In: Solar MHD: Theory and Observations, J. Leibacher, R. F. Stein, H. Uitenbroek (eds.), ASP Conf. Ser., vol. 354, p. 345

Schüssler, M., Vögler, A. 2008, A&A, 481, L5
Shapiro, A. I., Berdyugina, S. V., Fluri, D. M., Stenflo, J. O. 2007, A&A, 475, 349
Shelyag, S., Schüssler, M., Solanki, S. K., Berdyugina, S. V., Vögler, A. 2004, A&A, 427, 335
Shibata, K. 1983, PASJ, 35, 263
Socas-Navarro, H., Lites, B. W. 2004, ApJ, 616, 587
Spruit, H. C. 1976, Solar Phys., 50, 269
Stein, R. F., Bercik, D., Nordlund, Å. 2003, In: Current Theoretical Models and Future High Resolution Solar Observations: Preparing for ATST, A. A. Pevtsov, H. Uitenbroek (eds.), ASP Conf. Ser., vol. 286, p. 121
Steiner, O. 1991, A&A, 242, 290
Steiner, O. 2005, A&A, 430, 691
Steiner, O., Hauschildt, P. H., Bruls, J. 2001, A&A, 372, L13
Steiner, O., Vigeesh, G., Krieger, L., et al. 2007, Astronomische Nachrichten/AN, 328, 323
Steiner, O., Rezaei, R., Schaffenberger, W., Wedemeyer-Böhm, S. 2008, ApJ, 680, L85
Steiner, O., Rezaei, R., Schlichenmaier, R., et al. 2009, In: Second Hinode Science Meeting: Beyond Discovery – Toward Understanding, M. Cheung, B. Lites, T. Magara, J. Mariska, K. Reeves (eds.), ASP Conf. Ser., vol. 415, p. 67–70
Stenflo, J. O. 1982, Solar Phys., 80, 209
Stenflo, J. O., Keller, C. U., Gandorfer, A. 1998, A&A, 329, 319
Suematsu, Y. 1990, In: Progress of Seismology of the Sun and Stars, Y. Osaki, H. Shibahashi (eds.), Lecture Notes in Physics, Springer, Berlin, vol. 367, p. 211
Thomas, J. H. 1982, ApJ, 262, 760
Tritschler, A., Schmidt, W., Uitenbroek, H., Wedemeyer-Böhm, S. 2007, A&A, 462, 303
Trujillo Bueno, J., Shchukina, N., Asensio Ramos, A. 2004, Nat, 430, 326
Trujillo Bueno, J., Asensio Ramos, A., Shchukina, N. 2006, In: Solar polarization 4, R. Casini, B. W. Lites (eds.), ASP Conf. Ser., vol. 358, p. 269
Vigeesh, G., Hasan, S. S., Steiner, O. 2009, A&A (in press)
Vögler, A., Schüssler, M. 2007, A&A, 465, L43
Wedemeyer, S., Freytag, B., Steffen, M., Ludwig, H.-G., Holweger, H. 2004, A&A, 414, 1121
Wedemeyer-Böhm, S. 2008, A&A, 487, 399
Wedemeyer-Böhm, S., Lagg, A., Nordlund, Å. 2008, Sp. Sci. Rev., 171
Weiss, N. O. 1966, Roy. Soc. Lond. Proc. Ser. A, 293, 310
Wenzler, T., Solanki, S. K., Krivova, N. A. 2005, A&A, 432, 1057
Wiehr, E., Bovelet, B., Hirzberger, J. 2004, A&A, 422, L63
Zhugzhda, Y. D., Dzhalilov, N. S. 1982, A&A, 112, 16
Zhugzhda, Y. D., Dzhalilov, N. S. 1984a, A&A, 132, 45
Zhugzhda, Y. D., Dzhalilov, N. S. 1984b, A&A, 132, 52
Zhugzhda, Y. D., Dzhalilov, N. S. 1984c, A&A, 133, 333

The Evershed Effect with SOT/Hinode

K. Ichimoto and the SOT/Hinode Team

Abstract The Solar Optical Telescope onboard Hinode revealed the fine-scale structure of the Evershed flow and its relation to the filamentary structures of the sunspot penumbra. The Evershed flow is confined in narrow channels with nearly horizontal magnetic fields embedded in a deep layer of the penumbral atmosphere. It is a dynamic phenomenon with flow velocity close to the photospheric sound speed. Individual flow channels are associated with tiny upflows of hot gas (sources) at the inner end and downflows (sinks) at the outer end. SOT/Hinode also discovered "twisting" motions of penumbral filaments, which may be attributed to the convective nature of the Evershed flow. The Evershed effect may be understood as a natural consequence of thermal convection under a strong, inclined magnetic field. Current penumbral models are discussed in the lights of these new Hinode observations.

1 Introduction

Since its discovery by Evershed (1909), the Evershed effect has been one of the longstanding mysteries in solar physics. It is evident from recent high-resolution observations that the flow is closely related to the small-scale filamentary structures in penumbrae (for reviews see Solanki 2003; Thomas et al. 2004, 2008; Bellot Rubio 2007; Thomas 2010). The inclination of the penumbral magnetic field fluctuates in the azimuthal direction with the spatial scale of the penumbral filaments (interlocking comb structure), whereas the flow takes place in radial filaments, which have nearly horizontal magnetic field (Degenhardt and Wiehr 1991; Schmidt et al. 1992; Title et al. 1993). The flow vector is parallel to the magnetic fields (Bellot Rubio et al. 2004; Borrero et al. 2005).

To account for the filamentary structure of penumbrae, the following models have been proposed, that is, the embedded or rising flux tube model (Solanki and Mondavon 1993; Schlichenmaier et al. 1998) in which the Evershed flow channels

K. Ichimoto (✉)
Kwasan and Hida Observatories, Kyoto University, Japan

S.S. Hasan and R.J. Rutten (eds.), *Magnetic Coupling between the Interior and Atmosphere of the Sun*, Astrophysics and Space Science Proceedings, DOI 10.1007/978-3-642-02859-5_14,

are explained as rising flux tubes embedded in more vertical background magnetic fields in the penumbra, the gappy penumbral model (Spruit and Scharmer 2006; Scharmer and Spruit 2006) in which the bright penumbral filaments are regarded as manifestations of the protrusion of nonmagnetized, convecting hot gas into the oblique background fields of the penumbra, and the downward pumping model (Thomas et al. 2002; Weiss et al. 2004; Brummell et al. 2008) in which the penumbral fine structure is created by localized submergence of penumbral fields that are forced by the photospheric convection around the outer edge of penumbrae. There is still no consensus on the physical nature and origin of the penumbral fine structures.

Using highly stable time sequences of sunspot images and high precision spectropolarimetric data provided by the Solar Optical Telescope (SOT; Tsuneta et al. 2007; Suematsu 2007) onboard Hinode (Kosugi et al. 2007), Ichimoto et al. (2007a) confirmed the interlocking-comb penumbral structure and found that the Evershed flow preferentially takes place in bright filaments in the inner penumbra and in dark filaments in the outer penumbra. They also found the presence of a number of small patches with a vertical velocity component, upward motions distributed over the penumbra, and strong downward motions associated with magnetic polarity opposite to that of the sunspot in the mid and outer penumbra. The vertical motions may be regarded as the sources and sinks of the Evershed flow channels, though unequivocal identification of individual pairs was not reached.

Here we report results on the spatial distribution of vertical motions in penumbrae obtained from further analyses of the sunspot data taken by SOT. The characteristics of the Evershed flow are summarized, and the origin of the flow is discussed within the context of the present penumbral models.

2 Elementary Structure of the Evershed Flow

The spatial distribution of the vertical motions in sunspots were studied using SOT/SP data taken on 1 May 2007. The sunspot was located near disk center, with the viewing angle between the line of sight and the normal to the solar surface $\mu = 5.8°$. Stokes profiles of the pair of Fe I lines at 6,302 Å were fitted using the Milne–Eddington model to obtain vector magnetic fields.

Figure 1 shows the magnetic field inclination overlayed with contours showing the vertical motions. The blue contours show blueshift at $-0.8\,\mathrm{km\,s^{-1}}$ in the wing of Stokes-I of Fe I 6,301.5 Å and the red contours show $V/I_c = 0.01$ at Fe I 6,302.5 + 0.365 Å, representing strong downflow regions with opposite magnetic polarity to the spot. V and I_c are Stokes-V and the continuum intensity, respectively. Radial penumbral filaments with dark appearance in the inclination map are the channels with nearly horizontal magnetic field. It is remarkable that the upflow and downflow patches are aligned with the horizontal field channels that carry the Evershed flow, and that small-scale upflows are preferentially located near the inner ends, and downflows at the outer ends of the horizontal field channels. Thus, the upflow and downflow patches may be regarded as the sources and sinks of the Evershed

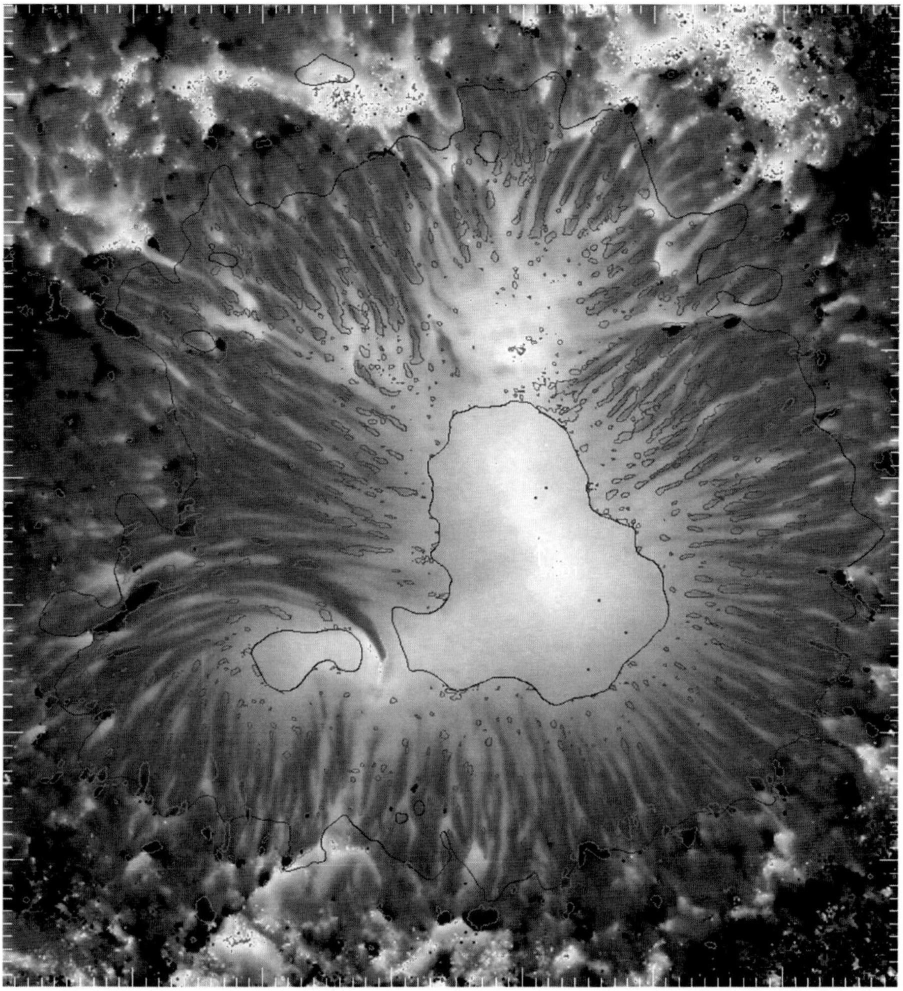

Fig. 1 Inclination of the magnetic field. This sunspot was observed by SP/SOT on 1 May 2007. *Top* is toward the disk center; the heliocentric distance is 5.8°. The *black* contours show the inner and outer boundaries of the penumbra. *Dark* channels in the penumbra represent nearly horizontal fields. Overlaid are contours showing the vertical motions, i.e., *blue* for upflow and *red* for downflow (color illustration are available in the on-line version)

flow elements, embedded in the deep penumbral photosphere. It is noted that the Stokes-V signal is well visible even at Fe I 6,302.5 Å + 0.454 Å (22 km s^{-1}). This suggests the presence of supersonic flows near the outer end of the flow channels (see Ichimoto et al. 2007a).

Figure 2 shows the continuum intensity of the sunspot overlayed with the same contours as in Fig. 1. The spatial correlation between bright grains and upflows is remarkable. It is obvious that the Evershed (up-)flow carries the energy to maintain

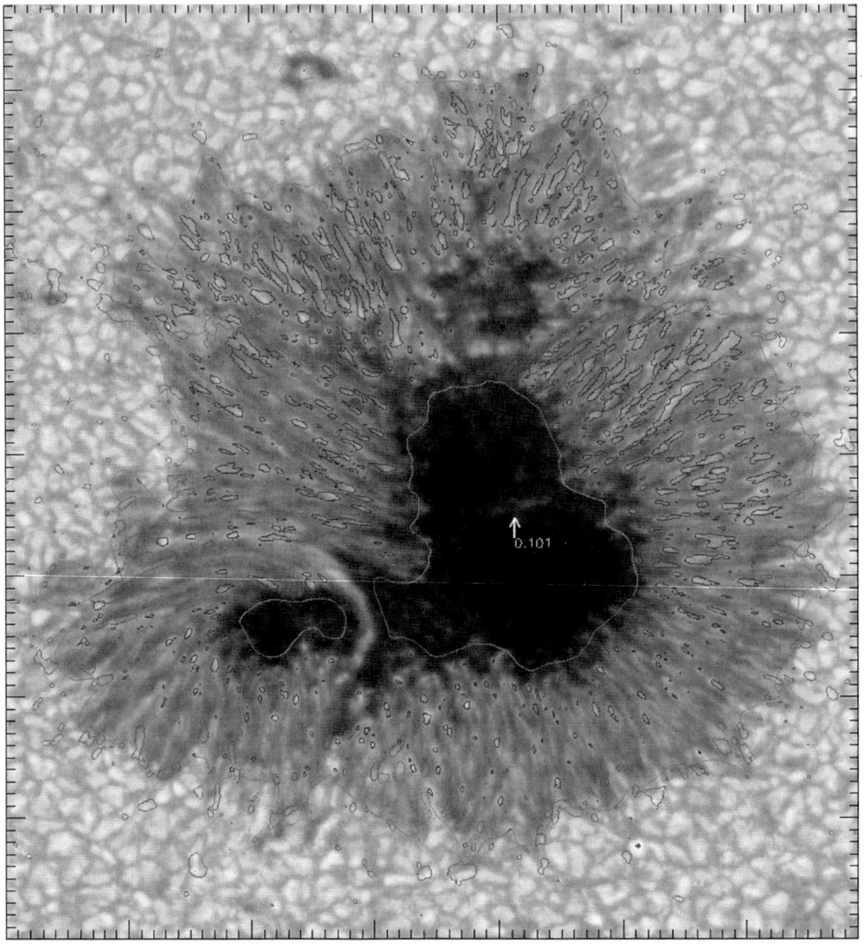

Fig. 2 The same sunspot in continuum intensity near $\lambda = 630$ nm, with the same contours as in Fig. 1

the penumbral brightness, although quantitative evaluation of the energy flux carried by the upflow is not straightforward because the line forming layer is likely well above the vigorously convecting layer.

3 Discussion

Ichimoto et al. (2007b) found a "twisting" motion in the leading part of the bright filaments in penumbrae. As the visibility of this twisting motion depends on the viewing angle, that is, it is seen only when the penumbral filaments are observed from their side, Ichimoto et al. (2007b) interpreted the apparent twisting feature as a manifestation of overturning convection in the source region of the Evershed flow. Zakharov et al. (2008) confirmed the twisting motions of penumbral filaments with ground-based observations and concluded that the twisting features indicate the presence of convective rolls in filaments with a nearly horizontal magnetic field.

From the analysis of the net circular polarization in penumbrae, Ichimoto et al. (2008) found evidence of positive correlation between the Evershed flow velocity and the magnetic field strength, spatially along the line of sight, in the sense that the flow velocity and the magnetic field strength both increase with depth in the flow channels. This supports the conclusion that the flowing gas is strongly magnetized. The Stokes-V Dopplershift or the Dopplershift of polarized spectral-line components in penumbrae serves as more direct evidence that the flowing gas is not field-free.

The properties of the Evershed effect so far obtained are summarized as follows:

- Each Evershed flow channel is associated with upflow (source) and downflow (sink) in its inner and outer ends
- The geometry of the magnetic field is consistent with low-lying arched tubes embedded in more inclined magnetic fields, that is, so-called interlocking-comb penumbral structure
- There is a very good spatial correlation between upflows and bright penumbral grains, which strongly suggests that the Evershed (up-)flow carries the energy for maintaining the penumbral brightness
- The source region of the Evershed flow channels shows hints of overturning convection
- The flow velocity (and the magnetic field strength) increase with depth in the visible layers of the flow channel
- The flow plasma is not field-free but magnetized.

These characteristics obviously suggest a convective origin for the Evershed effect. Let us consider the currently representative penumbral models, that is, the embedded-fluxtube model and the gappy model, in the light of these observational results. There is no observational evidence of the presence of a lower boundary of fluxtubes at least in the inner and middle penumbra. Thus, the picture of narrow fluxtubes in the embedded-fluxtube model is not supported by observation. On the other hand, the flowing gas is highly magnetized, which is an obvious contradiction between the field-free gap model and the observations. If the fluxtube model allows vertically elongated "flux tubes" (or slabs), and if the gap model discards the term "field free," then there is no fundamental difference between the two models. In both models, rising motions of hot gas are driven by buoyancy forces. These new observational results suggest that the Evershed effect can be understood as a natural consequence of thermal convection under strong, inclined magnetic fields.

Recent 3D numerical MHD simulations (Heinemann et al. 2007; Rempel et al. 2008) of sunspots have successfully reproduced the interlocking-comb structure of the magnetic fields in penumbrae with Evershed outflows along filaments with nearly horizontal magnetic field and overturning motions. Scharmer et al.

(2008) argued that the Evershed flow represents the horizontal flow component of overturning convection in penumbrae, based on numerical simulations. The velocity of the Evershed outflow in current numerical simulations is, however, only a few $\mathrm{km\,s^{-1}}$, which is much smaller than the observed one. The mechanism that drives the vigorous observed flow, especially at the outer end of the flow channel, is still to be identified. Siphon flow models (e.g., Thomas and Weiss 1992) predict supersonic flows near the outer footpoint of the flow channel (Montesinos and Thomas 1997), although the theory is based on a stationary model. A more dynamic picture is given by the downward pumping model (Weiss et al. 2004), in which stochastic flows may be driven by the convective collapse caused by the localized submergence of penumbral magnetic fields forced by the photospheric convection around the outer edge of penumbrae. The validity of these various models should be examined by investigating the temporal evolutions of the individual Evershed flow elements in further observations.

Acknowledgment Hinode is a Japanese mission developed and launched by ISAS/JAXA, with NAOJ as domestic partner and NASA (USA) and STFC (UK) as international partners. It is operated by these agencies in co-operation with ESA and NSC (Norway). This work was carried out at the NAOJ Hinode Science Center, which is supported by the Grant-in-Aid for Creative Scientic Research, The Basic Study of SpaceWeather Prediction from MEXT, Japan (Head Investigator: K. Shibata), generous donations from Sun Microsystems, and NAOJ internal funding.

References

Bellot Rubio, L. R. 2007, Highlights Spanish Astrophys. IV, 271
Bellot Rubio, L. R., Balthasar, H., Collados, M. 2004, A&A, 427, 319
Borrero, J. M., Lagg, A., Solanki, S. K., et al. 2005, A&A, 436, 333
Brummell, N. H., Tobias, S. M., Thomas, J. H., et al. 2008, ApJ, 686, 1454
Degenhardt, D., Wiehr, E. 1991, A&A, 252, 821
Evershed, J. 1909, MNRAS, 69, 454
Heinemann, T., Nordlund, A., Scharmer, G. B., et al. 2007, ApJ, 669, 1390
Ichimoto, K., Shine, R. A., Lites, B., et al. 2007a, PASJ, 59, 593
Ichimoto, K., Suematsu, Y., Tsuneta, S., et al. 2007b, Science, 318, 1597
Ichimoto, K., Tsuneta, S., Suematsu, Y., et al. 2008, A&A, 481, L9
Kosugi, T., Matsuzaki, K., Sakao, T., et al. 2007, Sol. Phys., 243, 3
Montesinos, B., Thomas, J. H., 1997, Nat, 390, 485
Rempel, M., Schuessler, M., Knölker, M. 2008, arXiv:0808.3294v2
Scharmer, G. B., Spruit, H. C. 2006, A&A, 460, 605
Scharmer, G. B., Nordlund, A., Heinemann, T., 2008, ApJ, 677, L149
Schlichenmaier, R., Jahn, K., Schmidt, H. U. 1998, A&A, 337, 897
Schmidt, W., Hofmann, A., Balthasar, H., et al. 1992, A&A, 264, L27
Solanki, S. K. 2003, Astron. Astrophys. Rev. 11, 153
Solanki, S. K., Montavon, C. A. P. 1993, A&A, 275, 283
Spruit, H. C., Scharmer, G. B. 2006, A&A, 447, 343
Suematsu, Y., Tsuneta, S., Ichimoto, K., et al. 2008, Sol. Phys., 249, 197
Thomas, J. H. 2010, In: Magnetic Coupling Between the Interior and the Atmosphere of the Sun, S. S. Hasan, R. J. Rutten (eds.), Astrophys. Space Sci. Procs., Springer, Heidelberg, these proceedings

Thomas, J. H., Weiss, N. O., 1992, in Sunspots; Theory and Observations, eds. J. H. Thomas, N. O. Weiss, Kluwer, Dordrecht, 3
Thomas, J. H., Weiss, N. O. 2004, ARAA, 42, 517
Thomas, J. H., Weiss, N. O. 2008, Sunspots and Starspots, Cambridge University Press, Cambridge
Thomas, J. H., Weiss, N. O., Tobias, S. M., et al. 2002, Nat, 420, 390
Title, A., Frank, Z., Shine, R., et al. 1993, ApJ, 403, 780
Tsuneta, S., Ichimoto, K., Katsukawa, Y., et al. 2008, Sol. Phys., 249, 167
Weiss, N. O., Thomas, J. H., Brummell, N. H., et al. 2004, ApJ, 600, 1073
Zakharov, V., Hirzberger, J., Riethmuller, T., et al. 2008, A&A, 488, L17

The Evershed Flow and the Brightness of the Penumbra

L.R. Bellot Rubio

Abstract The Evershed flow is a systematic motion of gas that occurs in the penumbra of all sunspots. Discovered in 1909, it still lacks a satisfactory explanation. We know that the flow is magnetized, often supersonic, and that it shows conspicuous fine structure on spatial scales of $0.2''$–$0.3''$, but its origin remains unclear. The hope is that a good observational understanding of the relation between the flow and the penumbral magnetic field will help us determine its nature. Here I review advances in the characterization of the Evershed flow and sunspot magnetic fields from high-resolution spectroscopic and spectropolarimetric measurements. Using this information as input for 2D heat transfer simulations, it has been demonstrated that hot Evershed upflows along nearly horizontal field lines are capable of explaining one of the most intriguing aspects of sunspots: the surplus brightness of the penumbra relative to the umbra. They also explain the existence of penumbral filaments with dark cores. These results support the idea that the Evershed flow is largely responsible for the transport of energy in the penumbra.

1 Introduction

The Evershed effect was discovered a century ago as a Doppler shift of spectral lines in sunspots away from disk center (Evershed 1909). The shift is to the red in the limb-side penumbra and to the blue in the center-side penumbra, and happens together with strong line asymmetries (e.g., Stellmacher and Wiehr 1980). Already from the very beginning, these signatures were explained in terms of a nearly horizontal outflow of gas – the Evershed flow. The motions that give rise to the Evershed effect represent the most conspicuous dynamical phenomenon observed in sunspots. Advances in instrumentation have allowed us to characterize them with increasing degree of detail, but their origin remains a fundamental problem in sunspot physics.

Another unsolved problem is the brightness of the penumbra, which radiates about 75% of the quiet Sun flux. This is significantly larger than the 20% emitted

L.R. Bellot Rubio (✉)
Instituto de Astrofísica de Andalucía, CSIC, Granada, Spain

S.S. Hasan and R.J. Rutten (eds.), *Magnetic Coupling between the Interior and Atmosphere of the Sun*, Astrophysics and Space Science Proceedings,
DOI 10.1007/978-3-642-02859-5_15,

by the umbra. In recent years, considerable efforts have been made to understand the energy transport in sunspots. Jahn and Schmidt (1994) proposed convection by interchange of magnetic flux tubes as the process that heats the penumbra, and Schlichenmaier et al. (1998) developed the moving tube model to simulate it. This mechanism was later found to be unable to explain the brightness of the penumbra (Solanki and Rüedi 2003; Schlichenmaier and Solanki 2003), but the moving tube model survived as a reasonably good, yet idealized, description of the penumbral magnetic field.

In the photosphere, the penumbra is a conglomerate of nearly horizontal field lines embedded in a stronger and more vertical field, as first proposed by Solanki and Montavon (1993) in their uncombed model. Strong gradients or discontinuities of the atmospheric parameters occur along the line of sight due to the vertical interlacing of different magnetic components. This characteristic organization of the field explains many features of the penumbra, including its filamentary appearance in continuum intensity and polarized light (see Schlichenmaier 2003; Bellot Rubio 2007; Borrero 2009; Schlichenmaier 2009; and Tritschler 2009 for reviews).

The Evershed flow occurs along the more horizontal field lines and is a global phenomenon, so it represents an excellent candidate to heat the penumbra. Based on the radiative cooling times of hot plasma embedded in a non-stratified atmosphere, Schlichenmaier and Solanki (2003) concluded that the Evershed flow would be able to provide the required amount of energy if the penumbral field lines return to the solar surface soon after they emerge into the photosphere. At that time, the existence of such field lines was unclear and many considered this result as a serious difficulty for the otherwise successful uncombed model.

The direct detection of small patches of magnetic fields diving back below the solar surface everywhere in the mid and outer penumbra (Ichimoto et al. 2007a; Sainz Dalda and Bellot Rubio 2008) removes this problem. Also, the radiative cooling times in a stratified atmosphere which is continuously heated by the Evershed flow may be quite different from those estimated by Schlichenmaier and Solanki (2003). With longer cooling times, a single flow channel would be able to heat larger penumbral areas than previously thought. These two considerations call for a reexamination of the Evershed flow as the mechanism responsible for the brightness of the penumbra.

Here, I summarize the properties of the Evershed flow as deduced from high-resolution spectroscopic and spectropolarimetric measurements (Sect. 2). Any serious attempt to model the energy transport in the penumbra has to include this information. Ruiz Cobo and Bellot Rubio (2008) used it to solve the heat transfer equation in a stratified atmosphere consisting of hot Evershed flows along nearly horizontal magnetic flux tubes. The calculations indicate that the tubes would be observed as bright penumbral filaments with a central dark lane. The filaments are heated by the flow over a length of about 3,000 km; together with their surroundings, they emit 50% of the quiet Sun intensity. These results suggest that the surplus brightness of the penumbra is a natural consequence of the Evershed flow (Sect. 3). Another mechanism proposed to explain the brightness of the penumbra is overturning convection (Sect. 4). However, this idea will remain speculative until measurements at 0.1″ demonstrate the existence of convective motions in penumbral filaments.

2 The Evershed Flow at High Spatial Resolution

Most of what we know about the Evershed flow has been learned from high-resolution filtergrams and spectropolarimetric measurements at 1″–2″ (see Tritschler 2009, and references therein). Despite their moderate angular resolution, the latter have resulted in a very detailed characterization of the flow, thanks to its relation with the more inclined fields of the penumbra, which allows inversion techniques to separate the different magnetic atmospheres that coexist in the pixel.

With the advent of instruments capable of reaching 0.3″, most notably the Universal Birefringent Filter at the Dunn Solar Telescope, the TRIPPEL spectrograph at the Swedish Solar Telescope, and the spectropolarimeter of the Solar Optical Telescope aboard Hinode, the results derived from Stokes inversions have been confirmed and often extended in a more direct way. As a consequence we now have a rather good understanding of the Evershed flow. In the next paragraphs, I will discuss the properties of the flow at high spatial resolution and its relation with the magnetic field of the penumbra based on these observations.

The Evershed Flow Occurs in the Dark Cores of Penumbral Filaments

Bright filaments in the inner penumbra consist of a central dark core surrounded by two lateral brightenings (Scharmer et al. 2002; Sütterlin et al. 2004). The distance between the lateral brightenings is typically 100 km, but in some cases it may be as large as 300 km, making these structures easy targets for 50 cm telescopes.

Bellot Rubio et al. (2005) performed spectroscopy of dark-cored penumbral filaments with the Solar Swedish Telescope. They used the TRIPPEL spectrograph to observe the Fe I 557.6 nm and 709.0 nm lines at a resolution of 0.2″. A bisector analysis of the data revealed that the Evershed flow is stronger in the dark cores, although the lateral brightenings also show non-negligible velocities. The velocities were found to increase with depth in the photosphere. The observed variation of the Doppler shift with position in the penumbra and heliocentric angle indicated that the Evershed flow is directed *upward* in the inner penumbra. Also, the Zeeman splitting of the Fe II 614.9 nm line was consistent with slightly weaker fields in the dark cores as compared with the lateral brightenings or the surrounding medium.

The observation that the Evershed flow occurs preferentially in the dark cores of penumbral filaments has been confirmed by Langhans et al. (2005, 2007), Rimmele and Marino (2006), and Rimmele (2008) using high-resolution filtergrams. The Fe I 630.25 nm magnetograms of Langhans et al. (2005) and van Noort and Rouppe van der Voort (2008) show the dark cores with weaker signals than the bright edges, suggesting more inclined fields. The field could also be weaker, but no definite conclusion can be made because of possible saturation effects (in the strong field regime, the amplitude of the circular polarization signal does not increase with the field strength). Interestingly, these measurements do not show negative polarities across the filaments, that is, the field in the dark cores and the lateral brightenings appears to have the same orientation (but see the remarks by Sánchez Almeida 2010). A similar conclusion has been reached by Bellot Rubio et al. (2007) from an analysis of Hinode spectropolarimetric measurements: the inclination of the magnetic field is slightly larger in the dark cores, but no change of polarity occurs with respect to the surroundings (at least at a resolution of 0.3″).

An intriguing feature revealed by high-resolution Dopplergrams (Hirzberger and Kneer 2001; Rouppe van der Voort 2002; Bellot Rubio et al. 2006) and spectropolarimetric measurements (Bellot Rubio et al. 2004; Ichimoto et al. 2008) is that there is no place in the penumbra where the velocities drop to zero, except where they should vanish because of projection effects. For example, in the inner penumbra the stronger flows occur in the bright filaments, but smaller velocities are also seen in between them. The origin of these motions remains unexplained.

The Evershed Flow Returns to the Solar Surface in the Middle Penumbra and Beyond

The large-scale geometry of the flow can be inferred from the azimuthal variation of the line-of-sight (LOS) velocity as a function of radial distance. The results of this analysis indicate that, on average, the flow is slightly inclined upward in the inner penumbra, then becomes horizontal at about 0.8 penumbral radii, and finally returns to the solar surface in the middle and outer penumbra with inclinations of 10°–30° to the horizontal (Schlichenmaier and Schmidt 2000; Bellot Rubio et al. 2003; Tritschler et al. 2004; Bellot Rubio et al. 2006, and references therein). This must be understood as an average behavior; of course, upflows also exist in the outer penumbra, but the downflows dominate the azimuthal average.

On smaller scales, Rimmele and Marino (2006) employed Fe I 557.6 nm filtergrams taken at the Dunn Solar Telescope to demonstrate that the Evershed flow emerges as a hot upflow in bright penumbral grains and quickly becomes horizontal along individual filaments. This result has been confirmed by Ichimoto et al. (2007a), who had the ingenious idea of using Stokes V maps in the far wings of Fe I 630.25 nm to identify strongly blueshifted or redshifted polarization signals. Figure 1 shows two such magnetograms for AR 10973 at −27.7 pm and +27.7 pm from line center, as observed by Hinode on 1 May 2007. The heliocentric angle of the spot was 6°. The signs have been reversed in the red-wing magnetogram so that black indicates the same polarity in the two panels. The structures visible in Fig. 1 represent strongly Doppler-shifted polarization signals. However, the strength of the signal does not inform about the velocity itself, as it also depends on other parameters such as the field inclination or the gas temperature.

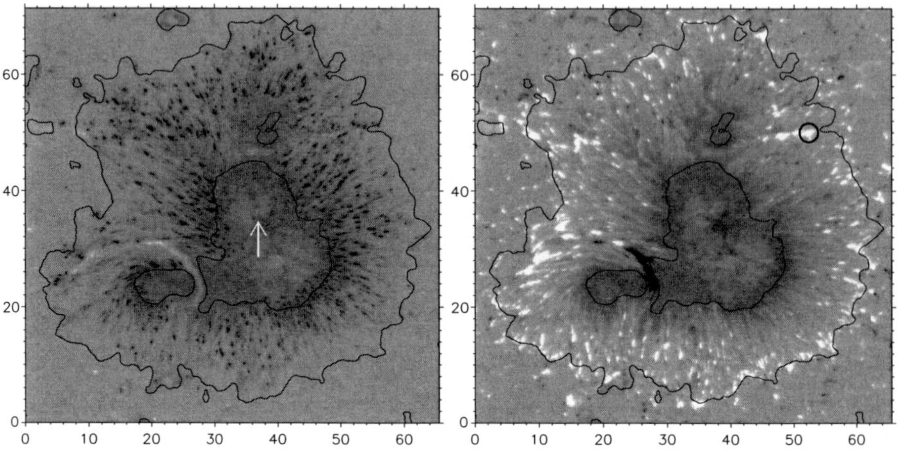

Fig. 1 Stokes *V* maps of AR 10973 at −27.7 pm (*left*) and +27.7 pm (*right*) from the center of the Fe I 630.25 nm line. The data were taken by the Hinode spectropolarimeter on 1 May 2007, with the spot at a heliocentric angle of 6°. *Black* represents the polarity of the spot in the two maps. The *circle* overplotted in the *right* panel marks the downflowing patch considered in Fig. 2. Distances are given in arcsec, and the *arrow* points to disk center

Figure 1 shows with unprecedented clarity the sources and sinks of the Evershed flow (for additional information, see the paper by Ichimoto 2010). The blue wing magnetogram is dominated by small patches of strong upflows located preferentially (but not exclusively) in the inner penumbra. The red wing magnetogram shows similar patches of downflows scattered all over the middle and outer penumbra. Strong downflows occur in the inner penumbra too, but they are less frequent. Westendorp Plaza et al. (1997) were the first to detect downflows in the penumbra, near the sunspot edge. The polarity of the magnetic signals in the downflowing patches is opposite to that of the spot. Thus, not only the flow, but also the magnetic field, returns to the solar surface. Observed at lower spatial resolution, this pattern of upflows/downflows would result in the average radial behavior mentioned above.

The flow field revealed by the Hinode magnetograms is highly organized. Presumably each upflow in the inner penumbra connects with a downflow in the mid or outer penumbra, although this needs to be verified by studying the evolution of the flow. The patches are larger than the angular resolution of the Hinode spectropolarimeter (0.3″), suggesting that they are resolved structures. Moreover, the downflows occur intermittently, not in every pixel as indicated by the inversions of Sánchez Almeida (2005). This is consistent with the idea that the Evershed flow is confined to discrete channels oriented radially; the far-wing magnetograms only show the relatively vertical inner and outer footpoints of these structures. The channels seem to be shorter than the width of the penumbra, although it is likely that their lengths vary as they evolve with time.

The Evershed Flow is Often Supersonic

The Evershed flow attains the largest velocities in the outer penumbra. Spectropolarimetry at 1″ has shown that the azimuthally averaged flow is nearly supersonic from the middle penumbra outward (Bellot Rubio et al. 2004). There have been reports of strongly Doppler-shifted line satellites in the penumbra (Wiehr 1995), and numerical models of the Evershed flow also predict supersonic velocities (Montesinos and Thomas 1997; Schlichenmaier et al. 1998). However, these velocities have not been detected directly until the advent of Hinode.

Figure 2 displays the Stokes spectra observed in one of the downflowing patches of Fig. 1 and a nearby pixel used as a reference (solid and dotted lines, respectively). The spectra represented by the solid lines are remarkable in a number of ways. To start with, the intensity profiles exhibit a very bright continuum and strongly tilted red wings. The tilt is maximum near the continuum, indicating that the flow velocity

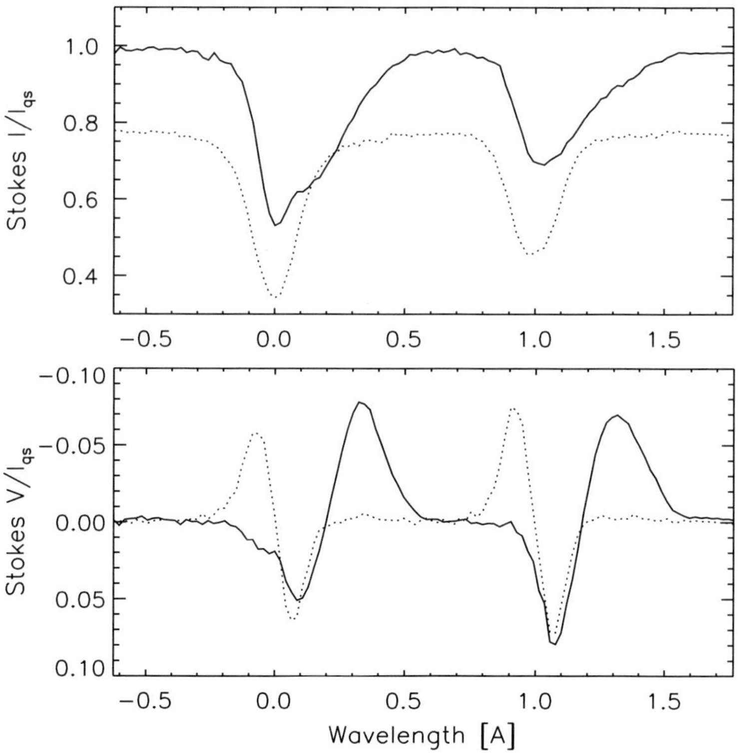

Fig. 2 Stokes I and V profiles observed in one of the downflowing patches of Fig. 1 and a nearby pixel at the same radial distance (*solid* and *dotted lines*, respectively). The exact position of the profiles is indicated in the *right* panel of Fig. 1 with a *circle*. The *dotted lines* represent typical signals in the outer penumbra, with the polarity of the spot and very small flows (the observations were taken close to disk center). The zero of the wavelength scale corresponds to the line-core position of the average quiet sun intensity profile, computed from pixels with Stokes V amplitudes smaller than 0.5% of the continuum intensity

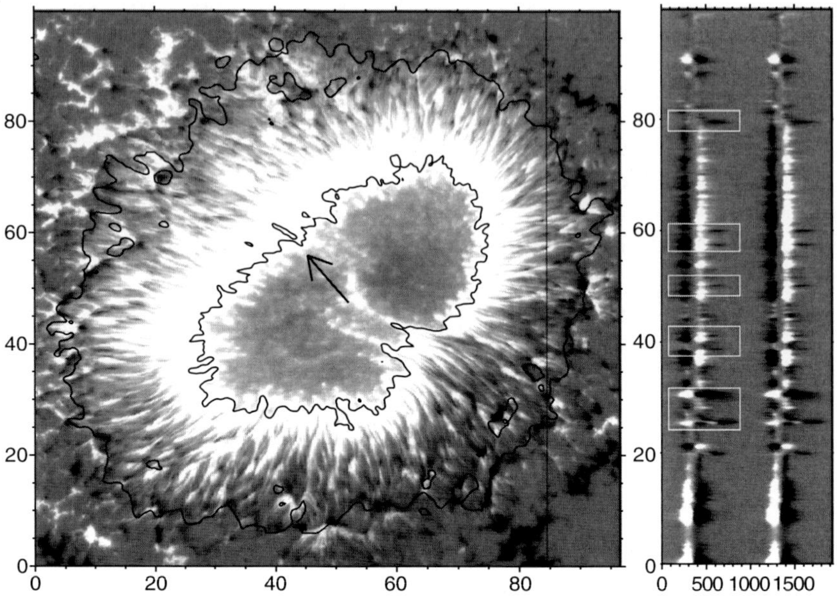

Fig. 3 *Left:* AR 10923 as observed with the Hinode spectropolarimeter on 14 November 2006 between 16:30 and 17:16 UT. The map shows the integral of the Stokes V profile of Fe I 630.25 nm over the *red* lobe. *Black* and *white* represent opposite magnetic polarities. The *arrow* indicates the direction to disk center. The heliocentric angle of the spot was 8°. *Right:* Circular polarization profiles of Fe I 630.15 and 630.25 nm along the slit indicated in the *left* panel. The *rectangles* show examples of pixels with opposite polarities and very strong Doppler shifts, both at the outer penumbral boundary and well within the penumbra (color illustration are available in the on-line version)

increases with depth in the atmosphere. The Stokes V profiles are normal and have two lobes. However, they are redshifted as a whole by 19.5 pm or $\sim$9 km s^{-1}. This LOS velocity is smaller than the true velocity because of projection effects, but it already exceeds the sound speed. The clear signatures of supersonic velocities in the spectra recorded by Hinode are a consequence of the high spatial resolution, which makes it possible to separate the flow channels from their surroundings.

In general, the downflowing patches observed in the middle and outer penumbra exhibit anomalous Stokes V profiles with three lobes (Fig. 3). They can be interpreted as the superposition of two regular signals. One of them has the polarity of the spot and is relatively unshifted, while the other has opposite polarity and shows a very strong redshift. The two signals do not necessarily come from different atmospheres in the pixel; at the resolution of Hinode, it is more likely that they are produced by two magnetic components stacked along the line of sight. We know that this configuration must exist in the penumbra because the observed Stokes V profiles have nonzero area asymmetries. The inversions of Jurčák et al. (2007) also demonstrate that models with strong gradients in the vertical direction are able to reproduce the complex profiles measured by Hinode without any horizontal interlacing of different magnetic components.

The Evershed Flow is Associated with the Weaker and More Inclined Fields of the Penumbra

A well-established observational result is that the Evershed flow happens along the more horizontal fields of the penumbra (see Tritschler 2009 for a review). Indeed, it has been demonstrated that the azimuthally averaged flow is parallel to the magnetic field vector all the way from the inner to the outer penumbra (Bellot Rubio et al. 2003). Near the umbra the inclined fields reside in the bright filaments (e.g., Jurčák et al. 2007), but at larger radial distances they tend to occur in dark structures. It is also accepted that the flow is associated with the weaker fields of the penumbra, except perhaps in the outer penumbra where the strength of the ambient field decreases rapidly (Bellot Rubio et al. 2004; Borrero et al. 2005, 2006; Tritschler et al. 2007; Beck 2008).

For the most part, these results have been obtained from the inversion of spectropolarimetric measurements that do not spatially resolve the different components of the penumbra. Title et al. (1993), Rimmele (1995), Langhans et al. (2005), and others investigated the relation between the flow and the magnetic field using high-resolution magnetograms where penumbral filaments are clearly distinguished, but their results cannot be considered complete because of the lack of linear polarization measurements.

Hinode has allowed us to study the magnetic and dynamic configuration of individual filaments with full Stokes polarimetry and complex atmospheric models. Jurčák et al. (2007), for instance, applied an uncombed model consisting of a single magnetic atmosphere with a flux tube occupying part of the line-forming region. Their inversions provide the variation of the physical parameters with height. Figure 4 shows a vertical cut crossing the inner penumbra of AR 10923 on 10 November 2006. The cut samples four bright penumbral filaments. As can be seen, the Evershed flow occurs deep in the photosphere at the position of the filaments. It is confined to narrow channels having weaker and more inclined fields than the ambient medium. This is the first time that the magnetic and thermodynamic structure of the flow channels is visualized simultaneously in the vertical and horizontal directions with high angular resolution. The results are less clear in the middle and outer penumbra where it becomes difficult to trace individual filaments, but the flow still seems to be associated with the more inclined fields (Jurčák and Bellot Rubio 2008). Similar results have been derived by Borrero et al. (2008) from inversions of Hinode measurements, confirming the overall picture of flow channels embedded in a stronger and more vertical field. As expected, the ambient field wraps around the flow channels, at least in the visible layers of the photosphere (Borrero et al. 2008).

The Evershed Flow is Magnetized

It should be clear by now that the Evershed flow is magnetized, but there are other arguments supporting this conclusion. One of them is the net circular polarization (NCP) observed in the penumbra (e.g., Sánchez Almeida and Lites 1992; Martínez Pillet 2000; Westendorp Plaza et al. 2001; Schlichenmaier and Collados

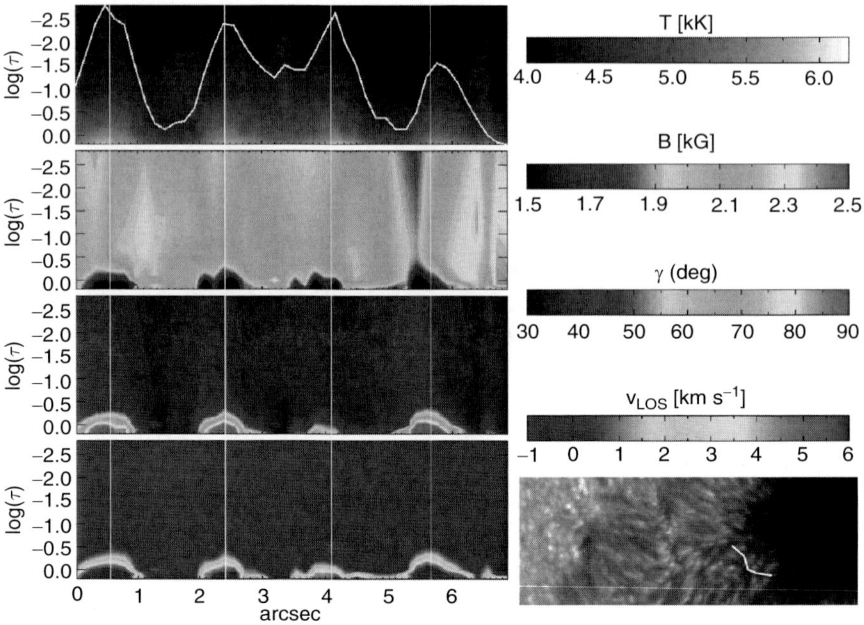

Fig. 4 Vertical stratification of atmospheric parameters along a cut in the inner penumbra of AR 10973 as observed by Hinode on 10 May 2006 at $\theta = 46^\circ$. The cut crosses four bright filaments of the limb-side penumbra (*lower right corner* of the figure). The *left* panels show, from *top* to *bottom*, temperature, field strength, field inclination (in the LOS reference frame), and LOS velocity as a function of optical depth, with *color bars* to the *right*. The *curve* overplotted in the temperature panel represents the continuum intensity along the cut. See Jurčák et al. (2007) for details (color illustration are available in the on-line version)

2002; Müller et al. 2002; Tritschler et al. 2007; Ichimoto et al. 2008). Nonzero NCPs can be produced by flows of field-free plasma, as happens, for instance, in the canopy of network and facular magnetic elements. However, this mechanism creates relatively small NCPs; the large values found in sunspots require the flow to occur in a magnetized atmosphere.

Another argument is the existence of strong Doppler shifts in the polarization profiles emerging from the penumbra. The shifts are induced by the Evershed flow, which has to be magnetized to be able to displace the Stokes signals. Figure 2 shows an extreme example: it is just impossible to explain this kind of profiles with field-free flows, because a magnetic field is required to generate the observed polarization signals through the Zeeman effect. The multi-lobed Stokes V profiles occurring all over the penumbra (and not only near the neutral line, as first pointed out by Bellot Rubio et al. 2002) provide more examples of Doppler shifts induced by flows of magnetized plasma. We have mentioned already that they can be interpreted as the superposition of two regular Stokes V profiles, one of which is displaced in wavelength with respect to the other. Only flows occurring in a magnetized environment can shift the polarization signals to the extent required to generate Stokes V spectra with three or more lobes (see, e.g., Fig. 1 of Bellot Rubio 2006).

Borrero and Solanki (2008) inverted Hinode measurements in terms of model atmospheres featuring vertical gradients of the parameters. Their results show that the field strength of structures associated with strong Evershed flows initially decreases with depth in high photospheric layers, but then increases as the continuum forming layers are approached. This behavior is incompatible with the presence of field-free plasma near or just below $\tau = 1$.

The Evershed Flow Continues Beyond the Sunspot Border as MMFs

Cabrera Solana et al. (2006) have shown that some moving magnetic features (MMFs) in the sunspot moat are generated by Evershed clouds, that is, patches of enhanced Evershed flows that move radially outward in the penumbra. Once in the moat, the Evershed clouds exhibit large velocities for some time. These findings support the idea that MMFs come from inside the spot (Sainz Dalda and Martínez Pillet 2005; Ravindra 2006; Kubo et al. 2007) and suggest that the Evershed flow is not caused by small-scale convection in a strong magnetic field, as it is also observed in the relatively field-free environment of the moat.

3 Heating of the Penumbra by the Evershed Flow

In a recent paper, Ruiz Cobo and Bellot Rubio (2008) investigated the ability of the Evershed flow to explain the brightness of the penumbra. To that end they solved the 2D heat transfer equation in a simple model of a flow channel embedded in a stratified atmosphere (Fig. 5). Consistent with the results described above, the simulated flow is radial, lies deep in the photosphere, and has weaker and more inclined fields than the ambient medium. The flow speed is 7 km s^{-1} at $y = 0$ km, but it becomes supersonic farther out.

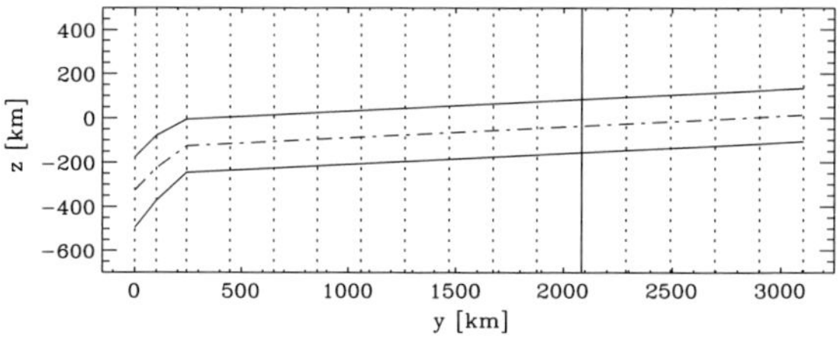

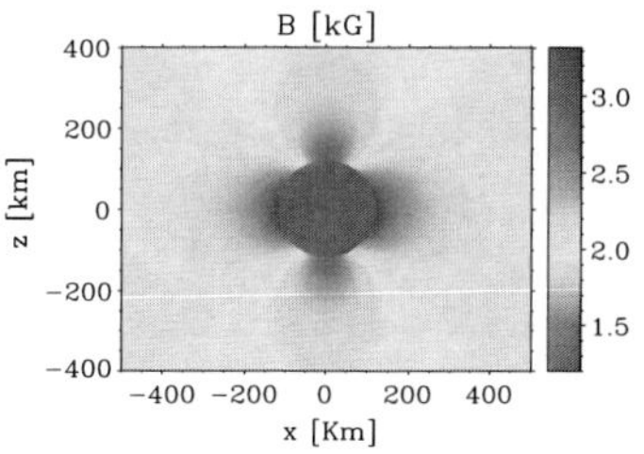

Fig. 5 Magnetic flux tube embedded in a background (umbral-like) atmosphere. The tube has a diameter of 240 km. This configuration is intended to represent bright filaments near the umbra/penumbra boundary. *Left:* Vertical cut in the radial (y) direction. The umbra is to the left and the outer penumbra to the right. The vertical axis indicates height in the photosphere. $z = 0$ km corresponds to $\tau = 1$. *Right:* Cross-section of the tube. Displayed in color is the distribution of the magnetic field strength. The tube has a weaker field than the background atmosphere. However, the ambient field is potential and therefore it shows strong spatial variations near the tube

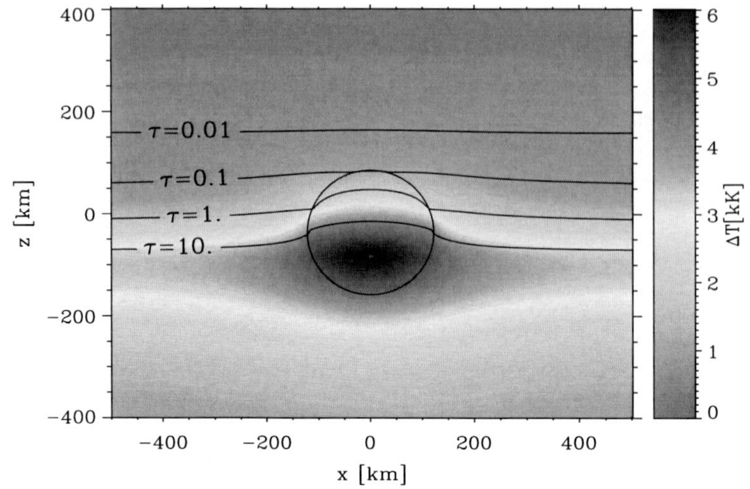

Fig. 6 Temperature perturbations induced by the Evershed flow in the *xz*-plane at $y = 2{,}083$ km (see Fig. 5). The temperature distribution without the flow is that of the cool umbral model of Collados et al. (1994). The *circle* represents the flow channel. The *solid lines* indicate constant Rosseland optical depths

The Evershed flow brings hot plasma from subsurface layers into the photosphere, leading to temperature enhancements inside and outside the flow channel. To determine the equilibrium temperature distribution, the stationary heat transfer equation was solved numerically in 17 vertical planes spanning more than 3,000 km in the radial direction. The temperatures obtained in this way ensure an exact balance between radiative losses and the energy supplied by the Evershed flow. As can be seen in Fig. 6, the flow induces temperature enhancements of up to 6,000 K. In the deepest photospheric layers accessible to the observations, around $\tau = 1$, the temperature of the plasma is increased by nearly 3,000 K.

The physical parameters obtained from the calculations were used to synthesize the Stokes profiles of the Fe I 630.25 nm line emerging from the model. The results are presented in Fig. 7. The three panels display a continuum image, a longitudinal magnetogram, and a Dopplergram. In continuum intensity, the flow channel would be observed as a filament with a bright head and a dark core. The lateral brightenings are separated by about 250 km. This filament is very similar to those observed in the inner penumbra and has a length of 3,000 km. The most important result, however, is that the Evershed flow heats the surroundings very efficiently: the average brightness in the box is approximately 50% of the quiet Sun value. Without the flow, it would be only 6%, corresponding to a very cool umbra. The average brightness is still a bit low, but it should be possible to increase it with small changes in the flow velocity and other simulation parameters.

The magnetogram of Fig. 7 shows weaker circular polarization signal in the dark core as compared with the lateral brightenings. This is exactly what is observed (Langhans et al. 2005, 2007; Bellot Rubio et al. 2007; van Noort and Rouppe van

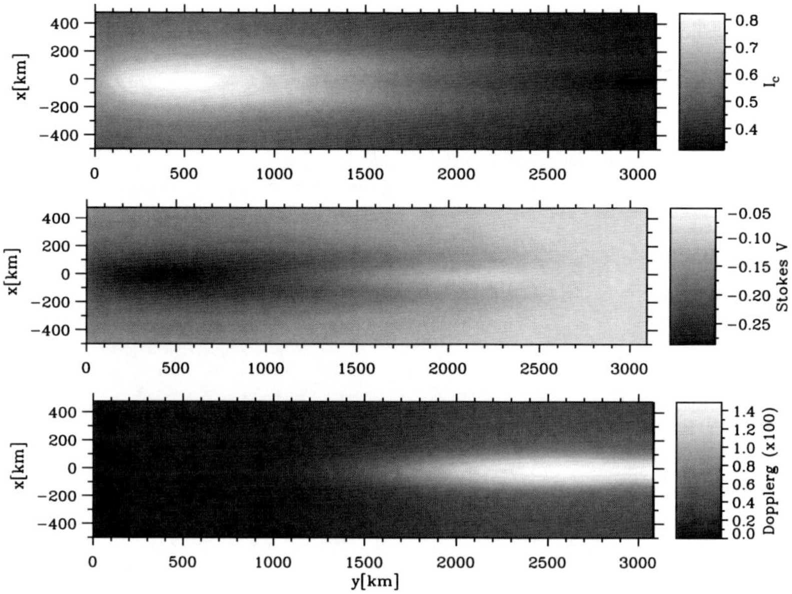

Fig. 7 Magnetic flux tube of Fig. 5 as observed through a 1 m telescope in the Fe I 630.25 nm line. *Top*: Continuum intensity at 630 nm. *Middle:* Stokes V map at +10 pm from line center. *Bottom:* Dopplergram calculated using Stokes I at ±15 pm from line center

der Voort 2008). Moreover, the Dopplergram indicates that the dark core harbors the largest velocities, in agreement with the findings of Bellot Rubio et al. (2005), Langhans et al. (2005), and Rimmele (2008).

A flow channel is capable of heating the atmosphere over a distance of 3,000 km (Fig. 7). Therefore, it does not need to return to the solar surface within the short lengths estimated by Schlichenmaier and Solanki (2003). The fundamental difference here is a much longer cooling time due to the fact that, in the stationary phase, the background atmosphere has been heated by the flow and is already hot. Thus, the temperature gradient between the flow channel and the surroundings is smaller than that considered by Schlichenmaier et al. (1999). This leads to longer cooling times and longer filaments.

According to the calculations of Ruiz Cobo and Bellot Rubio (2008), bright penumbral filaments are the manifestation of deep-lying, radial Evershed flows along nearly horizontal field lines. The dark cores are produced by the weaker fields associated with the flows, which increase the opacity and move the $\tau = 1$ surface to higher layers where the temperature is lower. The lateral brightenings represent the walls of the flow channel and their immediate surroundings. Thus, the dark core and the bright edges trace the same physical entity. This explains why they follow parallel trajectories and move with the same speed. The nonzero velocities observed in the bright edges can also be understood in a natural way: they are produced by the Evershed flow, but appear diluted because of the contribution of the ambient medium, which is essentially at rest. The penumbral filaments are deep-lying structures. The fact that the dark cores show up clearly in intensity measurements taken at the center of Zeeman-sensitive lines (e.g., Scharmer et al. 2008) does not imply that they are formed high in the atmosphere. They may appear dark in line-core images simply because they have weaker field strengths (which decreases the separation between the σ-components of the Zeeman pattern) and larger field inclinations (which increases the amplitude of the central π-component). The two effect work in the same direction, reducing the line-core intensity. The weaker and more inclined fields can be located deep in the photosphere and still be able to modify the line-core intensity due to the large width of the contribution functions.

4 Convection in Penumbral Filaments

Spruit and Scharmer (2006) and Scharmer and Spruit (2006) proposed the idea of field-free gaps protruding into the sunspot from below as an alternative mechanism to heat the penumbra. The gaps would be located beneath the bright penumbral filaments and would sustain overturning convection, carrying energy to the surface and producing the Evershed flow (for a detailed explanation, see the review by Scharmer 2008). Magnetoconvection simulations of sunspots, while still in their infancy, appear to support this picture (Heinemann et al. 2007). The gappy penumbral model is not free from problems, however, and radiative transfer calculations are urgently needed to confront it with spectropolarimetric measurements (Bellot Rubio 2007; Schlichenmaier 2009; Borrero 2009; Thomas 2010).

On the observational side, there have been some claims of the detection of convective flows in the penumbra. Using Hinode filtergrams, Ichimoto et al. (2007b) described a systematic twisting motion of brightness fluctuations in filaments located perpendicularly to the line of symmetry. The motion always occurs from the limb toward the observer, so the twist was interpreted as the manifestation of overturning convection within the filaments. The Doppler measurements of Zakharov et al. (2008) and Rimmele (2008) indicate the existence of weak upflows at the center of penumbral filaments and downflows on their sides. As in normal granular convection, the upflows would turn into downflows after releasing energy in the photosphere. This process of heat transport in the vertical direction would explain the brightness of the penumbra. The observed velocities are of the order of 1 $\mathrm{km\,s^{-1}}$, well below those associated with the Evershed flow. Interestingly, downflows at the periphery of penumbral filaments are also present in the MHD simulations of Rempel et al. (2009), albeit with larger velocities.

The question of whether overturning convection exists in the penumbra is far from settled, however. The best spectroscopic measurements of sunspots reach $0.2''$, but they do not provide evidence for overturning downflows in or around penumbral filaments, neither at disk center nor at heliocentric angles of $\sim$20° (Fig. 2 of Bellot Rubio et al. 2005). This is at odds with the observations of Zakharov et al. (2008) and Rimmele (2008). To reach a definite conclusion we need spectroscopic measurements at very high spatial resolution. With $0.1''$ it should be possible to determine the flow field across penumbral filaments, resolving internal fluctuations smaller than the width of the filaments themselves. Hopefully, this kind of observations will be provided soon by instruments like IMaX aboard SUNRISE or CRISP at the Swedish Solar Telescope.

5 Conclusions

The Evershed flow exhibits conspicuous fine structure at high angular resolution. It occurs preferentially in the dark cores of penumbral filaments, at least in the inner penumbra. The flow is magnetized and often supersonic, as demonstrated by the observation of Stokes V profiles shifted by up to $9\,\mathrm{km\,s^{-1}}$. At each radial distance, the flow is associated with the more inclined fields of the penumbra; in the inner penumbra this happens in the bright filaments, while in the outer penumbra the dark filaments have the largest inclinations. The flow is also associated with weaker fields (except perhaps near the edge of the spot).

High-resolution magnetograms by Hinode show the sources and sinks of the Evershed flow with unprecedented clarity, confirming earlier results from Stokes inversions at lower resolution: on average, the flow points upward in the inner penumbra, then becomes horizontal in the middle penumbra, and finally dives down below the solar surface in the outer penumbra. The Hinode observations reveal tiny patches of upflows concentrated preferentially in the inner penumbra and patches of downflows in the mid and outer penumbra; presumably they correspond to the ends of individual flow channels.

Recent numerical calculations by Ruiz Cobo and Bellot Rubio (2008) have demonstrated that Evershed flows with these properties are capable of heating the penumbra very efficiently, while reproducing many other observational features such as the existence of dark-cored penumbral filaments. This result strongly suggests that the radial Evershed flow is indeed responsible for the brightness of the penumbra.

At the same time, there have been observations of small-scale motions in penumbral filaments that could reflect the existence of overturning convection (Ichimoto et al. 2007b; Zakharov et al. 2008; Rimmele 2008). Convection is an essential ingredient of the field-free gap model proposed by Spruit and Scharmer (2006) and seems to occur also in MHD simulations of sunspots (Rempel et al. 2009). However, other spectroscopic observations at $0.2''$ do not show clear evidence for downflows in filaments near the umbra/penumbra boundary (Bellot Rubio et al. 2005).

It is important to clarify whether or not convection exists in the penumbra. To investigate this issue we need spectroscopic observations at $0.1''$. Narrow lanes of downflows should show up clearly in those measurements. Only then will it be possible to assess the contribution of overturning convection to the brightness of the penumbra and compare it with that of the supersonic Evershed flow. Ultimately, these efforts should reveal the primary mode of energy transport in the penumbra.

One possibility is that the two mechanisms operate at the same time. In fact, the strong vertical gradients of temperature observed within the flow channels (Fig. 6) may drive convective motions with upflows at the center of the filaments and downflows along their sides. The gas would not cross the field lines if the transverse component of the field is similar to that displayed in Fig. 1 of Borrero (2007). In any case, the velocities associated with this process have to be small. The superposition of the radial Evershed flow and overturning flows would result in two helical, outward-twisted motions, one on each half of the filament. The hot material would ascend at the center of the flow channel while being displaced radially outward by the dominant Evershed flow, and would descend along the filament edges after releasing its energy in the photosphere. Even in that case, the brightness of the penumbra would still be due to hot upflows channeled by nearly horizontal field lines.

Acknowledgment I thank Basilio Ruiz Cobo, Rolf Schlichenmaier, and Juan Manuel Borrero for clarifying discussions on the role of the Evershed flow in the heating of the penumbra. Financial support by the Spanish MICINN through projects ESP2006-13030-C06-02 and PCI2006-A7-0624 and by Junta de Andalucía through project P07-TEP-2687 is gratefully acknowledged.

References

Beck, C. 2008, A&A, 480, 825
Bellot Rubio, L. R. 2007, In: Highlights of Spanish Astrophysics IV, F. Figueras, J. M. Girart, M. Hernanz, C. Jordi (eds.), Springer, Dordrecht, p. 271
Bellot Rubio, L. R., Collados, M., Ruiz Cobo, B., Rodríguez Hidalgo, I. 2002, Nuovo Cimento C, 25, 543
Bellot Rubio, L. R., Balthasar, H., Collados, M., Schlichenmaier, R. 2003, A&A, 403, L47
Bellot Rubio, L. R., Balthasar, H., Collados, M. 2004, A&A, 427, 319
Bellot Rubio, L. R., Langhans, K., Schlichenmaier, R. 2005, A&A, 443, L7
Bellot Rubio, L. R. 2006, In: Solar Polarization 4, R. Casini, B.W. Lites (eds.), ASP Conf. Series, vol. 358, p. 107
Bellot Rubio, L. R., Schlichenmaier, R., Tritschler, A. 2006, A&A, 453, 1117
Bellot Rubio, L. R., Tsuneta, S., Ichimoto, K., et al. 2007, ApJ, 668, L91
Borrero, J. M. 2007, A&A, 471, 967
Borrero, J. M. 2009, Sci. China Ser. G, 52 (in press) [arXiv:0810.0080]
Borrero, J. M., Solanki, S. K. 2008, ApJ, 687, 668
Borrero, J. M., Lagg, A., Solanki, S. K., Collados, M. 2005, A&A, 436, 333
Borrero, J. M., Solanki, S. K., Lagg, A., Socas-Navarro, H., Lites, B. 2006, A&A, 450, 383
Borrero, J. M., Lites, B. W., Solanki, S. K. 2008, A&A, 481, L13
Cabrera Solana, D., Bellot Rubio, L. R., Beck, C., del Toro Iniesta, J. C. 2006, ApJ, 649, L41
Collados, M., Martínez Pillet, V., Ruiz Cobo, B., del Toro Iniesta, J. C., Vázquez, M. 1994, A&A, 291, 622
Evershed, J. 1909, MNRAS, 69, 454
Heinemann, T., Nordlund, Å., Scharmer, G. B., Spruit, H. C. 2007, ApJ, 669, 1390
Hirzberger, J., Kneer, F. 2001, A&A, 378, 1078
Ichimoto, K. 2010, In: Magnetic Coupling between the Interior and the Atmosphere of the Sun, S. S. Hasan, R. J. Rutten (eds.), Astrophys. Space Sci. Procs., Springer, Heidelberg, these proceedings [arXiv:0903.2605]
Ichimoto, K., Shine, R. A., Lites, B., et al. 2007a, PASJ, 59, 593

Ichimoto, K., Suematsu, Y., Tsuneta, S., et al. 2007b, Science, 318, 1597
Ichimoto, K., Tsuneta, S., Suematsu, Y., et al. 2008, A&A, 481, L9
Jahn, K., Schmidt, H. U. 1994, A&A, 290, 295
Jurčák, J., Bellot Rubio, L. R. 2008, A&A, 481, L17
Jurčák, J., Bellot Rubio, L., Ichimoto, K., et al. 2007, PASJ, 59, 601
Kubo, M., Shimizu, T., Tsuneta, S. 2007, ApJ, 659, 812
Langhans, K., Scharmer, G. B., Kiselman, D., Löfdahl, M. G., Berger, T. E. 2005, A&A, 436, 1087
Langhans, K., Scharmer, G. B., Kiselman, D., Löfdahl, M. G. 2007, A&A, 464, 763
Martínez Pillet, V. 2000, A&A, 361, 734
Montesinos, B., Thomas, J. H. 1997, Nature, 390, 485
Müller, D. A. N., Schlichenmaier, R., Steiner, O., Stix, M. 2002, A&A, 393, 305
Ravindra, B. 2006, Solar Phys., 237, 297
Rempel, M., Schüssler, M., Knölker, M. 2009, ApJ, 691, 640
Rimmele, T. 1995, A&A, 298, 260
Rimmele, T. 2008, ApJ, 672, 684
Rimmele, T., Marino, J. 2006, ApJ, 646, 593
Rouppe van der Voort, L. H. M. 2002, A&A, 389, 1020
Ruiz Cobo, B., Bellot Rubio, L. R. 2008, A&A, 488, 749
Sainz Dalda, A., Bellot Rubio, L. R. 2008, A&A, 481, L21
Sainz Dalda, A., Martínez Pillet, V. 2005, ApJ, 632, 1176
Sánchez Almeida, J. 2005, ApJ, 622, 1292
Sánchez Almeida, J. 2010, In: Magnetic Coupling Between the Interior and the Atmosphere of the Sun, S. S. Hasan, R. J. Rutten (eds.), Astrophys. Space Sci. Procs., Springer, Heidelberg, these proceedings [arXiv:0902.4532]
Sánchez Almeida, J. Lites, B. W. 1992, ApJ, 398, 359
Scharmer, G. B. 2008, Physica Scripta Volume T, 133, 014015
Scharmer, G. B., Spruit, H. C. 2006, A&A, 460, 605
Scharmer, G. B., Gudiksen, B. V., Kiselman, D., Löfdahl, M. G., Rouppe van der Voort, L. H. M. 2002, Nat, 420, 151
Scharmer, G. B., Narayan, H., Hillberg, T., et al. 2008, ApJ, 689, L69
Schlichenmaier, R. 2003, In: Current Theoretical Models and Future High Resolution Solar Observations: Preparing for ATST, A. A. Pevtsov, H. Uitenbroek (eds.), ASP Conf. Series, vol. 286, p. 211
Schlichenmaier, R. 2009, Space Science Reviews, 144, 213
Schlichenmaier, R., Collados, M. 2002, A&A, 381, 668
Schlichenmaier, R., Schmidt, W. 2000, A&A, 358, 1122
Schlichenmaier, R., Solanki, S. K. 2003, A&A, 411, 257
Schlichenmaier, R., Jahn, K., Schmidt, H. U. 1998, A&A, 337, 897
Schlichenmaier, R., Bruls, J. H. M. J., Schüssler, M. 1999, A&A, 349, 961
Solanki, S. K., Montavon, C. A. P. 1993, A&A, 275, 283
Solanki, S. K., Rüedi, I. 2003, A&A, 411, 249
Spruit, H. C., Scharmer, G. B. 2006, A&A, 447, 343
Stellmacher, G., Wiehr, E. 1980, A&A, 82, 157
Sütterlin, P., Bellot Rubio, L. R., Schlichenmaier, R. 2004, A&A, 424, 1049
Thomas, J. 2010, In: Magnetic Coupling between the Interior and the Atmosphere of the Sun, S. S. Hasan, R. J. Rutten (eds.), Astrophys. Space Sci. Procs., Springer, Heidelberg, these proceedings [arXiv:0903.4106]
Title, A. M., Frank, Z. A., Shine, R. A., et al. 1993, ApJ, 403, 780
Tritschler, A. 2009, In: Proceedings of the Second Hinode Science Meeting, M. Cheung et al. (eds.), ASP Conf. Series (in press) [arXiv: 0903.1300]
Tritschler, A., Schlichenmaier, R., Bellot Rubio, L. R. 2004, A&A, 415, 717
Tritschler, A., Müller, D. A. N., Schlichenmaier, R., Hagenaar, H. J. 2007, ApJ, 671, L85
van Noort, M. J., Rouppe van der Voort, L. H. M. 2008, A&A, 489, 429

Westendorp Plaza, C., et al. 1997, Nat, 389, 47
Westendorp Plaza, C., del Toro Iniesta, J. C., Ruiz Cobo, B., Martínez Pillet, V., Lites, B. W., Skumanich, A. 2001, ApJ, 547, 1130
Wiehr, E. 1995, A&A, 298, L17
Zakharov, V., Hirzberger, J., Riethmüller, T. L., Solanki, S. K., Kobel, P. 2008, A&A, 488, L17

A Topology for the Penumbral Magnetic Fields

J. Sánchez Almeida

Abstract We describe a scenario for the topology of the magnetic field in penumbrae that accounts for recent observations showing upflows, downflows, and reverse magnetic polarities. According to our conjecture, short narrow magnetic loops fill the penumbral photosphere. Flows along these arched field lines are responsible for both the Evershed effect and the convective transport. This scenario seems to be qualitatively consistent with most existing observations, including the dark cores in penumbral filaments reported by Scharmer et al. Each bright filament with dark core would be a system of two paired convective rolls with the dark core tracing the common lane where the plasma sinks down. The magnetic loops would have a hot footpoint in one of the bright filament and a cold footpoint in the dark core. The scenario fits in most of our theoretical prejudices (siphon flows along field lines, presence of overturning convection, drag of field lines by downdrafts, etc). If the conjecture turns out to be correct, the mild upward and downward velocities observed in penumbrae must increase upon improving the resolution. This and other observational tests to support or disprove the scenario are put forward.

1 Introduction

We are celebrating the centenary of the discovery by John Evershed (1909) of the effect now bearing his name. Photospheric spectral lines in sunspots are systematically shifted toward the red in the limb-side penumbra, and toward the blue in the center-side penumbra. A 100 years have passed and, despite the remarkably large number of works on the Evershed effect,[1] we still ignore how and why these line shifts are produced (see, e.g., the review paper by Thomas and Weiss 2004). Thus, the Evershed effect is among the oldest unsolved problems in astronomy. Although its study has never disappeared from the specialized literature, the Evershed effect has

J. Sánchez Almeida (✉)
Instituto de Astrofísica de Canarias, La Laguna, Tenerife, Spain

[1] The NASA Astrophysics Data System provides more than 1,400 papers under the keyword *penumbra*, 70 of them published during the last year.

S.S. Hasan and R.J. Rutten (eds.), *Magnetic Coupling between the Interior and Atmosphere of the Sun*, Astrophysics and Space Science Proceedings,
DOI 10.1007/978-3-642-02859-5_16, © Springer-Verlag Berlin Heidelberg 2010

undergone a recent revival triggered by the advent of new instrumentation (Scharmer et al. 2002; Kosugi et al. 2007), original theoretical ideas (Weiss et al. 2004; Spruit and Scharmer 2006), as well as realistic numerical simulations (Heinemann et al. 2007; Rempel et al. 2009). Unfortunately, this renewed interest has not come together with a renewal of the diagnostic techniques, that is, the methods and procedures that allow us to infer physical properties from observed images and spectra. Often implicitly, the observers assume the physical properties to be constant in the resolution element, a working hypothesis clearly at odds with the observations. Spectral line asymmetries show up even with our best spatial resolution (Ichimoto et al. 2007a; Sánchez Almeida et al. 2007, Sect. 2). This lack of enough resolution is not secondary. The nature of the Evershed flow has remained elusive so far because we have been unable to isolate and identify the physical processes responsible for the line shifts. Different measurements provide different ill-defined averages of the same unresolved underlaying structure, thus preventing simple interpretations and yielding the problems of consistency that plague the Evershed literature (e.g., non-parallelism between magnetic field lines and flows, Arena et al. 1990; violation of the conservation of magnetic flux, Sánchez Almeida 1998; non-parallelism between continuum filaments and magnetic field lines, Kàlmàn 1991).

Understanding the observed spectral line asymmetries complicates our analysis but, in reward, the asymmetries provide a unique diagnostic tool. They arise from sub-pixel variations of the magnetic fields and flows; therefore, by modeling and interpretation of asymmetries, one can get a handle on the unresolved structure. Although indirectly, such modeling allows us to surpass the limitations imposed by the finite resolution. The idea has tradition in penumbral research, starting from the discovery of the asymmetries almost 50 years ago (e.g., Bumba 1960; Grigorjev and Katz 1972). Sánchez Almeida (2005, hereinafter SA05) exploits the tool in a systematic study that encompasses a full round sunspot. The unresolved components found by SA05 inspire the topology for the penumbral magnetic fields proposed here. According to SA05, the asymmetries of the Stokes profiles[2] can be *quantitatively* explained if magnetic fields having a polarity opposite to the sunspot main polarity are common throughout the penumbra. The reverse polarity holds intense magnetic field aligned flows which, consequently, are directed downward. Counter-intuitive as it may be, the presence of such ubiquitous strongly redshifted reverse polarity has been directly observed with the satellite HINODE (Ichimoto et al. 2007a). This new finding supports the original SA05 results, providing credibility to the constraints that they impose on the magnetic fields and mass flows. The existence of such ubiquitous return of magnetic flux, together with a number of selected results from the literature, are assembled here to offer a plausible scenario for the penumbral magnetic field topology. Such exercise to piece together and synthesize information from different sources is confessedly speculative. It will not lead to

[2] We use Stokes parameters to characterize the polarization; I for the intensity, Q and U for the two independent types of linear polarization, and V for the circular polarization. The Stokes profiles are representations of I, Q, U, and V vs. wavelength for a particular spectral line. They follow well defined symmetries when the atmosphere has constant magnetic field and velocity (see, e.g., Sánchez Almeida et al. 1996).

a self-consistent MHD model for the penumbral structure. However, the exercise is worthwhile for a number of reasons. First, the compilation of observational results and theoretical arguments in Sect. 2 provides a unique brief-yet-restrictive list of constraints to be satisfied by any explanation of the Evershed effect deserving such name. This reference compilation will be useful even if the proposed magnetic topology turns out to be incorrect. Second, we will show how the presence of a reverse polarity is qualitatively consistent with all existing observations, including unobvious cases. Third, despite the amount of observational and theoretical papers on penumbrae, our understanding of the Evershed phenomenon is far from satisfactory. The solution to the riddle cannot be trivial, and it may require interpreting the observations with an alternative twist. Unconventional proposals are needed, and here we explore some of the possibilities. Finally, the proposal may help inspiring MHD numerical modelers.

This paper is structured as follows: it begins by summarizing a number of observations to constrain the topology of the magnetic field and velocity in penumbrae (Sect. 2). These observational results leave an important loose end. The reverse polarity found by Ichimoto et al. (2007a) does not show up in high spatial resolution observations with the Swedish Solar Telescope (SST; Langhans et al. 2005, 2007). Section 3 explains how the apparent inconsistency goes away if the dark cores in penumbral filaments found by Scharmer et al. (2002) correspond to the strongly redshifted reverse polarity regions spotted by HINODE. Once the difficulty has been cleared up, the actual scenario for the magnetic field topology is put forward in Sect. 4. Its predictions are qualitatively compared with observations in Sect. 5. Similarities and differences between this scenario and existing models for the penumbral structure and the Evershed flow are analyzed in Sect. 6, where we also suggest observational tests to confirm or disprove our conjecture.

2 Constraints on the Penumbral Structure

The bibliography on the magnetic structure of penumbrae is too extensive to be condensed in a few pages (see footnote No. 1). We refrain from giving an overview, and those readers interested in a more formal review should refer to, for example, Schmidt (1991); Thomas and Weiss (1992); Solanki (2003); Bellot Rubio (2004); or Thomas and Weiss (2004). Only the selected references that provide a framework for our proposal are introduced and discussed here. They include SA05 and Ichimoto et al. (2007a). Most of them are pure observational results, but several items involve theoretical arguments as well. The selection is obviously biased in the sense that some observations often bypassed are emphasized here, and vice versa. However, to the best of our knowledge, no potentially important constraint has been excluded.

1. The best penumbral images have a resolution of the order $0''.12$ (or 90 km on the Sun). They show many features at the resolution limit implying the existence of unresolved substructure. For example, the power spectrum of the penumbral images has signal down to the instrumental cutoff (Rouppe van der

Voort et al. 2004), and the width of the narrower penumbral filaments is set by the resolution of the observation (Scharmer et al. 2002; see also Fig. 1). This interpretation of the current observations should not be misunderstood. The penumbrae have structures of all sizes starting with the penumbra as a whole. However, the observations show that much of its observed structure is at the resolution set by the present technical limitations and, therefore, it is expected to be unresolved. This impression is corroborated by the presence of spectral line asymmetries as discussed in item 11.

2. The best penumbral images show *dark cores in penumbral filaments* (Scharmer et al. 2002). We prefer to describe them as dark filaments outlined by bright plasma. This description also provides a fair account of the actual observation (Fig. 1), but it emphasizes the role of the dark core. Actually, dark cores without a bright side are common, and the cores seldom emanate from a bright point (Fig. 1).

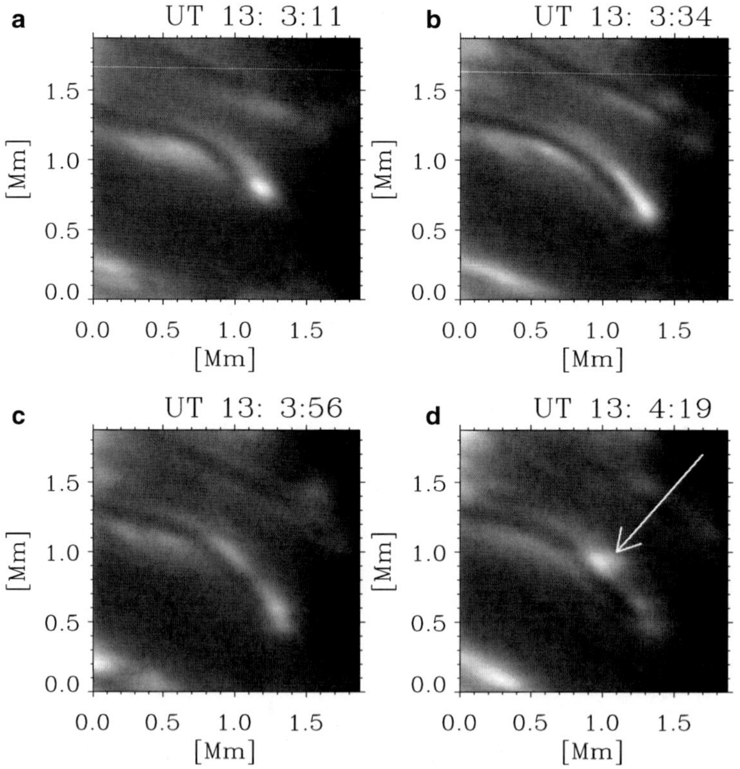

Fig. 1 Time evolution of one of the dark cores in penumbral filaments discovered by Scharmer et al. (2002). (The UT of observation is marked on *top* of each snapshot.) Note that one of the bright sides is partly missing in (**c**) and (**d**). Note also that the bright points are not on the dark filament but in a side. These two properties are common. The *arrow* indicates the emergence of a new bright point in a side of the preexisting dark filament. Note the narrowness of the bright filaments, and their large aspect ratio (length over width). The spatial scales are in Mm, and the angular resolution of the image is of the order of 0.09 Mm

The widths of the dark core and its bright boundaries remain unresolved, although the set formed by a dark core sandwiched between two bright filaments spans some 150–180 km across.

3. There is a local correlation between penumbral brightness and Doppler shift, so that bright features are blueshifted with respect to dark features (Beckers and Schröter 1969; Sánchez Almeida et al. 1993, 2007; Johannesson 1993; Schmidt and Schlichenmaier 2000). The correlation maintains the same sign in the limb-side penumbra and the center-side penumbra, a property invoked by Beckers and Schröter (1969) to conclude that it is produced by vertical motions. A positive correlation between vertical velocity and intensity is characteristic of the nonmagnetic granulation. The fact that the same correlation also exists in penumbrae suggests a common origin for the two phenomena, namely, convection.
4. The limb-side and center-side parts of a penumbra are slightly darker than the rest, an observational fact indicating that the bright penumbral filaments are elevated with respect to the dark ones (Schmidt and Fritz 2004). The behavior seems to continue down to the smallest structures. Dark cores are best seen where the low resolution penumbra is darkest according to Schmidt and Fritz (2004), that is, along the center-to-limb direction (e.g., Langhans et al. 2007; Ichimoto et al. 2007b). The two observations are probably connected, suggesting that dark cores are depressed with respect to their bright sides.
5. There is a local correlation between magnetic field inclination and horizontal velocity. The largest velocities are associated with the more horizontal fields (e.g., Title et al. 1993; Stanchfield et al. 1997).
6. The large horizontal motions occur in the dark penumbral filaments (e.g., Rüedi et al. 1999; Penn et al. 2003; Sánchez Almeida et al. 2007). This trend continues down to the dark cores in penumbral filaments (Langhans et al. 2005, 2007).
7. The observations on the correlation between magnetic field strength and brightness are contradictory. Some authors find the strongest field strengths associated with the darkest regions, and vice versa (c.f. Beckers and Schröter 1969; Hofmann et al. 1994). What seems to be clear is the reduced circular polarization signal existing in dark cores, which is commonly interpreted as a reduced field strength (Langhans et al. 2005, 2007). We show in Sect. 3 that such dimming of the circular polarization admits a totally different interpretation, consistent with an increase of field strength in dark cores.
8. Theoretical arguments indicate that the convective roll pattern should be the mode of convection for nearly horizontal magnetic fields (Danielson 1961; Hurlburt et al. 2000). The rolls have their axes along the magnetic field lines. Unfortunately, this is not what results from recent numerical simulations of magneto-convection in strong highly inclined magnetic fields (Heinemann et al. 2007; Rempel et al. 2009). Here the convection takes place as field-free plasma intrusions in a strong field background, resembling the gappy penumbra model by Spruit and Scharmer (2006). However, these numerical simulations may not be realistic enough. They are the first to come in a series trying to reduce the artificial diffusivities employed by the numerical schemes. It is unclear

whether such evolution will maintain the modes of convection. Moreover, even the present simulations hint at the existence of a convective roll pattern.

9. The stable sunspots are surrounded by a large annular convection cell called moat (Sheeley 1972). The moat presents an outflow which, contrarily to the commonly held opinion, has both radial and tangential velocities (Title 2003, private communication; Bonet et al. 2005). The tangential component is well organized so that it sweeps the plasma toward radial channels, creating a velocity pattern that resembles the convective rolls by Danielson (1961). Compare Fig. 7 in Bonet et al. (2005) or Fig. 11 in Bovelet and Wiehr (2003) with the rolls in Fig. 9a of García de la Rosa (1987).
10. The magnetic field lines can be dragged by the downdrafts of the granulation. Termed as *flux pumping* mechanism, this drag is modeled and studied by Weiss et al. (2004) and Brummell et al. (2008) to show that the vigorous sinking plumes of the granulation and mesogranulation easily pumps down magnetic fluxtubes outside the penumbra. It is conceivable that the same pumping by sinking plasma also operates within the magnetized penumbra.
11. The polarization of the spectral lines emerging from any sunspot has asymmetries and, consequently, it requires several unresolved velocities and magnetic fields to be produced (Bumba 1960; Grigorjev and Katz 1972; Golovko 1974; Sánchez Almeida and Lites 1992). The asymmetries show up even with the best spatial resolution achieved nowadays (150–250 km, Sánchez Almeida et al. 2007; Ichimoto et al. 2007a). Furthermore, part of the substructure producing asymmetries will never be resolved directly because it occurs along the line-of-sight (LOS). The spectral lines create net circular polarization (NCP), that is, the asymmetry of the Stokes V profiles is such that the integrated circular polarization of any typical spectral line is not zero. NCP can only be produced by gradients along the LOS and, therefore, within a range of heights smaller than the region where the lines are formed (say, 100 km). The NCP follows several general rules found by Illing et al. (1974a,b) and Makita (1986). Explaining them requires large gradients of magnetic field inclination along the LOS (Sánchez Almeida and Lites 1992). The gradients can be produced by the discontinuities occurring at the boundary of relatively large magnetic fluxtubes embedded in a background (Solanki and Montavon 1993), or as the cumulative effect of several smaller optically thin structures with varied magnetic field inclinations (Sánchez Almeida et al. 1996).
12. The sunspots seem to show upward motions in the inner penumbra and downward motions in the outer penumbra (e.g., Rimmele 1995; Schlichenmaier and Schmidt 2000; Bellot Rubio et al. 2003; Tritschler et al. 2004). However, this velocity pattern is inferred assuming uniform velocities in the resolution elements, an hypothesis inconsistent with observations (items 1 and 11).
13. Using a MISMA[3] framework, SA05 carries out a systematic fit of all Stokes profiles observed with $1''$-resolution in a medium sized sunspot. Line

[3] The acronym MISMA stands for MIcro-Structured Magnetic Atmosphere, and it was coined by Sánchez Almeida et al. (1996) to describe magnetic atmospheres having optically thin substructure.

asymmetries and NCP are reproduced (item 11). The resulting semi-empirical model sunspot provides both the large scale magnetic structure, as well as the small scale properties of the micro-structure. On top of a regular large scale behavior, the inferred small scale structure of the magnetic fields and flows is novel and unexpected. Some 30% of the volume is occupied by magnetic field lines that return to the sub-photosphere within the penumbral boundary. Mass flows are aligned with magnetic field lines; therefore, the field lines with the main sunspot polarity transport mass upward, while the reverse polarity is associated with high speed flows returning to the solar interior. This return of magnetic flux and mass toward the solar interior occurs throughout the penumbra, as opposed to previous claims of bending over and return at the penumbral border or beyond (item 12). The observed magnetic field strength difference between field lines pointing up and down can drive a siphon flow with the magnitude and sense of the Evershed flow. Within observational uncertainties, the mass transported upward is identical to the mass going downward.

14. The bright penumbral filaments are too long to trace individual streams of hot plasma. The original argument dates back to Danielson (1960), but here we recreate a recent account by Schlichenmaier et al. (1999). They estimate the length of a bright filament produced by hot plasma flowing along a magnetic fluxtube. The plasma cools down as it radiates away and so, eventually, the fluxtube becomes dark and transparent. An isolated loop would have a bright head whose length l is approximately set by the cooling time of the emerging plasma t_c times the velocity of the mass flow along the field lines U,

$$l \approx U t_c. \tag{1}$$

The cooling time depends on the diameter of the tube d, so that the thinner the tube the faster the cooling. For reasonable values of the Evershed flow speed ($U \approx 5\,\mathrm{km\,s^{-1}}$), and using the cooling time worked out by Schlichenmaier et al. (1999), the aspect ratio of the hot footpoint turns out to be of the order of one for a wide range of fluxtube diameters, that is,

$$l/d \approx 0.8\,(d/200\,\mathrm{km})^{0.5}. \tag{2}$$

Filaments must have $l/d >> 1$, and so, a hot plasma stream will show up as a bright knot rather than as a filament. In other words, the cooling of hot plasma moving along field lines cannot give rise to the kind of observed filaments (see Fig. 1). If arrays of hot plasma streams form the filaments, they must be arranged with their hot and cold footpoints aligned to give rise to the observed structures.

15. HINODE magnetograms of penumbrae obtained in the far wings of Fe I λ6302.5 Å show a redshifted magnetic component with a polarity opposite to the main sunspot polarity (Fig. 4 in Ichimoto et al. 2007a). The patches of opposite polarity are scattered throughout the penumbra. In addition, this reverse polarity is associated with extremely asymmetric Stokes V profiles

having three lobes (Ichimoto et al. 2007a, Fig. 5, and Ichimoto 2008, private communication). These asymmetric profiles are known in the traditional literature as cross-over effect, a term used along the paper.

3 Model MISMA for Penumbral Filaments with Dark Core

The set of constraints summarized in Sect. 2 leave an important loose end. The reverse polarity predicted by SA05 (item 13) and found by Ichimoto et al. (2007a, item 15) does not show up in SST magnetograms (Langhans et al. 2005, 2007, and Sect. 1). This fact poses a serious problem as SST has twice HINODE spatial resolution and, therefore, it should be simpler for SST to detect mixed polarities. Here we show how HINODE and SST observations can be naturally understood within the two component model MISMA by SA05, provided that the dark cores are associated with an enhanced contribution of the reverse polarity.

According to SA05, the component that contains most of the mass in each resolution element has the polarity of the sunspot and bears mild magnetic-field-aligned mass flows. It is usually combined with a minor component of opposite polarity and having large velocities. In $1''$-resolution observations, the outcoming light is systematically dominated by the major component, and the resulting Stokes profiles have rather regular shapes. An exception occurs in the so-called *apparent neutral line*, where the cross-over effect was discovered (see Sánchez Almeida and Lites 1992, and references therein). At the neutral line, the mean magnetic field vector is perpendicular to the LOS, and the contribution of the major component almost disappears due to projection effects. The improvement of angular resolution augments the chances of finding the minor component in excess, allowing the opposite polarity to show up. To illustrate the effect, several randomly chosen model MISMAs in SA05 were modified by increasing the fraction of atmosphere occupied by the minor component. One example is shown in Fig. 2. The resulting Stokes I, Q, and V profiles of Fe I $\lambda 6302.5$ Å are represented as solid lines in Fig. 2a–c, respectively. They model a point in the limb-side penumbra at $\mu = 0.95$ ($18°$ heliocentric angle). Note how Stokes I is redshifted and deformed, and how Stokes V shows the cross-over effect. Consequently, the improvement of spatial resolution with respect to traditional earth-based spectro-polarimetric observations naturally explains the abundance of cross-over Stokes V profiles found by HINODE. Figure 2a–c also show the case where the major component dominates (the dashed line). The strong asymmetries have disappeared, rendering Stokes V with antisymmetric shape and the sign of the dominant polarity. The magnetic field vector and the flows are identical in the two sets of Stokes profiles (Figs. 2e, f), although, we have globally decreased the temperature of the atmosphere giving rise to the asymmetric profiles to mimic dark features (see the Stokes I continua in Fig. 2a).

Understanding HINODE observations in terms of MISMAs also explains the lack of reverse polarity in SST magnetograms. Stokes V in reverse polarity regions shows cross-over effect (Fig. 5 in Ichimoto et al. 2007a, and Ichimoto 2008, private

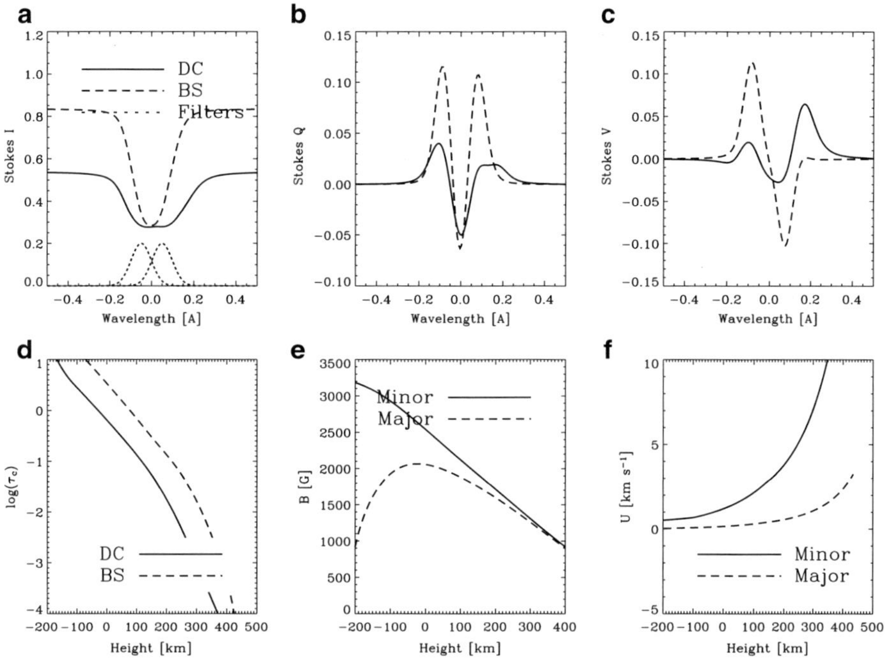

Fig. 2 (**a**) Stokes I profiles in one of the representative model MISMAs in SA05, which has been slightly modified to represent a dark core (*the solid line*), and its bright sides (*the dashed line*). They are normalized to the quiet Sun continuum intensity. (**b**) Stokes Q profiles. (**c**) Stokes V profiles. (**d**) Continuum optical depth τ_c vs. height in the atmosphere for the dark core and the bright sides, as indicated in the inset. (**e**) Magnetic field strength vs. height for the two magnetic components of the model MISMA. They are identical for the dark core and the bright sides. (**f**) Velocities along the magnetic field lines for the two magnetic components of the model MISMA. They are identical for the dark core and the bright sides

communication), that is, it presents two polarities depending on the sampled wavelength. It has the main sunspot polarity near line center, whereas the polarity is reversed in the far red wing (see the solid line in Fig. 2c). SST magnetograms are taken at line center ($\pm$50 mÅ), which explains why the reverse polarity does not show up. A significant reduction of the Stokes V signal occurs, though. Such reduction automatically explains the observed weakening of magnetic signals in dark cores (item 7 in Sect. 2), provided that the dark cores are associated with an enhancement of the opposite polarity, that is, if the cross-over profiles are produced in the dark cores. We have constructed images, magnetograms, and dopplergrams of a (naïve) model dark-cored filament that illustrate the idea. The filament is formed by a uniform 100 km wide dark strip, representing the dark core, bounded by two bright strips of the same width, representing the bright sides. The Stokes profiles of the dark core have been taken as the solid lines in Fig. 2a, c, while the bright sides are modelled as the dashed lines in the same figures. The color filters employed by Langhans et al. (2005, 2007) are approximated by Gaussian functions of 80 mÅ FWHM, and shifted $\pm$50 mÅ from the line center (see the dotted lines in Fig. 2a).

The magnetogram signals are computed from the profiles as

$$M = -\frac{\Delta\lambda}{|\Delta\lambda|}\int V(\lambda)f(\lambda-\Delta\lambda)\,\mathrm{d}\lambda \Big/ \int I(\lambda)f(\lambda-\Delta\lambda)\,\mathrm{d}\lambda, \tag{3}$$

with $f(\lambda)$ the transmission curve of the filter and $\Delta\lambda = -50\,\mathrm{m}\text{Å}$. Similarly, the Doppler signals are given by

$$D=-\frac{\Delta\lambda}{|\Delta\lambda|}\int I(\lambda)[f(\lambda-\Delta\lambda)-f(\lambda+\Delta\lambda)]\,\mathrm{d}\lambda \Big/ \int I(\lambda)[f(\lambda-\Delta\lambda)+f(\lambda+\Delta\lambda)]\,\mathrm{d}\lambda, \tag{4}$$

but here we employ the Stokes I profile of the nonmagnetic line used by Langhans et al. (2007; i.e., Fe I λ5576 Å). The signs of M and D ensure $M > 0$ for the main polarity of the sunspot, and also $D > 0$ for redshifted profiles. The continuum intensity has been taken as I at -0.4 Å from the line center. The continuum image of this model filament is shown in Fig. 3, with the dark core and the bright sides

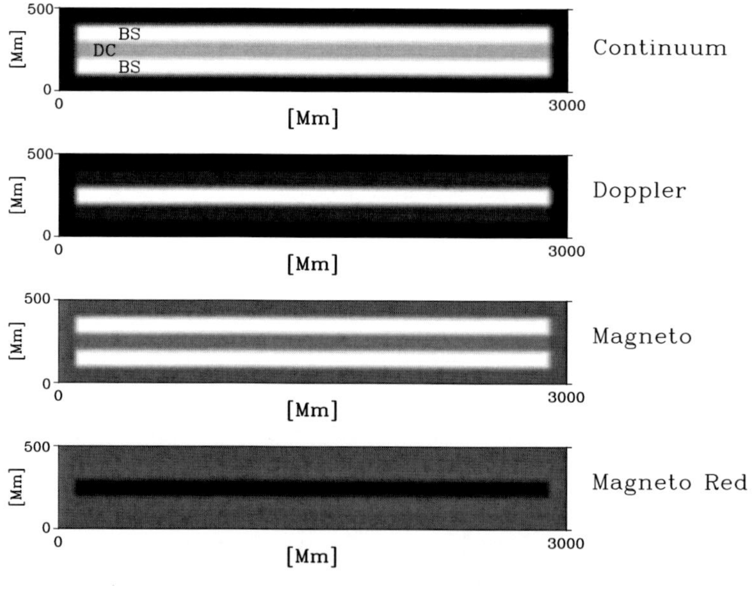

Fig. 3 Schematic modeling of SST observations of penumbral filaments by Langhans et al. (2005, 2007). A dark core (DC) surrounded by two bright sides (BS) is located in the limb-side penumbra of a sunspot at $\mu = 0.95$ (18° heliocentric angle). The three *top* images show a continuum image, a dopplergram, and a magnetogram, as labeled. The convention is such that both the sunspot main polarity and a redshift produce positive signals. The dark background in all images has been included for reference, and it represents signal equals zero. The fourth image (**Magneto Red**) corresponds to a magnetogram in the far red wing of Fe I λ6302.5 Å, and it reveals a dark core with a polarity opposite to the sunspot main polarity. The continuum image and the dopplergram have been scaled from zero (*black*) to maximum (*white*). The scaling of the two magnetograms is the same, so that their signals can be compared directly

marked as DC and BS, respectively. The dopplergram and the magnetogram are also included in the same figure. The dark background in all images indicates the level corresponding to no signal. In agreement with Langhans et al. observations, the filament shows redshifts ($D > 0$), which are enhanced in the dark core. In agreement with Langhans et al., the filament shows the main polarity of the sunspot ($M > 0$), with the signal strongly reduced in the dark core. Figure 3 (bottom) includes the magnetogram to be observed at the far red wing ($\Delta\lambda = 200\,\text{mÅ}$). The dark core now shows the reversed polarity ($M < 0$), while the bright sides still maintain the main polarity with an extremely weak signal. This specific prediction of the modeling is liable for direct observational test (Sect. 6).

Two final remarks are in order. First, the magnetogram signal in the dark core is much weaker than in the bright sides, despite the fact that the (average) magnetic field strength is larger in the core (see Fig. 2e, keeping in mind that the minor component dominates). Second, the model dark core is depressed with respect to the bright sides. Figure 2d shows the continuum optical depth τ_c as a function of the height in the atmosphere. When the two atmospheres are in lateral pressure balance, the layer $\tau_c = 1$ of the dark core is shifted by some 100 km downward with respect to the same layer in the bright sides. The depression of the observed layers in the dark core is produced by two effects; the decrease of density associated with the increase of magnetic pressure (e.g., Spruit 1976), and the decrease of opacity associated with the reduction of temperature (e.g., Stix 1991).

4 Scenario for the Small-Scale Structure of the Penumbra

Attending to the constraints presented in Sect. 2, penumbrae may be made out of short narrow shallow magnetic loops, which often return under the photosphere within the sunspot boundary (Fig. 4). One of the footpoints is hotter than the other (Fig. 5). The matter emerges in the hot footpoint, radiates away, cools down, and returns through the cold footpoint. The ascending plasma is hot, dense, and slowly moving. The descending plasma is cold, tenuous, and fast moving. The motions along magnetic field lines are driven by magnetic field strength differences between the two footpoints, as required by the siphon flow mechanism.

In addition to holding large velocities along field lines, the cold footpoint of each loop sinks down in a slow motion across field lines. In nonmagnetic convection, upflows are driven through mass conservation by displacing warm material around the downdrafts (Stein and Nordlund 1998; Rast 2003). The uprising hot material tends to emerge next to the downflows. If the same mechanism holds in penumbrae, the sinking of cold footpoints induces a rise of the hot footpoints physically connected to them, producing a backward displacement of the visible part of the loops (see Fig. 6). The sink of the cold footpoints could be forced by the drag of downdrafts in subphotospheric layers, in a magnetically modified version of the mechanism discussed in item 10 of Sect. 2.

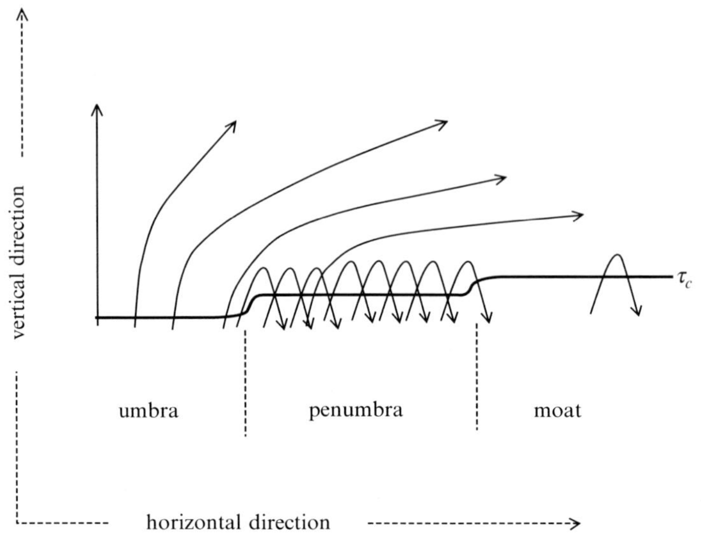

Fig. 4 Cartoon sketching the scenario for the penumbral magnetic field topology put forward in the paper. Magnetic field lines, represented as *solid lines* with *arrow* heads, return under the photosphere in the entire penumbra. The symbol τ_c stands for the continuum optical depth so that the *thick solid line* marked as $\tau_c = 1$ represents the base of the photosphere

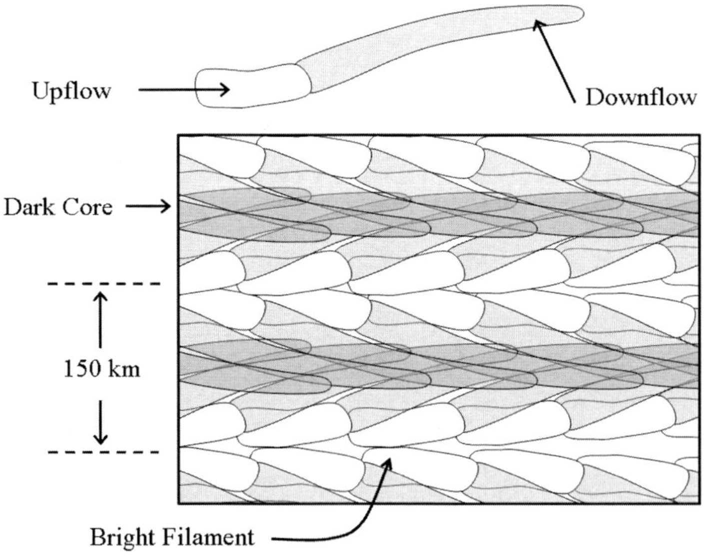

Fig. 5 A view from above of a small portion of penumbra (see the scale on the *left* hand side of the cartoon). Small magnetic loops like the one on *top* of the figure are averaged in our resolution element (the *rectangle*). They are so thin that various loops overlap along the LOS. The loops are arranged with the downflowing footpoints aligned forming a dark core. The hot upflows feeding a dark core give rise to two bright filaments. The Evershed flow is directed along the field lines toward the *right*

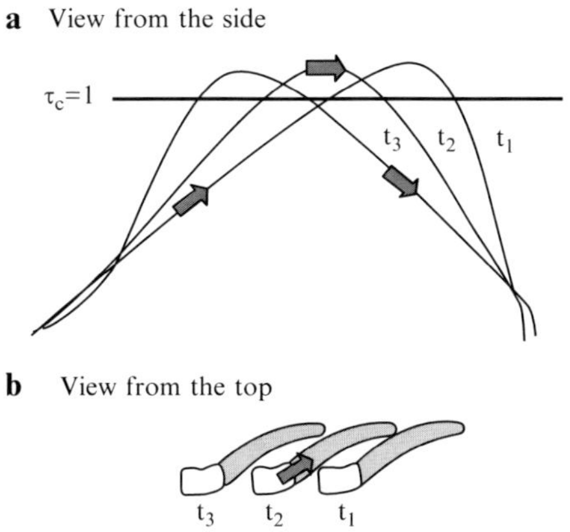

Fig. 6 Cartoon showing the time evolution of a single magnetic field line, with $t_1 < t_2 < t_3$ (the *thin solid lines*). (**a**) View from the side, with the vertical direction pointing upward. The *thick solid line* corresponds to base of the photosphere (continuum optical depth equals one). The Evershed flow is directed to the *right*. The *thick arrows* represent a parcel of fluid with the tip pointing in the direction of the plasma motion. (**b**) View from above of the same three instants. (Compare with Fig. 5, showing many field lines at a given time.) It only shows the fluid parcel at t_2 as it remains below the photosphere at t_1 and t_3

The properties of the loops (length, width, speed, and so on) should change with the position on the penumbra, but they are always narrow and so tend to be optically thin across field lines. One does not detect individual loops but assembles of them interleaved along the LOS. The cold legs of many different loops should be identified with the dark cores found by Scharmer et al. (2002); compare Figs. 1 and 5. The upflows of many loops account for the bright penumbral filaments. In this scenario, magnetic field lines are not exactly aligned with the penumbral bright and dark filaments, but the field lines diverge from the bright filaments and converge toward the dark filaments. The mean field is radial, though.

According to our scenario, bright penumbral filaments are associated with fields having the polarity of the sunspot, mild upflows, and relatively low field strengths. On the contrary, dark filaments are associated with fields whose polarity tends to be opposite to the sunspot polarity, have intense flows, high field strength, and they are more transparent than the bright filaments.

5 Qualitative Comparison with Observations

As we will discuss here, the scenario proposed in Sect. 4 fits in the observations in Sect. 2. Our model penumbra is formed by narrow loops whose length is typically smaller than the penumbral size (Fig. 4), and whose width is not spatially resolved. The smallness of the physical scales is in agreement with items 1 and 11. (These and the rest of numbers refer to the labels in Sect. 2.) Magnetic field lines bend over and return under the photosphere over the entire penumbra, as required by items 13 and 15. The loops have a hot footpoint with upward motion and a cold footpoint with downward motion, in agreement with the local correlation between brightness and upward velocity observed in penumbrae (item 3). The downflows are expected to be faster than the upflows as they are accelerated by the magnetic field strength difference between the two footpoints, an image that fits in well the observations showing the largest velocities to be associated with the dark penumbral components (item 6).

We identify the dark cores found by Scharmer et al. (2002, item 2) with cold footpoints of many loops, as sketched in Fig. 5. Dark cores trace downdrafts engulfing cold footpoints (item 10). The bright filaments around the dark cores would be naturally explained by the presence of the downflows, as it happens with the enhanced brightness at the borders of the granules in nonmagnetic convection. Mass conservation induces an upflow of hot material around the downdrafts (Rast 1995, 2003; Stein and Nordlund 1998). The same mechanism would produce the upraise of hot (magnetized) material around the dark cores, forming two bright filaments outlining each core (item 2; Fig. 1). The hot magnetized material would eventually cool down and sink into the dark core to restart the process. In other words, a dark core would be the downdraft of two paired convective rolls, resembling those proposed long ago by Danielson (item 9). In this case, however, the magnetic field lines are not exactly horizontal, and the plasma has a large velocity component along the field lines. Note that these hypothetical convective rolls reproduce the expected mode of convective transport in highly inclined magnetic fields (see item 8, including the comment on the recent numerical simulations of penumbrae which seem to disfavor this mode). Moreover, a pattern of motions similar to these convective rolls occurs in the moat surrounding the sunspot (item 9), and it is conceivable that it continues within the sunspot.

The existence of small scale convective upflows and downflows does not contradict the systematic upward motions in the inner penumbra and downward motions in the outer penumbra found by various authors (see item 12). Most observational techniques employed so far assume uniform velocities in the resolution element. When spatially unresolved upflows and downflows are interpreted as a single resolved component, the measured velocity corresponds to an ill-defined mean of the actual velocities. The contribution of upflows and downflows to such mean is not proportional to the mass going up and down. It depends on the physical properties of the upflows and downflows, as well as on the method employed to measure. The mean vertical flux of mass inferred by SA05 is zero (item 13); however, the local averages are biased,[4] showing net upflows in the inner penumbra and net downflows in the outer penumbra, in agreement with item 12.

[4] The effect is similar to the convective blueshift of the spectral lines formed in the granulation, whose existence does not imply a net uplifting of the quiet photosphere.

Our scenario with overlaying loops of various velocities and inclinations accounts for the observed Stokes asymmetries, including the rules for the NCP mentioned in item 11.

The bright filaments are more opaque than the dark cores (Sect. 3), and they tend to block the light coming from the dark cores when the filaments are observed sideways. This depression of the dark cores explains why they are elusive in the penumbra perpendicular to the center-to-limb direction, as well as why penumbrae are slightly darker in the line along the center-to-limb direction (item 4).

The length of the bright filaments is not set by the cooling time of individual fluxtubes, which avoids the difficulty posed in item 14. It is given by the length of the dark core.

Does the model account for the penumbral radiative flux? The radiative flux emanating from penumbrae F is some 75% of the flux in the quiet Sun. To balance this loss with energy transported by convection, the vertical velocity U_z must satisfy (e.g., Spruit 1987; Stein and Nordlund 1998),

$$U_z \approx F/(\rho\,\alpha\,\epsilon) \approx 1\,\mathrm{km\,s^{-1}}, \tag{5}$$

with ρ the density, α the fraction of atmospheric volume occupied by upward motions, and ϵ the energy per unit mass to be radiated away. As the physical conditions in penumbrae are similar to those of the quiet Sun, the U_z accounting for F must be similar too, rendering the speed in the right-hand-side of equation (5). Unfortunately, the observed upward vertical velocities are one order of magnitude smaller than the requirement set by equation (5)[5] (items 3 and 13). The discrepancy can be explained if an observational bias underestimates the true velocities. Such bias is to be expected because the velocity structure remains unresolved (items 1 and 11). Removing the bias involves resolving the structure both along and across the LOS, in particular, the cross-over effect Stokes V profiles associated with the reverse polarity must be properly interpreted to retrieve realistic velocities.

Why does the low-density plasma of the cold footpoints sink rather than float? We have been arguing by analogy with the nonmagnetic convection, where the (negative) buoyancy forces in the intergranular lanes drive the sinking of cold plasma and the rising of hot material around it. The plasma tends to sink down due to its enhanced density as compared to the hot upwelling plasma. The scenario for the penumbral convection discussed above does not reproduce this particular aspect of the granular convection. The descending footpoint has reduced density as compared to the upflowing footpoint. The density in the descending leg is lower than that in the ascending leg, and one may think that the descending plasma is buoyant. However, the density of the cold leg has to be compared to the local density in the downdraft, which can easily be larger than the downdraft density. Recall that the

[5] This discrepancy between the required and observed velocities was used to discard the transport of energy by convection in penumbrae (Spruit 1987), leading to the concept of shallow penumbra by Schmidt et al. (1986).

downdrafts have low temperature and high magnetic field strengths, two ingredients that naturally produce low densities in magneto-hydrostatic equilibrium (see, e.g., Sánchez Almeida 2001).

Why are the dark cores radially oriented? According to our conjecture, the dark cores are not tracing individual field lines. They are faults in the global structure of the sunspot where downflow motions are easier. Such discontinuities of the global magnetic structure would be favored if they are aligned with the mean magnetic field, as it happens with interchange instabilities (Parker 1975; Meyer et al. 1977; Jahn and Schmidt 1994). The dark cores would be oriented along the direction of the mean penumbral magnetic field.

6 Discussion and Tests

We have described a penumbral magnetic field formed by short magnetic loops most of which return to the sub-photosphere within the sunspot boundary (Fig. 4). Matter flowing along magnetic field lines would give rise to the Evershed effect. This flow along field lines would also be responsible for the convective transport of heat in penumbrae. Cold downflows of many loops are aligned in a sort of lane of downdrafts whose observational counterpart would be the dark cores in penumbral filaments found by Scharmer et al. (2002). To some extent, the scenario resembles the convective rolls put forward by Danielson some 50 years ago, except that (1) the mass also flows along field lines, and (2) the mass may not re-emerge after submergence. It is also akin to the interchange convection (e.g., Schlichenmaier 2002), where flows along field lines transport heat from below and give rise to the Evershed effect phenomena. In our case, however, the upraise of the hot tubes is induced by the presence of downdrafts and the need to satisfy mass conservation (as in the nonmagnetic granular convection). The submergence of the cool footpoints of the loops happening in the downdrafts may be due to flux pumping, as numerical models have shown to occur in the nonmagnetic downdrafts outside penumbrae (Weiss et al. 2004). Our scenario also owes some properties to the siphon flow model (Meyer and Schmidt 1968; Thomas and Montesinos 1993). A gas pressure difference between the loop footpoints drives the flows, but this difference is set by the magnetic field strength difference between the hot and the cold penumbral structures, rather than from field strength differences between the penumbra and magnetic concentrations outside the sunspot. Finally, the model that emerges has features from the return flux model of Osherovich (1982), where the field lines forming umbra and penumbra differ because the former are open while the latter return to the photosphere. The return of magnetic flux is included in our scenario, except that the return of penumbral field lines occurs throughout the penumbra, rather than outside the sunspot border.

This scenario seems to be compatible with most observations existing in the literature, in particular, with the ubiquitous downflows and the reverse magnetic flux found by SA05 and Ichimoto et al. (2007a) (see Sect. 5). However, two difficulties remain. First, the upflows and downflows measured so far are not fast enough to transport the radiative flux emerging from penumbrae (Sect. 5). Second, the reverse polarity does not show up in SST observations, which nevertheless provide the highest spatial resolution available at present (Sect. 3). We believe that the two difficulties are caused by our still insufficient resolution. Solving the problem is not only a question of improving the resolution across the LOS (e.g., by enlarging the telescope diameter), but the resolution along the LOS seems to be critical. The finding by Ichimoto et al. (2007a) that the reverse polarity coincides with very asymmetric Stokes profiles with NCP strongly suggests that measuring the true velocities and polarities demands resolution along the LOS. In other words, the observed polarization cannot be correctly interpreted if one assumes the light to come from an atmosphere with uniform properties.

Several specific tests will allow us to confirm or falsify the scenario. (1) Diffraction limited high resolution spectropolarimetry with 1 m class telescopes must show a correlation between brightness and velocity with an amplitude of the order of $1\,\mathrm{km\,s^{-1}}$. As we point out above, a pure brute force approach may not suffice to render the right amplitude, and modeling the observed asymmetries seems to be required. (2) SST magnetograms taken in the far red wing of Fe I 6302.5 Å must show dark cores with reverse polarity (Sect. 3). (3) The molecular spectra trace the cold penumbral component (e.g., Penn et al. 2003). They should show a global downflow of the order of $1\,\mathrm{km\,s^{-1}}$, no matter the angular resolution of the observation. (4) Local correlation tracking techniques applied to sequences of penumbral images should reveal proper motions of bright features moving toward dark filaments. Such motions are masked by the tendency of the bright plasma to cool down and become dark. A global trend is to be expected, though. Such trend seems to be present in high resolution SST images (Márquez et al. 2006), although the result requires independent confirmation.

Acknowledgment In a paper for nonspecialists, Priest (2003) speculates *Are the bright filaments doubly convective rolls with a dark core that are cooling and sinking?* This question seems to advance the scenario advocated here. In addition, the scenario is sketched in an unpublished work by Sánchez Almeida (2004), and it is also mentioned in Sánchez Almeida (2006). Using arguments differing from those introduced here, Zakharov et al. (2008) have recently claimed evidence for convective rolls in penumbrae. The image in Fig. 1 is courtesy of the Institute for Solar Physics of the Royal Swedish Academy of Sciences. It was obtained with the SST operated in the Spanish Observatorio del Roque de Los Muchachos (La Palma). This work has partly been funded by the Spanish Ministry of Science and Technology, project AYA2007-66502, as well as by the EC SOLAIRE Network (MTRN-CT-2006-035484). I thank the SOC of the Evershed centenary meeting for this opportunity to expound unconventional ideas.

References

Arena, P., Landi Degl'Innocenti, E., Noci, G. 1990, Solar Phys., 129, 259
Beckers, J. M., Schröter, E. H. 1969, Solar Phys., 10, 384
Bellot Rubio, L. R. 2004, Rev. Mod. Astron., 17, 21

Bellot Rubio, L. R., Balthasar, H., Collados, M., Schlichenmaier, R. 2003, A&A, 403, L47
Bonet, J. A., Márquez, I., Muller, R., Sobotka, M., Roudier, T. 2005, A&A, 430, 1089
Bovelet, B., Wiehr, E. 2003, A&A, 412, 249
Brummell, N. H., Tobias, S. M., Thomas, J. H., Weiss, N. O. 2008, ApJ, 686, 1454
Bumba, V. 1960, Izv. Crim. Astrophys. Obs., 23, 253
Danielson, R. E. 1960, AJ, 70, 343
Danielson, R. E. 1961, ApJ, 134, 289
Evershed, J. 1909, MNRAS, 69, 454
García de la Rosa, J. I. 1987, In: The Role of Fine-Scale Magnetic Fields on the Structure of the Solar Atmosphere, E.-H. Schröter, M. Váquez, A. A. Wyller (eds.), Cambridge University Press, Cambridge, p. 140
Golovko, A. A. 1974, Solar Phys., 37, 113
Grigorjev, V. M., Katz, J. M. 1972, Solar Phys., 22, 119
Heinemann, T., Nordlund, Å., Scharmer, G. B., Spruit, H. C. 2007, ApJ, 669, 1390
Hofmann, J., Deubner, F.-L., Fleck, B., Schmidt, W. 1994, A&A, 284, 269
Hurlburt, N. E., Matthews, P. C., Rucklidge, A. M. 2000, Solar Phys., 192, 109
Ichimoto, K., Shine, R. A., Lites, B., et al. 2007a, PASJ, 59, 593
Ichimoto, K., Suematsu, Y., Tsuneta, S., et al. 2007b, Science, 318, 1597
Illing, R. M. E., Landman, D. A., Mickey, D. L. 1974a, A&A, 35, 327
Illing, R. M. E., Landman, D. A., Mickey, D. L. 1974b, A&A, 37, 97
Jahn, K., Schmidt, H. U. 1994, A&A, 290, 295
Johannesson, A. 1993, A&A, 273, 633
Kàlmàn, B. 1991, Solar Phys., 135, 299
Kosugi, T., Matsuzaki, K., Sakao, T., et al. 2007, Solar Phys., 243, 3
Langhans, K., Scharmer, G., Kiselman, D., Löfdahl, M., Berger, T. E. 2005, A&A, 436, 1087
Langhans, K., Scharmer, G. B., Kiselman, D., Löfdahl, M. G. 2007, A&A, 464, 763
Makita, M. 1986, Solar Phys., 106, 269
Márquez, I., Sánchez Almeida, J., Bonet, J. A. 2006, ApJ, 638, 553
Meyer, F., Schmidt, H. U. 1968, Mitt. Astron. Gesellschaft Hamburg, 25, 194
Meyer, F., Schmidt, H. U., Weiss, N. O. 1977, MNRAS, 179, 741
Osherovich, V. A. 1982, Solar Phys., 77, 63
Parker, E. N. 1975, Solar Phys., 40, 291
Penn, M. J., Cao, W. D., Walton, S. R., Chapman, G. A., Livingston, W. 2003, ApJ, 590, L119
Priest, E. 2003, Physics World, 16(2), 19
Rüedi, I., Solanki, S. K., Keller, C. U. 1999, A&A, 348, L37
Rast, M. P. 1995, ApJ, 443, 863
Rast, M. P. 2003, ApJ, 597, 1200
Rempel, M., Schüssler, M., Knölker, M. 2009, ApJ, 691, 640
Rimmele, T. R. 1995, ApJ, 445, 511
Rouppe van der Voort, L. H. M., Löfdahl, M. G., Kiselman, D., Scharmer, G. B. 2004, A&A, 414, 717
Sánchez Almeida, J. 1998, ApJ, 497, 967
Sánchez Almeida, J. 2001, ApJ, 556, 928
Sánchez Almeida, J. 2004, astro-ph/0811.4319
Sánchez Almeida, J. 2005, ApJ, 622, 1292
Sánchez Almeida, J. 2006, In: Solar Polarization 4, R. Casini, B. W. Lites (eds.), ASP Conf. Ser., 358, ASP, San Francisco, 13
Sánchez Almeida, J., Lites, B. W. 1992, ApJ, 398, 359
Sánchez Almeida, J., Martínez Pillet, V., Trujillo Bueno, J., Lites, B. W. 1993, In: The Magnetic and Velocity Fields of Solar Active Regions, H. Zirin, G. Ai, H. Wang (eds.), ASP Conf. Ser., 46, ASP, San Francisco, p. 192
Sánchez Almeida, J., Landi Degl'Innocenti, E., Martínez Pillet, V., Lites, B. W. 1996, ApJ, 466, 537
Sánchez Almeida, J., Márquez, I., Bonet, J. A., Domínguez Cerdeña, I. 2007, ApJ, 658, 1357

Scharmer, G. B., Gudiksen, B. V., Kiselman, D., Löfdahl, M. G., Rouppe van der Voort, L. H. M. 2002, Nat, 420, 151
Schlichenmaier, R. 2002, Astron. Nachr., 323, 303
Schlichenmaier, R., Schmidt, W. 2000, A&A, 358, 1122
Schlichenmaier, R., Bruls, J. H. M. J., Schüssler, M. 1999, A&A, 349, 961
Schmidt, H. U. 1991, Geophys. Astrophys. Fluid Dyn., 62, 249
Schmidt, H. U., Spruit, H. C., Weiss, N. O. 1986, A&A, 158, 351
Schmidt, W., Fritz, G. 2004, A&A, 421, 735
Schmidt, W., Schlichenmaier, R. 2000, A&A, 364, 829
Sheeley, N. R. 1972, Solar Phys., 25, 98
Solanki, S. K. 2003, A&A Rev., 11, 153
Solanki, S. K., Montavon, C. A. P. 1993, A&A, 275, 283
Spruit, H. C. 1976, Solar Phys., 50, 269
Spruit, H. C. 1987, In: The Role of Fine-Scale Magnetic Fields on the Structure of the Solar Atmosphere, E.-H. Schröter, M. Váquez, A. A. Wyller (eds.), Cambridge University Press, Cambridge, p. 199
Spruit, H. C., Scharmer, G. B. 2006, A&A, 447, 343
Stanchfield, D. C. H., Thomas, J. H., Lites, B. W. 1997, ApJ, 477, 485
Stein, R. F., Nordlund, Å. 1998, ApJ, 499, 914
Stix, M. 1991, The Sun, Springer, Berlin
Thomas, J. H., Montesinos, B. 1993, ApJ, 407, 398
Thomas, J. H. Weiss, N. O. 1992, In: Sunspots. Theory and Observations, J. H. Thomas, N. O. Weiss (eds.), NATO ASI Ser., 375, Kluwer, Dordrecht, p. 3
Thomas, J. H., Weiss, N. O. 2004, ARA&A, 42, 517
Title, A. M., Frank, Z. A., Shine, R. A., et al. 1993, ApJ, 403, 780
Tritschler, A., Schlichenmaier, R., Bellot Rubio, L. R., the KAOS Team. 2004, A&A, 415, 717
Weiss, N. O., Thomas, J. H., Brummell, N. H., Tobias, S. M. 2004, ApJ, 600, 1073
Zakharov, V., Hirzberger, J., Riethmüller, T. L., Solanki, S. K., Kobel, P. 2008, A&A, 488, L17

Theoretical Models of Sunspot Structure and Dynamics

J.H. Thomas

Abstract Recent progress in theoretical modeling of a sunspot is reviewed. The observed properties of umbral dots are well reproduced by realistic simulations of magnetoconvection in a vertical, monolithic magnetic field. To understand the penumbra, it is useful to distinguish between the inner penumbra, dominated by bright filaments containing slender dark cores, and the outer penumbra, made up of dark and bright filaments of comparable width with corresponding magnetic fields differing in inclination by some 30° and strong Evershed flows in the dark filaments along nearly horizontal or downward-plunging magnetic fields. The role of magnetic flux pumping in submerging magnetic flux in the outer penumbra is examined through numerical experiments, and different geometric models of the penumbral magnetic field are discussed in the light of high-resolution observations. Recent, realistic numerical MHD simulations of an entire sunspot have succeeded in reproducing the salient features of the convective pattern in the umbra and the inner penumbra. The siphon-flow mechanism still provides the best explanation of the Evershed flow, particularly in the outer penumbra where it often consists of cool, supersonic downflows.

1 Introduction

Understanding the structure and dynamics of a sunspot poses a formidable challenge to magnetohydrodynamic theory. The marvelous details revealed in high-resolution observations of sunspots have shown how very complex a sunspot is, but have also stimulated real progress in theoretical modeling.

Here I review recent advances on some important theoretical issues concerning sunspots, including the following questions. Is the overall near-surface structure of a sunspot best described as a monolithic (but inhomogeneous) magnetic flux tube or as a cluster of individual flux tubes? What is the nature of magnetoconvection

J.H. Thomas (✉)
Department of Mechanical Engineering and Department of Physics and Astronomy, University of Rochester, USA

S.S. Hasan and R.J. Rutten (eds.), *Magnetic Coupling between the Interior and Atmosphere of the Sun*, Astrophysics and Space Science Proceedings,
DOI 10.1007/978-3-642-02859-5_17,

in a sunspot, and how does it produce the umbral dots and the filamentary intensity pattern in the penumbra? What causes the complicated interlocking-comb configuration of the magnetic field in the penumbra? How do we explain the significant differences between the inner and outer penumbra? What causes the Evershed flow in the penumbra? How do the outflows along the dark penumbral cores in bright filaments in the inner penumbra relate to the stronger and downward plunging Evershed flows in the outer penumbra?

This review is of necessity selective, and some important topics will not be discussed at all (for example, sunspot seismology, which is well covered by Rajaguru and Hanasoge in this volume). For a broader coverage of both theory and observations of sunspots, see the recent book by Thomas and Weiss (2008) and the reviews by Solanki (2003) and Thomas and Weiss (2004).

2 Umbral Magnetoconvection

In a broad sense, there are two competing models of the structure of a sunspot below the solar surface: a monolithic, but inhomogeneous, magnetic flux tube, or a tight cluster of smaller flux tubes separated by field-free plasma (Parker 1979). One way in which we might distinguish between these two models is to examine the form of convective energy transport in the umbra, and in particular the mechanism that produces the bright umbral dots.

In the monolithic model, the umbral dots are thought to correspond to slender, hot, rising plumes that form within the ambient magnetic field and penetrate into the stable surface layer, spreading horizontally and sweeping magnetic flux aside (flux expulsion), thereby producing a small, bright region with a weakened magnetic field. This picture is supported by several idealized model calculations involving both Boussinesq and fully compressible magnetoconvection (see the reviews by Proctor 2005 and Thomas and Weiss 2008). In the cluster model, convection is imagined to be effectively suppressed in the magnetic flux tubes but unimpeded in the nearly field-free regions around them, where the convection penetrates upward into the visible layers to form bright regions. In that case, however, we might reasonably expect to see a bright network enclosing dark features, rather than the observed pattern of bright, isolated umbral dots on a dark background (e.g., Knobloch and Weiss 1984). The essential differences between the monolith and cluster models are that in the cluster model the weak-field gaps are permanent and are connected to the field-free plasma surrounding the sunspot, whereas in the monolithic model the gaps are temporary and are embedded within the overall flux tube, isolated from the surroundings of the spot.

Recently, Schüssler and Vögler (2006) carried out realistic numerical simulations of umbral magnetoconvection in the context of the monolithic model, assuming an initially uniform vertical magnetic field. They study three-dimensional compressible magnetoconvection within a realistic representation of an umbral atmosphere, including partial ionization effects and radiative transfer. Their model

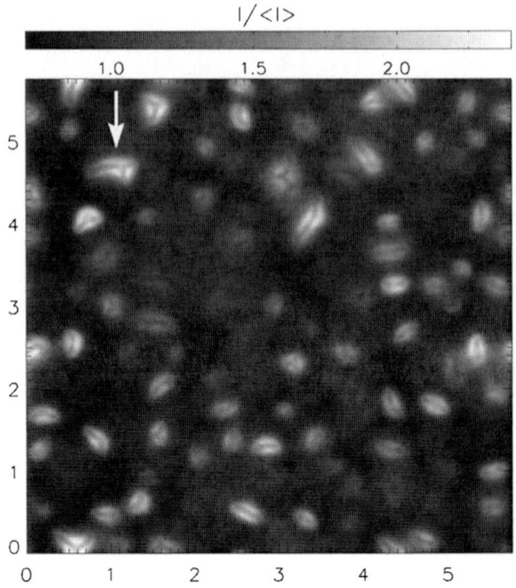

Fig. 1 The pattern of vertically emerging surface intensity in a realistic numerical simulation of umbral magnetoconvection (From Schüssler and Vögler 2006)

reproduces all of the principal observed features of umbral dots (see, e.g., Bharti et al. 2007). The results show an irregular pattern of slender, isolated plumes of width 200–300 km and lifetime around 30 min. An individual plume achieves a peak upward velocity of about 3 km s^{-1} before decelerating (by buoyancy braking) and spreading laterally as it meets the stable surface layer, greatly reducing the local magnetic field strength. Figure 1 shows a snapshot of the emerging intensity at the surface corresponding to optical depth $\tau_{500} = 1$ (which is elevated above the rising plumes). Note that the plumes are generally oval rather than circular in shape, and they have dark streaks along their major axes. These dark streaks are absorption features caused by the local increase of density and pressure associated with buoyancy braking of the plumes (cf. Sect. 6). The dark streaks have been seen in Hinode observations (Bharti et al. 2009).

While the results of Schüssler and Vögler do not necessarily rule out the cluster model, they do provide strong support for the monolithic model, in the sense that they show that umbral dots arise naturally as a consequence of magnetoconvection in a space-filling, vertical magnetic field. The magnetic flux is partially expelled from the plume regions to allow convective motions to occur, but these regions are not entirely field free and, more importantly, they are isolated within the overall flux bundle and not in contact with field-free plasma below, as they would be in the cluster model.

3 The Inner and Outer Penumbra

In understanding the structure of the penumbra, it is useful to distinguish between the *inner* and the *outer* penumbra (Brummell et al. 2008). The boundary between them is somewhat arbitrary, but it may be conveniently defined as the line separating inward-moving and outward-moving grains in the bright filaments, lying at about 60% of the radial distance between the inner and outer edges of the penumbra and dividing the penumbra into roughly equal surface areas (Sobotka et al. 1999; Sobotka and Sütterlin 2001; Márquez et al. 2006). This pattern may be understood as a transition from isolated, vertical convective plumes in the umbra to elongated, roll-like convective structures in the outer penumbra, as a consequence of the increasing inclination (to the local vertical) of the magnetic field. The moving bright grains are then traveling patterns of magnetoconvection in an inclined magnetic field, with the motion switching from inward to outward at some critical inclination angle of the magnetic field.

The inner penumbra is dominated by bright filaments containing slender dark cores (Scharmer et al. 2002; Langhans et al. 2007) and has relatively small azimuthal variations in the inclination of the magnetic field. The field in a dark core is slightly more inclined than the field in its bright surroundings, by some 4–10°. A dark core typically originates at a bright feature near the umbra, where there is an upflow that bends over into an outflow along the inclined magnetic field in the core.

The outer penumbra, on the other hand, is made up of dark and bright filaments of comparable width, with corresponding magnetic fields differing significantly in inclination, by 20–30° or more, the more horizontal field being in the dark filaments. The Evershed flow is stronger in the outer penumbra and is generally concentrated in the dark filaments, along nearly horizontal and often downward-plunging magnetic fields, with the flow velocity and the magnetic field well aligned all along the filament. One of the most intriguing features of the outer penumbra is the presence of "returning" magnetic flux, that is, field lines with inclinations greater than 90° that plunge back below the solar surface. There is now overwhelming observational evidence for a substantial amount of returning magnetic flux in the outer penumbra, in several high-resolution polarimetric studies based on different inversion schemes (e.g., Westendorp Plaza et al. 2001; Bellot Rubio et al. 2004; Borrero et al. 2004; Langhans et al. 2005; Ichimoto et al. 2007, 2009; Beck 2008; Jurčák and Bellot Rubio 2008).

The outer edge of the penumbra is quite ragged, with prominent dark filaments protruding into the surrounding granulation. The proper motions of granules in the moat surrounding a spot show convergence along radial lines extending outward from the protruding dark filaments (Hagenaar and Shine 2005), providing evidence for submerged magnetic flux extending outward from the spot. This submerged magnetic field is presumably held down, in opposition to its inherent buoyancy, by magnetic flux pumping, as described in the next section.

4 The Formation and Maintenance of the Penumbra

One of the important challenges for sunspot theory is to explain how the filamentary penumbra forms and its magnetic field acquires the observed interlocking comb structure with downward-plunging field lines in the outer penumbra, and how this structure is maintained. Eventually this whole process may be amenable to direct numerical simulation (see Sect. 6 below), but for now we can only speculate based on less ambitious models of specific aspects of the process.

The following scenario has been suggested for the formation of a fully fledged sunspot with a penumbra (Thomas et al. 2002; Weiss et al. 2004). The development of a solar active region begins with the emergence of a fragmented magnetic flux tube into the photosphere. The emergent flux is quickly concentrated into small, intense magnetic elements which can accumulate in the lanes between granules and mesogranules to form small pores. Some of these pores and magnetic elements may then coalesce to form a sunspot. Simple models show that, as a growing pore accumulates more magnetic flux, the inclination (to the local vertical) of the magnetic field at its outer boundary increases until it reaches a critical value, whereupon a convectively driven fluting instability sets in and a penumbra forms. The fluting of the magnetic field near the outer boundary of the sunspot's flux tube brings the more horizontal spokes of field into greater contact with the granulation layer in the surroundings, and then downward magnetic pumping of this flux by the granular convection further depresses this magnetic field to form the "returning" magnetic fields (inclination greater than 90°) seen in the outer penumbra. The transition between a pore and a sunspot shows hysteresis, in the sense that the largest pores are bigger than the smallest sunspots; this may be explained by the flux-pumping mechanism, which can keep the fields in the dark filaments submerged even when the total flux in a decaying spot is less than that at which the transition from pore to spot occurred.

We have demonstrated the efficacy of the process of magnetic flux pumping by granular convection through a series of idealized numerical experiments (Thomas et al. 2002; Weiss et al. 2004), most recently for a more realistic, arched magnetic field configuration that accounts more accurately for the magnetic curvature forces (in addition to the buoyancy forces) opposing the downward pumping (Brummell et al. 2008). We solve the equations governing three dimensional, fully compressible, nonlinear magnetoconvection in a rectangular box, consisting of two layers: an upper, superadiabatic layer of vigorous convection representing the granulation layer, and a lower, marginally stable or weakly superadiabatic layer representing the rest of the convection zone. The simulation is run without a magnetic field until a statistically steady state is reached, and then a strong magnetic field is introduced, in the form of a purely poloidal (x–z), double arched magnetic field, and the gas density is adjusted to maintain pressure equilibrium. The calculation proceeds and we examine the effect of the convection in redistributing the magnetic flux.

Figure 2 shows the state of the magnetic field shortly after it was introduced (scaled time $t = 0.5$) and at a few later stages, the last stage ($t = 42.8$) being after a new quasi-steady statistical state has been reached. Here we see that a significant fraction of the large-scale magnetic field is pumped rapidly downward out of the upper granulation layer and concentrated mostly in the upper part of the lower, more quiescent convective layer. These new numerical experiments demonstrate that the downward pumping by turbulent granular convection is indeed able to overcome the combined effects of the magnetic buoyancy force and the curvature

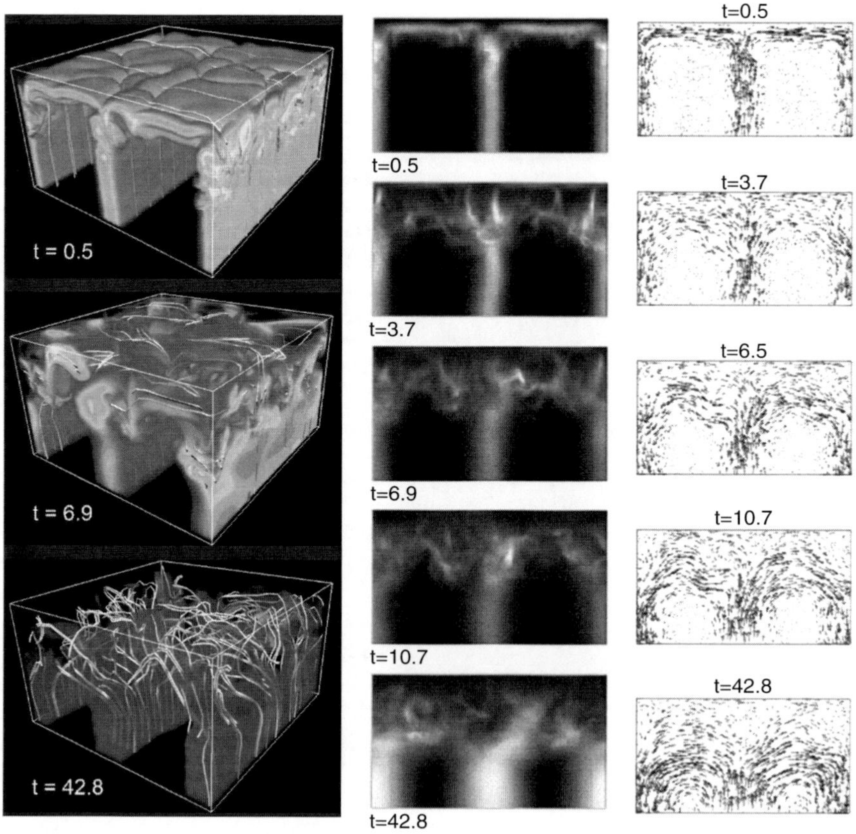

Fig. 2 Numerical simulation of downward magnetic flux pumping of penumbral magnetic fields by granular convection in the surroundings. The panels show volume renderings of magnetic energy density (*left*), x-averaged magnetic energy density (*middle*), and the x-averaged vector magnetic field at different times during the run. (Here x is the direction perpendicular to the page.) The initial arched magnetic field configuration is still distinctly visible in the uppermost plots at $t = 0.5$ (From Brummell et al. 2008)

force due to magnetic tension, and thus to submerge much of the initial, nearly horizontal magnetic flux beneath the granulation layer, as we propose in the scenario presented above.

5 The Magnetic Field Configuration in the Penumbra

Here we consider some geometric models that have been proposed for the observed interlocking-comb structure of the penumbral magnetic field. The scenario described in Sect. 4 for the formation of the penumbra and the returning flux tubes through flux pumping leads us to a magnetic field configuration in the outer

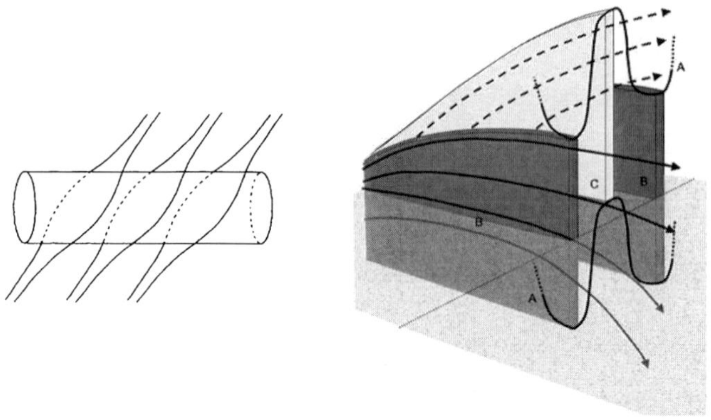

Fig. 3 Two simple models of the penumbral magnetic field configuration. *Left panel:* Sketch of the magnetic field configuration in the "uncombed" penumbral model of Solanki and Montavon (1993), with an ambient magnetic field wrapping around a thin horizontal flux tube (dark filament). *Right panel:* Schematic diagram of the "interleaved sheet" model of the outer penumbra (Brummell et al. 2008.), with a fluted magnetopause (A) and slabs of nearly horizontal magnetic field (B, dark filaments) extending downward to some depth below the surface and separated by a slab of less steeply inclined magnetic field (C, bright filament)

penumbra roughly as depicted in the right-hand panel of Fig. 3 (Thomas et al. 2006; Brummell et al. 2008). This configuration, which we might describe as an "interleaved sheet" model, has vertical sheets of nearly horizontal magnetic field (dark filaments) interleaved between sheets of more vertical magnetic field (bright filaments). In this picture, the sheets of horizontal field extend downward below the visible surface to a depth of, say, 5 Mm (A simple estimate gives the depth of penetration equal to one-quarter of the width of the penumbra: Brummell et al. 2008).

Another geometric model, with a longer history, is the "uncombed" penumbral model[1] of Solanki and Montavon (1993), depicted in the left-hand panel of Fig. 3. In this model the more horizontal component of the penumbral magnetic field is represented by horizontal magnetic flux tubes, of nearly circular cross-section, embedded in a more vertical background magnetic field that wraps around these tubes. Scharmer and Spruit (2006) pointed out that the magnetic tension forces in the background magnetic field will tend to compress a circular flux tube in the horizontal direction, causing it to expand upward at the top and downward at the bottom, perhaps indefinitely. Borrero et al. (2006) then argued that buoyancy forces will halt this squeezing process, leaving a flux tube of tall, narrow cross-section. If the vertical elongation of the flux tube is significant, the configuration begins to look much like the interleaved sheet model depicted in the right-hand panel of Fig. 3, and these two models are then not very different.

---

[1] Sometimes the term "uncombed" is used more generally to describe the observed penumbral field configuration, but here I use the term specifically to represent the geometric model proposed by Solanki and Montavon (1993).

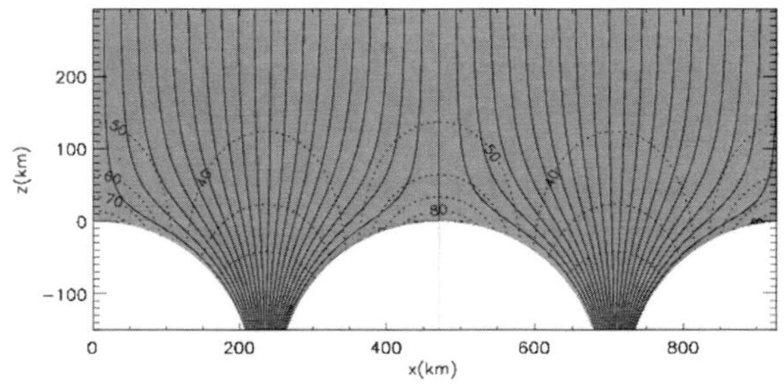

Fig. 4 The potential magnetic field configuration in the "gappy penumbra" model of Spruit and Scharmer (2006). Shown here are the magnetic field lines (*solid lines*) projected onto a vertical (x–z) plane perpendicular the axis (y-axis) of a penumbral filament, along with contours (*dotted lines*) of constant inclination of the field in the y–z plane

A quite different model of the penumbral magnetic field, the "gappy penumbra" model of Spruit and Scharmer (2006; Scharmer and Spruit 2006), is based on the cluster model of a sunspot. It postulates field-free, radially aligned gaps in the magnetic field below the visible surface of the penumbra, protruding into a potential magnetic field configuration. The gaps are assumed to extend indefinitely downward, allowing the field-free convection in the gaps to carry the bulk of the upward heat flux in the penumbra. Figure 4 shows the proposed magnetic field configuration. The gaps themselves represent the bright penumbral filaments, while the intervening regions of strong magnetic field represent the dark filaments. As can be seen from the contours of constant inclination in Fig. 4, the magnetic field is more nearly horizontal above the bright filaments (the gaps) and more nearly vertical (here 45°) above the dark filaments. However, this magnetic field configuration is in direct contradiction with numerous observations that show that the field is more horizontal in the dark filaments (e.g., Rimmele 1995; Stanchfield et al. 1997; Westendorp Plaza et al. 2001; Langhans et al. 2005), including very recent spectropolarimetric observations from Hinode by Jurčák and Bellot Rubio (2008) and by Borrero and Solanki (2008). The last authors also examined the vertical stratification of magnetic field strength in the penumbra and found that it is inconsistent with the existence of regions void of magnetic field at or just below the $\tau_{500} = 1$ level. While the gappy penumbra model itself contains no flows, Spruit and Scharmer suggest that the Evershed flow occurs along the (very restricted) region of nearly horizontal field just above the center of the gap. At least in the outer penumbra, this is in conflict with numerous observations that show that the flow is concentrated in the dark filaments. It seems, then, that the gappy penumbra is incompatible with observations.

Spruit and Scharmer (2006) also suggested that the observed narrow dark cores running along the center of bright filaments in the inner penumbra can be understood as an effect of the increased opacity due to increased gas pressure in the field-free gaps. This important suggestion seems to be basically correct, although the field-free gaps are not necessary: dark cores also form as opacity effects in the case of magnetoconvection in a strong-field region, as shown in the simulations of umbral dots discussed in Sect. 2 above and in the simulations of penumbral filaments discussed in the next section.

6 Numerical Simulations of a Sunspot

Any attempt to perform a direct numerical MHD simulation of an entire sunspot faces serious computational difficulties: the simulation must represent a very large structure while still resolving fine-scale features and even smaller scale diffusive effects; it must cope with a wide range of values of the Alfvén speed and plasma beta; and the computation must be carried out long enough to reach a relaxed, quasi-steady state. In spite of these formidable problems, there have been very recently impressive attempts by two groups to model an entire sunspot by direct, realistic simulations including radiative transfer (Heinemann et al. 2007; Rempel et al. 2009). These efforts are surely just the beginning of a new and fruitful approach to sunspot theory.

Both groups model a large section of a sunspot in a rectangular box. They each introduce a two-dimensional, vertically spreading, initial magnetic field into a state of fully developed nonmagnetic convection representing the upper convection zone and a stable atmospheric layer above it. The calculations continue for several hours of real (solar) time, through a dynamic adjustment phase, until a quasi-steady state is attained. The results show the formation of filamentary structures resembling those in the inner penumbra of a real sunspot, including bright filaments containing central dark cores.

In very recent work, Rempel et al. have extended their simulations to model an entire circular sunspot within a rectangular box. Figure 5 shows a snapshot of the surface intensity pattern and magnetic field in this beautiful simulation.

The simulations reproduce most of the important features of the bright penumbral filaments found in the inner penumbra. Figure 6 shows a blowup of a single bright penumbral filament produced in the rectangular sunspot simulation of Rempel et al. (2009). The continuum intensity pattern shows an elongated bright filament with a dark central core and a bright "head", which migrates inward toward the umbra during the lifetime of the filament. The dark core is produced as an opacity effect due to buoyancy braking of the upflow, much as in the simulations of umbral dots discussed in Sect. 2. The magnetic field (not shown here) is weaker and more inclined in the filament than in its immediate surroundings.

The pattern of vertical velocity shows roll-like convection along the filament with an inclined upflow along the central axis of the filament (i.e., along the dark core) and inclined downflows (return flows) along the sides of the filament. Correspondingly, the radial (x) component of the velocity is outward along the axis of the filament, with a peak value of about 2 $\mathrm{km\,s^{-1}}$, and inward and along the sides of the filament. The return flow is in regions with stronger and less inclined magnetic field, so the horizontal component is smaller in magnitude than that in the outflow at the same optical depth; as a result, the radial inflows and outflows do not cancel when averaged in the y-direction, but instead show an average outflow of about 1 $\mathrm{km\,s^{-1}}$.

The simulated penumbral filaments are slender structures with a width of a few hundred km and a depth of about 2 Mm. The filaments form and remain embedded within an overall region of strong magnetic field, and they are well isolated from the field-free convection beneath the penumbra. As Rempel et al. emphasize, these ces between the inner and outer penumbra, and it is them.

netic flux by turbulent granular convection offers producing the returning magnetic flux in the outer

aved sheet" models of the penumbral magnetic field ite similar, in view of the squeezing effect on the ombed model.

el for the penumbral magnetic field configuration is ns.

of an entire sunspot have succeeded in reproducing umbra. However, they do not reproduce the structure its horizontal and returning magnetic fields and fast s along arched channels.

s in the inner penumbra are well reproduced in these nvection (not interchange convection). Magnetic flux convective plumes, but the resulting "gaps" are not r plasma and hence are fundamentally different from umbra" model. The simulations reproduce the central aments, as an opacity effect due to buoyancy braking flows seen in these cores.

ll provides the best description of the Evershed flow in noving-tube model describes the transient nature of the produce returning flux tubes and downflows. A thin- g the best features of these two models is suggested.

Hasan for making it possible for me to attend this meeting, and esults and figures prior to their publication. I also thank my col- Tobias, and Nigel Weiss, with special thanks to Nigel Weiss for d issues covered in this review.

H., Collados, M. 2004, A&A, 427, 319
N. A. 2007, ApJ, 665, L79
. N. A., Jain, R. 2009, MNRAS, 393, 65
lanki, S. K. 2006, in Solar Polarization 4, eds. R. Casini and B. W. 19
2008, ApJ, 687, 668
Bellot Rubio, L. R., Lagg, A., Mathew, S. K. 2004, A&A, 422, 1093
l., Thomas, J. H., Weiss, N. O. 2008, ApJ, 686, 1454
Rubio, L. R., Collados, M. 2001, ApJ, 549, L139
2005, ApJ, 635, 659
n Scharmer, G. B., Spruit, H. C. 2007, ApJ, 669, 1390
, tes, B. W., et al. 2007, PASJ, 59, S593

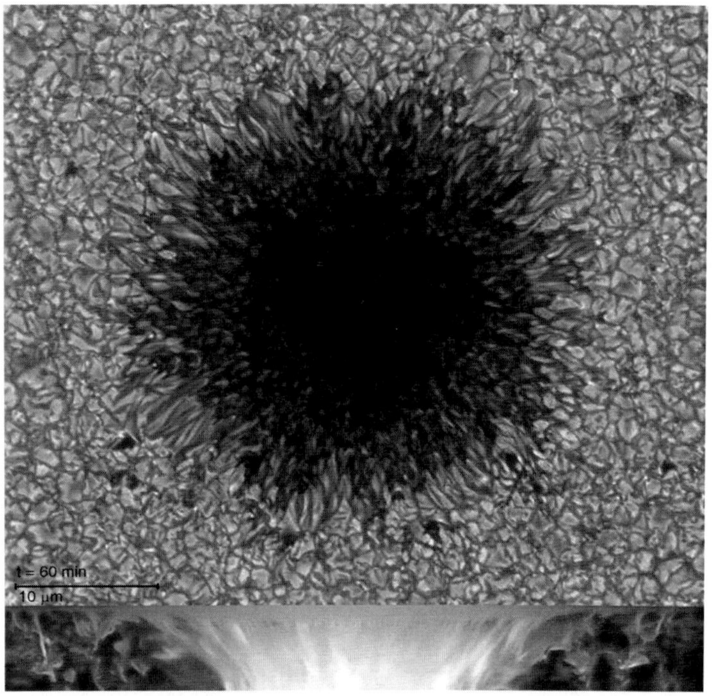

Fig. 5 Numerical simulation of a circular sunspot, following the same method as in the slab model of Rempel et al. (2009). Shown here are (*above*) a snapshot of the surface intensity and (*below*) the corresponding values of $|B|^{1/2}$ on a vertical slice through the center of the spot, depicted on a gray scale (Courtesy of Matthias Rempel)

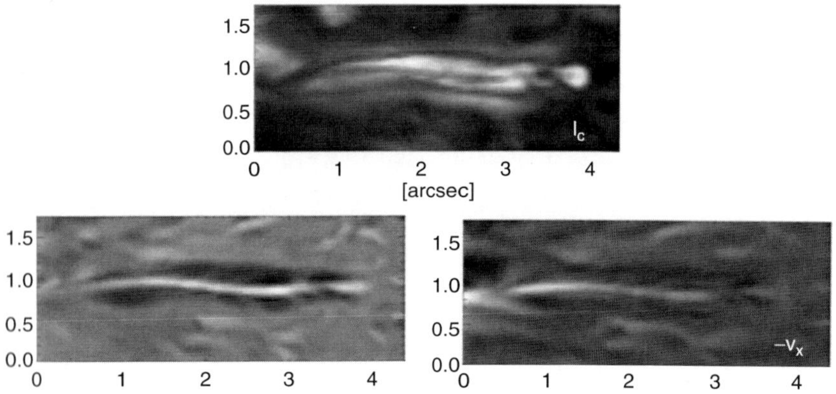

Fig. 6 Enlarged view of a single bright penumbral filament produced in the simulation of Rempel et al. (2009). The umbra lies to the right of this filament. The *upper* panel shows a surface continuum intensity image at wavelength 630 nm, and the *lower* panels show vertical velocity v_z (*left*) and horizontal velocity v_x (*right*), where the x-axis is parallel to the *bottom* of the panels

weak-field "gaps" formed within the overall magnetic field by the convection are fundamentally different from those proposed in the "gappy penumbra" model of Spruit and Scharmer, which are protrusions of the exterior field-free plasma into the penumbra as envisioned in the cluster model (see Sect. 5). At a fundamental level, the simulations discussed in this section are based on the monolithic model and they support that model by producing results that match observations of the inner penumbra; they lend no support for the "gappy penumbra."

Rempel et al. also point out that their results do not support the "moving tube" model of Schlichenmaier, Jahn and Schmidt (1998): vertical heat transport takes place all along the simulated filaments, not just along separate, thin flux tubes, and the movement of the filament "heads" inward toward the penumbra is due to a propagation of the magnetoconvective pattern rather than the bodily motion of an individual thin flux tube.

In both simulations discussed above, the overall extension of the penumbra is rather small and the inclination of the magnetic field in the outer part of the penumbra is generally much less than that found in a real sunspot. Thus, as both groups admit, the simulations so far seem to reproduce only the inner penumbra. One reason for this is that the periodic boundary conditions employed effectively place another sunspot of the same magnetic polarity nearby, on either side of the simulated spot. This hinders the formation of nearly horizontal fields in the outer penumbra. (Indeed, observations show that sunspots often do not form a penumbra in a sector near another spot of the same polarity.) This could be remedied, for example, by using periodic boundary conditions like those of Brummell et al. (2008), which produce a row of spots of alternating polarity (see Fig. 2).

7 The Evershed Flow

Since the occasion for this meeting is the centennial of John Evershed's discovery, it seems appropriate to close with some remarks about theoretical interpretations of the Evershed flow. The flow occurs along arched, elevated flow channels. Recent results from Hinode support this picture. Ichimoto et al. (2007) find that the Evershed downflows in the outer penumbra have the flow velocity vector and magnetic field vector well aligned, at an angle of about 30° to the solar surface. Jurčák and Bellot Rubio (2008) find that the average inclination of the magnetic field associated with the Evershed flow channels increases from 85° to 105° in going from the middle to the outer penumbra, quite consistent with the earlier results of Langhans et al. (2005) from the Swedish Solar Telescope.

The arched nature of the flow channels and the strong, often supersonic, field-aligned downflows in the outer penumbra are well reproduced in the siphon flow model (e.g., Montesinos and Thomas 1997). The "moving tube" model of Schlichenmaier et al. (1998) does not produce this configuration: it has no returning flux or downflow, but instead has all of the flow continuing radially outward along the elevated magnetic canopy. Schlichenmaier (2002) did find a class of super-

Alfvénic, serpentine sol
returning flux tube, but th
artifact of the outer boun
is gravitationally unstable
seems to have been ignore
to be invoked as a possibl
et al. 2007; Sainz Dalda ar

The numerical simulati
ward horizontal velocity co
Fig. 6), which might explai
penumbra, although it is no
core are not observed. Howe
of the Evershed flow, as clai
peak outward velocity is onl
a few filaments is only abou
the outer penumbra. The simu
flow speeds of 7–16 km s^{-1},
that are observed in dark filar
et al. 2001; del Toro Iniesta et

The supersonic, cool Evers
flow model (Montesinos and
description of the Evershed flo
puted so far have all been stead
of flows. A thin-flux-tube mod
model (arched, returning flux tu
tube model (transient flows, hea
all of the salient features of the E

In a broad sense the Evershe
nomenon. Even in the models ba
or the siphon-flow model – the f
tube produced by some combinati
pressure) or convective collapse
returning flux is produced by turbu
ities increase and the numerical si
convection in a sunspot and its imm
flow and the returning flux tubes to

8 Conclusions

The principal conclusions of this revi

- The observed properties of umbra
tions of magnetoconvection in a ve
need to invoke a cluster model.

Ichimoto, K., SOT/Hinode Team 2010, in Magnetic Coupling between the Interior and the Atmosphere of the Sun, eds. S. S. Hasan and R. J. Rutten, Astrophys. Space Sci. Procs., Springer, Heidelberg, these proceedings
Jurčák, J., Bellot Rubio, J. L. 2008, A&A, 481, L17
Knobloch, E., Weiss, N. O. 1984, MNRAS, 207, 203
Langhans, K., Scharmer, G. B., Kiselman, D., Löfdahl, M. G., Berger, T. E. 2005, A&A, 436, 1087
Langhans, K., Scharmer, G. B., Kiselman, D., Löfdahl, M. G. 2007, A&A, 464, 763
Márquez, I., Sánchez Almeida, J., Bonet, J. A. 2006, ApJ, 638, 553
Montesinos, B., Thomas, J. H. 1997, Nat, 390, 485
Parker, E. N. 1979, ApJ, 230, 905
Penn, M. J., Cao, W. D., Walton, S. R., Chapman, G. A., Livingston, W. 2003, ApJ, 590, L119
Proctor, M. R. E. 2005, in Fluid Dynamics and Dynamos in Astrophysics and Geophysics, eds. A. M. Soward, C. A. Jones, D. W. Hughes, and N. O. Weiss, 235
Rempel, M., Schüssler, M., Knölker, M. 2009, ApJ, 691, 640
Rimmele, T. R. 1995, ApJ, 445, 511
Sainz Dalda, A., Bellot Rubio, L. R. 2008, A&A, 481, L21
Scharmer, G. B., Spruit, H. C. 2006, A&A, 460, 605
Scharmer, G. B., Gudiksen, B. V., Kiselman, D., Löfdahl, M. G., Rouppe van der Voort, L. H. M. 2002, Nat, 420, 151
Scharmer, G. B., Nordlund, Å., Heinemann, T. 2008, ApJ, 677, L149
Schlichenmaier, R. 2002, Astron. Nachr., 323, 303
Schlichenmaier, R., Jahn, K., Schmidt, H. U. 1998, A&A, 337, 897
Schlichenmaier, R., Müller, D. A. N., Beck, C. 2007, in Modern Solar Facilities – Advanced Solar Science, eds. F. Kneer, K. G. Puschmann, and A. D. Wittmann, Göttingen Universitätsverlag, 233
Schüssler, M., Vögler, A. 2006. ApJ, 641, L73
Sobotka, M., Sütterlin, P. 2001, A&A, 380, 714
Sobotka, M., Brandt, P. N., Simon, G. W. 1999, A&A, 348, 621
Solanki, S. K. 2003, A&A Rev., 11, 153
Solanki, S. K., Montavon, C. A. P. 1993, A&A, 275, 283
Spruit, H. C., Scharmer, G. B. 2006, A&A, 447, 343
Stanchfield, D. C. H. II, Thomas, J. H., Lites, B. W. 1997, ApJ, 477, 485
Thomas, J. H. 2005, A&A, 440, L29
Thomas, J. H., Weiss, N. O. 2004, ARA&A, 42, 517
Thomas, J. H., Weiss, N. O. 2008, Sunspots and Starspots, Cambridge University Press
Thomas, J. H., Weiss, N. O., Tobias, S. M., Brummell, N. H. 2002, Nat, 420, 390
Thomas, J. H., Weiss, N. O., Tobias, S. M., Brummell, N. H. 2006, A&A, 452, 1089
Weiss, N. O, Thomas, J. H., Brummell, N. H., Tobias, S. M. 2004, ApJ, 600, 1073
Westendorp Plaza, C., del Toro Iniesta, J. C., Ruiz Cobo, B., Martínez Pillet, V. 2001, ApJ, 547, 1148

Convection and the Origin of Evershed Flows

Å. Nordlund and G.B. Scharmer

Abstract Numerical simulations have by now revealed that the fine scale structure of the penumbra in general and the Evershed effect in particular is due to overturning convection, mainly confined to gaps with strongly reduced magnetic field strength. The Evershed flow is the radial component of the overturning convective flow visible at the surface. It is directed outwards – away from the umbra – because of the broken symmetry due to the inclined magnetic field. The dark penumbral filament cores visible at high resolution are caused by the "cusps" in the magnetic field that form above the gaps. Still remaining to be established are the details of what determines the average luminosity of penumbrae, the widths, lengths, and filling factors of penumbral filaments, and the amplitudes and filling factors of the Evershed flow. These are likely to depend at least partially also on numerical aspects such as limited resolution and model size, but mainly on physical properties that have not yet been adequately determined or calibrated, such as the plasma beta profile inside sunspots at depth and its horizontal profile, the entropy of ascending flows in the penumbra, etc.

1 Introduction

Recently – and in just the right time for the Evershed centenary – the first realistic three-dimensional radiation-magnetohydrodynamics models of sunspots and sunspot penumbrae have become available (Heinemann et al. 2007; Scharmer et al. 2008; Rempel et al. 2009). This has provided the opportunity to resolve the longstanding debate about the origin of the penumbral fine structure (Danielson 1961; Mamadazimov 1972; Schmidt et al. 1986; Solanki and Montavon 1993; Jahn and Schmidt 1994; Martens et al. 1996; Schlichenmaier et al. 1998a,b, 1999; Martínez Pillet 2000; Thomas et al. 2002; Bellot Rubio et al. 2003; Schlichenmaier and Solanki 2003; Schmidt and Fritz 2004; Borrero et al. 2004; Bellot Rubio

Å. Nordlund (✉)
Niels Bohr Institute, University of Copenhagen, Denmark

G.B. Scharmer
Institute for Solar Physics, Royal Swedish Academy of Sciences, Stockholm, Sweden

S.S. Hasan and R.J. Rutten (eds.), *Magnetic Coupling between the Interior and Atmosphere of the Sun*, Astrophysics and Space Science Proceedings,
DOI 10.1007/978-3-642-02859-5_18,

et al. 2004; Weiss et al. 2004; Thomas and Weiss 2004; Tildesley and Weiss 2004; Borrero et al. 2005; Spruit and Scharmer 2006; Thomas et al. 2006; Scharmer and Spruit 2006; Brummell et al. 2008) by "looking the horse in the mouth."

When combined with unavoidable requirements from basic physics, the evidence from even this first generation of numerical models is sufficient to establish the basic mechanisms at work. The models illustrate, for example, that the large luminosity of the penumbra, of the order of 75–95% of that of the photosphere (Solanki 2003, Fig. 3.2), must be essentially due to convective heat transport, which is able to carry nearly as much heat to the solar surface in penumbrae as in the surrounding photosphere, even in the presence of the strong penumbral magnetic field.

The numerical models still have shortcomings, as discussed in more detail below. In part, the shortcomings may be due to the choice of parameters and boundary conditions for the models. Additionally, the limited numerical resolution of the models may also be important. Indeed, modeling entire sunspots is exceedingly demanding in computational resources and even the most highly resolved models (c.f. Rempel et al. 2009) have mesh spacings that are only a few times smaller than the thickness of penumbral filaments.

As sunspots are not necessarily round, one can also choose to model only a narrow strip, stretching in one direction from an umbra to a surrounding photosphere, and assumed to be periodic in the perpendicular horizontal direction (Heinemann et al. 2007; Rempel et al. 2009). Such a model may be considered to represent a piece of a sunspot with a section of more or less straight penumbra filaments. There are abundant examples of such sunspots, as well as of "umbrae-without-penumbra" and "penumbrae-without-umbrae," and thus one should expect to see essentially the same phenomena in such models as in models of entire sunspots. The much reduced size requirement in one direction can be utilized to increase the resolution in the other two directions, with given amounts of computing resources.

As we show below, the situation can also be modeled, with essential features reproduced, in even smaller patches, with accordingly higher possible spatial resolution, by making also the longer of the two horizontal directions periodic. Such models represent local rectangular patches of penumbrae, and include neither an umbra nor a piece of photosphere. The one major drawback of this type of model is that it cannot allow for the inclination of the penumbra surface. The setup does allow for an arbitrary inclination of the penumbra magnetic field, and in fact requires the inclination to be specified, at least as an initial condition. These types of models are thus ideal for exploring the parameter space spanned by magnetic field inclination and by lower boundary plasma beta and entropy, and may be used to disentangle the relative importance of each of these factors, at spatial resolutions that could not be achieved in the other two types of models.

The three classes of models are complementary, in that the more localized ones can achieve higher spatial resolution, and thus can address resolution issues more effectively than global models, which on the other hand can be compared more directly with observations.

Below we briefly summarize and discuss some of the findings from the numerical models, attempt to identify reasons for remaining problems, and indicate directions for future work.

2 Recent Modeling Results

Heinemann et al. (2007) were the first to perform realistic numerical simulations of sunspots, where one could see signs of the formation of penumbra filaments. The qualifier "realistic" is here used to indicate MHD-models that include an ionizing equation of state representative of solar conditions and radiative energy transfer with an H^{-}-like continuum opacity. Those requirements are essential for making sure that the model behavior near the optical depth unity, where the surface luminosity is determined, is at least qualitatively similar to that of the real Sun. Gray radiative transfer was used in the Pencil Code results on which the Heinemann et al. (2007) paper was based on, but similar experiments using the Copenhagen Stagger Code that includes non-gray radiative energy transfer confirm the results. The use of non-gray (binned) opacity influences only layers higher up in the atmosphere, whose behavior is not essential to understand the penumbra fine structure (although it may be important for computing realistic synthetic spectral line diagnostics).

The Heinemann et al. (2007) experiments consisted of a rectangular box of size $12 \times 6 \times 3$ Mm, with an isotropic grid spacing of approximately 24 km; a vertical cross section is shown in Fig. 1. As users of numerical MHD-simulations are the first to realize, this is by no means equivalent to having a "physical resolution" of 24 km; as a rule-of-thumb a feature resolved by less than about ten grid points is significantly affected by numerical resolution, and hence the penumbral small scale

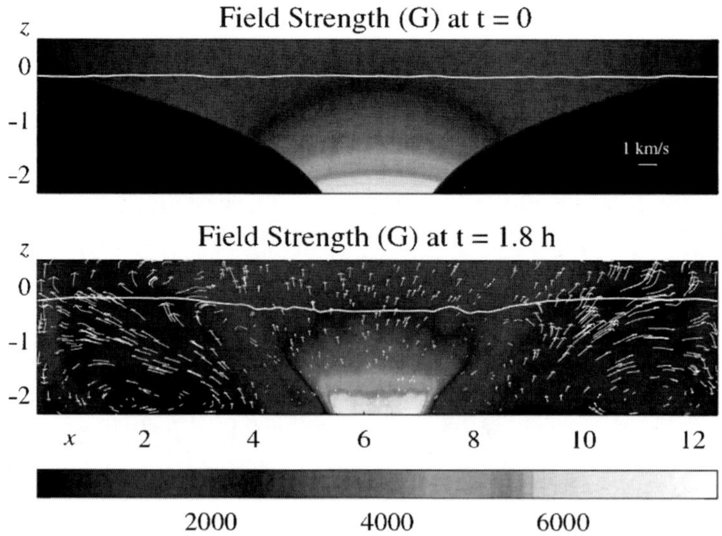

Fig. 1 Cross sections of the Heinemann et al. (2007) experiment, showing the initial state (*top*), where the magnetic field has a potential shape corresponding to a constant beta as a function of height, and the state after 1.8 solar hours (*bottom*), where the gas pressure has been significantly reduced in the surface layers, because of the cooling of the surface layers associated with the strong magnetic field. The *arrows* shown in the *bottom* panel illustrate the "moat flow"; a mean sub-surface radial flow driven by the reduced cooling under the spot

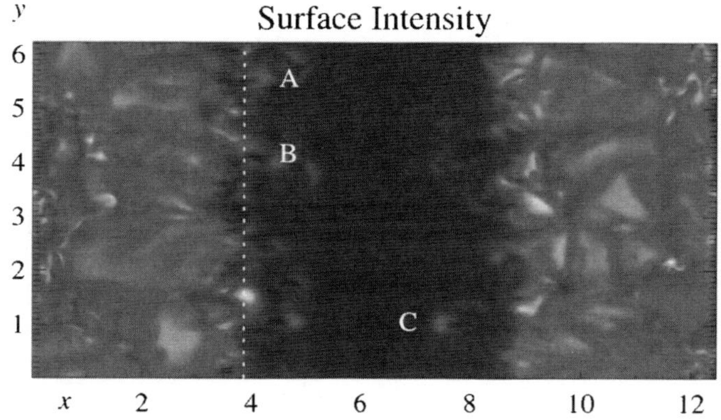

Fig. 2 Image of the surface radiation intensity from the Heinemann et al. (2007) experiment, showing a number of short penumbral filaments with dark cores. Three of the locations are marked, and are further illustrated in Fig. 3. Note the small scale bright (magnetic) features outside the spot – in animations they can be seen drifting out from the spot, carried by the moat flow shown in Fig. 1

features seen in the Heinemann et al. (2007) models are indeed only marginally resolved. One can nevertheless recognize a number of properties reminiscent of solar penumbrae and their surroundings: bright filaments with "heads" that move towards the umbra, with dark cores (cf. Fig. 2) and systematic radial outflows of several km s^{-1} in the penumbral filaments, and a "moat flow" carrying magnetic flux features away from the spot. Both the speeds of inward migration, the magnitude of the Evershed-like outflows, and the moat flow speeds are similar to observed values. Those agreements could in principle, given the limited spatial size and resolution of the model, be fortuitous, but could also be indications of underlying physical processes that are so robust that they start becoming established even when the numerical resolution is marginal. As illustrated by the subsequent work by Rempel et al. (2009), the latter indeed seems to be the case.

As pointed out by Heinemann et al. (2007), the filamentary features seen at the surface correspond to local convective flow channels (cf. Fig. 3), similar to the umbral dots seen in the numerical simulations by Schüssler and Vögler (2006), and in light bridge models (Nordlund 2006, cf. Fig. 4). The analogy is additional evidence of robust physical processes at work. The main difference is probably just a matter of the inclination of the magnetic field, which brakes the symmetry and forces a preferred direction onto the overturning part of the convective flow.

The basic mechanism at work is thermal convection, modified by the presence of a strong and inclined magnetic field. The luminosity of the penumbra, which visually is so clearly attributed to the densely packed penumbral filaments, must be due to convective heat transport up to the visible surface, with the penumbral filaments being the top layer manifestation of that heat transport, much as the granulation pattern in the surrounding photosphere is the surface manifestation of the even more efficient convective heat transport there.

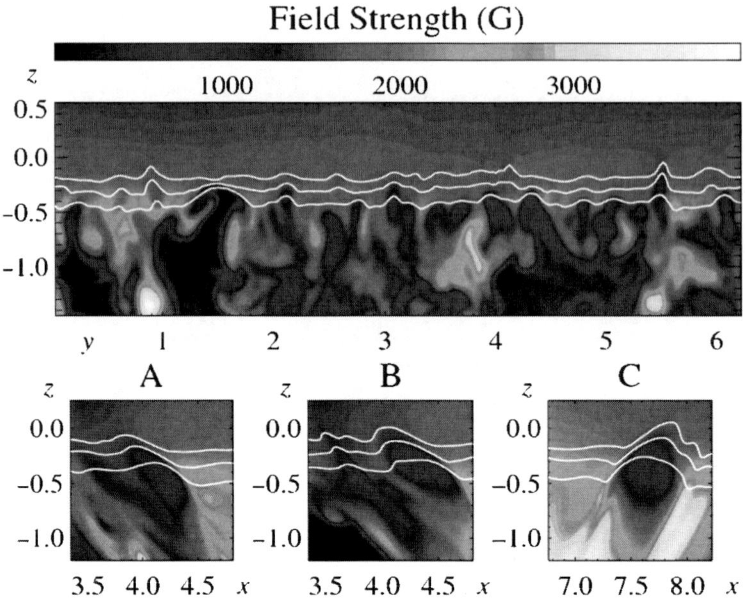

Fig. 3 Cross sections showing the magnetic field strength along the *dotted line*, and across the three features indicated in Fig. 2. These are migrating penumbral features, with strong upflows on their umbral sides and downflows on their penumbral sides

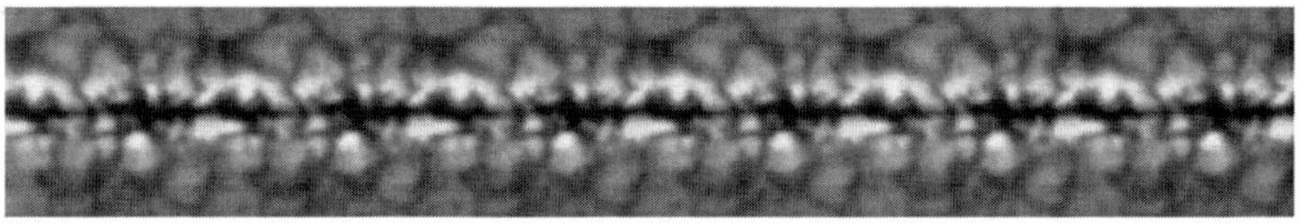

Fig. 4 Light bridge formed by a "gap" in an umbral-strength magnetic field. As a consequence of approximate horizontal pressure equilibrium the gap is cusp-shaped, closing at the *top* at some height in the visible photosphere. Because of the correspondingly higher density of the gas inside the cusp and the overall drop of temperature with height, a dark filament-like structure appears in the emergent intensity (cf. Nordlund 2006)

In the penumbra, the overturning and cooling of ascending hot gas is constrained to happen primarily in the radial direction; the generally radial orientation of the magnetic field strongly disfavors motions in the azimuthal direction. As each parcel of ascending hot gas must be exposed to surface cooling for about the same time as in the granulation pattern in the surrounding photosphere, and as the velocity magnitudes are constrained by the requirement to transport nearly the same amount of energy to the surface, the trajectories of parcels of gas as they overturn at the surface must be about as long in the surrounding photosphere; that is, paths that extend above the surface for several megameters horizontally. But as the motions are constrained to occur mainly in the radial direction, there is less horizontal expansion and more "crowding" of the paths.

From these general considerations one concludes that already the energy transport requirements and the constraints imposed by the magnetic field in a natural way leads to the occurrence of bright, densely packed filaments with lengths of at least several megameters. To the extent that the models do not yet produce all of these properties (e.g., too short and non-space-filling filaments), one can then also conclude that the energy transport (and hence penumbral brightness) is likely to be too small, for which one should then seek remedies.

However, the features that are already present in the current models do perform the expected type of convective energy transport, just at a lower level of intensity, and with a smaller filling factor. One can thus already study the basic mechanisms at work, and perhaps then also conjecture about how conditions should be changed in order for the models to more closly match the observed properties of solar penumbrae.

2.1 Evershed Results

Scharmer et al. (2008) investigated the nature of the Evershed-like flows in the penumbral filaments seen in the Heinemann et al. (2007) models, looking in particular for explanations of the inward migration and for the outward, Evershed-like flows. They found the inward migration to be due to a pattern-motion, where the "head" represents a strong convective upflow location, which is able to push aside the penumbral magnetic field, and also to push material that then quickly cools up along the inclined magnetic field lines. The cooler and heavier material results in the bending down of field lines, and allows the convective flow to overturn and return down below the surface. That process, the ascent of hot gas that cools at the surface and then descends, is – from a "heat engine" point of view – essentially the same as the convective process in normal solar granulation (cf. Stein and Nordlund 1998; Nordlund 2008) (Fig. 5).

The filaments seen in the Heinemann et al. (2007) simulations are quite short, which is perhaps not too surprising, given the limitations in size and numerical resolution. The corresponding features in the models of Rempel et al. (2009) are longer, but still rather short compared to real penumbral filaments. In both cases, the filaments are also much less space-filling than in real solar penumbrae. As the size constraints have been lifted in the Rempel et al. (2009) work, the length and filling factor issues should be seen as indications that some other setup property common to both works is responsible; for example, the limited numerical resolution. Rempel et al. (2009) also argue that the Evershed-like flow seen in their model is too weak relative to real sunspots, but this may simply be a reflection of the too small filling factor. As illustrated by Fig. 6 of their paper, the peak flow speeds in the filaments reach approximately 6 km s^{-1}, which is certainly in the right ball park (cf. Solanki 2003) (Fig. 6).

Likely factors, which could be responsible for both the lack of space filling and the shortness of the filaments, are the basic strength and inclinations of the magnetic field in the penumbral part of the models. These properties depend on details of the

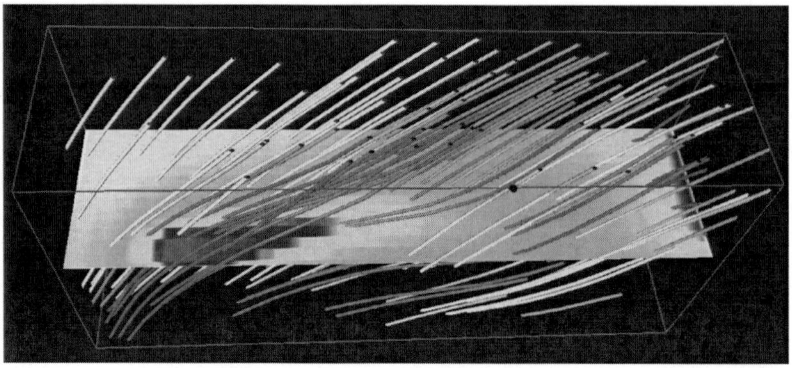

Fig. 5 This snapshot of a 3D visualization (Scharmer et al. 2008) shows the vector magnetic field, color coded with magnetic field strength, with a cutting plane containing a color coded image of temperature. *Purple* to *blue* represents large values, *orange* to *red* represent small values (color illustrations are available in the on-line version). The size of the box is approximately 1200 × 340 × 550 km (color illustration are available in the on-line version)

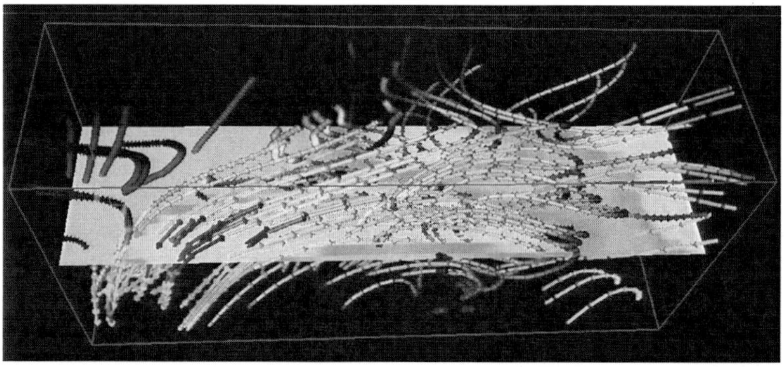

Fig. 6 The vector velocity field, color coded with velocity magnitude, and with a cutting plane color-coded image showing the strength of the vertical magnetic field component (color illustration are available in the on-line version)

boundary conditions applied at the base of the model – mainly the amplitude and profile of the plasma beta (ratio of gas to magnetic pressure) across the base of the sunspot model. Already a visual inspection of Fig. 1 and of Fig. 2 of Rempel et al. (2009) indicates that the average model magnetic fields are not sufficiently inclined, as compared to what is implied by Fig. 3.2 of Solanki (2003). Likely remedies that should be tried include decreasing the sunspot beta value at the bottom boundary, adjusting the horizontal beta profile, and/or reducing the sensitivity to boundary conditions by extending the models to larger depths. Future model results will show if any of these suggestions pans out, or if there are yet other properties of the models that need to be improved (e.g., using even higher spatial resolution, and continuing for even longer time spans, sufficient for the model to forgive and forget oversimplified initial conditions).

3 Local Penumbra Models

As mentioned in the introduction, one of the compromises one can choose to make to maximize the available spatial resolution is to restrict the total size of the model so much that it represents only a small piece of the penumbra. To avoid having to specify necessarily awkward radial boundary conditions, one can choose periodic boundary conditions also in the radial direction.

Such models necessarily lack one seemingly central aspect of real penumbrae, in that the sloping of the penumbra surface (e.g., of optical depth unity) cannot be represented. That may also be seen as an advantage, as in the alternative, more normal set-ups, which include an umbra and a photosphere, one cannot avoid having a sloping penumbra. By using local models one can thus figure out if the sloping plane is essential for the formation of the penumbral fine structure, or if perhaps – as in the case of rectangular vs. round sunspot models – this is a side issue that is not of central importance.

The main advantage with the local models is that they allow a much larger spatial resolution. A rectangular model of a penumbral stripe such as the Heinemann et al. (2007) model needs to be at least as wide as a few granules in order to do a reasonable job in representing the external photosphere. At the same time it needs to be at least as "long" as the diameter of a sunspot, and then some, to allow for the surrounding photosphere and to prevent neighboring periodic images of the spot from distorting the spot magnetic field too much. The Heinemann et al. (2007) model, which was 12×6 Mm wide, thus represents the low end of the possible size range of these models.

A local and doubly periodic model that covers only a patch of the penumbra, on the other hand, needs only to be as wide as a few penumbral filaments (say, about 1 Mm), and long enough to contain a section of a penumbral filament (which was argued above to be of the order of the size of a granule, so a size of about 6 Mm would seem to be enough). To summarize, one can use a horizontal area more than an order of magnitude smaller in local models, with a corresponding gain possible in the spatial resolution. Figure 7 shows an example of such a model, with a grid spacing of about 3 km.

The filamentary structures visible in Fig. 7 display a pattern motion towards the penumbra, consistent with the behavior found in similar, but more idealized setups that have been investigated in the past (cf. Hurlburt et al. 2000). This demonstrates that pattern motions similar to those observed in the inner parts of real penumbra can occur also without a "sloping plane" penumbra surface. Systematic studies of the effects of varying the strength and inclination of the magnetic field in such models will be helpful in disentangling the interplay between factors that influence the lengths, widths, and filling factors of small scale penumbral structures.

From the various results mentioned earlier, one can draw some conclusions relevant to the debate about mechanisms (or mechanism labels!) responsible for the creation of penumbral filaments. A major label issue has for example been whether to refer to the penumbral structures as "flux tubes" or as "gaps."

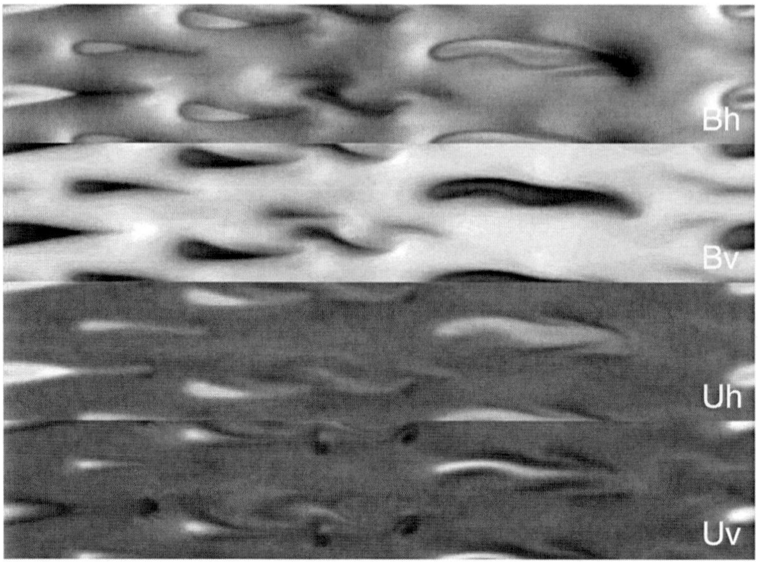

Fig. 7 A stack of horizontal cross sections near mean optical depth unity from a local penumbra model with 3 km grid spacing. The panels show *gray* scale images of the horizontal and vertical magnetic field strengths and velocity amplitudes

To some extent such issues become moot when detailed models become available, or at most they turn into "what-did-I-say" attempts to paste the labels onto the detailed models. It is, however, relevant to remark that the flows that occur in penumbral filaments (and this includes the Evershed flow) are flows that – at least locally – dominate over the magnetic fields. The overturning convective flows are able to push aside the magnetic field, weakening it sufficiently so that it no longer prevents the flow, but allows it. In this respect, the situation is more adequately labeled as characterized by magnetic "gaps," rather than one characterized by magnetic "flux tubes," in as much as "flux tubes" traditionally have been thought of as magnetic field channels that constrain the motions of gas to be along the magnetic field in the structures, rather than flowing relatively freely into and out of them, carrying a relatively weak magnetic field along. In the penumbra, the filaments do represent structures that are "open" (at the bottom) rather than "closed," and where the topology of the flow determines the topology of the magnetic field rather than the other way around. Of course, the gaps are not entirely "field free," but they fulfil essentially the same function as the idealized field-free gaps originally proposed by Spruit and Scharmer (2006) and Scharmer and Spruit (2006).

With respect to the "siphon flow" type of models (Montesinos and Thomas 1989, 1993, 1997; Thomas and Montesinos 1990, 1991; Thomas 2005), it appears safe to conclude that they are not directly relevant for penumbral fine structure. It is possible that localized siphon flows do occur, for example, along field lines that happen to connect locations inside the penumbra with strongly evacuated flux concentrations outside the sunspot, but these must be rare, and cannot be a significant contributing effect in the creation and maintenance of penumbral fine structure.

The idea that downward flux pumping *outside* the sunspot could contribute to creating and maintaining the penumbra fine structure by "pulling down" field lines, and hence helps explain the occurrence of nearly horizontal field lines in the penumbra (Thomas et al. 2002; Weiss et al. 2004) also does not work, or at least is not needed. However, *inside* the penumbra, pulling down of field lines by convection is certainly relevant and operative, and near the periphery of penumbrae some aspect of such a process may be relevant as well (cf. Brummell et al. 2008).

An interesting question, which can be more easily addressed with the local type of penumbra models that allow maximizing the numerical resolution, is whether and to what extent the reduction of the magnetic field strength in the gaps depends on the numerical resolution. Should one expect, for example, that the magnetic field strength inside the gaps is reduced to a larger and larger extent as the numerical resolution is increased in the models? Only future experiments can address that question quantitatively, but two fundamental aspects deserve to be mentioned already before the results of such investigations are in:

- To the extent that the flows are of finite duration, and their source regions are not entirely field free (both assumptions are unavoidably satisfied in the penumbral context), there is a definitive limit to how weak the magnetic fields can get, in that even with "frozen-in" field lines the magnetic flux density is reduced only by an amount corresponding to the perpendicular expansion of the flow from the "source" point to the point of observation.
- Moreover, a general conclusion from other situations with very low diffusivities is that some sort of turbulence typically develops, resulting in large scale statistics that become essentially independent of the scale at which diffusivity sets in. To the extent that this happens here, one should not expect a continued dependence on the numerical resolution.

4 Conclusions and Concluding Remarks

In conclusion, with access to realistic three-dimensional numerical modeling of sunspots, we are now in an excellent position to finally answer the intriguing questions raised by Evershed's observations a 100 years ago (Evershed 1909a,b). Accurate and detailed matches and comparisons with observations must await the next generation of improved models, but the current set of models already allow a number of conclusions regarding the physical mechanisms at work in the penumbra:

- The basic mechanism, responsible for creating the penumbra fine structure as well as the Evershed flow, is thermal convection, modified by the presence of a strong and inclined magnetic field, with the penumbral filaments being the top layer manifestation of the heat transport, analogous to the convection pattern in the surrounding photosphere.

- In the penumbra, the overturning and cooling of ascending hot gas is constrained to happen primarily in the radial direction, with the inclination of the magnetic field braking the symmetry and causing the overturning to happen as an outward directed mainly radial flow – this is the physical background for the Evershed flow.
- The convective motions are able to force open "gaps" in the magnetic field. These gaps are of course not "field-free," but the magnetic field strengths are reduced enough to allow the overturning convective flows to occur.
- Energy transport requirements dictate that ascending hot parcels of gas spend similar intervals of time, with similar flow speeds, while cooling in the optically visible surface layers as they do in the photospheric granulation pattern. Individual sections of penumbral filaments must therefore have lengths similar to granulation scales, and there both upflows and downflows must occur, intertwined, in the penumbral filaments.

With qualitative and semi-quantitative agreement of the results of realistic numerical simulations from two independent groups (Heinemann et al. 2007; Rempel et al. 2009), and with quantitative results moving in the right direction with improvements in the numerical resolution and in the horizontal and vertical extent of the models, the results that have already been gleaned from the first generation of models are likely to hold up to future scrutiny. But it will indeed be interesting to see models improving to the point where direct, quantitative comparisons can be made between observations and synthetic diagnostic produced from the models.

The intricate correlations between magnetic field strength and inclination, on the one hand, and velocity, temperature, and gas density, on the other hand, makes it virtually impossible to uniquely "invert" even very detailed observations. Indeed, the more detailed the observations are (in the spatial, temporal, and spectral domains), the more parameters it would take to represent the physical properties of inversion type models.

On the other hand, with access to detailed numerical models, one can compute synthetic spectral lines and related diagnostics such as synthetic magnetograms and Doppler maps, and compare both their qualitative features and more quantitative "finger prints," such as average spectral line shapes and other statistical measures. Each and every such quantitative measure represents a "collapse" of information, making it possible to compare unique signatures of the complex reality with analogous results from (necessarily always less complex) numerical models. If accurate matches are obtained, one may still argue that this does not constitute formal proof of correctness of the models, but to the extent that the models adhere closely to the physical properties (equation of state, realistic radiative energy transfer, etc.) of the situation they are representing, one would be hard put to argue that a good match is inconclusive.

Acknowledgment We thank the conference organizers for a very good meeting and the editors for excellent instructions and an amazing patience. Discussions and collaborations with Henk Spruit and Tobias Heinemann are gratefully acknowledged. The work of ÅN is supported by the Danish Natural Science Research Council. Computing resources were provided by the Danish Center for Scientific Computing.

References

Bellot Rubio, L. R., Balthasar, H., Collados, M., Schlichenmaier, R. 2003, A&A, 403, L47
Bellot Rubio, L. R., Balthasar, H., Collados, M. 2004, A&A, 427, 319
Borrero, J. M., Solanki, S. K., Bellot Rubio, L. R., Lagg, A., Mathew, S. K. 2004, A&A, 422, 1093
Borrero, J. M., Lagg, A., Solanki, S. K., Collados, M. 2005, A&A, 436, 333
Brummell, N. H., Tobias, S. M., Thomas, J. H., Weiss, N. O. 2008, ApJ, 686, 1454
Danielson, R. E. 1961, ApJ, 134, 289
Evershed, J. 1909a, The Observatory, 32, 291
Evershed, J. 1909b, MNRAS, 69, 454
Heinemann, T., Nordlund, Å., Scharmer, G. B., Spruit, H. C. 2007, ApJ, 669, 1390
Hurlburt, N. E., Matthews, P. C., Rucklidge, A. M. 2000, Solar Phys., 192, 109
Jahn, K., Schmidt, H. U. 1994, A&A, 290, 295
Mamadazimov, M. 1972, Solar Phys., 22, 129
Martens, P. C. H., Hurlburt, N. E., Title, A. M., Acton, L. W. 1996, ApJ, 463, 372
Martínez Pillet, V. 2000, A&A, 361, 734
Montesinos, B., Thomas, J. H. 1989, ApJ, 337, 977
Montesinos, B., Thomas, J. H. 1993, ApJ, 402, 314
Montesinos, B., Thomas, J. H. 1997, Nat, 390, 485
Nordlund, Å. 2006, In: Solar MHD Theory and Observations: A High Spatial Resolution Perspective, J. Leibacher, R. F. Stein, H. Uitenbroek (eds.), ASP Conf. Ser., vol. 354, p. 353
Nordlund, Å. 2008, Physica Scripta Vol. T, 133, 014002
Rempel, M., Schüssler, M., Knölker, M. 2009, ApJ, 691, 640
Scharmer, G. B., Spruit, H. C. 2006, A&A, 460, 605
Scharmer, G. B., Nordlund, Å., Heinemann, T. 2008, ApJ, 677, L149
Schlichenmaier, R., Solanki, S. K. 2003, A&A, 411, 257
Schlichenmaier, R., Jahn, K., Schmidt, H. U. 1998a, ApJ, 493, L121+
Schlichenmaier, R., Jahn, K., Schmidt, H. U. 1998b, A&A, 337, 897
Schlichenmaier, R., Bruls, J. H. M. J., Schüssler, M. 1999, A&A, 349, 961
Schmidt, H. U., Spruit, H. C., Weiss, N. O. 1986, A&A, 158, 351
Schmidt, W., Fritz, G. 2004, A&A, 421, 735
Schüssler, M., Vögler, A. 2006, ApJ, 641, L73
Solanki, S. K. 2003, A&A Rev., 11, 153
Solanki, S. K., Montavon, C. A. P. 1993, A&A, 275, 283
Spruit, H. C., Scharmer, G. B. 2006, A&A, 447, 343
Stein, R. F., Nordlund, A. 1998, ApJ, 499, 914
Thomas, J. H. 2005, A&A, 440, L29
Thomas, J. H., Montesinos, B. 1990, ApJ, 359, 550
Thomas, J. H., Montesinos, B. 1991, ApJ, 375, 404
Thomas, J. H., Weiss, N. O. 2004, ARA&A, 42, 517
Thomas, J. H., Weiss, N. O., Tobias, S. M., Brummell, N. H. 2002, Nat, 420, 390
Thomas, J. H., Weiss, N. O., Tobias, S. M., Brummell, N. H. 2006, A&A, 452, 1089
Tildesley, M. J., Weiss, N. O. 2004, MNRAS, 350, 657
Weiss, N. O., Thomas, J. H., Brummell, N. H., Tobias, S. M. 2004, ApJ, 600, 1073

The Magnetic Field of Solar Spicules

R. Centeno, J. Trujillo Bueno, and A. Asensio Ramos

Abstract Determining the magnetic field of solar spicules is vital for developing adequate models of these plasma jets, which are thought to play a key role in the thermal, dynamic, and magnetic structure of the chromosphere. Here we report on magnetic spicule properties in a very quiet region of the off-limb solar atmosphere, as inferred from new spectropolarimetric observations in the He I 10830 Å triplet. We have used a novel inversion code for Stokes profiles caused by the joint action of atomic level polarization and the Hanle and Zeeman effects (HAZEL) to interpret the observations. Magnetic fields as strong as 40 G were unambiguously detected in a very localized area of the slit, which may represent a possible lower value of the field strength of organized network spicules.

1 Introduction

The first observational evidence of solar spicules came in the drawings of Father Angelo Secchi in the late nineteenth century. He recorded the shape of these off-limb jet-like structures and listed some of their properties. Spicules can be described as rapidly evolving chromospheric plasma jets, protruding outside the solar limb into the corona. It is thought that they constitute an important ingredient of the mass balance of the solar atmosphere, since they are estimated to carry about 100 times the mass of the solar wind. After the spicular material is shot up (with typical apparent velocities around 25 km s^{-1}) and has reached its maximum height, it returns to the surface along the same path or a different one. However, many spicules do not seem

R. Centeno (✉)
High Altitude Observatory, Boulder, USA
and
Instituto de Astrofísica de Canarias, La Laguna, Spain

J. Trujillo Bueno
Instituto de Astrofísica de Canarias, La Laguna, Spain
and
Consejo Superior de Investigaciones Científicas, Spain

A. Asensio Ramos
Instituto de Astrofísica de Canarias, La Laguna, Spain

S.S. Hasan and R.J. Rutten (eds.), *Magnetic Coupling between the Interior and Atmosphere of the Sun*, Astrophysics and Space Science Proceedings,
DOI 10.1007/978-3-642-02859-5_19,

to retract, but rather fade away (see De Pontieu et al. 2007). The average direction of spicules deviates from the vertical, reaching typical heights of 6,500–9,500 km. Observations yield densities about (3×10^{-13} g cm^{-3}) and temperatures in the range 5,000–15,000 K that seem to be constant with height (see Beckers 1972 for an early review of spicule properties).

All classical spicule models make use of a magnetic flux tube that expands from the photosphere all the way up into the corona as their main ingredient (e.g., Sterling 2000, De Pontieu et al. 2004). An injection of energy into the flux tube is required to launch the material and to raise it up to heights of several thousand kilometers. Although the various models seem to explain some of the observational aspects of quiet-Sun spicules, they all fail to reproduce one or other observed parameter. One of the key impediments is our poor knowledge of the magnetic properties of spicules. We need more and better observations to constrain the models, and this is what motivates the present investigation. We want to find reliable constraints of some of the physical aspects of spicules, focusing, in particular, on understanding the magnetic field topology and its behavior along the length of the spicule.

We use new spectropolarimetric measurements of the He I 10830 Å multiplet beyond the limb to infer the magnetic properties of quiet-Sun spicules. The information was retrieved by inverting the observed Stokes profiles (caused by atomic level polarization and the Hanle and Zeeman effects), as was first done by Trujillo Bueno et al. (2005). These authors inferred strengths of 10 G in quiet-Sun spicules at an atmospheric height of 2,000 km. They point out, however, that significantly stronger fields could also be present (as indicated by larger Stokes-V signals detected during another observing run). The possibility of magnetic fields significantly larger than 10 G was also tentatively suggested by the He I D_3 measurements of López Ariste and Casini (2005), although these spicules emanated from active plage. More detailed spectropolarimetric observations of spicules in the He I D_3 multiplet were carried out by Ramelli et al. (2006), who found $B \approx 10$ G in quiet Sun and $B \approx 50$ G in more active areas. The presented He I 10830 Å spectropolarimetric observations of quiet-Sun spicules provide an unambiguous demonstration that the magnetic field strength of some spicules can be significantly large.

2 Observations

Observations were carried out with the Tenerife Infrared Polarimeter (TIP, Martínez Pillet et al. 1999) at the German Vacuum Tower Telescope (Tenerife, Spain) on 17 August 2008. The TIP instrument allowed us to measure (almost) simultaneously the full Stokes vector of the 10,830 Å spectral region for all the points along the spectrograph slit, with spectral and spatial samplings of 11 mÅ and 0.17″, respectively. Standard data reduction routines were applied to all the data-sets, encompassing dark current and flat-field correction of the images as well as polarization calibration and cross-talk correction of the Stokes profiles.

We placed the slit 2″ off, and parallel to, the visible South limb, crossing a forest of spicules. There, we carried out several time series with the slit at fixed distances to the visible limb. The seeing conditions were not optimal and the off-limb pointing rendered the Adaptive Optics system inoperable. However, the atmospheric conditions were very stable (no wind), warranting image stability with no coelostat vibrations during the runs. Each data-set of 50-min duration was averaged in time to obtain a large S/N ratio. Spectral and spatial pixel binning were performed for the same purpose, maintaining sufficient sampling ($\approx$0.7″).

3 Analysis

Solar magnetic fields leave their fingerprints on the emergent polarization patterns of spectral lines that form in the solar atmosphere. This occurs through the Hanle and Zeeman effects.

The spectral line polarization produced by the Zeeman effect is a consequence of the wavelength shifts between the π and σ components of the atomic transitions, as the energy levels split due to the presence of a magnetic field. This splitting is normally proportional to the magnetic field strength and the Landé factor of the level. Typically, fields of 100 G or more are needed to be able to observe the signature of the transverse Zeeman effect on the Stokes Q and U profiles of a spectral line, while much weaker resolved fields are enough to produce measurable Stokes-V signals via the longitudinal Zeeman effect. However, when the magnetic field is too weak and/or when there are mixed magnetic polarities within the spatio-temporal resolution element, the circular polarization produced by the longitudinal Zeeman effect tends to be negligible.

Fortunately, even in the absence of magnetic fields, measurable polarization signals in a spectral line occur if there are population imbalances among the magnetic sub-levels of the atom. The key mechanism that produces this so-called atomic level polarization in the solar atmosphere is the anisotropic illumination of the atoms. Such "optical pumping" needs no magnetic field to operate and it is very effective in generating atomic level polarization when the depolarizing rates from elastic collisions are low. Structures such as chromospheric spicules are subject to the center-to-limb (CLV) variation of the photospheric illumination, receiving more radiation from the plasma that is directly underneath them than from the sides.

The Hanle effect is the modification of the atomic level polarization due to the presence of a magnetic field inclined with respect to the axis of symmetry of the radiation field. It is sensitive to weaker magnetic fields than those needed to induce a measurable Zeeman polarization signal and it does not tend to cancel out when mixed polarities are present (see Trujillo Bueno 2005). The observational signatures of the Hanle effect in the 90° scattering geometry of our observations are a reduction of the linear polarization amplitude and a rotation of the direction of linear polarization, with respect to the unmagnetized case.

The formation of the He I 10830 Å triplet is sensitive to both the Zeeman and Hanle effects. We have taken advantage of this fact to "measure" the magnetic field in spicules.

3.1 Detection of Zeeman-Induced Stokes V

The left-most panel of Fig. 1 shows time-averaged intensity as a function of wavelength and position along the slit. The bright and fainter vertical strips correspond to the red and blue components of the He multiplet, respectively. Stokes I provides physical and thermodynamical information: the damping, the Doppler width, the optical depth, and the macroscopic velocity of the plasma. Combining it with the information carried by Stokes Q and U (second and third panels, respectively), one can infer the magnetic field orientation. However, in the Hanle saturation regime (which is above about 8 G for this multiplet), linear polarization is barely responsive to the magnetic field strength, hindering its determination.

One of the most striking findings in this particular observation was the clear detection of a Zeeman-induced Stokes-V signature (right-most panel of Fig. 1). The circular polarization signal was, in many cases, large enough that it allowed us to pin down magnetic field strengths beyond the Hanle saturation value. The antisymmetric Stokes V profile must be produced by a net line-of-sight (LOS) component of the magnetic field, $B_{\rm LOS}$, that would be fully resolved. However, cancelation effects due to the unresolved magnetic structure in our spatio-temporal resolution element (or along the line-of-sight) make the inferred $B_{\rm LOS}$ a lower limit for the field strength. For the profiles in Fig. 2 (indicated by the horizontal lines in Fig. 1), the inferred value is $B_{\rm LOS} \approx 25$ G, provided by HAZEL.

How do we interpret this signal? In a picture where the spicules are oriented arbitrarily along the line of sight, we would expect the $B_{\rm LOS}$ to cancel, producing no net Zeeman Stokes V. However, these data would seem to imply a preferred direction of the magnetic field. Along the slit there are areas of strong and weak Stokes-V signals, suggesting that, in the latter cases, the magnetic field is conspiring to minimize (or, at least, reduce) the net LOS component.

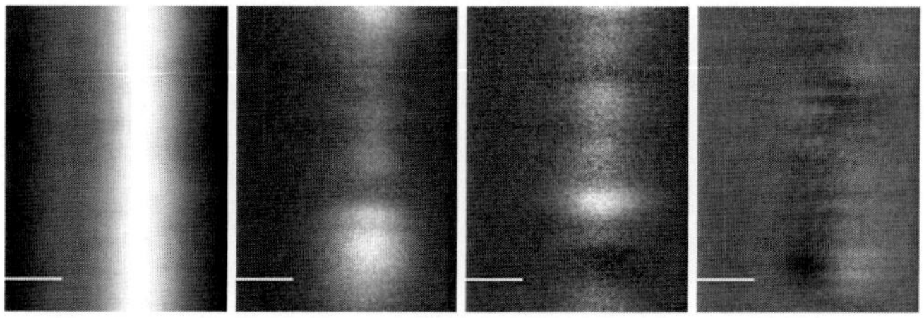

Fig. 1 From left to right, maps of Stokes I, Q, U, and V. The x-axis represents wavelength (increasing to the right), the y-axis the position along the slit (which is about 80″ long)

3.2 Inversions

To determine the magnetic field strength and other physical quantities from the observations, we inverted the full Stokes vector for every position along the slit using the inversion code HAZEL (see Asensio Ramos et al. 2008) to complete this task. HAZEL accounts for the physical ingredients and mechanisms operating in the generation of polarized light in this kind of observations: optical pumping, atomic level polarization, and the Hanle and Zeeman effects. Radiative transfer is computed in a constant-property slab that is permeated by a deterministic magnetic field. The slab is located at height h above the visible solar surface, and is illuminated by the CLV of the photospheric continuum. The slab's optical thickness, τ, accounts for the integrated number of emitters and absorbers along the line of sight, taking care of the collective effect of having several spicules interposed along the path (although we cannot prescribe how many).

Figure 2 is an example in which our measurement shows a sizable Stokes-V profile, at the location shown in Fig. 1. The open circles represent the observed profiles while the solid line shows the best fit (in a χ^2 sense) obtained from a HAZEL inversion. The inferred magnetic field strength is 36 G, with inclination 38.6° from the solar local vertical and azimuth −2.2° with respect to the LOS. The magnetic field orientation is very well constrained by the observed Stokes Q and U profiles. Except for the 180° and the Van-Vleck ambiguities (see Asensio Ramos et al. 2008 and references therein), a good fit is only possible in a very narrow range of values. However, the field strength is well determined only when the Stokes-V signal is present.

We applied this inversion procedure to all the pixels along the slit and both datasets, deriving the magnetic field for all the spatial positions at two heights above the

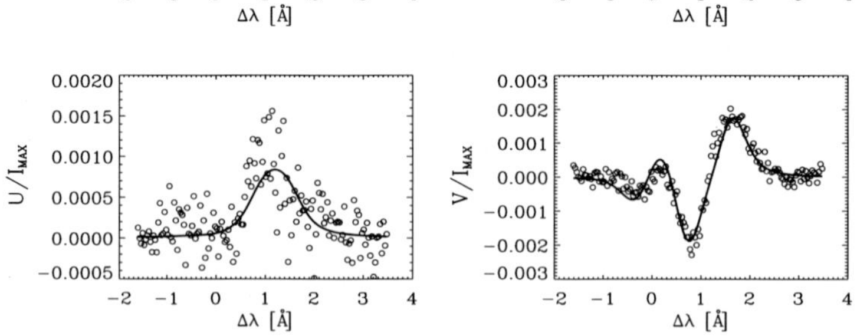

Fig. 2 Best fit produced by HAZEL for a set of Stokes profiles for one position along the slit that has a sizable Stokes-V signal, with $B = 36\,\mathrm{G}$, $\theta_B = 38.6°$, and $\chi_B = -2.2°$. The reference direction for Q is the tangential to the closest solar limb

visible limb. From these inversion we were able to trace the magnetic field vector and construct a reliable picture of its behavior along the spicules. Variations of the field strength and topology were detected, which we will describe in detail in forthcoming publications. Likewise, new observations will be carried out in 2009 to complement this preliminary work.

4 Conclusion

We carried out spectropolarimetric measurements of quiet Sun spicules in the He I 10830 Å triplet, detecting clear Stokes V signals that allow us to infer magnetic field strengths beyond the Hanle saturation regime. Values as high as 40 G were found in localized regions of the slit, which may correspond to organized network spicules or perhaps a macro-spicule.

We determined the magnetic field vector of all the pixels along the slit at two heights from the South limb, detecting spatial variations in the magnetic field strength and orientation. We plan to pursue this investigation further with new observations in the 10,830 Å multiplet complemented with other useful data, such as Hα or Ca II filtergrams and He D_3 spectropolarimetry.

Acknowledgment The National Center for Atmospheric Research is sponsored by the US National Science Foundation. Financial support by the Spanish Ministry of Science through project AYA2007-63881 and by the European Commission via the SOLAIRE network (MTRN-CT-2006-035484) are gratefully acknowledged.

References

Asensio Ramos, A., Trujillo Bueno, J., Landi Degl'Innocenti, E. 2008, ApJ, 683, 542
Beckers, J. M. 1972, ARA&A, 10, 73
De Pontieu, B., Erdélyi, R., James, S. P. 2004, Nat, 430, 536
De Pontieu, B., McIntosh, S., Hansteen, V. H., et al. 2007, PASJ, 59, 655
López Ariste, A., Casini, R. 2005, A&A, 436, 325
Martínez Pillet, V., Collados, M., Sánchez Almeida, J., et al. 1999, In: High Resolution Solar Physics: Theory, Observations, and Techniques, T. R. Rimmele, K. S. Balasubramaniam, R. R. Radick (eds.), ASP Conf. Ser., vol. 183, p. 264
Ramelli, R., Bianda, M., Merenda, L., Trujillo Bueno, T. 2006, ASP Conf. Ser., 358, 448
Sterling, A. C. 2000, Solar Phys., 196, 79
Trujillo Bueno, J. 2005, In: The Dynamic Sun: Challenges for Theory and Observations, ESA-SP, p. 600
Trujillo Bueno, J., Merenda, L., Centeno, R., Collados, M., Landi Degl'Innocenti, E. 2005, ApJ, 619, L191

Three-Dimensional Magnetic Reconnection

C.E. Parnell and A.L. Haynes

Abstract The importance of magnetic reconnection as an energy release mechanism in many solar, stellar, magnetospheric and astrophysical phenomena has long been recognised. Reconnection is the only mechanism by which magnetic fields can globally restructure, enabling them to access a lower energy state. Over the past decade, there have been some major advances in our understanding of three-dimensional reconnection. In particular, the key characteristics of 3D magnetohydrodynamic (MHD) reconnection have been determined. For instance, 3D reconnection (1) occurs with or without nulls, (2) occurs continuously and continually throughout a diffusion region and (3) is driven by counter rotating flows. Furthermore, analysis of resistive 3D MHD magnetic experiments have revealed some intriguing effects relating to where and how reconnection occurs. To illustrate these new features, a series of constant-resistivity experiments, involving the interaction of two opposite-polarity magnetic sources in an overlying field, are considered. Such a simple interaction represents a typical building block of the Sun's magnetic atmosphere. By following the evolution of the magnetic topology, we are able to explain where, how and at what rate the reconnection occurs. Remarkably, there can be up to five energy release sites at any one time (compared to one in the potential case) and the duration of the interaction increases (more than doubles) as the resistivity decreases (by a factor of 16). The decreased resistivity also leads to a higher peak ohmic dissipation and more energy being released in total, as a result of a greater injection of Poynting flux.

1 Introduction

Magnetic reconnection is a fundamental plasma physics process that is essential to many phenomena on the Sun, such as solar flares, CMEs, coronal heating, nano/microflares, X-ray bright points, explosive events and the solar dynamo. It is also an extremely important mechanism in the magnetosphere, where it plays a

C.E. Parnell (✉) and A.L. Haynes
School of Mathematics and Statistics, University of St Andrews, Scotland, UK

S.S. Hasan and R.J. Rutten (eds.), *Magnetic Coupling between the Interior and Atmosphere of the Sun*, Astrophysics and Space Science Proceedings,
DOI 10.1007/978-3-642-02859-5_20, © Springer-Verlag Berlin Heidelberg 2010

key role in linking the magnetic fields from the Sun and Earth and in powering flux transfer events and substorms. It is also very important in many astrophysics applications such as accretion discs, stellar flares and coronæ, astrophysical jets and stellar dynamos.

Magnetic reconnection is important for two key reasons: first, it is a mechanism by which energy stored in a magnetic field may be rapidly released and converted into thermal and kinetic energy, causing heating, bulk plasma motions and the acceleration of particles. Second, it is the mechanism by which global restructuring of the magnetic field may take place. Indeed, it is this restructuring that facilitates the release of energy by allowing the magnetic field to access a lower energy state.

The entire surface of the Sun is threaded by magnetic fields that are directed both into and out of the Sun. The magnetic field is clumped into numerous photospheric flux features that range from large sunspots with fluxes of about 10^{20} Mx (Schrijver and Harvey 1994) down to tiny intranetwork fields with just 10^{16} Mx or less (Wang et al. 1995). These photospheric features, both small and large, are the feet of magnetic loops that are intermingled and expand to fill the whole of the solar atmosphere (Fig. 1).

Supergranular and granular flows, overshoots of convection cells from below the Sun's surface, drive these photospheric flux features towards downflow regions at the convergence of three or more cells. These photospheric motions result in the following behaviours being displayed by the flux features (Schrijver et al. 1997): (1) emergence, in which pairs of features, with equal, but opposite-polarity flux, appear; (2) cancellation, the disappearance (generally through submergence) of equal amounts of flux from a pair of opposite-polarity features; (3) coalescence, the merging of two like-polarity features creating a larger feature and (4) fragmentation, the splitting of a large feature into two or more smaller features. Clearly, all these changes in photospheric flux cause changes to the intermingled magnetic loops in the atmosphere above, leading to the redistribution of flux between features – that is, these flows and flux changes drive magnetic reconnection (Close et al. 2004, 2005).

To estimate the time it takes to completely redistribute all the flux between the features during solar minimum, Close et al. (2004, 2005) considered a 12 h series of high-resolution MDI magnetograms in which all photospheric magnetic features

Fig. 1 Potential magnetic field extrapolation from a quiet-sun magnetogram (Close et al. 2004)

were identified and then tracked in time. Their birth mechanism (emergence or fragmentation) was noted, as was their death mechanism (cancellation or coalescence). Potential field extrapolations were then used to determine the connectivity of the photospheric flux features. By assuming that the evolution of the field went through a series of equi-potential states, the observed connectivity changes were coupled with the birth and death information of the features to determine the coronal flux recycling/reconnection time. Remarkably, it was found that during solar minimum the total flux in the solar corona completely changes all its connections in just 1.4 h (Close et al. 2004, 2005), a factor of ten times faster than the time it takes for all the flux in the quiet-Sun photosphere to be completely replaced (Schrijver et al. 1997; Hagenaar et al. 2003).

Clearly, reconnection operates on a wide range of scales from kinetic to MHD. The micro-scale physics at the kinetic scales governs the portioning of the released energy into its various new forms and plays a role in determining the rate of reconnection. MHD (the macro-scale physics) determines where the reconnection takes place and, hence where the energy is deposited, and also effects the reconnection rate. In this paper, we focus on macro-scale effects, and investigate the behaviour of three-dimensional (3D) reconnection using MHD numerical experiments.

Two-dimensional (2D) reconnection has been studied in detail and is relatively well understood, especially in the solar and magnetospheric contexts. Over the past decade, our knowledge of 3D reconnection has significantly improved (Lau and Finn 1990; Priest and Démoulin 1995; Démoulin et al. 1996; Priest and Titov 1996; Birn et al. 1998; Longcope 2001; Hesse et al. 2001; Pritchett 2001; Priest et al. 2003; Linton and Priest 2003; Pontin and Craig 2006; De Moortel and Galsgaard 2006a,b; Pontin and Galsgaard 2007; Haynes et al. 2007; Parnell et al. 2008). It is abundantly clear that the addition of the extra dimension leads to many differences between 2D and 3D reconnection. In Sect. 2, we first review the key characteristics of both 2D and 3D reconnection. Then, in Sect. 3, we consider a series of 3D MHD experiments in order to investigate where, how and at what rate reconnection takes place in 3D. The effects of varying resistivity and the resulting energetics of these experiments are discussed in Sect. 4. Finally, in Sect. 5, we draw our conclusions.

2 Characteristics of 2D and 3D Reconnection

A comparison of the main properties of reconnection in 2D and 3D highlight the significant differences that arise due to the addition of the extra dimension (Table 1). In 2D, magnetic reconnection can only occur at X-type nulls. Here, pairs of field lines with different connectivities, say $A \rightarrow A'$ and $B \rightarrow B'$, are reconnected at a single point to form a new pair of field lines with connectivities $A \rightarrow B'$ and $B \rightarrow A'$. Hence, flux is transferred from one pair of flux domains into a different pair of flux domains. The fieldline mapping from $A \rightarrow A'$ onto $A \rightarrow B'$ is discontinuous and

Table 1 A comparison of the main characteristics of reconnection in 2D and 3D

2D reconnection	3D reconnection
1. Must occur at X-type null points	1. Can occur at null points or in the absence of null points
2. Occurs at a single point	2. Occurs continually and continuously throughout diffusion region volume – not at a single point
3. Pairs of field lines break and recombine into two new pairs of field lines	3. Pairs of field lines or even pairs of surfaces break, but do not recombine into two new pairs of field lines or surfaces
4. Discontinuous field line mapping	4. Continuous or discontinuous field line mapping
5. Stagnation type flow	5. Counter-rotating flows

jumps at the instant of reconnection itself. There is been a considerable body of work on 2D reconnection and a good review of this work can be found in Priest and Forbes (2000).

In 3D, magnetic reconnection can occur both at 3D nulls, but more commonly it will occur in a null-less region of magnetic field, for instance, in a hyperbolic flux tube (Galsgaard et al. 2003; Linton and Priest 2003; Pontin et al. 2005; Aulanier et al. 2006; De Moortel and Galsgaard 2006a,b; Wilmot-Smith and De Moortel 2007), in an elliptic flux tube (Wilmot-Smith and Priest 2007) or near a separator (Longcope and Cowley 1996; Galsgaard and Nordlund 1997; Galsgaard et al. 2000; Haynes et al. 2007; Parnell et al. 2008). Here, the diffusion region where reconnection occurs is not a single point, but is a finite volume.

Many, but not all, of the above situations depend on the fact that field lines from flux domains with two different connectivities, say $A \rightarrow A'$ and $B \rightarrow B'$, are reconnected to form field lines in two new flux domains with connectivities $A \rightarrow B'$ and $B \rightarrow A'$, exactly as in the 2D case. In 3D, however, it is generally not possible to identify pairs of field lines (or even pairs of surfaces) that reconnect to form new pairs of field lines (or surfaces). Instead, reconnection will occur continually and continuously throughout the finite diffusion region, converting flux from two domains into flux in two other domains. A consequence of having continual and continuous reconnection is that in 3D, the field line mapping can be continuous between pre- and post-reconnected field lines. The theory behind this behaviour is explained in Hornig and Priest (2003) and is illustrated very nicely using numerical experiments in Pontin et al. (2005) and Aulanier et al. (2006).

Using theoretical arguments, Hornig and Priest (2003) determined that counter rotating flows are an essential ingredient of 3D reconnection. Wilmot-Smith and De Moortel (2007) and Parnell et al. (2009) have both found counter-rotating flows about the reconnection sites in their numerical experiments.

3 3D Magnetic Interactions

To determine how reconnection occurs in 3D, we investigated a simple interaction that may be considered as a basic building block of the Sun's complex coronal magnetic field. The interaction involves two opposite-polarity magnetic features which are driven past each other in an overlying field (Galsgaard et al. 2000; Parnell and Galsgaard 2004; Galsgaard and Parnell 2005; Haynes et al. 2007, 2009; Parnell et al. 2008, 2009). One would imagine that the resulting magnetic interaction is trivial, after all there are only four sources: $P1$ and $N1$ – the two opposite-polarity sources prescribed on the base, and $P\infty$ and $N\infty$ – the sources that produce the overlying magnetic field (Fig. 2a). However, the interaction turns out to be much more complicated than one initially imagines.

3.1 Potential 3D Interaction

For comparative purposes, we first discuss the details of the potential 3D interaction (Haynes et al. 2007). The topology of the initial magnetic field involves just three flux domains: one domain linking $P1$ to $N\infty$, another linking $P\infty$ to $N1$ and the overlying field which links $P\infty$ to $N\infty$, as seen in Fig. 2b. In each phase, throughout the evolution, there are two null points: one positive and the other negative. Both of these nulls are situated on the base. From each of these null points a separatrix-surface emanates. In Fig. 2b, which shows the topology of the *initial phase*, the separatrix surfaces are mapped out by the field lines lying in them. The blue (red) field lines show the separatrix surface originating from the positive (negative) null.

The two opposite-polarity magnetic sources $P1$ and $N1$ are driven in an anti-parallel manner (Fig. 2a). In this subsection, we assume that they evolve through a

Fig. 2 (**a**) Sketch of the experimental setup showing the two opposite-polarity sources, $P1$ and $N1$ and the overlying field due to $P\infty$ and $N\infty$. The *arrows* indicate the direction of advection of $P1$ and $N1$. (**b**) A three-dimensional view of the initial magnetic field showing fieldlines in the positive (*blue*) and negative (*red*) separatrix surfaces (color illustration are available in the on-line version)

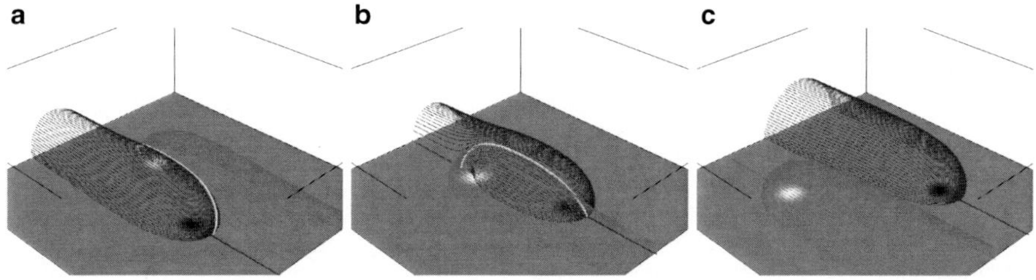

Fig. 3 Three-dimensional views of the potential magnetic topology evolution during the interaction of two opposite-polarity features in an overlying field: (**a**) single-separator closing phase; (**b**) single-separator opening phase; and (**c**) final phase. Field lines lying in the separatrix surfaces from the positive (*blue*) and negative (*red*) nulls are shown. The *yellow lines* indicate the separators (color illustration are available in the on-line version)

series of equi-potential states. This means that the different flux domains interact (reconnect) the moment the separatrix surfaces come into contact. Hence, the first change to a new magnetic topology (new phase) starts as soon as the flux domains from $P1$ and $N1$ come into contact. When this happens, a new flux domain and a separator (yellow curve) are created (Fig. 3a). We call this phase the *single-separator closing phase*, because the reconnection at this separator transfers flux from the open $P1 - N\infty$ and $P\infty - N1$ domains to the newly formed closed, $P1 - N1$, domain and the overlying, $P\infty - N\infty$, domain.

When the sources $P1$ and $N1$ reach the point of closest approach, all the flux from them has been completely closed and they are fully connected. This state was reached via a global separatrix bifurcation. As they start moving away from each other, the closed flux starts to re-open and a new phase is entered (Fig. 3b). Again, there is still only one separator, but reconnection at this separator now re-opens the flux from the sources (i.e., flux is transferred from the closed, $P1 - N1$, and overlying, $P\infty - N\infty$, domains to the two newly formed re-opened, $P1 - N\infty$, and, $P\infty - N1$, domains). This is known as the *single-separator re-opening phase*.

Eventually, the two sources $P1$ and $N1$ become completely unconnected from each other, leaving them each just connected to a single source at infinity, and surrounded by overlying field (Fig. 3c). In this phase, the *final phase*, there are no separators and there is no reconnection. The field is basically the same as that in the initial phase, but the two sources ($P1$ and $N1$) and their associated separatrix surfaces and flux domains have swapped places.

To visualize the above flux domains, and therefore the magnetic evolution more clearly, we plot 2D cuts taken in the $y = 0.5$ planes (Fig. 4). In the three frames of this figure, there are no field lines lying in the plane. Instead, the thick and thin lines show the intersections of the positive and negative separatrix surfaces, respectively, with the $y = 0.5$ plane. Where these lines cross there will be a separator threading the plane, shown by a diamond. These frames clearly show the numbers of flux domains and separators during the evolution of the equi-potential field. They are useful as they enable us to easily determine the direction of reconnection at each separator by looking at which domains are growing or shrinking.

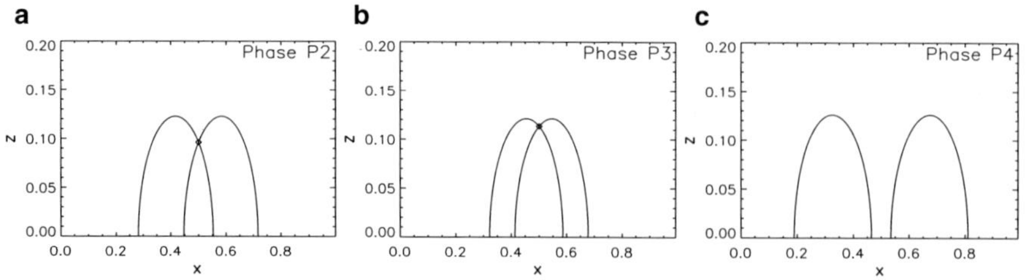

Fig. 4 Cross-sectional cuts ($y = 0.5$ plane) showing the potential magnetic topology evolution during the interaction of two opposite-polarity features in an overlying field. A cut through the positive separatrix surface is shown by a *thick line* whilst the cut through the negative one is given by a *thin line*. *Diamonds* indicate where the separators intersect the plane

Here, the evolution involves $0 \rightarrow 1 \rightarrow 1 \rightarrow 0$ separators during the four topological phases found. However, such an equi-potential evolution can only arise in a situation where there is perfect, instantaneous reconnection, that is, in the limit that the magnetic resistivity, η, tends to ∞. This is not very realistic and so, in the subsection below, we consider what happens when we have a more realistic finite resistivity.

3.2 3D Resistive MHD Interaction

A series of experiments with different constant resistivity (constant-η) values was considered to see how the evolution of the magnetic topology of the above magnetic interaction varied with varying resistivity (Haynes et al. 2009). A 3D resistive MHD code was used to model these interactions. The initial magnetic setup was exactly that used in the potential case (Fig. 2b) and the sources were driven in the same manner (Fig. 2a). Details of the code used and the setup can be found in Haynes et al. (2009).

For each constant-η experiment, the evolution of the resulting magnetic topology was determined in the same way as above. The different phases found were numbered and described in terms of the numbers of flux domains and separators found, as well as the direction of reconnection at each separator. Table 2 details the topological evolution found in each constant-η experiment.

As described earlier, the equi-potential interaction simply evolves through four phases. From Table 2, however, it is clear that in the case of finite resistivity, there are always more than four phases. Furthermore, it is also clear that as the resistivity decreases, the number of phases increases, as does the overall duration of the interaction. Below we describe the evolution of the $\eta_0/16$ magnetic topology, as this experiment covers all the phases found.

The magnetic topology of the $\eta_0/16$ experiment evolves through seven different topological phases. Figure 5 shows a frame from each of the first six phases. The seventh phase is the same as the final phase seen in the potential evolution and shown in Fig. 3c. For clarity, we also include cross-sections of the magnetic

Table 2 The start times of each of the phases through which the magnetic topology of the various constant resistivity experiments evolve

		Phases (No. separators : No. flux domains)							
Res.	S	1 (0:3)	2 (2:5)	3 (1:4)	4 (5:8)	5 (3:6)	6 (1:4)	7 (0:3)	R_T
Pot.	0	0.0	–	0.45	–	–	4.11	7.76	2.0
η_0	4.8×10^3	0.0	–	1.50	–	6.04	7.02	10.3	2.31
$\eta_0/2$	9.8×10^3	0.0	–	1.78	–	6.46	8.79	11.7	2.68
$\eta_0/4$	2.0×10^4	0.0	1.92	2.07	–	6.89	10.9	13.6	3.01
$\eta_0/8$	3.9×10^4	0.0	2.21	2.35	7.17	7.32	14.2	16.0	3.47
$\eta_0/16$	7.9×10^4	0.0	2.35	2.92	7.60	7.88	18.9	19.2	3.94

Each phase is numbered, with the number of separators and numbers of flux domains given in brackets next to the phase number. S is the average maximum Lundquist number of each experiment. The average mean Lundquist number is a factor of 8 smaller than this value. R_{T} is the number of times that the total flux in a single source reconnects. $\eta_0 = 5 \times 10^{-4}$

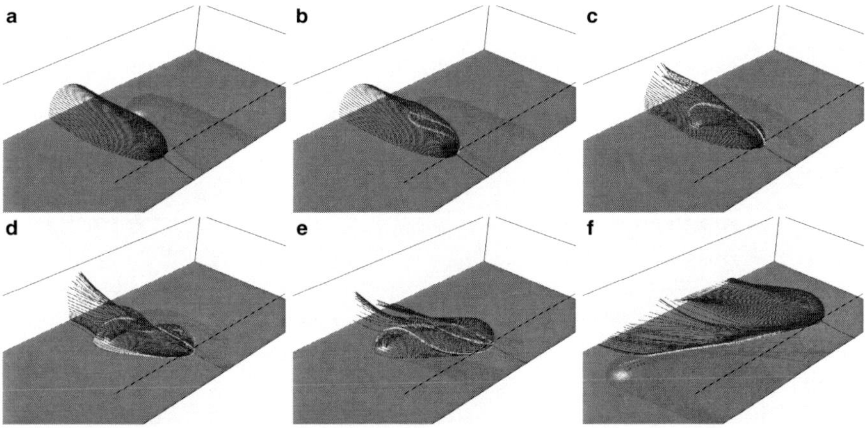

Fig. 5 Three-dimensional views of the magnetic topology evolution during the $\eta_0/16$ constant-η interaction of two opposite-polarity features in an overlying field. Fieldlines in the separatrix surfaces from the positive (*blue*) and negative (*red*) are shown. The *yellow lines* indicate the separators (color illustration are available in the on-line version)

skeleton ($y = 0.5$ cuts) for each of these six frames in Fig. 6. From these two figures, it is clear that the separatrix surfaces intersect each other multiple times giving rise to multiple separators. Also, the filled contours of current in these cross-sections clearly demonstrate that the current sheets in the system are all threaded by a separator. Hence, the number of reconnection sites is governed by the number of separators in the system.

Figures 5a and 6a show the magnetic topology towards the end of the *initial phase*, when the sources $P1$ and $N1$ are still unconnected. To enter a new phase reconnection must occur, producing closed flux. Closed flux connects $P1$ to $N1$ and so must be contained within the two separatrix surfaces, hence these separatrix surfaces must overlap. In the potential situation, the surfaces first overlapped in photosphere

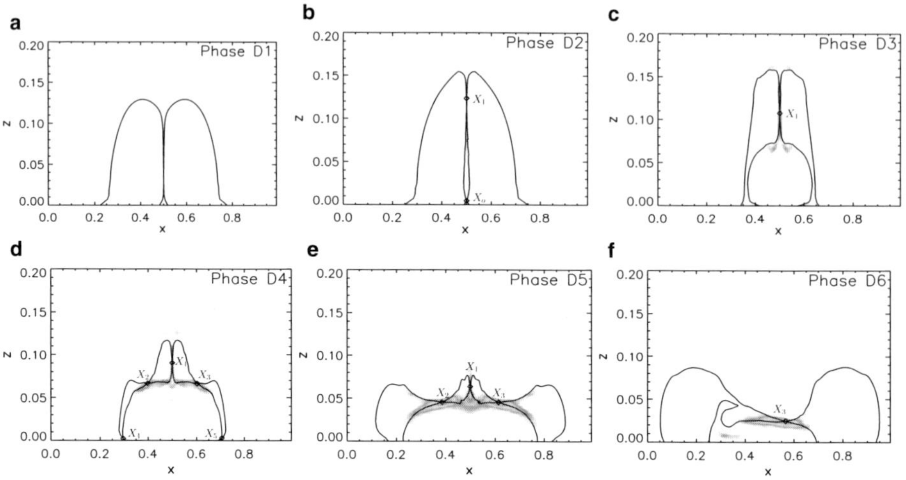

Fig. 6 The $y = 0.5$ cross-sectional cuts through the frames in Fig. 5 showing the magnetic topology evolution. Cut though the positive (negative) separatrix surface is shown by a *thick (thin) line*. *Diamonds* indicate where the separators intersect the plane. The coloured contours, *white – blue – green – red – yellow*, represent increasing current density (color illustration are available in the on-line version)

(at the base) and so a single intersection (separator) line between the surfaces was found. However, in general, when two surfaces intersect, they will cross in two locations, forming two separators. Figures 5a and 6a show that the two surfaces come closest together above the photosphere (base) and so when they first intersect a new pair of separators (X_0 and X_1) is created (via a global double-separator bifurcation (GDSB)) along with two new flux domains: one containing closed flux; the other, trapped below this, containing overlying field (Figs. 5b and 6b). This is known as the *double-separator hybrid phase*. It is called a hybrid phase, because flux is both closed and re-opened during this phase: the top separator, X_1, closes the flux from the sources, while reconnection at the lower separator, X_0, re-opens this closed flux again, with the reopened flux being transferred into the original open flux domains, $P1 - N\infty$ and $P\infty - N1$. The reconnection at the lower separator is very weak and, since the flux in the trapped overlying field domain is small, it soon runs out of flux to reconnect and so disappears, leading to the start of another phase.

Figures 5c and 6c show the magnetic topology in this phase, which involves four flux domains and a single separator, X_1, at which flux is closed. This is equivalent to the *single-separator closing phase* that occurred during the potential evolution. During this phase the closed flux domain increases in size, while the two original open flux domains decrease in size. From Fig. 6c, it is clear that the growing closed flux domain pushes the left-hand side of the negative separatrix surface towards the left-hand side of the positive separatrix surface. Similarly, the right-hand side of the positive separatrix surface is pushed towards the right-hand side of the negative separatrix surface. When these surfaces touch a new phase begins.

The new phase starts when flux starts to be re-opened, and this occurs following the occurrence of another two GDSBs, creating four new separators ($X_2 - X_5$)

and four new domains. The new separators and domains are created as the inner separatrix surface sides bulge out through the sides of the outer separatrix surfaces. These new separators and flux domains can be clearly seem in Figs. 5d and 6d. In total there are eight flux domains and five separators. This phase is called the *quintuple-separator hybrid phase*, as flux is both closing and re-opening during this phase. The central separator is separator X_1 and reconnection here is still closing flux. Reconnection at separators X_2 and X_3 (the two upper side separators) is re-opening flux and so filling the two new flux domains below these separators and the original open flux domains above them. At the two lower side separators, X_4 and X_5, flux is being closed. Below these two separators are two new flux domains, which have been pinched off from the two original open flux domains. Above them are the new re-opened flux domains. It is the flux from these domains that is converted at X_4 and X_5 into closed flux and overlying flux. These lower side separators do not last long and disappear as soon as the flux in the domains beneath them is used up, which leads to the main reopening flux phase.

The next phase is called the *triple-separator hybrid phase*, and is a phase that occurs in all the constant-η experiments (Figs. 5e and 6e). There is a total of six flux domains and three separators: the central separator (X_1) where flux is closed; the side separators (X_2 and X_3) where flux is re-opened.

The above phase ends, and a new phase starts, when the flux in one of the original open flux domains is used up. This leads to the destruction of separators X_1 and X_2 via a GDSB, leaving just separator X_3, which continues to re-open the remain closed flux (Figs. 5f and 6f). This phase is the same as the *single-separator re-opening phase* seem in the equi-potential evolution and it ends once all the closed flux has been reopened. The final phase, as has already been mentioned, is the same as that in Figs. 3c and 4c and involves no reconnection.

3.3 Recursive Reconnection and Reconnection Rates

From Table 2, it is clear that there are three main phases involving reconnection in each of the constant-η experiments: the *single-separator closing phase* (phase 3), the *triple-separator hybrid phase* (phase 5) and the *single-separator re-opening phase* (phase 6). Figure 7a shows a sketch of the direction of reconnection at the separator

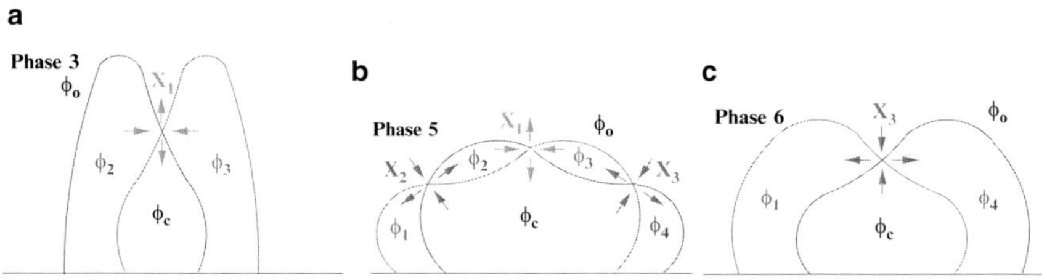

Fig. 7 Sketch showing the direction of reconnection at (**a**) the separator, X_1 in phase 3, (**b**) each of the separators, X_1–X_3 in phase 5 and (**c**) the separator, X_3, in phase 6

(X_1) in phase 3. In this phase, the rate of reconnection across X_1 can be simply calculated from the rate of change of flux in anyone of the four flux domains (flux in domains: ϕ_c – closed, ϕ_o – overlying, ϕ_2 – original positive open, ϕ_3 – original negative open). Hence, the rate of reconnection at X_1 during this phase, α_1, is given by

$$\alpha_1 = \frac{d\phi_c}{dt} = -\frac{d\phi_2}{dt} = -\frac{d\phi_3}{dt} = \frac{d\phi_o}{dt}.$$

Figure 7b illustrates the direction of reconnection at each of the three separators during phase 5. Here, once again the flux is being closed at the central separator (X_1) but at the two outer separators (X_2 and X_3) it is being re-opened. This overlapping of the two reconnection processes allows flux to both close and then re-open multiple times, that is, to be *recursively reconnected.* There are some interesting consequences from this recursive reconnection, which are discussed below.

Here, the rate of reconnection at the separators X_2 and X_3 can be simply determined and is equal to

$$\alpha_2 = \frac{d\phi_1}{dt}, \quad \text{and} \quad \alpha_3 = \frac{d\phi_4}{dt},$$

where ϕ_1 and ϕ_4 are the fluxes in the new re-opened negative and positive flux domains, respectively. The rate of reconnection at X_1 is slightly harder to determine since every flux domain surrounding this separator is losing, as well as gaining flux. The rate of reconnection, α_1, during this phase equals

$$\alpha_1 = \frac{d\phi_1}{dt} - \frac{d\phi_2}{dt} = \frac{d\phi_4}{dt} - \frac{d\phi_3}{dt}.$$

Figure 7c illustrates the direction of reconnection at the separator X_3 during phase 6, the *single separator re-opening phase*. Here, the rate of reconnection α_3 at separator X_3 is simply equal to

$$\alpha_3 = -\frac{d\phi_c}{dt} = \frac{d\phi_4}{dt} = \frac{d\phi_1}{dt} = -\frac{d\phi_o}{dt}.$$

For each experiment, it is possible to calculate the global rate of reconnection in the experiment, α,

$$\alpha = \sum_{i=0}^{5} \alpha_i,$$

where $\alpha_i = 0$ when the separator X_i does not exist. Plots of the global reconnection rate, α, against time for each experiment are shown in Fig. 8, with the start and end of each phase labelled. From these graphs, we note the following points: (1) as the value of η decreases, the instantaneous reconnection rate falls, with the peak rate in the η_0 experiment some 2.4 times greater than the peak rate in the $\eta_0/16$ experiment, and (2) as η decreases, the overall duration of the interaction increases.

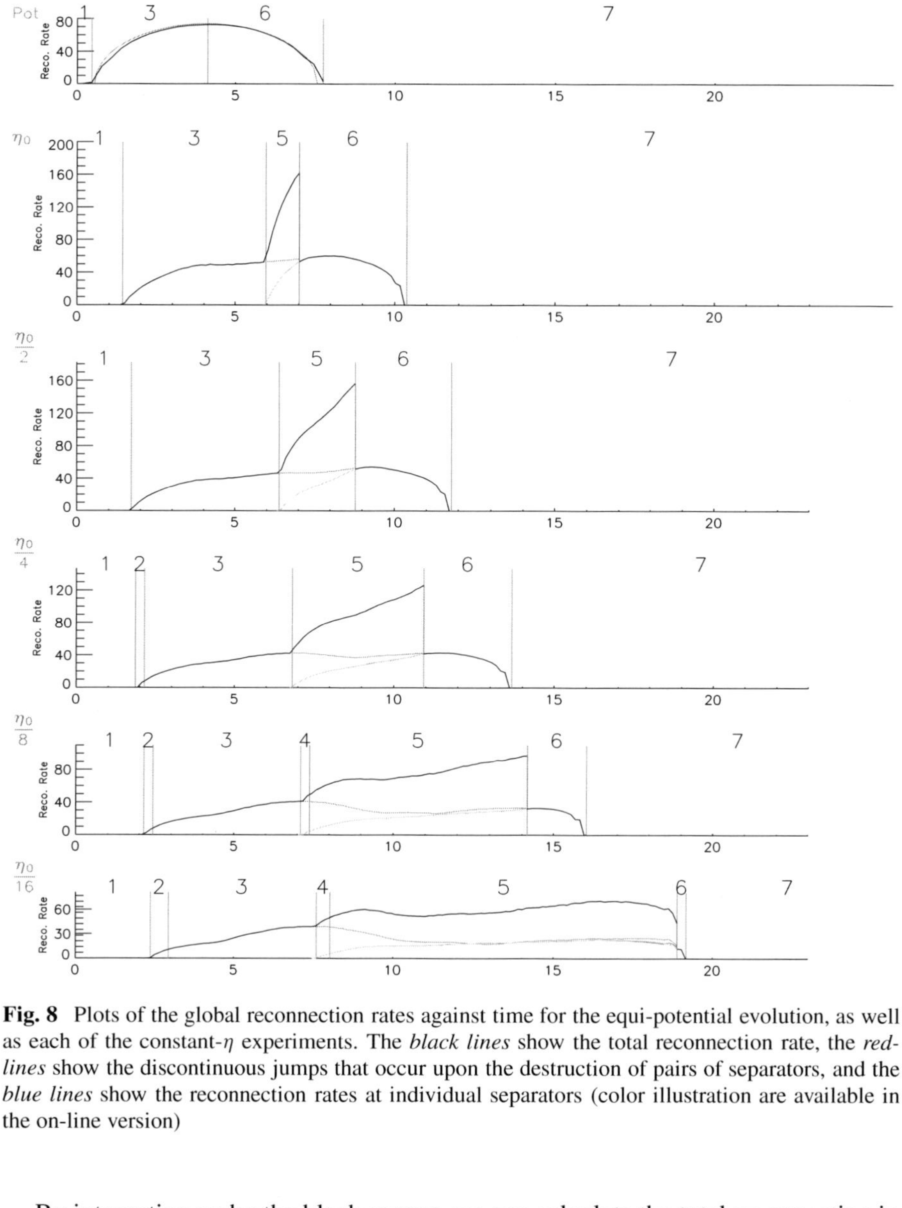

Fig. 8 Plots of the global reconnection rates against time for the equi-potential evolution, as well as each of the constant-η experiments. The *black lines* show the total reconnection rate, the *red-lines* show the discontinuous jumps that occur upon the destruction of pairs of separators, and the *blue lines* show the reconnection rates at individual separators (color illustration are available in the on-line version)

By integrating under the black curves, we can calculate the total reconnection in each experiment,

$$R_{\mathrm{T}} = \int_0^\infty \alpha \, \mathrm{d}t.$$

These numbers are given in the right-hand column of Table 2 and it is clear that R_{T} increases as η decreases ($R_{\mathrm{T}}(\eta_0/16)/R_{\mathrm{T}}(\mathrm{pot}) = 1.97$).

Where does this "extra flux" come from that is reconnected? In the potential case, there is no overlap between the closing and reopening phases and the total flux from one source is simply reconnected twice through the following evolution:

$$\text{open} \quad \rightarrow \quad \text{closed} \quad \rightarrow \quad \text{reopened.}$$

In the case of finite η, there is an overlap between the closing and re-opening reconnection processes, which permits recursive reconnection, hence the flux now evolves in the following manner:

$$\text{open} \quad \rightarrow \quad \text{closed} \quad \rightleftarrows \quad \text{reopened.}$$

Thus, there is no actual extra flux, instead re-opened flux is closed and re-opened again multiple times during phase 5.

From Fig. 8 we see that phase 3, which only closes the flux, lasts for about the same duration in each of the constant-η experiments, although the flux closed during this time decreases as η decreases. Phase 5, however, increases in length considerably as η decreases, permitting more and more recursive reconnection. In contrast, phase 6, the simple reopening phase, decreases in duration, such that in the $\eta_0/16$ case it lasts less than half an Alfvén crossing time.

During phase 5, there is a rapid increase in the global reconnection rate (Fig. 8). The rapidity of this increase decreases as η decreases. The upper blue line plotted during this phase shows the rate of reconnection at separator X_1, while the lower blue line shows the rate of reconnection at both separators X_2 and X_3. When these blue lines meet, all three separators are reconnecting at exactly the same rates. As the system is symmetric, the rates of reconnection at all three separators are always essentially the same at the end of phase 5, at which point two of the separators coincide and destroy each other. Hence, the discontinuous jump at the end phase 5 (red line) always involves the global rate of reconnection dropping by a factor of 3.

4 Energetics

We have seen that the instantaneous global rate of reconnection drops as η decreases, but the total flux reconnected increases. So what are the implications for the rate of heating and the total amount of free magnetic energy of the interaction?

As η decreases, the total amount of free magnetic energy in the system increases (Fig. 9a). Although, not shown in this graph, we find that all experiments eventually relax down to the same magnetic energy. The extra energy in the small constant-η experiments is injected through the base as Poynting flux (Fig. 9b). This is because, as η decreases, the onset of reconnection is delayed (Fig. 8) and so the magnetic field from the sources becomes more horizontal. Therefore, the normal component of the Poynting flux, $(E_x B_y - E_y B_x)/\mu$, increases.

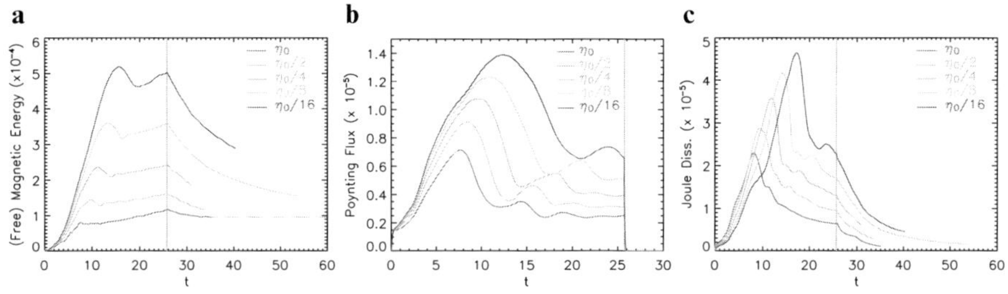

Fig. 9 (**a**) Free magnetic energy, (**b**) Poynting flux and (**c**) ohmic dissipation against time for the constant-η experiments. The *green vertical line* indicates the end of the driving of the sources (color illustration are available in the on-line version)

Figure 9c indicates that as η decreases, not only does the total energy released from the system increase, but the peak rate of joule dissipation also increases. This means that, although the instantaneous rate of reconnection in the system decreases, the rate of heating increases. This is because the heating from the reconnection (ohmic heating) is proportional to ηj^2. So the decrease in η is offset by a far greater increase in the current density. The currents are built up due to the delay in onset of reconnection caused by the small value of η.

5 Conclusions

Research over the last decade has revealed that the properties of 3D magnetic reconnection are quite different to those of 2D reconnection. In particular, 3D reconnection can occur in both null and non-null regions. Furthermore, it does not occur at a single point, but occurs continuously and continually throughout the diffusion volume.

Through the detailed analysis of a simple 3D magnetic interaction, forming a basic building block of the Sun's magnetic atmosphere, we have investigated some of the interesting consequences of 3D reconnection. In particular, we have demonstrated that the reconnection within a simple magnetic flux system is not trivial, but is instead highly complex. The key to revealing the details of the interaction was understanding the evolution of the magnetic topology, that is, determining the evolution of the magnetic skeleton. This allowed us to determine the following:

- The location of the reconnection/energy release sites
- The direction of the reconnection processes that occurred at these sites
- The rates of reconnection at these sites and also the global reconnection rate.

An unexpected discovery from the magnetic skeleton analysis was the process of recursive reconnection, where the same flux can be reconnected multiple times. It is interesting to consider what structures permit the recursive reconnection to take place. Recursive reconnection is possible because several separators link the same two null points: the nulls are said to be *multiply connected* (Parnell 2007). Multiply connected nulls go hand-in-hand with multiply-connected source pairs. They can occur in all types of magnetic fields: potential (Parnell 2007), force-free, and non force-free, as seen here.

The process of recursive reconnection leads to the following important consequences:

- *Greater spread of energy*: Recursive reconnection involves many separators at which reconnection is occurring, and hence many energy release sites. This leads to a better/wider distribution of energy which is important, for instance, for heating the corona. The more energy release sites there are, the easier it is to maintain a background temperature of over a million degrees across almost the whole surface.
- *Repeated heating of the plasma*: As flux is repeatedly reconnected, it is possible for certain parts of the plasma to have injections of energy in rapid succession over a short period of time enabling it to potentially get significantly hotter than other regions of plasma around it. Furthermore, this repeated reconnection of certain parts of flux may explain why, in some circumstances, you can observe solar flare loops before they reconnect when you would normally expect them to be cool and so invisible in X-rays.
- *Longer period of heating*: As flux is repeatedly closed and re-opened many times, the entire heating process lasts longer than if it simply closed and re-opened just once. This again is useful for coronal heating and also solar flares.

Finally, it is interesting to observe that the rate of reconnection is not proportional to the rate of ohmic heating. In fact, here, as η decreases, so does the rate of reconnection, but the rate and peak value of ohmic heating increases. Further work needs to be undertaken to see how these two rates are related in general.

References

Aulanier, G., Pariat, E., Démoulin, P., Devore, C. R. 2006, Solar Phys., 238, 347
Birn, J., Hesse, M., Schindler, K. 1998, J. Geophys. Res., 103, 6843
Close, R. M., Parnell, C. E., Longcope, D. W., Priest, E. R. 2004, ApJ, 612, L81
Close, R. M., Parnell, C. E., Priest, E. R. 2005, Geophys. Astrophys. Fluid Dynamics, 99, 513
De Moortel, I., Galsgaard, K. 2006a, A&A, 451, 1101
De Moortel, I., Galsgaard, K. 2006b, A&A, 459, 627
Démoulin, P., Priest, E. R., Lonie, D. P. 1996, J. Geophys. Res., 101, 7631
Galsgaard, K., Nordlund, Å. 1997, J. Geophys. Res., 102, 231
Galsgaard, K., Parnell, C. E. 2005, A&A, 439, 335
Galsgaard, K., Parnell, C. E., Blaizot, J. 2000, A&A, 362, 395
Galsgaard, K., Priest, E. R., Nordlund, Å. 2000, Solar Phys., 193, 1
Galsgaard, K., Titov, V. S., Neukirch, T. 2003, ApJ, 595, 506
Hagenaar, H. J., Schrijver, C. J., Title, A. M. 2003, ApJ, 584, 1107
Haynes, A. L., Parnell, C. E., Galsgaard, K., Priest, E. R. 2007, Royal Soc. Lond. Proc. Ser. A, 463, 1097
Haynes, A. L., Parnell, C. E., Galsgaard, K. 2009, Solar Phys. (submitted)
Hesse, M., Kuznetsova, M., Birn, J. 2001, J. Geophys. Res., 106, 29831

Hornig, G., Priest, E. R. 2003, Phys. Plasma, 10, 2712
Lau, Y.-T., Finn, J. 1990, ApJ, 350, 672
Linton, M. G., Priest, E. R. 2003, ApJ, 595, 1259
Longcope, D. W. 2001, Phys. Plasmas, 8, 5277
Longcope, D. W., Cowley, S. C. 1996, Phys. Plasmas, 3, 2885
Parnell, C. E. 2007, Solar Phys., 242, 21
Parnell, C. E., Galsgaard, K. 2004, A&A, 428, 595
Parnell, C. E., Haynes, A. L., Galsgaard, K. 2008, ApJ, 675, 1656
Parnell, C. E., Haynes, A. L., Galsgaard, K. 2009, J. Geophys. Res. (in press)
Pontin, D. I., Craig, I. J. D. 2006, ApJ, 642, 568
Pontin, D. I., Galsgaard, K. 2007, J. Geophys. Res., 112, 3103
Pontin, D. I., Galsgaard, K., Hornig, G., Priest, E. R. 2005, Phys. Plasmas, 12, 052307
Priest, E. R., Démoulin, P. 1995, J. Geophys. Res., 100, 23443
Priest, E. R., Titov, V. S. 1996, Phil. Trans. Roy. Soc. Lond., 355, 2951
Priest, E. R., Forbes, T. G. 2000, Magnetic reconnection, Cambridge University Press
Priest, E. R., Hornig, G., Pontin, D. I. 2003, J. Geophys. Res., 108, 1285
Pritchett, P. L. 2001, J. Geophys. Res., 106, 25961
Schrijver, C. J., Harvey, K. L. 1994, Solar Phys., 150, 1
Schrijver, C. J., Title, A. M., van Ballegooijen, A. A., Hagenaar, H. J., Shine, R. A. 1997, ApJ, 487, 424
Wang, J., Wang, H., Tang, F., Lee, J. W., Zirin, H. 1995, Solar Phys., 160, 277
Wilmot-Smith, A. L., De Moortel, I. 2007, A&A, 473, 615
Wilmot-Smith, A. L., Priest, E. R. 2007, Phys. Plasmas, 14, 102903

Signatures of Coronal Heating Mechanisms

P. Antolin, K. Shibata, T. Kudoh, D. Shiota, and D. Brooks

Abstract Alfvén waves created by sub-photospheric motions or by magnetic reconnection in the low solar atmosphere seem good candidates for coronal heating. However, the corona is also likely to be heated more directly by magnetic reconnection, with dissipation taking place in current sheets. Distinguishing observationally between these two heating mechanisms is an extremely difficult task. We perform 1.5-dimensional MHD simulations of a coronal loop subject to each type of heating and derive observational quantities that may allow these to be differentiated. This work is presented in more detail in Antolin et al. (2008).

1 Introduction

The "coronal heating problem," that is, the heating of the solar corona up to a few hundred times the average temperature of the underlying photosphere, is one of the most perplexing and unresolved problems in astrophysics to date. Alfvén waves produced by the constant turbulent convective motions or by magnetic reconnection

P. Antolin (✉)
Kwasan Observatory, Kyoto University, Japan
and
The Institute of Theoretical Astrophysics, University of Oslo, Norway

K. Shibata
Kwasan Observatory, Kyoto University, Japan

T. Kudoh
National Astronomical Observatory of Japan, Japan

D. Shiota
The Earth Simulator Center, Japan Agency for Marine-Earth Science and Technology (JAMSTEC), Japan

D. Brooks
Space Science Division, Naval Research Laboratory, USA
and
George Mason University, USA

S.S. Hasan and R.J. Rutten (eds.), *Magnetic Coupling between the Interior and Atmosphere of the Sun*, Astrophysics and Space Science Proceedings,
DOI 10.1007/978-3-642-02859-5_21,

in the lower and upper solar atmosphere may transport enough energy to heat and maintain a corona (Uchida and Kaburaki 1974). A possible dissipation mechanism for Alfvén waves is mode conversion. This is known as the Alfvén wave heating model (Hollweg et al. 1982; Kudoh and Shibata 1999).

Another promising coronal heating mechanism is the nanoflare reconnection heating model, first suggested by Parker (1988), who considered coronal loops being subject to many magnetic reconnection events, releasing energy impulsively and sporadically in small quantities of the order of 10^{24} erg or less ("nanoflares"), uniformly along loops. It has been shown that both these candidate mechanisms can account for the observed impulsive and ubiquitous character of the heating events in the corona (Katsukawa and Tsuneta 2001; Moriyasu et al. 2004). How then can we distinguish observationally between both heating mechanisms when these operate in the corona?

We propose a way to discern observationally between Alfvén wave heating and nanoflare reconnection heating. The idea relies on the fact that the distribution of the shocks in loops differs substantially between the two models, due to the different characteristics of the wave modes they produce. As a consequence, X-ray intensity profiles differ substantially between an Alfvén-wave heated corona and a nanoflare-heated corona. The heating events obtained follow a power-law distribution in frequency, with indices that differ significantly from one heating model to the other. We thus analyze the link between the power-law index of the frequency distribution and the operating heating mechanism in the loop. We also predict different flow structures and different average plasma velocities along the loop, depending on the heating mechanism and its spatial distribution.

2 Signatures for Alfvén Wave Heating

Alfvén waves generated at the photosphere, due to nonlinear effects, convert into longitudinal modes during propagation, with the major conversion happening in the chromosphere. An important fraction of the Alfvénic energy is also converted into slow and fast modes in the corona, where the plasma β parameter can get close to unity sporadically and spontaneously. The resulting longitudinal modes produce strong shocks that heat the plasma uniformly. The result is a uniform loop satisfying the RTV scaling law (Rosner et al. 1974; Moriyasu et al. 2004), which is, however, very dynamic (Table 1). Synthetic Fe XV emission lines show a predominance of red shifts (downflows) close to the footpoints (Fig. 1). Synthetic XRT intensity profiles show spiky patterns throughout the corona. Corresponding intensity histograms show a distribution of heating events, which stays roughly constant along the corona, and which can be approximated by a power law with index steeper than -2, an indication that most of the heating comes from small dissipative events (Hudson 1991).

Table 1 Observational signatures for coronal heating mechanisms

Heating model	Mean & maximum velocities (km s^{-1})	Doppler velocities (Fe XV) (km s^{-1})	Intensity flux pattern	Mean power law index
Alfvén wave	$\langle v \rangle \sim 50$ $v_{max} > 200$	Red shifts ~ 3	Bursty everywhere	$\langle \delta \rangle < -2$
Footpoint nanoflare	$\langle v \rangle \sim 15$ $v_{max} > 200$	Blue shifts ~ 7	Bursty close to footpoints	$-1.5 > \langle \delta \rangle > -2$ decreases
Uniform nanoflare	$\langle v \rangle \sim 5$ $v_{max} < 40$	Blue shifts ~ 1	Flat everywhere	$\langle \delta \rangle \sim -1$ decreases

We consider Alfvén wave heating, nanoflare-reconnection heating with heating events concentrated at the footpoints ("footpoint nanoflare") or uniformly distributed in the corona ("uniform nanoflare"). Second column: mean and maximum flow velocities. Third column: Doppler velocities derived from synthesized Fe XV 284.16 Å emission line. The loops are assumed to be observed from above (disk center). Fourth column: shape of the intensity flux time series. Fifth column: mean over the power law indexes obtained from intensity histograms for many positions along the loop from the footpoint to the apex.

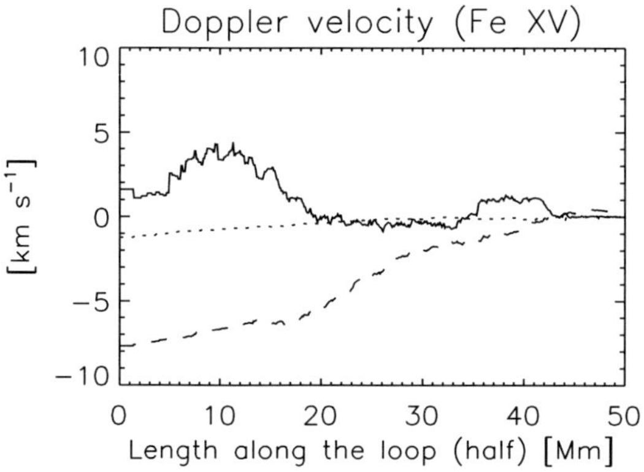

Fig. 1 Doppler velocities from synthesis of Fe XV 284.16 Å emission with respect to distance along the first half of the loop. The loop is assumed to be at disk center. The CHIANTI atomic database was used for calculation of the line profiles (Dere et al. 1997; Landi et al. 2006). The *solid*, *dashed*, and *dotted curves* correspond, respectively, to Alfvén wave heating, footpoint nanoflare heating, and uniform nanoflare heating. Positive velocities correspond to redshifts (downflows)

3 Signatures for Nanoflare Reconnection Heating

The nanoflare reconnection model consists of artificial injections of energy, simulating nanoflares that can be distributed towards the footpoints of the loop ("footpoint nanoflare heating") or uniformly along the corona ("uniform nanoflare heating"). Strong slow shocks created by such heating events are only obtained in the first case, close to the footpoints. Fast dissipation of the slow shocks by thermal conduction leads to only weak shocks near the apex of the loop. We obtain only weak shocks in the case of uniform nanoflare heating. For both cases, the mean flow speeds are considerably smaller than in the Alfvén wave heating model (Table 1). Synthetic Fe XV emission lines show, mainly for footpoint heating, a predominance of blueshifts (upflows) close to the footpoints (cf. Fig. 1), which may match observations of active regions (Hara et al. 2008). In this case, spiky patterns result in XRT intensity profiles close to the footpoints and a flattening of the profile at the apex. The corresponding power-law index of the heating events distribution decreases closer to the apex. In the case of uniform nanoflare heating, the intensity profiles are not spiky but rather uniform in time due to the stronger dissipation by thermal conduction. It results in a very shallow average power-law index for the distribution of heating events.

4 Conclusion

The observable differences between the two coronal heating mechanisms are summarized in Table 1. The power-law index of the heating distribution is found to be sensible to the location of the heating along coronal loops and to the heating mechanism itself. Different flow patterns are also obtained. Downflows of hot plasma are present in the Alfvén-wave heating model, while hot upflows are obtained for nanoflare reconnection heating. Footpoint nanoflare heating seems to match the observations in active region loops better. Are Alfvén wave heating or uniform nanoflare heating more adequate for quiet-Sun loops?

Acknowledgment The numerical calculations were carried out on Altix3700 BX2 at YITP in Kyoto University. We thank the conference organizers for a very good meeting and the editors for excellent instructions.

References

Antolin, P., Shibata, K., Kudoh, T., Shiota, D., Brooks, D. 2008, ApJ, 688, 669
Dere, K.P., Landi, E., Mason, H.E., Monsignori Fossi, B.C., Young, P.R. 1997, A&AS, 125, 149
Hara, H., Watanabe, T., Harra, L.K., Culhane, J.L., Young, P., Mariska, J.T., Doschek, G.A. 2008, ApJ, 678, 67
Hollweg, J.V., Jackson, S., Galloway, D. 1982, Solar Phys., 75, 35
Hudson, H.S. 1991, Solar Phys., 133, 357
Katsukawa, Y., Tsuneta, S. 2001, ApJ, 557, 343
Kudoh, T., Shibata, K. 1999, ApJ, 514, 493
Landi, E., Del Zanna, G., Young, P.R., Dere, K.P., Mason, H.E., Landini, M. 2006, ApJS, 162, 261
Moriyasu, S., Kudoh, T., Yokoyama, T., Shibata, K. 2004, ApJ, 601, L107
Parker, E.N. 1988, ApJ, 330, 474
Rosner, R., Tucker, W.H., Vaiana, G.S. 1978, ApJ, 220, 643
Uchida, Y., Kaburaki, O. 1974, Solar Phys., 35, 451

Waves in Polar Coronal Holes

D. Banerjee

Abstract The fast solar wind originates from polar coronal holes. Recent observations from SoHO suggest that the solar wind is flowing from funnel-shaped magnetic fields anchored in the lanes of the magnetic network at the solar surface. Using the spectroscopic diagnostic capability of SUMER on SoHO and of EIS on HINODE, we study waves in polar coronal holes, in particular their origin, nature, and acceleration. The variation of the width of spectral lines with height above the solar surface supplies information on the properties of waves as they propagate out of the Sun.

1 Introduction

Recent data from Ulysses show the importance of the polar coronal holes, particularly at times near solar minimum, for the acceleration of the fast solar wind. Acceleration of the quasi-steady, high-speed solar wind emanating from large coronal holes requires energy addition to the supersonic region of the flow. It has been shown theoretically that Alfvén waves from the sun can accelerate the solar wind to these high speeds. Until now, this is the only mechanism that has been shown to enhance the flow speed of a basically thermally driven solar wind to the high flow speeds observed in interplanetary space. The Alfvén speed in the corona is quite large, so Alfvén waves can carry a significant energy flux even for a small wave energy density. These waves can therefore propagate through the corona and the inner solar wind without increasing the solar wind mass flux substantially, and deposit their energy flux to the supersonic flow. For this mechanism to work, the wave velocity amplitude in the inner corona must be 20–30 km s^{-1}.

Waves can be detected using the oscillatory signatures they impose on the plasma (density changes, plasma motions). Another method of detecting waves is to examine the variation they produce in line widths measured from spectral lines. There have been several off-limb spectral line observations performed to search

D. Banerjee (✉)
Indian Institute of Astrophysics, Bangalore, India

S.S. Hasan and R.J. Rutten (eds.), *Magnetic Coupling between the Interior and Atmosphere of the Sun*, Astrophysics and Space Science Proceedings,
DOI 10.1007/978-3-642-02859-5_22,

for Alfvén wave signatures. Measurements of ultraviolet Mg X line widths made during a rocket flight showed an increase of width with height to a distance of 70 000 km, although the signal to noise was weak (Hassler et al. 1990). With the 40-cm coronagraph at the Sacramento Peak Observatory, Fe X profiles in a coronal hole showed an increase of line width with height (Hassler and Moran 1994). The SUMER ultraviolet spectrograph (Wilhelm et al. 1995) on board SoHO has allowed further high-resolution, spatially resolved measurements of ultraviolet coronal line widths, which have been used to test for the presence of Alfvén waves (Doyle et al. 1998; Banerjee et al. 1998).

The SUMER instrument was used to record the off-limb, height-resolved spectra of a Si VIII density-sensitive line pair, in an equatorial coronal region (Doyle et al. 1998) and a polar coronal hole (Banerjee et al. 1998). The measured variation of the line width with density and height supports undamped wave propagation in low coronal holes, as the Si VIII line widths increase with higher heights and lower densities (see Fig. 1). This was the first strong evidence for outwardly propagating undamped Alfvén waves in coronal holes, which may contribute to coronal hole heating and the high-speed solar wind. We revisit the subject here with the new EIS instrument on HINODE and compare with our previous results as recorded by SUMER/SoHO.

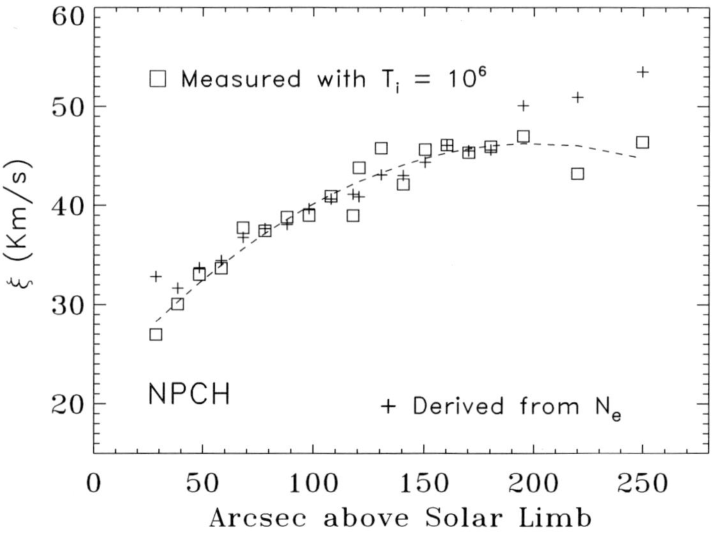

Fig. 1 The nonthermal velocity derived from Si VIII SUMER observations, using $T_{ion} = 1\ 10^6$ K. The *dashed curve* is a second-order polynomial fit. The *plus symbols* correspond to theoretical values (Banerjee et al. 1998)

2 Observation and Results

We observed the North polar coronal hole with EIS onboard Hinode, on and off the limb with the 2″ slit on 10 October 2007. Raster scans were made during over 4 h, constituting 101 exposures with an exposure time of 155 s and covering an area of 201.7″ × 512″. All data have been reduced and calibrated with the standard procedures in the SolarSoft (SSW)[1] library. For further details see Banerjee et al. (in preparation). The spectral line profile of an optically thin coronal emission line results from the thermal broadening caused by the ion temperature T_i as well as broadening caused by small-scale unresolved nonthermal motions. The expression for the FWHM is

$$\mathrm{FWHM} = \left[W_\mathrm{inst}^2 + 4\ln 2 \left(\frac{\lambda}{c}\right)^2 \left(\frac{2k\,T_\mathrm{i}}{M_\mathrm{i}} + \xi^2\right)\right]^{1/2}, \tag{1}$$

where T_i, M_i, and ξ are, respectively, the ion temperature, ion mass, and nonthermal velocity, while W_inst is the instrumental line width.

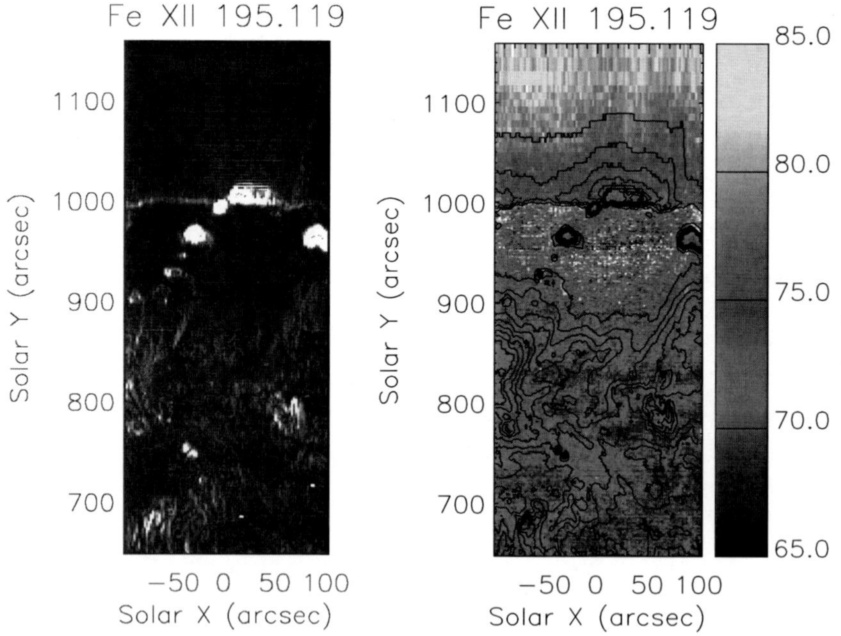

Fig. 2 Fe XII 195 Å intensity (*left*) and FWHM (*right*) maps of the North polar coronal hole

[1] http://www.lmsal.com/solarsoft/

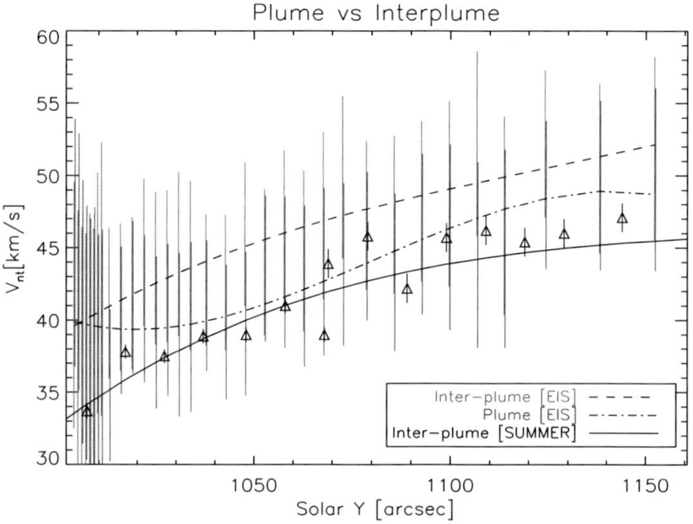

Fig. 3 Variation of nonthermal velocity with height for Fe XII 195 Å along a polar plume and interplume. The *solid curve* corresponds to the nonthermal velocity derived from Si VIII 1445.75 Å from SUMER (Banerjee et al. 1998). The *dashed curve* is a second-order polynomial fit

Figure 2 shows maps of line intensity and width (FWHM) for the North polar coronal hole. The EIS density diagnostics provide density maps of the corona in high spatial resolution from isolated emission lines, which we will detail in Banerjee et al. (in preparation). Inspection of Fig. 2 reveals the coronal hole boundary, fine scale structures, bright points within coronal holes, plume structures, and interplume lanes, of which the physical properties will be discussed in Pérez-Suárez et al. (in preparation). Here, I concentrate only on one plume and interplume lane off the limb, and compare our results with the previous SUMER results. To study the variation of the FWHM with height, we focus our attention to $X = 22$ as a representative location for the plume and on $X = 57$ for the interplume (see the off-limb part in Fig. 2). From the measured FWHM and using (1), we calculate the nonthermal velocities at different altitudes, plotted in Fig. 3, where the triangles represent results from our previous SUMER study (Banerjee et al. 1998).

3 Conclusion

The observational detection of Alfvén waves has gained momentum with the launch of HINODE. The recent detections of low-frequency (<5 mHz) propagating Alfvénic motions in the corona (Tomczyk et al. 2007) and the chromosphere (De Pontieu et al. 2007b) and their relationship with spicules observed at the solar limb (De Pontieu et al. 2007a) with the Solar Optical Telescope (SOT; Tsuneta et al.

2008) on Hinode (Kosugi et al. 2007) have widened the interest on this subject. In Alfvén waves, heavy ions follow the electrons and protons in their wave motion; the width of the spectral line from such ions should therefore increase with decreasing density, that is, increasing distance from the Sun, if the wave broadening is larger than the thermal broadening. As the wave velocity amplitude in the very inner corona must be 20–30 km s^{-1}, broadening of spectral lines from elements as O, Si, Mg, and Fe, increasing with radial distance in the inner corona, should be seen if MHD waves from the sun play an important role in accelerating high speed streams from this region.

All previous observations of emission-line broadening increase with altitude are consistent with the propagation of linear undamped Alfvén waves in open field regions with outward decreasing density. Banerjee et al. (1998) studied Si VIIII line profiles with SUMER in off-limb northern polar hole regions. They recorded line broadening up to 110,000 km (150$''$ off-limb) followed by line-width leveling-off up to 220,000 km height (see Fig. 1). Further out, there was a faint hint of width fall-off, but inconclusive. O'Shea et al. (2003) measured the variation of Mg X 624 AA line widths from CDS above the north polar limb and found an initial linear increase with altitude, supporting the interpretation of linear undamped Alfvén waves propagating outward in open-field regions. They also noted a turnover at a particular altitude where the line widths suddenly decreased or leveled off, which may be attributed to wave dissipation or nonlinear behavior.

In the present study we see a similar trend of linear increase of the line widths up to a certain height, with differences between plume and interplume structures. Thus, our EIS observations reconfirm the detection of Alfvén waves in the polar region, but with a difference between plume and interplume. These data will be useful to determine whether plumes or interplumes are the preferred channel for the acceleration of the fast solar wind. More detail and a comparison with linearly propagating Alfvén waves within an expanding magnetic flux tube will be presented in Banerjee et al. (in preparation).

Acknowledgment This work was supported by Royal Society/British Council and STFC grant PP/D001129/1. The SUMER project is financially supported by DLR, CNES, NASA, and the ESA PRODEX program (Swiss contribution). Hinode is a Japanese mission developed and launched by ISAS/JAXA, with NAOJ as domestic partner and NASA (USA) and STFC (UK) as international partners. It is operated by these agencies in co-operation with ESA and NSC (Norway).

References

Banerjee, D., Pérez-Suárez, D., Doyle, J. G., in preparation
Banerjee, D., Teriaca, L., Doyle, J. G., Wilhelm, K. 1998, A&A, 339, 208
De Pontieu, B., McIntosh, S., Hansteen, V. H., et al. 2007a, PASJ, 59, 655
De Pontieu, B., McIntosh, S. W., Carlsson, M., et al. 2007b, Science, 318, 1574
Doyle, J. G., Banerjee, D., Perez, M. E. 1998, Solar Phys., 181, 91
Hassler, D. M., Moran, T. G. 1994, Sp. Sci. Rev., 70, 373
Hassler, D. M., Rottman, G. J., Shoub, E. C., Holzer, T. E. 1990, ApJ, 348, L77

Kosugi, T., Matsuzaki, K., Sakao, T., et al. 2007, Solar Phys., 243, 3
O'Shea, E., Banerjee, D., Poedts, S. 2003, A&A, 400, 1065
Pérez-Suárez, D., Banerjee, D., Doyle, J. G., in preparation
Tomczyk, S., McIntosh, S. W., Keil, S. L., et al. 2007, Science, 317, 1192
Tsuneta, S., Ichimoto, K., Katsukawa, Y., et al. 2008, Solar Phys., 249, 167
Wilhelm, K., Curdt, W., Marsch, E., et al. 1995, Solar Phys., 162, 189

MHD Wave Heating Diagnostics

Y. Taroyan and R. Erdélyi

Abstract Analyzing the structure of solar coronal loops is crucial to our understanding of the processes that heat and maintain the coronal plasma at multimillion degree temperatures. The determination of the physical parameters of coronal loops remains both an observational and theoretical challenge. A novel diagnostic technique for quiescent coronal loops based on the analysis of power spectra of Doppler-shift time series is developed and proposed to test on real data.

We point out that the analysis of the power spectra allows *distinction* between uniformly heated loops from loops heated near their footpoints. We also argue that it becomes possible to estimate the average energy of a single heating event.

Through examples of synthetic and direct SoHO/SUMER and Hinode/EIS observations of waves, the applicability of the method is demonstrated successfully.

1 The State of the Art

The heating mechanism(s) of the solar corona is (are) a mystery in spite of the multitude of efforts spanning over half a century. It is clear now that the ubiquitous magnetic field of the atmosphere plays a key role in the observed multi-million temperature of the coronal plasma. High-resolution ground and space-based observations provide countless evidence of structuring of the atmospheric magnetic field, both in closed (e.g., loops) and in open (e.g., plumes) format. The determination of the physical parameters of coronal magnetic loops remains both an observational and a theoretical challenge. A novel diagnostic technique for quiescent coronal loops based on the analysis of power spectra of Doppler shift time series is developed and the technique was presented at this meeting.

It is assumed that the loop is heated randomly both in space and time by small-scale discrete impulsive events of an unspecified nature. The loop evolution is then characterized by longitudinal motions caused by the random heating events taking

Y. Taroyan and R. Erdélyi (✉)
Solar Physics and Space Plasma Research Centre, Department of Applied Mathematics, University of Sheffield, UK

S.S. Hasan and R.J. Rutten (eds.), *Magnetic Coupling between the Interior and Atmosphere of the Sun*, Astrophysics and Space Science Proceedings,
DOI 10.1007/978-3-642-02859-5_23,

place in the loop. These random motions can be represented as a superposition of the normal modes of the loop, that is, its standing acoustic wave harmonics (see Taroyan 2008). The idea is borrowed from helioseismology where a similar approach resulted in the advanced understanding of the physical mechanisms and the physical state of the solar interior.

We demonstrated that the wavelet analysis of the power spectra of EUV Doppler and intensity signals allows the *unique distinction* between uniformly heated loops from loops heated near their footpoints. We also derived an estimate of the average energy of a single heating event that took place in the test loop analyzed. To show the applicability, viability, and robustness of the technique, through a couple of further examples of synthetic and direct SoHO/SUMER and Hinode/EIS observations of waves, the method was probed successfully.

A full paper by Taroyan and Erdélyi (submitted) with the theoretical and observational details will appear in Space Sci. Rev.

Acknowledgment We thank the conference organizers for the very good meeting and the excellent hospitality. Y.T. thanks the Leverhulme Trust for financial support. R.E. acknowledges M. Kéray for patient encouragement and NSF, Hungary (OTKA, ref. no. K67746).

References

Taroyan, Y. 2008, In Waves and Oscillations in the Solar Atmosphere: Heating and Magneto-Seismology, R. Edélyi, C. A. Mendoza-Briceño (eds.), Procs. IAU Symposium vol. 247, p. 184

Taroyan, Y., Erdélyi, R., Space Sci. Rev., submitted

Coronal Mass Ejections from Sunspot and Non-Sunspot Regions

N. Gopalswamy, S. Akiyama, S. Yashiro, and P. Mäkelä

Abstract Coronal mass ejections (CMEs) originate from closed magnetic field regions on the Sun, which are active regions and quiescent filament regions. The energetic populations such as halo CMEs, CMEs associated with magnetic clouds, geoeffective CMEs, CMEs associated with solar energetic particles and interplanetary type II radio bursts, and shock-driving CMEs have been found to originate from sunspot regions. The CME and flare occurrence rates are found to be correlated with the sunspot number, but the correlations are significantly weaker during the maximum phase compared to the rise and declining phases. We suggest that the weaker correlation results from high-latitude CMEs from the polar crown filament regions that are not related to sunspots.

1 Introduction

Coronal mass ejections (CMEs) are the most energetic phenomena in the solar atmosphere and represent the conversion of stored magnetic energy into plasma kinetic energy and flare thermal energy. The transient nature of CMEs contrasts them from the solar wind, which is a quasi steady plasma flow. Once ejected, CMEs travel through the solar wind and interact with it, often setting up fast-mode MHD shocks, which in turn accelerate charged particles to very high energies. CMEs often

N. Gopalswamy (✉)
NASA Goddard Space Flight Center, Greenbelt, Maryland, USA

S. Akiyama and P. Mäkelä
NASA Goddard Space Flight Center, Greenbelt, Maryland, USA
and
The Catholic University of America, Washington, USA

S. Yashiro
NASA Goddard Space Flight Center, Greenbelt, Maryland, USA
and
The Catholic University of America, Washington, USA
and
Interferometrics, Herndon, USA

S.S. Hasan and R.J. Rutten (eds.), *Magnetic Coupling between the Interior and Atmosphere of the Sun*, Astrophysics and Space Science Proceedings,
DOI 10.1007/978-3-642-02859-5_24, © Springer-Verlag Berlin Heidelberg 2010

propagate far into the interplanetary (IP) medium impacting planetary atmospheres and even the termination shock of the heliosphere. The magnetic fields embedded in CMEs can merge with Earth's magnetic field, resulting in intense geomagnetic storms, which have serious consequences throughout the geospace and even for life on Earth. Thus, CMEs represent magnetic coupling at various locations in the heliosphere. Active regions on the Sun, containing sunspots and plages, are the primary sources of CMEs. Closed magnetic field regions such as quiescent filament regions also cause CMEs. These secondary source regions can occur at all latitudes, but during the solar maximum, they occur prominently at high latitudes, where sunspots are not found. This paper summarizes the properties of CMEs as an indicator of solar activity in comparison with the sunspot number.

2 Summary of CME Properties

Figure 1 illustrates a CME as a large-scale structure moving in the corona and the associated soft X-ray flare. The CME observations were made by the Solar and Heliospheric Observatory (SOHO) Mission's Large Angle and Spectrometric Coronagraph (LASCO). The CME is clearly an inhomogeneous structure with a well defined leading edge (LE) followed by a dark void and finally an irregular bright core. The core is nothing but an eruptive prominence normally observed in Hα or microwaves, but here it is observed in the photospheric light Thomson-scattered by the prominence. Prominence eruptions and flares have been known for a long time before the discovery of CMEs in the early 1970s. Several coronagraphs have operated since then and have accumulated a wealth of information on the properties of CMEs (see, e.g., Hundhausen 1993; Gopalswamy 2004; Kahler 2006). Here

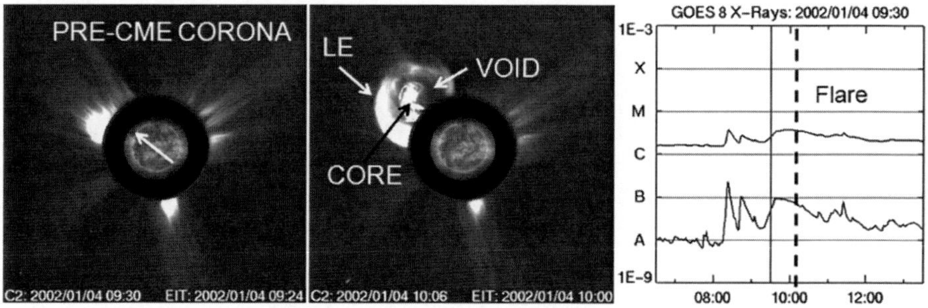

Fig. 1 Example of a CME originating from near the northeast limb of the sun (pointed by *arrow*) as a distinct structure into the pre-CME corona. The CME roughly fills the northeast quadrant of the sun. The three primary structures of the CME, viz., the leading edge (LE), which is curved like a loop in 2D projection, the dark void, and the structured prominence core are indicated by *arrows*. The plot to the *right* shows the GOES soft X-ray flare associated with the CME. The vertical *solid line* marks the LASCO frame at 09:30 UT (pre-CME corona) and the *dashed line* marks the frame with the CME at 10:06 UT

we summarize the statistical properties of CMEs detected by SOHO/LASCO and compiled in a catalog (Gopalswamy et al. 2009c):

- The CME speed is obtained by tracking the leading edge until it reaches the edge of the LASCO field of view (FOV, extending to about 32 $R_\odot$). Some CMEs become faint before reaching the edge of the FOV and others go farther. Therefore, the CME speed we quote here is an average value within the LASCO FOV. Since the height–time measurements are made in the sky plane, the speed is a lower limit. Figure 2 shows that the speed varies over two orders of magnitude from 20 km s^{-1} to more than 3,000 km s^{-1}, with an average value of 466 km s^{-1}.
- The CME angular width is measured as the position angle extent of the CME in the sky plane. Figure 2 shows the width distribution for all CMEs and for CMEs with width >30°. The narrow CMEs ($W < 30°$) were excluded because the manual detection of such CMEs is highly subjective (Yashiro et al. 2008b). The apparent width ranges from <5° to 360° with an average value of 41° (60° when CMEs wider than 30° are considered). There is actually a correlation between

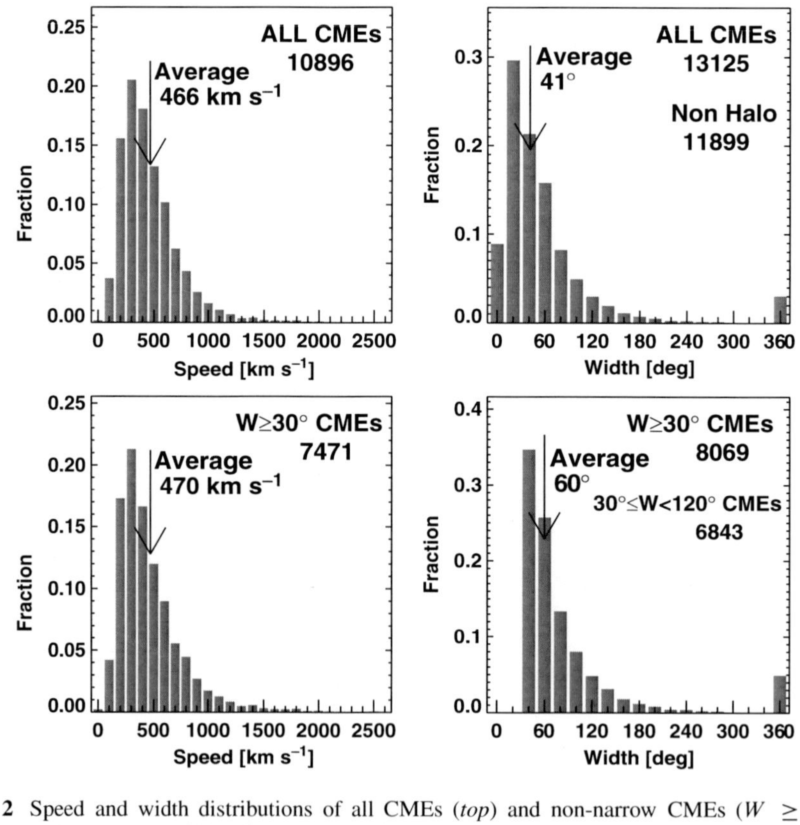

Fig. 2 Speed and width distributions of all CMEs (*top*) and non-narrow CMEs ($W \geq 30°$; *bottom*). The average width of non-narrow CMEs is calculated using only those CMEs with $W \geq 30°$

CME speed (V km s^{-1}) and width (W in degrees), indicating that faster CMEs are generally wider: $V = 360 + 3.64\ W$ (Gopalswamy et al. 2009a).

- CMEs with the above-average speeds decelerate due to coronal drag, while those with speeds well below the average accelerate. CMEs with speeds close to the average speed do not have observable acceleration. This is because the average CME speed is close to the slow solar wind speed.
- The CME mass ranges from 10^{12} to $>10^{16}$ g with an average value of 10^{14} g. Wider CMEs generally have a greater mass content (M): $\log M = 12.6 + 1.3 \log W$ (Gopalswamy et al. 2005). From the observed mass and speed, one can see that the kinetic energy ranges from 10^{27} to $>10^{33}$ erg, with an average value of 5.4×10^{29} erg.
- The daily CME rate averaged over Carrington rotation periods ranges from <0.5 (solar minimum) to >6 (solar maximum). The average speed increases from about 250 km s^{-1} during solar minimum to >550 km s^{-1} during solar maximum (see Fig. 3).
- CMEs moving faster than the coronal magnetosonic speed drive shocks, which accelerate solar energetic particles (SEPs) to GeV energies. The shocks also accelerate electrons, which produce nonthermal radio emission (type II radio bursts) throughout the inner heliosphere.
- The CME eruption is accompanied by solar flares whose intensity in soft X-rays is correlated with the CME kinetic energy (Hundhausen 1997; Yashiro and Gopalswamy 2009).
- There is a close temporal and spatial connection between CMEs and flares: CMEs move radially away from the eruption region, except for small deviations that depend on the phase of the solar cycle (Yashiro et al. 2008a). However, more than half of the flares are not associated with CMEs.

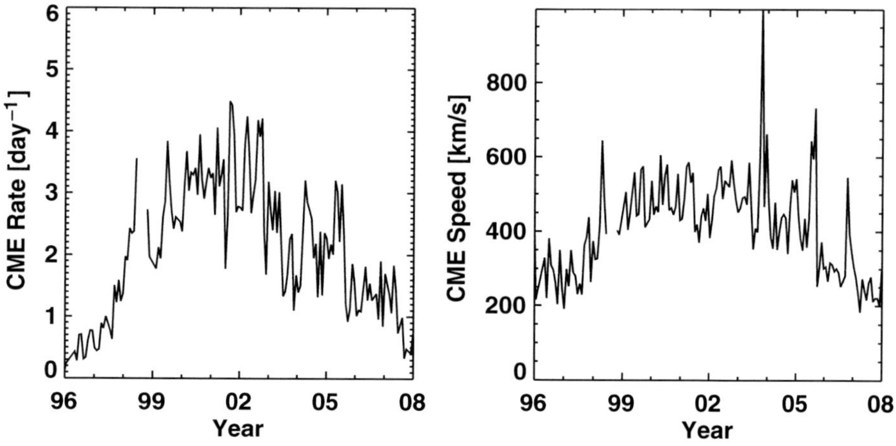

Fig. 3 The daily CME rate (for CMEs with $W \geq 30°$) and the mean CME speed plotted as a function of time showing the solar cycle variation. The occasional spikes are due to super-active regions

- CMEs are comprised of multithermal plasmas containing coronal material at a temperature of a few times 10^6 K and prominence material at about $8,000$ K in the core. In-situ observations show high charge states within the CME, confirming the high temperature in the eruption region due to the flare process associated with the CME (Reinard 2008).
- CMEs originate from closed field regions on the Sun, which are active regions, filament regions, and transequatorial interconnecting regions.
- Some energetic CMEs move as coherent structures in the heliosphere all the way to the edge of the solar system.
- Theory and IP observations suggest that CMEs contain a fundamental flux rope structure identified with the void structure in Fig. 1 (see e.g., Gopalswamy et al. 2006; Amari and Aly 2009). The prominence core is thought to be located at the bottom of this flux rope. The frontal structure is the material piled up at the leading edge of the flux rope.
- CMEs are often associated with EUV waves, which may be fast mode shocks when the CME is fast enough (Neupert 1989; Thompson et al. 1998).
- Coronal dimmings are often observed as compact regions located on either side of the photospheric neutral line, which are thought to be the feet of the erupting flux rope (Webb et al. 2000).

3 Special Populations of CMEs

In this section, we consider several subsets of CMEs that have significant consequences in the heliosphere: halo CMEs, SEP-producing CMEs, CMEs associated with IP type II radio bursts, CMEs associated with shocks detected in situ, CMEs detected at 1 AU as magnetic clouds, and non-cloud ICMEs.

3.1 Halo CMEs

CMEs appearing to surround the occulting disk in coronagraphic images are known as halo CMEs (Howard et al. 1982). Halo CMEs are like any other CMEs, except that they move predominantly toward or away from the observer. Figure 4 illustrates this geometrical effect: the same CME is observed from two viewpoints by the Solar Terrestrial Relations Observatory (STEREO) coronagraphs separated by an angle of 50°. In the SOHO data, 3.6% of all CMEs were found to be full halos, while CMEs with $W \geq 120°$ account for 11% (Gopalswamy 2004). Coronagraph occulting disks block the solar disk and the inner corona, so observations in other wavelengths such as Hα, microwave, X-ray, or EUV are needed to determine whether a CME is frontsided or not. Figure 5 shows the source locations of halo CMEs defined as the heliographic coordinates of the associated flares. Clearly, most of the sources are concentrated near the central meridian. Halos within a central meridian distance

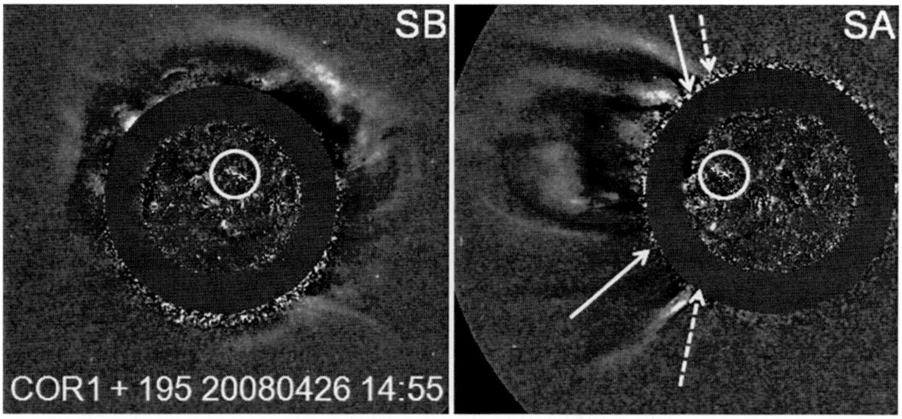

Fig. 4 Two images of the 26 April 2008 CME observed simultaneously by the two STEREO spacecraft separated by about 50°. Clearly, the CME appears as a halo in the view of the Behind spacecraft (SB), which trailed Earth by about 25°. On the other hand, the CME appears as a normal CME in the view of the Ahead spacecraft (SA), which was ahead of Earth by about 25°. In the SA view, the CME is confined to the angular extent bounded by the two *solid arrows*. The *dashed arrows* indicate the disturbances surrounding the CME, which are likely to be shocks for fast CMEs. In the SB view, it is difficult to see the inner and outer structures, so what is seen is likely to be the disturbances surrounding the CME. The solar source of the CME is shown encircled in the difference image taken by STEREO's extreme ultraviolet imager (EUVI). The elongated feature is the post eruption arcade, which is normally taken as the solar source of the CME. The halo CME is roughly symmetric around the solar source in SB view, while the CME is mostly to the east in the SA view. Thus a top view produces a halo CME, while a broadside view produces a normal CME. Occasionally, the CME can become a halo even in the broadside view; such CMEs are known as "limb halos" because the disturbances can be seen on the limb opposite to the solar source. Such limb halos are generally the fastest events

(CMD) of 45° are known as disk halos, while those with $45° < \mathrm{CMD} \leq 90°$ are known as limb halos. Halos occurring close to the central meridian appear as symmetric halos in the coronagraph images. Halos originating from close to the limb appear asymmetric (see Gopalswamy et al. 2003a). The outermost structures seen in most of the halos are likely to be the disturbances (shocks) surrounding the main body of the CME (Sheeley et al. 2000). When the halo CMEs erupt on the visible face of the Sun, they are highly likely to impact Earth and cause intense geomagnetic storms. About 70% of the halos have been found to be geoeffective ($Dst \leq -50$ nT). Disk halos are more geoeffective because of their direct impact on Earth, while the limb halos are less geoeffective due to the glancing impact they deliver. More details on halo CMEs can be found in Gopalswamy et al. (2007).

The speed distribution in Fig. 5 shows that the halos are high-speed CMEs (the average speed is about 1,090 km s^{-1}, more than two times the average speed of all CMEs). The halo CMEs are also associated with flares of greater X-ray importance (Gopalswamy et al. 2007). Halo CMEs observed from a single spacecraft have no width information. However, higher-speed CMEs have been found to be wider from a set of CMEs erupting close to the limb (Gopalswamy et al. 2009a).

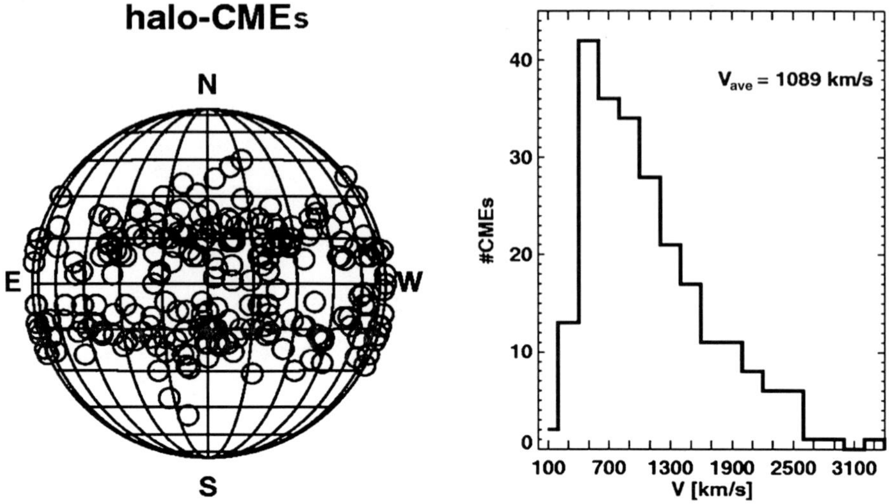

Fig. 5 Source locations (*left*) and the speed distribution (*right*) of halo CMEs recorded by SOHO during 1996–2007. Note the heavy concentration of the sources close to the central meridian. The average speed of halos (V_{ave}) is more than two times the average speed of the general population

Furthermore, wider CMEs are more massive (Gopalswamy et al. 2005). Therefore, one can conclude that halos are more energetic on the average. This is the reason that a large fraction of halos are important for space weather as they drive shocks that produce SEPs and produce geomagnetic storms.

3.2 CMEs Resulting in ICMEs

CMEs in the IP medium are known as IP CMEs or ICMEs for short. ICMEs are generally inferred from single point observations from the solar wind plasma, magnetic field, and composition signatures as the ICME blows past the observing spacecraft (see Gosling 1996 for a discussion on the identification of ICMEs). ICMEs can be traced back to the frontside of the Sun by looking at 5 days of solar observations preceding the ICME arrival at Earth. When ICMEs have a flux rope structure with depressed proton temperature compared to the pre-ICME solar wind, they are known as magnetic clouds (MCs) or IP flux ropes (Burlaga et al. 1981). CMEs resulting in MCs generally originate close to the disk center. CMEs resulting in non-cloud ICMEs generally originate at CMD $> 30°$, but there are many exceptions. Shock-driving ICMEs are easier to identify because shocks are well-defined discontinuities observed in the solar wind. Sometimes, IP shocks are observed without a discernible ICME. In these cases, the CMEs are ejected at large angles to the Sun–Earth line, and so the ICME part misses Earth. Thus as one goes from the disk center to the limb, one encounters solar sources of MCs close to the disk center, those of non-clouds ICMEs, and finally shocks without drivers (Gopalswamy 2006). This suggests the IP manifestation of CMEs depends on the observer–Sun–CME angle. Assuming that CMEs reaching 1 AU have an average width of 60°, one can see that about one third of such CMEs should be MCs and the remaining two-thirds should be non-cloud ICMEs (excluding the pure shock cases). This is generally the case on the average (e.g., Burlaga 1995), but there are notable exceptions: (1) the fraction of MCs is very high during the rise phase of the solar cycle compared to the maximum phase (Riley et al. 2006), (2) there are many non-cloud ICME sources close to the disk center, (3) some pure-shock sources are close to the disk center during the declining phase. These exceptions can be explained as the effect of external influences (Gopalswamy et al. 2009b).

ICMEs reaching Earth are highly likely to cause magnetic storms, provided they contain south-pointing magnetic field either in the ICME portion or in the sheath portion or both. In fact, about 90% of the large geomagnetic storms are due to ICME impact on Earth's magnetosphere. The remaining 10% of the large storms are caused by corotating interaction regions (CIRs), resulting from the interaction between fast and slow solar wind streams (see e.g., Zhang et al. 2007).

3.3 Shock-Driving CMEs

CMEs driving fast mode MHD shocks can be directly observed in the solar wind. Occurrence of type II radio bursts at the local plasma frequency in the vicinity of the observing spacecraft (Bale et al. 1999) is strong evidence that the radio bursts are produced by electrons accelerated at the shock front by the plasma emission mechanism first proposed by Ginzburg and Zhelezniakov (1958). The frequency of type II burst emission is related to the plasma density in the corona, so high frequency (about 150 MHz) type II bursts are indicative of shocks accelerating electrons near the Sun. CMEs must have speeds exceeding the local fast-mode speed in order to drive a shock. CMEs associated with metric type II bursts have a speed of about 600 km s^{-1}, while those producing type II bursts at decameter-hectometric (DH) wavelengths have an average speed exceeding 1,100 km s^{-1}. Type II bursts with emission components from metric to kilometric wavelengths are associated with the fastest CMEs (average speed about 1,500 km s^{-1}). Thus type II bursts are good indicators of shock-driving CMEs (Gopalswamy et al. 2005). Here we take type II bursts at DH wavelengths to be indicative of CMEs driving IP shocks. However, not all DH type II bursts are indicative of shocks detected in situ. This is mainly because CMEs originating even behind the limbs can produce type II bursts due to the extended nature of the shock and the wide beams of the radio emission, but these shocks need not reach Earth. On the other hand, there are some shocks detected at 1 AU, which are not associated with DH type II bursts; shocks have to be of certain threshold strength to accelerate electrons.

A subset of shock-driving CMEs are associated with solar energetic particle (SEP) events detected near Earth. Naturally, the associated CMEs form a subset of those producing DH type II bursts because the same shocks accelerate electrons and ions. In fact, all the major SEP events are associated with DH type II bursts (Gopalswamy 2003; Cliver et al. 2004), but only about half of the DH type II bursts have SEP association (Gopalswamy et al. 2008). CMEs associated with SEP events have the highest average speed (about 1,600 km s^{-1}).

3.4 Comparing the Properties of the Special Populations

Table 1 compares the speed and width information of the special population of CMEs discussed earlier. The lowest average speed is for MC-associated CMEs and the highest speed is for SEP-producing CMEs. The cumulative speed distribution of all CMEs is shown in Fig. 6 with the lowest and highest speeds in Table 1 marked. Even the lowest speed (782 km s^{-1} for MCs) in Table 1 is well above the average speed of all CMEs. The average speed of the SEP-producing CMEs is the highest (1,557 km s^{-1}). All the other special populations have their average speeds between these two limits.

The fraction of halo CMEs in a given population is an indicator of the energy of the CMEs, because halo CMEs are more energetic on the average owing to their higher speed and larger width. The majority of CMEs in all special populations are halos. If partial halos are included, the fraction becomes more than 80% in each population. Even the small fraction of non-halo CMEs ($W < 120°$) have an above-average width. The large fraction of halos in each population implies that there is a high degree of overlap among the populations, that is, the same CME appears in various subgroups.

From Fig. 6 one can see that the number of CMEs with speeds >2,000 km s^{-1} is exceedingly small. In fact, only two CMEs are known to have speeds exceeding 3,000 km s^{-1} among the more than 13,000 CMEs detected by SOHO during 1996–2007. This implies a limit to the speed that CMEs can attain of about $4,000\,\mathrm{km\,s^{-1}}$. For a mass of about 10^{17} g, a 4,000 km s^{-1} CME would possess a kinetic energy of 10^{34} erg. Active regions that produce such high energy CMEs must possess a free energy of at least 10^{34} erg to power the CMEs. It has been estimated that the free energy in active regions is of the order of the potential field energy and that the total magnetic energy in the active region is about twice the potential field energy (Mackay et al. 1997; Metcalf et al. 1995; Forbes 2000; Venkatakrishnan and Ravindra 2003). The potential field energy depends on the

Table 1 Speed and width of the special populations of CMEs

	Halos	MCs	Non-MCs	Type IIs	Shocks	Storms	SEPs
Speed (km s^{-1})	1,089	782	955	1,194	966	1,007	1,557
% Halos	100	59	60	59	54	67	69
% Partial halos	–	88	90	81	90	91	88
Non-halo width (°)	–	55	84	83	90	89	48

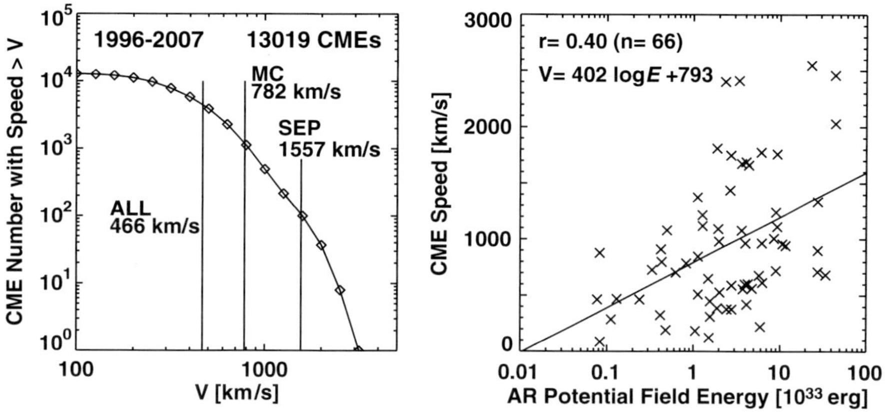

Fig. 6 *Left*: the cumulative number distribution of CMEs as a function of speed (V). The average speed of all CMEs (466 km s^{-1}), CMEs resulting in MCs (782 km s^{-1}), and CMEs producing SEPs (1,557 km s^{-1}) are indicated. Note that the number of CMEs with speeds >3,000 km s^{-1} is negligibly small. *Right*: scatter plot between active-region (AR) potential energy (E) and V for a set of 66 CMEs that had significant impact on Earth. These CMEs either produced large geomagnetic storms or became MCs. E was calculated using the AR flux (ϕ) and area A measured when the AR was close to the central meridian (CMD $\leq 45°$) as follows: $E = \phi^2/8\pi\sqrt{A}$. The scatter plot shows a weak correlation ($r = 0.40$). The regression line is also shown on the plot

size and the average magnetic field strength in the active regions. Figure 6 shows a scatter plot between the potential field energy and the CME speed for a set of CMEs that produced large geomagnetic storms or arrived at Earth as MCs. The correlation is weak but there is certainly a trend that faster CMEs arise from regions of higher potential field energy, as previously shown by Venkatakrishnan and Ravindra (2003). The limiting speed of the CMEs can thus be traced to the maximum energy that can be stored in solar active regions. The largest reported active region size is about 5,000 millionths of a solar hemisphere (Newton 1955), and the highest magnetic field strength observed in sunspots is about 6,100 G (Livingston et al. 2006). Combining these two, one can estimate an upper limit of 10^{36} erg for the potential field energy and hence the free energy that can be stored in an active region. Usually, a single CME does not exhaust all the free energy in the active region. Note that some stars have much larger spots and can release a lot more energy than in solar eruptions (Byrne 1988).

3.5 Solar Sources of the Special Populations

The lowest speed of the MC-associated CMEs and the highest speed of the SEP-associated CMEs shown in Table 1 may be partly due to projection effects because the speeds were measured in the sky plane. To see this, we have shown the distributions of the solar sources of the special populations in Fig. 7. The solar sources

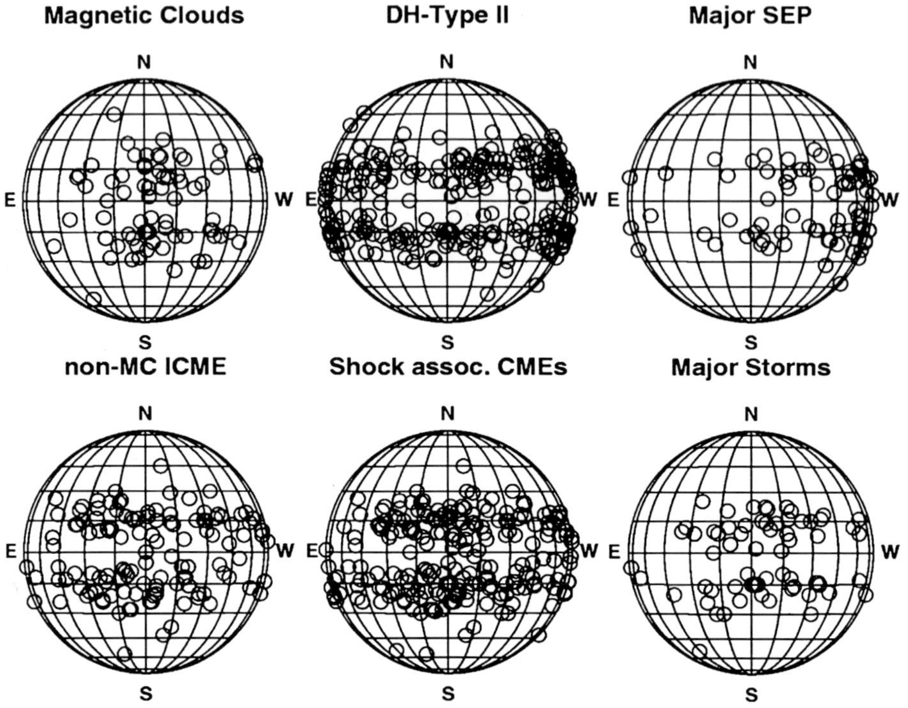

Fig. 7 Heliographic coordinates of the solar sources of the special populations

are taken as the heliographic coordinates of the associated Hα flares from the Solar Geophysical Data. For events with no reported flare information, we have taken the centroid of the post eruption arcade from EUV, X-ray, or microwave images as the solar source. CMEs associated with MCs generally originate from the disk center, so they are subject to projection effects; the SEP-associated CMEs are mostly near the limb, so the projection effects are expected to be minimal. Note that the speed difference between MC- and SEP-associated CMEs is similar to that of disk and limb halo CMEs (933 km s^{-1} vs. 1,548 km s^{-1}; see Gopalswamy et al. 2007). It is also possible that the SEP associated CMEs are the fastest because they have to drive shocks and accelerate particles.

The solar source distributions in Fig. 7 reveal several interesting facts: (1) Most of the sources are at low latitudes with only a few exceptions during the rise phase. (2) The MC sources are generally confined to the disk center, but the non-cloud ICME sources are distributed at larger CMD. There is some concentration of the non-MC sources to the east of the central meridian. (3) Subsets of MCs and non-MC ICMEs are responsible for the major geomagnetic storms, so the solar sources of storm-associated CMEs are also generally close to the central meridian. The slight higher longitudinal extent compared to that of MC sources is due to the fact that some storms are produced by shock sheaths of some fast CMEs originating at larger CMD. (4) The solar sources of CMEs producing DH type II bursts have nearly uniform distribution in longitude, including the east and west limbs. There are also sources behind the east and west limbs that are not plotted. The radio emission can reach the observer from large angles owing to the wide beam of the radio bursts. (5) The sources of SEP-associated CMEs, on the other hand, are confined mostly to the western hemisphere with a large number of sources close to the limb. In fact, there are also many sources behind the west limb, not plotted here (see Gopalswamy et al. 2008a for more details). This western bias is known to be due to the spiral structure of the IP magnetic field along which the SEPs have to propagate before being detected by an observer near Earth. Typically, the longitude W70 is well connected to an Earth observer. An observer located to the east is expected to detect more particle events from the CMEs that produce DH type II bursts but located on the eastern hemisphere. There are a few eastern sources producing SEPs, but these are generally low-intensity events from very fast CMEs. (6) The shock sources are quite similar to the DH type II sources, except for the limb part. As the associated CMEs need to produce a shock signature at Earth, they are somewhat restricted to the disk. Occasional limb CMEs did produce shock signatures at Earth, but these are shock flanks. Comparison with DH type II sources reveals that many shocks do not produce radio emission probably due to the low Mach number (Gopalswamy et al. 2008b).

It is also interesting to note that the combined MC and non-cloud ICME source distribution is similar to those of halo CMEs and the ones associated with shocks at 1 AU. Even the sources of the SEP associated CMEs are similar to the halos originating from the western hemisphere of the Sun because of the requirement of magnetic connectivity to the particle detector.

4 Solar Cycle Variation

CMEs originating close to the disk center and in the western hemisphere have important implications to the space environment of Earth because of the geomagnetic storms and the SEP events they produce. Source regions of CMEs come close to the disk center in two ways: (1) the solar rotation brings active regions to the central meridian, and (2) the progressive decrease in the latitudes where active regions emerge from beneath the photosphere (the butterfly diagram). The effect due to the solar rotation is of short-term because an active region stays in the vicinity of the disk center only for 3–4 days during its disk passage. To see the effect of the butterfly diagram, we need to plot the solar sources of as a function of time.

Figure 8 shows the latitude distribution of the solar sources of the special populations as a function of time during solar cycle 23. Sources corresponding to the three phases of the solar cycle are distinguished using different symbols: the rise phase starts from the beginning of the cycle in 1996 to the end of 1998. The maximum phase is taken from the beginning of 1999 to the middle of 2002. The time of completion of the polarity reversal of the solar polar magnetic fields is considered as the end of the solar maximum phase and the beginning of the declining phase. The boundary between phases is not precise, but one can see the difference

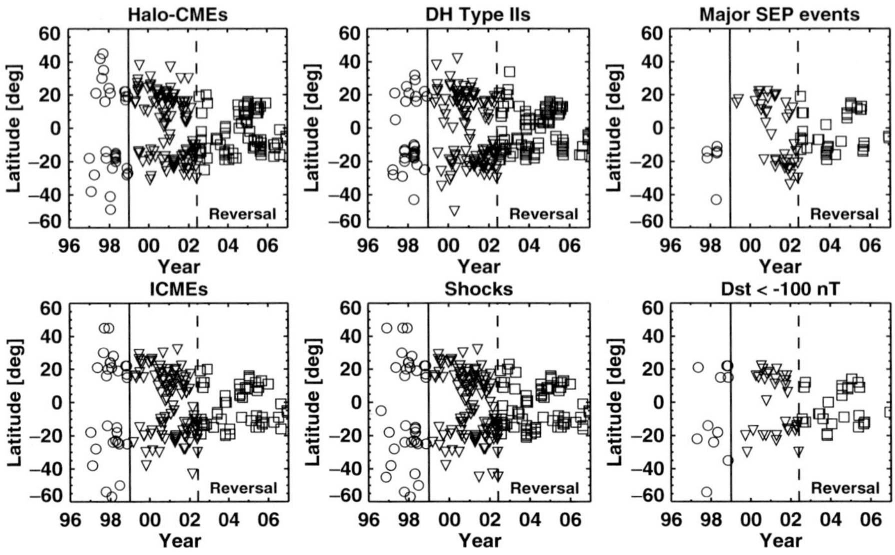

Fig. 8 The source latitudes of the special populations plotted as a function of time: halo CMEs, CMEs detected in situ as ICMEs, CMEs producing DH type II bursts, CMEs driving IP shocks, CMEs producing large SEP events, and CMEs resulting in large geomagnetic storms ($Dst \leq -100$ nT). The *solid vertical line* (end of the year 1998) divides the rise and maximum phases. The *dashed vertical line* (end of May 2002) divides the maximum and declining phases. This line also marks the end of polarity reversal at solar poles (i.e., beginning of the new solar magnetic cycle). CMEs from the three phases of solar cycle 23 are also distinguished by different symbols (rise – *open circles*, maximum – *triangles*, and declining – *squares*)

in the levels of activity and the change in latitude between phases. We have combined the MCs and non-MC ICMEs into a single group as ICMEs. As we noted before, there is a close similarity between the ICME and halo CME sources because many frontside halo CMEs become ICMEs (see the left column in Fig. 8). There are clearly ICMEs without corresponding halos, which means some ICMEs are due to non-halo CMEs. During the rise phase there are several sources at latitudes higher than the ones at which sunspots emerge (about 40°). What is striking is that such high-latitude CMEs produced an observable signature at Earth. Case studies (Gopalswamy et al. 2000, Gopalswamy and Thompson 2000) and statistical studies (Plunkett et al. 2001; Gopalswamy et al. 2003b; Cremades et al. 2006) have shown that CMEs during the solar minimum get deflected towards the equator because of the strong global dipolar field of the Sun. In the declining phase, a different type of deflection occurs: eruptions occurring near low latitude coronal holes tend to be deflected away from coronal holes. Occasionally, such deflections push CMEs toward or away from the Sun–Earth line (Gopalswamy et al. 2009c).

Despite the longitudinal differences presented in Fig. 8, we see a close latitudinal similarity of the solar sources of the special populations. The sources clearly follow the sunspot butterfly diagram. This means the special populations originate only from active regions, where one expects to have higher free energy needed to power these energetic CMEs.

4.1 Solar Sources of the General Population

In contrast to the solar sources of the special populations discussed above, the general CME population is known to occur at all latitudes during solar maxima (Hundhausen 1993; Gopalswamy et al. 2003b). Figure 9 illustrates this using the latitude distributions of prominence eruptions (PEs) and the associated CMEs. Note that these CMEs constitute a very small sample because they are chosen based on their association with PEs detected by the Nobeyama radioheliograph (Nakajima et al. 1994), which is a ground based instrument operating only about 8 h per day. Nevertheless, the observations provide accurate source information for the CMEs and the sample is not subject to projection effects. One can clearly see a large number of high latitude CMEs between the years 1999 and 2003, with a significant north-south asymmetry in the source distributions. These high-latitude CMEs are associated with polar crown filaments, which migrate toward the solar poles and completely disappear by the end of the solar maximum. The cessation of high-latitude CME activity has been found to be a good indicator of the polarity reversal at solar poles (Gopalswamy et al. 2003c). Low-latitude PEs may be associated with both active regions and quiescent filament regions, but the high-latitude CMEs are always associated with filament regions. One can clearly see that the high-latitude CMEs have no relation to the sunspot activity because the latter is confined to latitudes below 40°. Comparing Figs. 8 and 9, we can conclude that the special populations are primarily an active region phenomenon. It is interesting that the high-latitude CMEs occur only during the period of maximum sunspot number (SSN), but are not directly related to the sunspots.

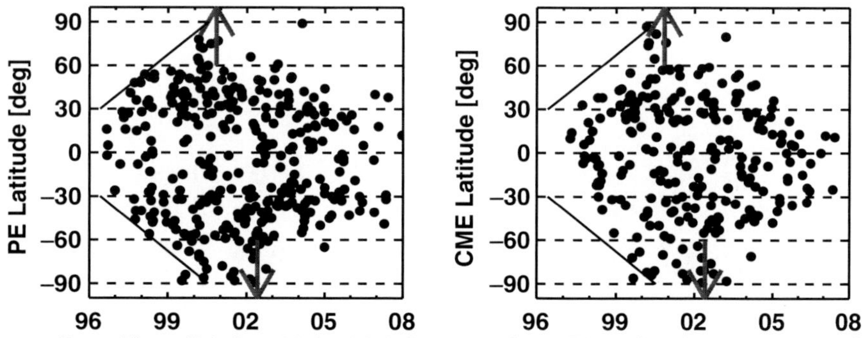

Fig. 9 Latitude of prominence eruptions (PEs) and those of the associated CMEs shown as a function of time. The *up* and *down arrows* denote, respectively, the times when the polarity in the north and south solar poles reversed. Note that the high-latitude CMEs and PEs are confined to the solar maximum phase and their occurrence is asymmetric in the northern and southern hemispheres. PEs at latitudes below 40° may be from active regions or quiescent filament regions, but those at higher latitudes are always from the latter

4.2 Implications to the Flare: CME Connection

The difference in the latitude distributions of CMEs (no butterfly diagram) and flares (follow the sunspot butterfly diagram) coupled with the weak correlation between CME kinetic energy and soft X-ray flare size (Hundhausen 1997) has been suggested as evidence that CMEs are not directly related to flares. However, this depends on the definition of flares. If flares are defined as the enhanced electromagnetic emission from the structures left behind after CME eruptions, one can find flares associated with all CMEs – both at high and at low latitudes. This is illustrated using Fig. 10, which shows the solar source locations of flares reported in the Solar Geophysical Data. During 2004 January to 2007 March, the GOES Soft X-ray Imager (SXI) provided the solar sources of all flares, including the weak ones that can be found at all latitudes, similar to the source distribution shown in Fig. 9 for PEs. On the other hand, if we consider only larger flares (X-ray importance >C3.0), we see that the flares follow the sunspot butterfly diagram. This is quite consistent with the fact that the solar sources of the special populations of CMEs follow the sunspot butterfly diagram because these CMEs are associated with larger flares. For example, the median size of flares associated with halo CMEs is M2.5, an order of magnitude larger than the median size of all flares (C1.7) during solar cycle 23 (Gopalswamy et al. 2007). Thus, CMEs seem to be related to flares irrespective of the origin in active regions or quiescent filament regions. There are in fact several new indicators of the close connection between CMEs and flares: CME speed and flare profiles (Zhang et al. 2001), CME and flare angular widths (Moore et al. 2007), CME magnetic flux in the IP medium and the reconnection flux at the Sun (Qiu et al.

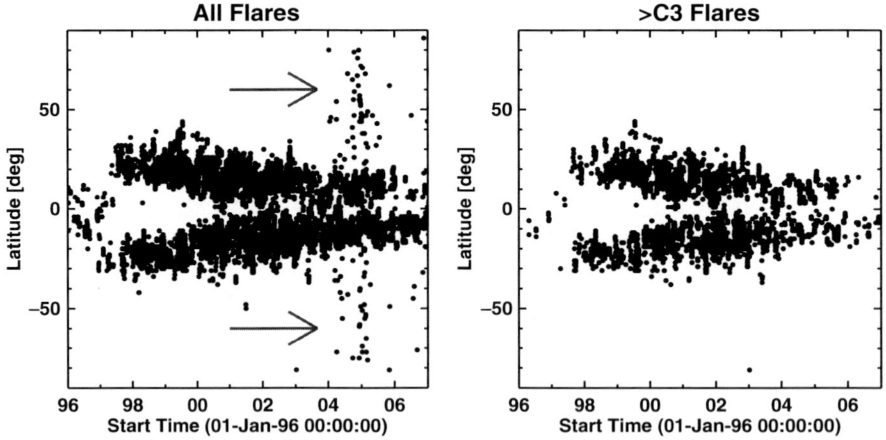

Fig. 10 Flare locations reported in the solar geophysical data plotted as a function of time for all flares (*left*) and larger flares (soft X-ray importance >C3.0) (*right*). The *arrows* point to the weak flares from higher latitudes. Note that the GOES Soft X-ray imager provided solar source locations of flares only during January 2004 to March 2007, so there is no information on the high-latitude flare locations for other times

2007), and the CME and flare positional correspondence (Yashiro et al. 2008a). The close relationship between flares and CMEs does not contradict the fact that more than half of the flares are not associated with CMEs. This is because the stored energy in the solar source regions can be released to heat the flaring loops with no mass motion.

5 Sunspot Number and CME Rate

The above discussion made it clear that the high-latitude CMEs do not follow the sunspot butterfly diagram but occur during the period of maximum solar activity. This should somehow be reflected in the relation between CME and sunspot activities. To see this, we have plotted the daily CME rate (R) as a function of the daily sunspot number (SSN) in Fig. 11. There is an overall good correlation between the two types of activity, which has been known for a long time (Hildner et al. 1976; Webb et al. 1994; Cliver et al. 1994; Gopalswamy et al. 2003a). The SOHO data yield a relation $R = 0.02\,\mathrm{SSN} + 0.9$ (correlation coefficient $r = 0.84$), which has a larger slope compared to the one obtained by Cliver et al. (1994): $R = 0.011\,\mathrm{SSN} + 0.06$. The higher rate has been attributed to the better

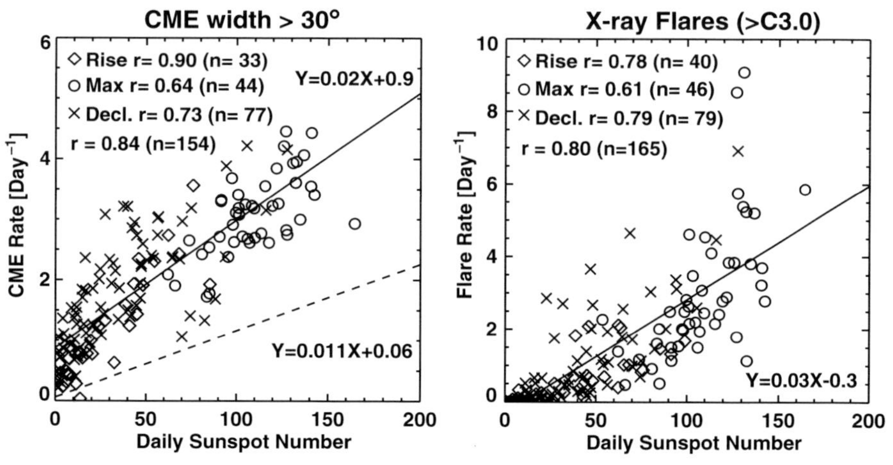

Fig. 11 Correlation of the daily sunspot number with the daily CME rate (*left*) and the daily flare rate (*right*). All numbers are averaged over Carrington rotation periods (27.3 days). The number of rotations (n) is different for the CME and flare rates because of CME data gaps. Rates in different phases of the solar cycle are shown by different symbols. The correlation coefficients are also shown for individual phases as well as for the entire data set (the *solid lines* are the regression lines). In the CME rate, only CMEs wider than 30° are used to avoid subjectivity in CME identification. In the CME plot, the *dashed line* corresponds to the regression line ($\mathrm{Y} = 0.011\mathrm{X} + 0.06$) obtained by Cliver et al. (1994) for CMEs from the pre-SOHO era. In the flare rate, only flares of importance >C3.0 are included

dynamic range and wider field of view of the SOHO coronagraphs compared to the pre-SOHO coronagraphs. However, when the CMEs are grouped according to the phase of the solar cycle, the correlation becomes weak during the maximum phase ($r = 0.64$) compared to the rise ($r = 0.90$) and declining ($r = 0.73$) phases. We attribute this diminished correlation to the CMEs during the maximum phase that are not associated with sunspots (see also Gopalswamy et al. 2003a). Note that we have excluded narrow CMEs ($W < 30°$) because manual detection of such CMEs is highly subjective.

A similar scatter plot involving the daily flare rate as a function of SSN reveals a similar trend in terms of the overall correlations ($r = 0.80$) and the individual phases: rise ($r = 0.78$), maximum ($r = 0.61$), and declining ($r = 0.79$). In particular, the weak correlation between the flare rate and SSN during the maximum phase is striking. When all the flares are included, the overall correlation diminished only slightly ($r = 0.76$) mainly due to the weaker correlation during the maximum phase ($r = 0.46$) because the correlation remained high during the rise ($r = 0.85$), and declining ($r = 0.79$) phases. Interestingly, flare rate vs. SSN correlations are very similar to the CME rate vs. SSN correlations, including the weaker correlation during the maximum phase. This needs further investigation by separating the flares into high and low-latitude events.

6 Summary and Conclusions

In this paper, we studied several subsets of CMEs that have significant consequences in the heliosphere: halo CMEs, SEP-producing CMEs, CMEs associated with IP type II radio bursts, CMEs associated with shocks detected in situ in the solar wind, CMEs detected at 1 AU as magnetic clouds, and non-cloud ICMEs. The primary common property of these special populations is their above-average energy, which helps them propagate far into the IP medium. Most of the CMEs in these subsets are frontside halo CMEs. Notable exceptions are the IP type II bursts and large SEP events. IP type II bursts can be observed from CMEs from behind the east and west limbs because the shocks responsible for the radio emission are more extended than the driving CMEs, and the radio emission is wide beamed. SEP events are also observed from behind the west limb for the same reason (extended shock) and the fact that the SEPs propagate from the shock to the observer along the spiral magnetic field lines. Another common property of the special populations is that they follow the sunspot butterfly diagram. This suggests that the energetic CMEs originate mostly from the sunspot regions, where large free energy can be stored to power the energetic CMEs. Quiescent filament regions are the other source of CMEs, not related to the sunspots, and hence do not follow the sunspot butterfly diagram. During the maximum phase of the solar activity cycle, the quiescent filament regions occur in high abundance at high latitudes, resulting in higher rate of CMEs from there. As the high-latitude CMEs are not related to the sunspots, the correlation between daily CME rate and sunspot number is weak in the maximum phase. A similar weak correlation was found between the flaring rate and sunspot number during the maximum phase. This result further confirms the close relation between flares and CMEs irrespective of the source region: sunspot regions or quiescent filament regions.

Acknowledgment This work is supported by NASA's LWS program.

References

Amari, T., Aly, J.-J. 2009, in Universal Heliophysical Processes, eds. N. Gopalswamy and D. F. Webb, 212
Bale, S. D., Reiner, M. J., Bougeret, J.-L., et al. 1999, Geophys. Res. Lett., 26, 1573
Burlaga, L. F. 1995, Int. Ser. Astron. Astrophys., 3, 3
Burlaga, L., Sittler, E., Mariani, F., Schwenn, R. 1981, J. Geophys. Res., 86, 6673
Byrne, P. B. 1988, Irish Astron. J., 18, 172
Cliver, E. W., St. Cyr, O. C., Howard, R. A., McIntosh, P. S. 1994, in IAU Colloq. 144: Solar Coronal Structures, 83
Cliver, E. W., Kahler, S. W., Reames, D. V. 2004, ApJ, 605, 902
Cremades, H., Bothmer, V., Tripathi, D. 2006, Adv. Space Res., 38, 461
Forbes, T. G. 2000, J. Geophys. Res., 105, 23153
Ginzburg, V. L., Zhelezniakov, V. V. 1958, Sov. Astron., 2, 653
Gopalswamy, N. 2003, Geophys. Res. Lett., 30, SEP 1-1
Gopalswamy, N. 2004, in The Sun and the Heliosphere as an Integrated System, eds. G. Poletto and S. T. Suess, Astrophys. Space Sci. Lib., 317, 201
Gopalswamy, N. 2006, Space Sci. Rev., 124, 145
Gopalswamy, N., Thompson, B. J. 2000, J. Atmosph. Solar-Terrestr. Phys., 62, 1457
Gopalswamy, N., Hanaoka, Y., Hudson, H. S. 2000, Adv. Space Res., 25, 1851
Gopalswamy, N., Lara, A., Yashiro, S., Howard, R. A. 2003a, ApJ, 598, L63
Gopalswamy, N., Lara, A., Yashiro, S., Nunes, S., Howard, R. A. 2003b, in Solar Variability as an Input to the Earth's Environment, ed. A. Wilson, ESA-SP, 535, 403
Gopalswamy, N., Shimojo, M., Lu, W., et al. 2003c, ApJ, 586, 562
Gopalswamy, N., Aguilar-Rodriguez, E., Yashiro, S., et al. 2005, J. Geophys. Res., 110, A12S07
Gopalswamy, N., Mikić, Z., Maia, D., et al. 2006, Space Sci. Rev., 123, 303
Gopalswamy, N., Yashiro, S., Akiyama, S. 2007, J. Geophys. Res., 112, 6112
Gopalswamy, N., Yashiro, S., Akiyama, S., et al. 2008a, Annales Geophys., 26, 3033
Gopalswamy, N., Yashiro, S., Xie, H., et al. 2008b, ApJ, 674, 560
Gopalswamy, N., Dal Lago, A., Yashiro, S., Akiyama, S. 2009a, Central European Astrophysical Bulletin, J. Geophys. Res., 114, A00A22
Gopalswamy, N., Mäkelä, P., Xie, H., Akiyama, S., Yashiro, S. 2009b, J. Geophys. Res. (in press)
Gopalswamy, N., Yashiro, S., Michalek, G., et al. 2009c, Earth Moon and Planets, 204, 295
Gosling, J. T. 1996, ARA&A, 34, 35
Hildner, E., Gosling, J. T., MacQueen, R. M., et al. 1976, Solar Phys., 48, 127
Howard, R. A., Michels, D. J., Sheeley, Jr., N. R., Koomen, M. J. 1982, ApJ, 263, L101
Hundhausen, A. J. 1993, J. Geophys. Res., 98, 13177
Hundhausen, A. J. 1997, in Coronal Mass Ejections, AGU Geophysical Monograph, 99, Am. Geophys. Union, Washington, 1
Kahler, S. W. 2006, in Solar Eruptions and Energetic Particles, AGU Geophys. Monograph, 165, 21
Livingston, W., Harvey, J. W., Malanushenko, O. V., Webster, L. 2006, Solar Phys., 239, 41
Mackay, D. H., Gaizauskas, V., Rickard, G. J., Priest, E. R. 1997, ApJ, 486, 534
Metcalf, T. R., Jiao, L., McClymont, A. N., Canfield, R. C., Uitenbroek, H. 1995, ApJ, 439, 474
Moore, R. L., Sterling, A. C., Suess, S. T. 2007, ApJ, 668, 1221

Nakajima, H., Nishio, M., Enome, S., et al. 1994, IEEE, 82, 705
Neupert, W. M. 1989, ApJ, 344, 504
Newton, H. W. 1955, Vistas Astron., 1, 666
Plunkett, S. P., Thompson, B. J., St. Cyr, O. C., Howard, R. A. 2001, J. Atmosph. Solar-Terrestr. Phys., 63, 389
Qiu, J., Hu, Q., Howard, T. A., Yurchyshyn, V. B. 2007, ApJ, 659, 758
Reinard, A. A. 2008, ApJ, 682, 1289
Riley, P., Schatzman, C., Cane, H. V., Richardson, I. G., Gopalswamy, N. 2006, ApJ, 647, 648
Sheeley, N. R., Hakala, W. N., Wang, Y.-M. 2000, J. Geophys. Res., 105, 5081
Thompson, B. J., Plunkett, S. P., Gurman, J. B., et al. 1998, Geophys. Res. Lett., 25, 2465
Venkatakrishnan, P., Ravindra, B. 2003, Geophys. Res. Lett., 30, SSC 2-1
Webb, D. F., Howard, R. A. 1994, J. Geophys. Res., 99, 4201
Webb, D. F., Lepping, R. P., Burlaga, L. F., et al. 2000, J. Geophys. Res., 105, 27251
Yashiro, S., Michalek, G., Akiyama, S., Gopalswamy, N., Howard, R. A. 2008a, ApJ, 673, 1174
Yashiro, S., Michalek, G., Gopalswamy, N. 2008b, Annales Geophys., 26, 3103
Yashiro, S., Gopalswamy, N. 2009, in Universal Heliophysical Processes, eds. N. Gopalswamy and D. F. Webb, 233
Zhang, J., Dere, K. P., Howard, R. A., Kundu, M. R., White, S. M. 2001, ApJ, 559, 452
Zhang, J., Richardson, I. G., Webb, D. F., et al. 2007, J. Geophys. Res., 112, A12103

CME Observations from STEREO

N. Srivastava

Abstract Coronal mass ejections (CMEs) are spectacular ejections of material from the Sun as seen in the coronal field of view. Regular observations are possible with both ground-based and space-based coronagraphs. I present our current understanding of CMEs based on multi-wavelength observations from ground-based instruments as well as from space missions such as SoHO. Based on the continuous and multi-wavelength observations of CMEs from SoHO over a period of more than a solar cycle, the physical properties of CMEs are described. Recent observations of CMEs with the SECCHI coronagraphs, namely COR1 and COR2, aboard the twin STEREO spacecrafts A and B are also presented. STEREO surpasses previous missions by providing a 3D view of CME structure from two vantage points. Applications of STEREO observations to 3D reconstructions of the leading edge of CMEs are described.

1 Introduction

Coronal mass ejections (CMEs) were first observed by OSO-7 in 1971 (Tousey 1973). Since then, a number of space-based and ground-based coronagraphs have been regularly recording images of the corona and CMEs. These observations have led to a fairly good general understanding of CMEs. They were traditionally observed in white light or continuum images by space-based coronagraphs such as Solwind, the Solar Maximum Mission (SMM), and LASCO/C2&C3, and with ground-based instruments such as the MK III and MK IV coronagraphs at Mauna Loa. However, CMEs have also been recorded in coronal emission lines for, for example, Fe XIV or Fe X, such as with the Norikura observatory and with LASCO-C1 onboard SoHO. Table 1 gives the field of view of various coronagraphs. SoHO was the first successful space mission with multiple instruments onboard to record various aspects of transient activity at multiple wavelengths. The Extreme Ultraviolet telescope (EIT) onboard SoHO not only can track down the source region of a CME,

N. Srivastava (✉)
Udaipur Solar Observatory, Physical Research Laboratory, Udaipur, India

S.S. Hasan and R.J. Rutten (eds.), *Magnetic Coupling between the Interior and Atmosphere of the Sun*, Astrophysics and Space Science Proceedings,
DOI 10.1007/978-3-642-02859-5_25, © Springer-Verlag Berlin Heidelberg 2010

Table 1 Comparison of different space-based coronagraphs used to record CMEs

Coronagraph	Year	Field of view	Resolution
OSO-7	1971–1973	3.0–10.0 $R_\odot$	3′
Skylab	1973–1974	2.0–6.0 $R_\odot$	5″
Solwind/P78-I	1979–1985	3.0–10.0 $R_\odot$	Same as OSO
SMM	1980 1984–1989	1.6–6.0 $R_\odot$	30″
LASCO			
C1	1995–1998	1.1–3.0 $R_\odot$	11.2″
C2	Since 1995	2.0–6.0 $R_\odot$	23.2″
C3		3.7–32.0 $R_\odot$	112″
SECCHI	Since 2006		
COR1		1.4–4.0 $R_\odot$	7.5″
COR2		2.0–15.0 $R_\odot$	14″

but SoHO also can record CME propagation through a field of view of 1–32 $R_\odot$ with the Large Angle Spectrometric COronagraphs (LASCO) at an improved time cadence (Delaboudinière et al. 1995; Brueckner et al. 1995). These continuous, multi-wavelength, and high time-cadence images have led to a better understanding of CME initiation and propagation. The details are presented in various reviews (Hudson et al. 2007; Schwenn et al. 2007).

An important research area in which SoHO observations proved to be extremely useful is space weather prediction. A large number of studies have been undertaken in this direction, in particular, for predicting the arrival time of CMEs at the Earth through continuous monitoring of the Sun and CME tracking. This helps in predicting the time of commencement of geomagnetic storms, which are normally expected to occur when a high-speed CME is directed towards the Earth and its magnetic field is oriented southward so as to ensure an effective coupling between the Earth's magnetic field and the propagating magnetic cloud. However, most prediction schemes used plane-of-sky speeds as proxies of radial speeds, which has led to large errors in the estimated arrival times (Schwenn et al. 2005). This error can be attributed to the lack of capability of SoHO to measure the radial speeds of CMEs. The problem is expected to be overcome with new observations from STEREO, which comprises twin spacecrafts with identical sets of instruments (Howard et al. 2008). The main goal of STEREO is to improve our understanding of the 3D structure of CMEs with accurate and direct estimations of their true speeds and propagation directions.

In what follows, statistical properties of CMEs are highlighted on the basis of SoHO observations and new observations from the SECCHI coronagraphs onboard the STEREO mission are discussed. In particular, I showcase the results of 3D reconstruction of a CME leading edge, using SECCHI-COR1 and COR2 observations, and discuss implications for space-weather prediction.

2 What Have We Learnt from LASCO Observations?

The three LASCO coronagraphs (Brueckner et al. 1995) onboard SoHO provide a nested field of view and have tracked a large number of CMEs from their launch at the solar surface. These observations confirm that CMEs generally have a three-part structure. They have a bright leading edge, a dark cavity believed to have a high magnetic field, followed by a bright and intense knot, which mainly comprises prominence material. A large number of earthward-directed CMEs were recorded by LASCO. These are observed as full halos in which brightness enhancement is seen in an angular span of 360° around the occulter, while in the case of a partial halo, the brightness is seen over an angular span of more than 120°. The kinematics of full and partial halos have also been studied (Wang et al. 2002; Zhang et al. 2003; Zhao and Webb 2003; dal Lago et al. 2004). These studies were aimed at inferring the travel time of the CMEs to the Earth. The estimated arrival time of a CME at the Earth is an important input for forecasting the time of occurrence of the resulting geomagnetic storms, provided there is a strong southward component of the propagating magnetic cloud.

A huge dataset has been collected with the LASCO coronagraphs over more than a solar cycle. It allows us to study CME properties over a large time period. The rate of occurrence of CMEs has been found to be 0.3/day during the solar minimum around 1996, and it slowly increased to 5–6 CMEs/day during the solar maxima (Fig. 1). Yashiro et al. (2004) studied the variation of angular or apparent width of all CMEs that occurred during 1996–2003 and found that all non-halo CMEs have an average angular width of 47° (left-hand panel of Fig. 2). The kinematics of the CMEs recorded by LASCO have also been studied by Yashiro et al. (2004). They measured the projected plane-of-sky speeds of CMEs and found that these lie in the range 10–3,000 km s^{-1}, the average being 487 km s^{-1}(right-hand panel in Fig. 2). They also found that the average speed varies with the solar cycle, the average plane-of-sky speeds in the descending phase being lower than in the minimum.

The white-light coronal images have also been used to measure the density of CMEs by estimating their excess brightness. The density values thus obtained were

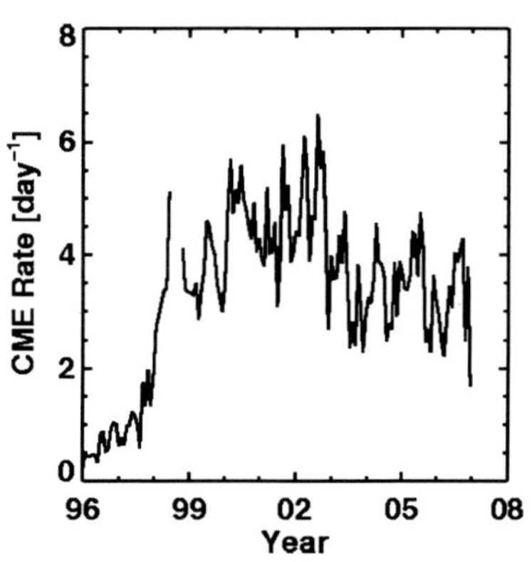

Fig. 1 The rate of occurrence of LASCO CMEs averaged over a full solar cycle, i.e., 1996–2007 (adapted from Gopalswamy et al. 2009)

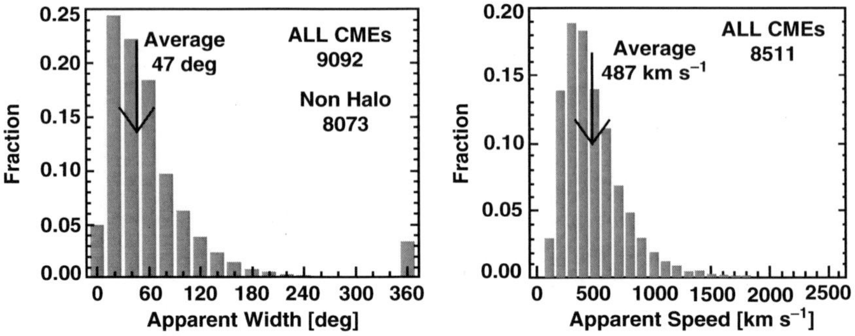

Fig. 2 *Left*: average width of all non-halo (angular width 0–120°) LASCO CMEs. *Right*: the projected speed of both halo and limb CMEs as recorded by the LASCO coronagraphs

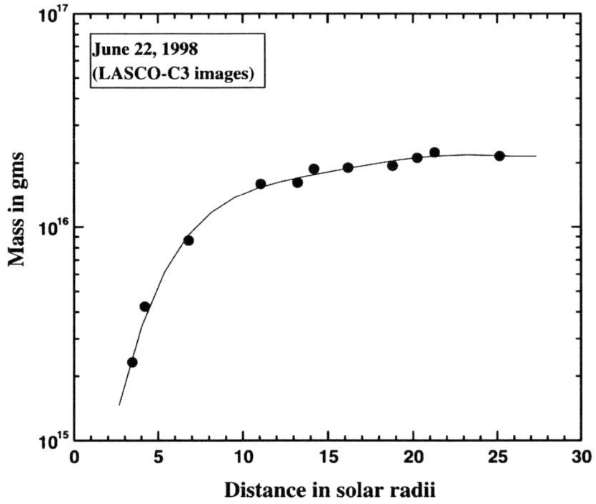

Fig. 3 The variation of total mass of the CME observed on 21–22 June 1998, with distance to the center of the sun (adapted from Srivastava et al. 2000)

used to derive the mass of CMEs, which ranges between 10^{15} and 10^{16} gm. The average CME mass is smaller than the pre-SoHO values because LASCO could record low-mass CMEs owing to its high sensitivity. Usually, most CMEs show an initial increase in mass before reaching a constant value, as shown in Fig. 3. Yashiro et al. (2004) found that the mass estimates are uncertain by a factor of two. Using the estimated values of density and the measured values of the plane-of-sky speeds of the CMEs, the kinetic energies of CMEs were found to lie in the range 10^{31}–10^{32} erg. A comparison of statistical properties of white-light CMEs recorded by LASCO and recorded by earlier space-borne coronagraphs, namely, Solwind and SMM, is given in Table 2.

A number of studies have been made on height-time profiles of various types of CMEs, indicating that there is broad spectrum of profiles. At one end of the speed spectrum, there are CMEs that rise gradually with slow speed (50–100 km s^{-1}) over a period of several hours. They reach a terminal speed of 300–400 km s^{-1}, equivalent to that of the slow solar wind speed at $20\,R_\odot$ (Srivastava et al. 1999;

Table 2 Average CME properties

Parameter	LASCO	Solwind/SMM
Observing duty cycle	81.7%	66.5%
Kinetic energy (erg)	2.6×10^{30}	3.5×10^{30}
Average mass (gm)	1.4×10^{15}	4.1×10^{15}
Mass flux (gm/day)	2.7×10^{15}	7.5×10^{15}
Average speed (km s^{-1})	487	349
Speed range (km s^{-1})	$10 - 3{,}000$	$80 - 1{,}042$
Rate of occurrence (CME/day)		
Cycle minimum	$0.31 - 0.77$	0.5
Cycle maximum	$1.75 - 3.11$	$5 - 6$
Angular width (°)	47	40

Srivastava et al. 2000). These are generally associated with eruptive filaments or prominences. At the other end of the speed spectrum are fast events, with speeds greater than 600 km s^{-1}, which undergo maximum acceleration in the lower corona. Such CMEs are generally associated with flares (Zhang et al. 2004).

3 CME Observations from SECCHI/STEREO Coronagraphs

The twin STEREO spacecrafts were launched during the solar minimum period in October 2006, when there was low expectation of the occurrence of CMEs. A preliminary study shows that the rate of CMEs observed with STEREO soon after its launch was 1 CME/day, higher than the rate of LASCO CMEs recorded during solar minimum. COR1 data show that about 353 CMEs have been recorded until 30 January 2009, as shown at the COR1 website http://cor1.gsfc.nasa.gov/docs/prelim__events. The rate decreased to 0.5 CME/day in 2008, but shows a rising trend since then. Some of these CMEs could also be tracked in the outer corona with the COR2 and HI coronagraphs. As the STEREO observations provide simultaneous images of the corona from two vantage points, that is, from "ahead" and "behind" spacecrafts, they are useful to study the 3D structure of CMEs. Prior to the launch of STEREO, different techniques were employed to derive the 3D structure of solar features using SoHO data (Pizzo and Biesecker 2004; Inhester 2006). The CME propagation properties were also derived by applying a cone model to the LASCO images (Zhao et al. 2002; Michalek et al. 2003; Michalek 2006). Other techniques that have been used for 3D reconstruction are based on polarization measurements of the white light corona (Moran and Davila 2004; Dere et al. 2005). Based on the findings of Schwenn et al. (2005) that the ratio between lateral expansion and radial propagation of CMEs is a constant, estimations of radial speeds, and hence the arrival time of CMEs at the Earth, were made.

Recently, with the launch of the twin spacecrafts STEREO A and B, disk observations of the solar atmosphere in extreme ultraviolet wavelengths (EUVI) and coronal observations in white light using the SECCHI coronagraphs from two vantage points simultaneously became available. This was used to study 3D structure by reconstruction of solar features such as flare loops and CMEs. Using stereo pair images, one can also determine the true speeds and the directions of the leading edge and prominence of a CME. These are extremely valuable for space weather predictions, as one cannot only estimate the true speeds and propagation direction of a CME in the corona, but also the exact arrival time at the Earth. A number of studies in this direction using STEREO/SECCHI and EUVI images have been made recently. These studies are mainly based on tie-pointing reconstruction of STEREO images (Mierla et al. 2009; Mierla et al. 2009; Srivastava et al. 2009). The technique has proven to be extremely successful when applied to the leading edge of white light CMEs (Mierla et al. 2009; Mierla et al. 2009) and disk filament and loops (Gissot et al. 2008; Aschwanden et al. 2008). Essentially, the tie-pointing technique for reconstructing CMEs is based on epipolar geometry, wherein the position of the two STEREO spacecrafts A and B and a point on the solar surface define a plane called the epipolar plane. The STEREO mission's plane is a special epipolar plane, passing through the Sun's center and the two spacecrafts with its normal oriented towards the ecliptic North direction. The projections of all epipolar planes in the spacecraft's images are seen as epipolar lines in one stereoscopic image that passes through the same epipolar line in the other stereoscopic image. Tie-pointing involves finding a one-to-one correspondence of a feature in both the stereo images along equal epipolar lines, calculating the line-of-sight ray that belongs to the respective images, and eventually constraining the rays to lie on the same epipolar plane (Trucco and Verri 1998).

A quick method based on tie-pointing is the height-time method for the 3D reconstruction of CME features (Mierla et al. 2009). This is based on estimating the projected or "plane-of-sky" speeds of selected moving features of CMEs. Mierla et al. (2009) estimated the true heights, speeds, and directions of the leading edges of three CMEs using COR1 coronagraphic images. An example of 3D reconstruction of one of the CMEs dated 20 May 2007 studied in their paper is shown in Fig. 4. By applying the height-time method to COR1 images, Mierla et al. (2009) found that the 20 May 2007 CME was located at ecliptic longitude and latitude of around $-2°$ and $-27°$ (i.e., south of the ecliptic), respectively. The plane-of-sky speeds as measured from the COR1 A and COR1 B coronagraphs were estimated to be approximately 242 and 253 km s^{-1}. The true speed was estimated to be approximately 548 km s^{-1}. In this contribution, we extended the analysis to COR2 images, which cover a field of view from 2 to 15 $R_{\odot}$. Using the stereo pair images of the COR2 coronagraphs and applying height-time analysis on the same feature as in the COR1 images, we obtained the projected speed as approximately 298 and 250 km s^{-1} in the A and B images, respectively (Fig. 5). The reconstructed speed was estimated at 544 km s^{-1}. The average ecliptic longitude and latitude were estimated to be $-2°$ and $-28°$, respectively. A comparison of the reconstruction parameters obtained using different techniques for the leading edge of the 20 May 2007 CME is given in Table 3. The projected speeds (V_{proj}) of the leading edge are given in the third column; the reconstructed parameters, that is, the reconstructed speed (V_{rec}), the ecliptic longitude (λ), and the ecliptic latitude (θ) of the identified feature along

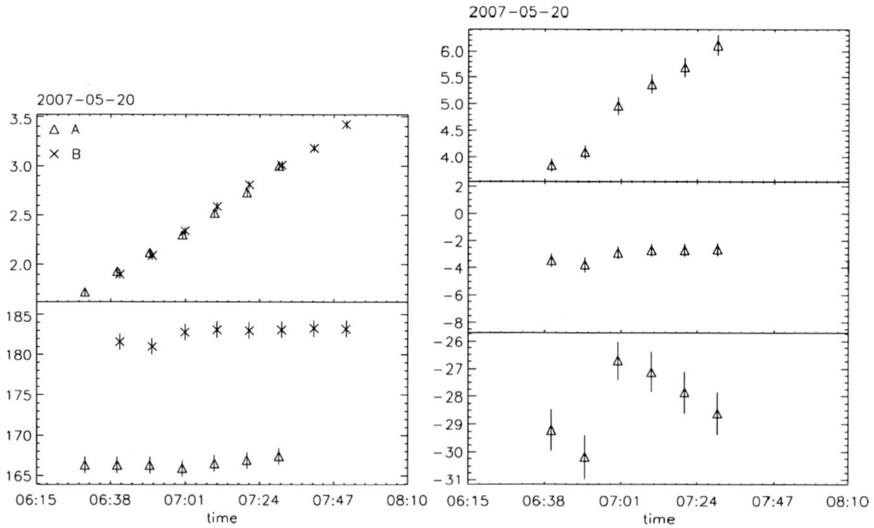

Fig. 4 *Left*: projected height (*top*) and position angle (*bottom*) of an identifiable feature along the leading edge of the 20 May 2007 CME, as observed in the COR1A and B coronagraphs. The position angle of the selected feature is measured counterclockwise from the North ecliptic pole. *Right*: true height (*top*), ecliptic longitude (*middle*), and ecliptic latitude (*bottom*) for the same feature, obtained using the height-time technique (adapted from Mierla et al. 2009)

the leading edge, are given in the fourth column. The table shows that the results from the different techniques are in good agreement. This implies that the leading edge moves at almost constant speed, with only small deflections in the direction of propagation.

A magnetic cloud was found to be associated with the CME of 20 May 2007 (Kilpua et al. 2009). This cloud arrived at the STEREO A spacecraft at 00:26 UT on 23 May. Thus, the actual travel time of the CME to the Earth was 68 h. In addition, the measured speed of the magnetic cloud at STEREO A was 535 km s^{-1}. Assuming that this speed is the average speed at which the CME traveled towards the Earth, the predicted travel time is about 74 h, which is equal to the actual arrival time within the measurement errors. It should be emphasized that use of the projected speeds would yield large errors in the estimation of the travel time. Thus, 3D reconstruction using STEREO pair images has an important bearing on space weather prediction (Srivastava et al. 2009).

4 Summary and Conclusion

SECCHI observations have surpassed the capabilities of SoHO observations. With new views of CMEs obtained by applying reconstruction techniques to the stereo image pairs, our knowledge of the 3D structure and propagation of CMEs will

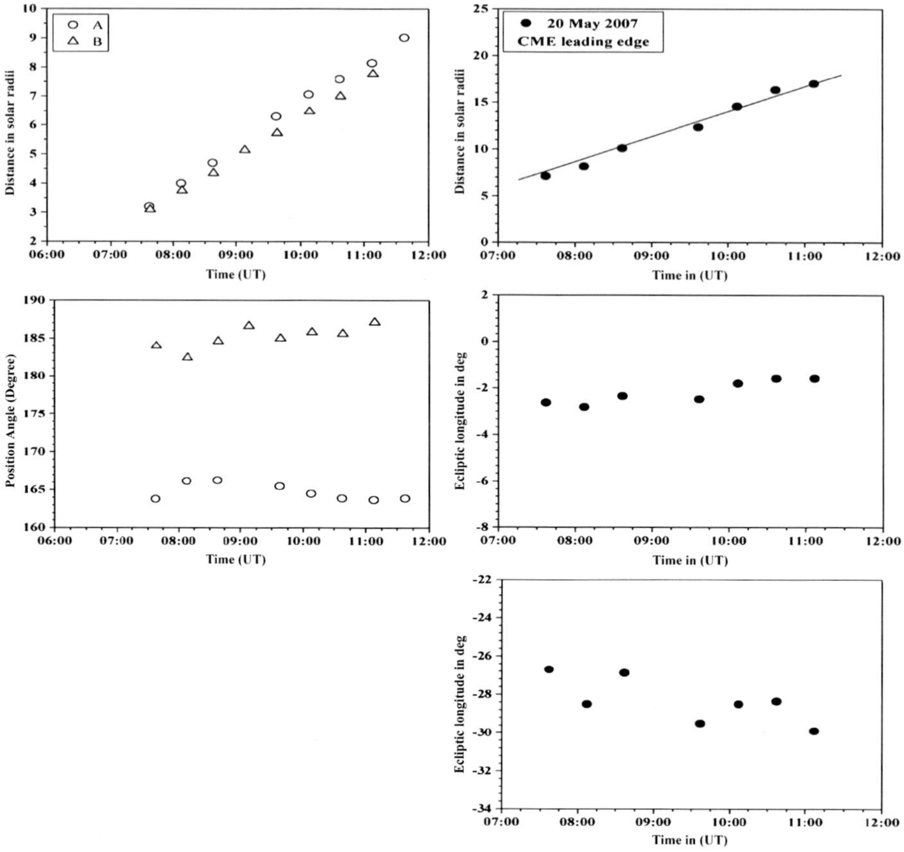

Fig. 5 *Left*: projected height (*top*) and position angle (*bottom*) of a selected feature along the leading edge of the 20 May 2007 CME, as observed in COR2A and B coronagraphs. *Right*: true height (*top*), ecliptic longitude (*middle*), and ecliptic latitude (*bottom*) for the same feature, obtained using the height-time technique for reconstruction

Table 3 Reconstruction of the leading edge of the 20 May 2007 CME

Method	Data	Projected V_{proj} ($km\,s^{-1}$)		Reconstruction V_{rec} ($km\,s^{-1}$)	λ (deg)	θ (deg)	Remarks
Height-time	COR1 A&B	242	253	548	−2	−27	Mierla et al. (2008)
	COR2 A&B	295	250	544	−2	−28	this paper
Tie-pointing	COR1 A&B COR2 A&B			510	11	−30	Srivastava et al. (2009)

advance considerably. Also, with reliable estimations of true CME speeds and propagation directions, the solar community is poised to achieve unprecedented success in space weather prediction.

Acknowledgment The author acknowledges the help of STEREO/SECCHI consortia for providing the data used and presented in this paper. The author also thanks M. Mierla (Royal Observatory of Belgium) and B. Inhester of MPS, Germany, for fruitful discussions on CME reconstruction using COR1 and COR2 images.

References

Aschwanden, M., Wlser, J.-P., Nitta, N. V., Lemen, J. R. 2008, ApJ, 679, 827

Brueckner, G. E., Howard, R. A., Koomen, M. J., Korendyke, C. M., Michels, D. J., Moses, J. D., et al. 1995, Solar Phys., 162, 357

dal Lago, A., Vieira, L. E. A., Echer, E., Gonzalez W. D., 2004, Solar Phys., 222, 323

Delaboudiniére, J.-P., Artzner, G. E., Brunaud, J., Gabriel, A. H., Hochedez, J. F., Millier, F., Song, X. Y., Au, B., Dere, K. P., Howard, R. A., and 18 coauthors, Solar Phys., 162, 291

Dere, K. P., Wang, X., Howard, R. 2005, ApJ, 620, L119

Gissot, S. F., Hochedez, J.-F., Chainais, P., Antoine, J.-P., Solar Phys., 252, 397

Gopalswamy, N., Yashiro, S., Michalek, G., Stenborg, G., Vourlidas, A., Freeland, S., Howard, R. 2009, Earth, Moon and Planets, doi 10.1007/s11038-008-9282-7

Hapgood, M. A. 1992, Plan. Space Science, 40, 711

Howard, R. A., Moses, J. D., Vourlidas, A., Newmark, J. S., Socker, D. G., Plunkett, S. P., et al. 2008, Space Sci. Rev., 136, 67

Hudson et al. 2007, in Coronal Mass Ejections, eds. H. Kunow, N.U. Crooker, J.A. Linker, R. Schwenn, and R. Von Steiger, Space Science Series of ISSI 2007, 13

Inhester, B. 2006, Publ. Int. Space Sci. Inst., astro-ph/0612649 (in press)

Kaiser, M. L., Kucera, T. A., Davila, J. M., St. Cyr, O. C., Guhathakurta, M., Christian, E. 2008, Space Sci. Rev., 136, 5

Kilpua, E. K. J., Liewer, P. C., Farrugia, C., Luhmann, J. G., Moest, C., Li, Y., Liu, Y., Lynch, B. J., Russell, C. T., Vourlidas, A., Acuna, M. H., Galvin, A. B., Larosn, D., Sayvaud, J. A. 2009, Solar Phys. 254, 325

Michalek, G. 2006, Solar Phys., 237, 101

Michalek, G., Gopalswamy, N., Yashiro, S. 2003, ApJ, 584, 472

Mierla, M. Davila, J., Thompson, W., Inhester, B., Srivastava, N., Kramar, M., St. Cyr. O. C., Stenborg, G, Howard, R. A. 2008, Solar Phys., 252, 385

Mierla, M., Inhester, B., Marque, C., Rodriguez, L., Gissot, S., Zhukov, A., Berghmans, D., Davila, J. 2009, Solar Phys., doi 10.1007/s11207-009-9416-8

Moran, T.G., Davila, J. 2004, Science, 305, 66

Pizzo, V.J., Biesecker, D.A. 2004, Geophys. Res. Lett., 31, L21802

Schwenn, R., Dal Lago, A., Huttunen, E., Gonzalez, W. D., 2005, Ann. Geophys., 23, 1033

Schwenn, et al. 2007, in Coronal Mass Ejections eds. H. Kunow, N. U. Crooker, J.A. Linker, R. Schwenn, and R. Von Steiger, Springer, Berlin, 137

Srivastava, N., Schwenn, R., Inhester, B., Stenborg, G., and Podlipnik, B. 1999, in Solar Wind 9, AIP Conf. Proc. 471, 115

Srivastava, N., Schwenn, R., Inhester, B., Martin, S. F., Hanaoka, Y. 2000, ApJ, 534, 468

Srivastava, N., Inhester, B., Mierla, M., Podlipnik, B. 2009, Solar Phys., doi 10.1007/s11207-009-9423-9

St. Cyr, O. C., Howard, R. A., Sheeley, N. R., Plunkett, S. P., Michels, D. J., Paswaters, S. E., Koomen, M. J., Simnett, G. M., Thompson, B. J., Gurman, J. B., Schwenn, R., Webb, D. F., Hildner, E., Lamy, P. L. 2000, J. Geophys. Res., 105, 18169

Thompson, W.T. 2006, A&A, 449, 791

Thompson, W. T., Davila, J. M., Fisher, R. R., Orwig, L. E., Mentzell, J. E., Hetherington, S.E. 2003, in Innovative Telescopes and Instrumentation for Solar Astrophysics, eds. Keil, S. L., Avakyan, S. V., SPIE, 4853, 1

Tousey R. 1973, The Solar Corona, in Space Research XIII 1973, eds. M. J. Rycroft, S. K. Runcorn, 713, Akademie-Verlag, Berlin
Trucco, E., Verri, A. 1998, Introductory Techniques for 3-D Computer Vision, Prentice Hall
Vourlidas, A., Buzasi, D., Howard, R. A., Esfandiari, E. in: Solar Variability: From Core to Outer Frontiers. The 10th European Solar Physics Meeting, ed. A. Wilson, ESA SP-506, 91
Wang, Y. M., Ye, P. Z., Wang, S., Zhou, G. P., Wang, J. X. 2002, J. Geophys. Res., 107, 2
Xie, H., Ofman, L., Lawrence, G. 2004, J. Geophys. Res., 109, A03109
Yashiro, S., Gopalswamy, N., Michalek, G., St. Cyr, O. C., Plunkett, S. P., Rich, N. B., Howard, R. A. 2004, J. Geophys. Res., 109, A07105
Yurchyshyn, V., Wang, H., Abramenko, V. 2003, Adv. Sp. Res., 32, 1965
Zhang, J., Dere, K. P., Howard, R. A., Bothmer, V. 2003, ApJ, 582, 520
Zhang, J., Dere, K. P., Howard, R. A., Vourlidas, A., 2004, ApJ, 604, 420
Zhao, X. P., Webb, D. F. 2003, J. Geophys. Res., 108, 4
Zhao, X. P., Plunkett, S. P., Liu, W. 2002, J. Geophys. Res., 107, 1223

Low-Frequency Radio Observations of Coronal Magnetic Fields

R. Ramesh, S.M. Sonnett, and C. Kathiravan

Abstract We report observations of circularly polarized noise storm emission from the solar corona at 77 and 109 MHz during the period 11–18 August 2006 (Carrington Rotation 2046) with the recently commissioned East-West one-dimensional radio polarimeter at the Gauribidanur observatory, near Bangalore. Two-dimensional imaging observations at 77 MHz around the same epoch with the radioheliograph at the observatory revealed that the radio source was associated with active region AR 10904 and co-rotated with it. The radial distance of the corresponding 77 MHz plasma level derived independently from the observed rotation was about 1.1 $R_{\odot}$. The average magnetic field value that follows from the observed circularly polarized emission at this plasma level and radial distance is about 4 G.

1 Introduction

Despite its fundamental importance, it is difficult to directly measure coronal magnetic fields through optical observations due to a variety of reasons (Lin et al. 2000 and references therein). The direct measurement of magnetic fields in the solar photosphere through Zeeman splitting of Fraunhofer lines is a well-established and reliable technique. In the absence of similar method for the upper layers of the solar atmosphere, considerable efforts are spent in extrapolating the observed solar-surface field distribution under the assumption that it is potential or force-free. It is possible to obtain information on coronal magnetic fields through low-frequency radio observations of circularly polarized radio emission from noise storms and transient burst emission [Dulk and McLean 1978 and references therein] as the observed emission originates there. Noise storms in particular are very useful candidates as they are the most frequently observed solar activity in the above frequency range and

R. Ramesh (✉) and C. Kathiravan
Indian Institute of Astrophysics, Bangalore, India

S.M. Sonnett
Institute for Astronomy, University of Hawaii, Manoa, USA

S.S. Hasan and R.J. Rutten (eds.), *Magnetic Coupling between the Interior and Atmosphere of the Sun*, Astrophysics and Space Science Proceedings,
DOI 10.1007/978-3-642-02859-5_26,

they are usually circularly polarized. The emission consists of occasional, short-lived (0.1–1 s), narrowband radio enhancements ("noise storm bursts"), superimposed on frequently observed continuous, slowly varying, long lasting (hours–days), broadband background emission called the "noise storm continuum," [Elgarøy (1977) and references therein]. Note that linear polarization, if present at the source region in the solar corona, is washed out at lower frequencies due to Faraday rotation of the plane of polarization between the source and the observer ((Gragnard and McLean 1973)).

2 Observations

The radio data reported were obtained at 77 and 109 MHz with the recently commissioned polarimeter at the Gauribidanur observatory (Ramesh et al. 2008). Figures 1 and 2 show the Stokes I and V output from the solar corona at 77 and 109 MHz, observed on 11 August 2006. There is clear evidence of circularly polarized emission from the Sun at both 77 and 109 MHz. Similar emission from the Sun was observed on the following days also, up to 18 August 2006. As the 77 MHz frequency is also used with the Gauribidanur radioheliograph (GRH) (GRH; Ramesh et al. 1998), we inspected the two-dimensional radioheliograms (in Stokes I) obtained with the GRH around the same epoch to identify the emission source(s) that were responsible for the observed polarization. There was an isolated, discrete source of intense emission (brightness temperature $T_b \geq 10^8$ K) on 11 August 2006 in the south-east quadrant (Fig. 3). Because of issues related to the dynamic range of the GRH, emission from the background corona [$T_b \approx 10^5\ K$ at 77 MHz, McLean and Labrum 1985] is not seen in this image (Ramesh et al. 1999). The emission site persisted in images from the GRH over the next few days and lasted until 18 August 2006. Interestingly, it co-rotated with the Sun and was located close to its West limb on 18 August 2006 (Fig. 4). We found that its average rotation rate, from

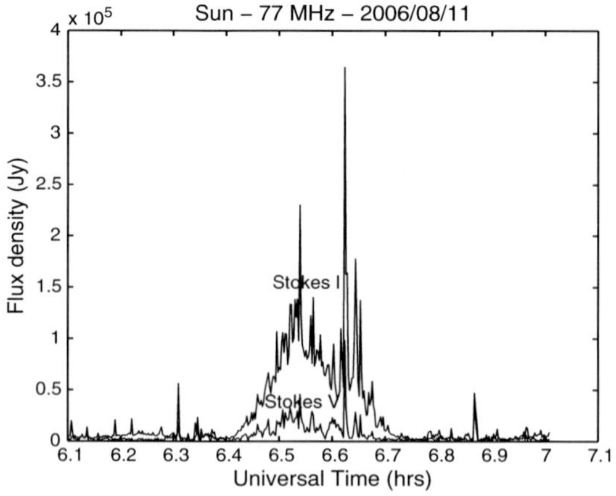

Fig. 1 Stokes I and V observations around solar transit through the local meridian at Gauribidanur, at 77 MHz on 11 August 2006 with an integration time of 512 ms. The "spikes" on top of the background continuum are individual noise storm bursts

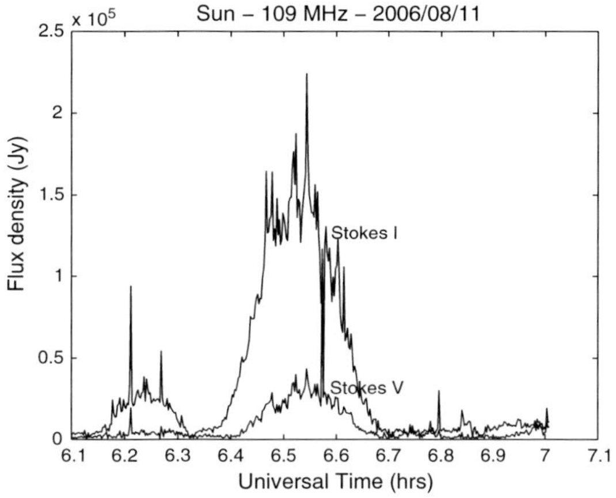

Fig. 2 Same as Fig. 1, but at 109 MHz. The hump between 6.1 and 6.3 UT, to the left of the Stokes *I* deflection, is a sidelobe artifact

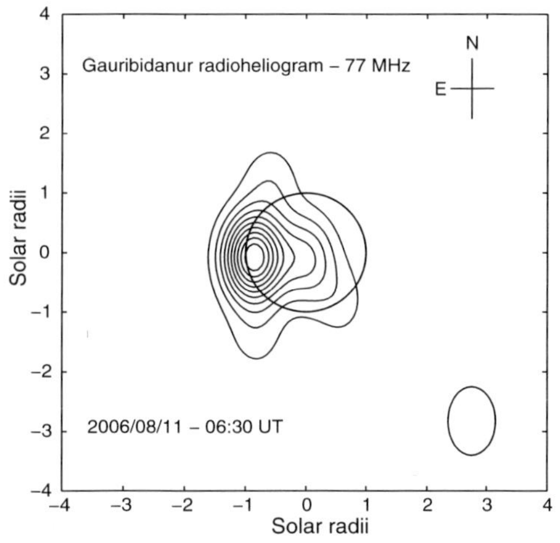

Fig. 3 GRH observations of 11 August 2006 at 77 MHz. The peak value is $T_b \approx 2.6 \times 10^8$ K. The open circle at the center represents the limb of the solar photosphere, the ellipse near the bottom-right corner the GRH "beam" at 77 MHz

the radioheliograms, was about 4.1′/day. This corresponds to $r \approx 1.1\,R_\odot$ in the solar atmosphere for the emission region (Ramesh et al. 2000).

A comparison of the radioheliograms from 11 and 18 August 2006 with the ones observed on 10 and 19 August 2006 indicates that the discrete source of intense emission was present only during the interval 11–18 August 2006. We calculated the spectral index of the Stokes *I* emission observed with the polarimeter; its average value was −0.84, indicating that the observed emission is nonthermal in nature. As radio noise storms are the only long-duration (days) events in the solar atmosphere that belong to this category [McLean and Labrum 1985 and references therein], we argue that the circularly polarized emission observed with the new Gauribidanur radio polarimeter must be due to noise storm emission. The estimated average degree of circular polarization was $V/I \approx 0.53$ at 77 MHz and 0.64 at 109 MHz, which supports this claim (Kai 1962).

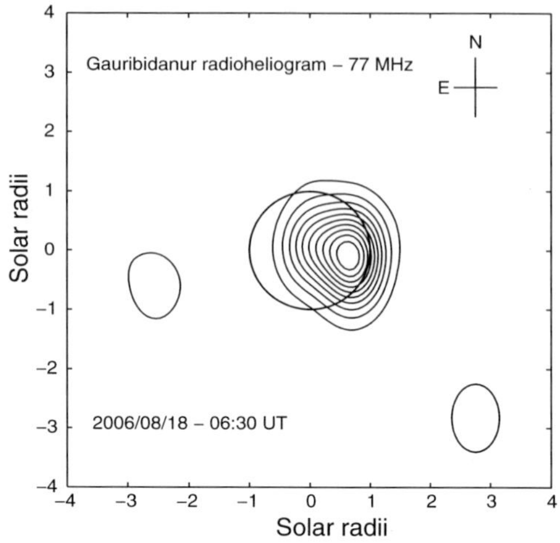

Fig. 4 Same as Fig. 3, but for 18 August 2006. The peak value is $T_b \approx 3.1 \times 10^8$ K. The isolated contour beyond the east limb is due to local interference

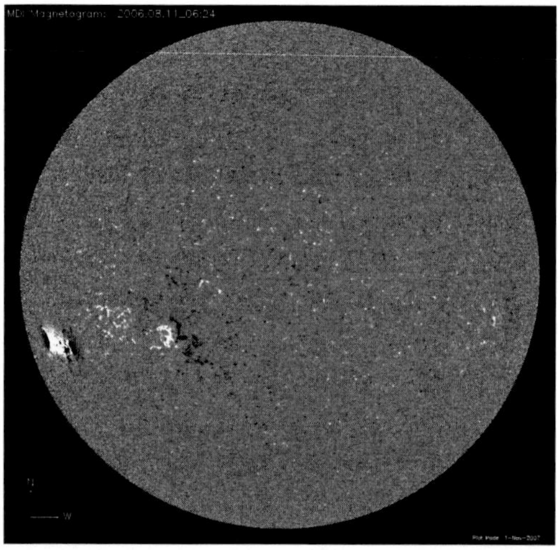

Fig. 5 SoHO-MDI magnetogram obtained on 11 August 2006. The bright region located close to the east limb is AR 10904. Other than AR 10903 (located to the west of AR 10904), there are no dominant magnetic regions on the solar disk

3 Results and Discussion

In general, any transient or long-duration phenomenon observed in the solar corona should have its origin in corresponding activities at lower levels in the atmosphere. This is particularly true for radio noise storms as it is well established that these are closely associated with sunspot groups in the photosphere (e.g., Elgarøy 1977). We therefore inspected SoHO-MDI magnetogram images to identify the photospheric counterpart of the observed circular polarization at 77 and 109 MHz. Figures 5 and 6 show these images for 11 and 18 August 2006 (CR 2046). A comparison with Figs. 3 and 4 indicates that the bright magnetic region AR 10904 (S14 E63) located close to the east/west limb of the Sun on 11/18 August 2006, respectively, must be primarily

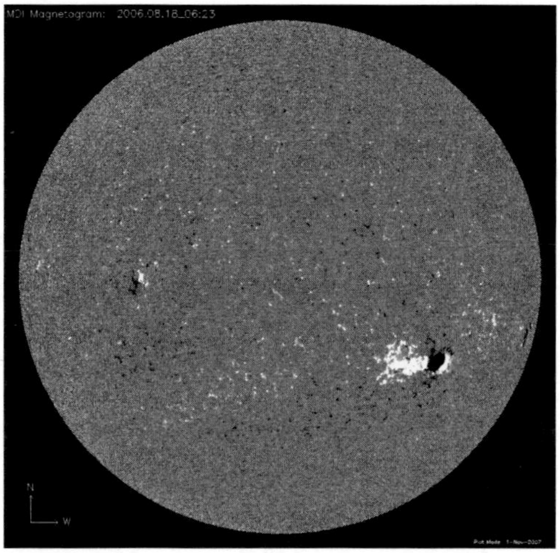

Fig. 6 Same as Fig. 5, but obtained on 18 August 2006. AR 10904 had rotated close to the west limb in this image

responsible for the appearance of the Stokes V output in Figs. 1 and 2 observed on 11 August 2006 and the subsequent days, up to 18 August 2006. Active region AR 10903 (S12 E26) located to the west of AR 10904 (see Fig. 5) can be ruled out as the cause for the observed noise storm emission because (1) AR 10903 was on the visible solar hemisphere even prior to the appearance of AR 10904 at the east limb while no detectable noise storm emission or circular polarization was observed, and (2) we observed significant circular polarization on 18 August 2006 when AR 10903 was not anymore on the disk but had passed over the west limb of the Sun.

Relating the observations at the photosphere and corona can de done assuming magnetic flux conservation by setting $B_p A_p = B_c A_c$ (Kai 1970; Lee 2007), where the magnetic field B_c corresponds to the observed circularly polarized radio emission at 77 MHz and B_p is the SoHO-MDI line-of-sight magnetic field for AR 10904. The average value of the latter during 11–18 August 2006 was about 3,040 G. The area A_p of AR 10904 was about $22'' \times 32''$ while the GRH beam size was taken for the coronal area A_c at 77 MHz (about $600'' \times 900''$). The latter is close to the expected size (about $660''$) of noise storm continuum sources at 77-MHz ((Sundaram and Subramanian 2004)). With these numbers we obtain $B_c \approx 4\,\text{G}$ at $r = 1.1\,R_\odot$.

We are presently deriving the coronal field strength more rigorously; the results will be reported elsewhere.

Acknowledgment The estimation of photospheric magnetic flux values from SoHO-MDI data is due to K. Nagaraju at the Indian Institute of Astrophysics (IIA). S.M. Sonnett was a visiting intern at IIA, Bangalore during June–July 2007. She was funded by the US National Science Foundation's Office of International Science and Education under Grant Number 0654111: International Research Experience for Students, managed by the US National Solar Observatory's Global Oscillation Network Group Program in coordination with the IIA. We thank the Scientific Organizing Committee for the opportunity to participate in the meeting and present our work.

References

Dulk, G. A., McLean, D. J. 1978, Solar Phys., 57, 279
Elgarøy, Ø. 1977, Solar Noise Storms, Pergamon Press
Gragnard, R. J. -M., McLean, D. J. 1973, Solar Phys., 29, 149
Kai, K. 1962, PASJ, 14, 1
Kai, K. 1970, Solar Phys., 11, 456
Lee, J. W. 2007, Space Sci. Rev., 133, 73
Lin, H., Penn, M. J., Tomczyk, S. 2000, ApJ, 541, L83
McLean, D. J., Labrum, N. R. 1985, Solar Radio Physics, Cambridge University Press
Ramesh, R., Subramanian, K. R., SundaraRajan, M. S., Sastry, Ch. V. 1998, Solar Phys., 181, 439
Ramesh, R., Subramanian, K. R. Sastry, Ch. V. 1999, A&AS, 139, 179
Ramesh, R., Subramanian, K. R., Sastry, Ch. V. 2000, Astrophys. Lett. Comm., 40, 93
Ramesh, R., Kathiravan, C., SundaraRajan, M. S., Indrajit V. Barve, Sastry, Ch. V. 2008, Solar Phys., 253, 319
Sundaram, G. A. S., Subramanian, K. R. 2004, ApJ, 605, 948

Evolution of Near-Sun Solar Wind Turbulence

P.K. Manoharan

Abstract This paper presents a preliminary analysis of the turbulence spectrum of the solar wind in the near-Sun region $R < 50\,R_\odot$, obtained from interplanetary scintillation measurements with the Ooty Radio Telescope at 327 MHz. The results clearly show that the scintillation is dominated by density irregularities of size about 100–500 km. The scintillation at the small-scale side of the spectrum, although significantly less in magnitude, has a flatter spectrum than the larger-scale dominant part. Furthermore, the spectral power contained in the flatter portion rapidly increases closer to the Sun. These results on the turbulence spectrum for $R < 50\,R_\odot$ quantify the evidence for radial evolution of the small-scale fluctuations ($\leq$50 km) generated by Alfvén waves.

1 Introduction

The solar wind is highly variable and inhomogeneous, and exhibits fluctuations over a wide range of spatial and temporal scales. The properties of these fluctuations, as they move outward in the solar corona, are controlled by the presence of both waves and turbulence (e.g., Coleman 1968, Belcher & Davis 1971). However, their relative contributions to the heating and acceleration of the solar wind have yet to be assessed fully (Tu & Marsch 1995, Harmon & Coles 2005).

Radio scattering and scintillation experiments measure density fluctuations, which are related to the wave field, density fluctuations, and magnetic turbulence (e.g., Higdon 1986, Montgomery et al. 1987). The density fluctuation spectrum roughly follows a Kolmogorov power law in the spatial scale range 100–1,000 km, at distances well outside the solar wind acceleration region. However, nearer to the Sun the spectrum tends to be flat (e.g., Woo & Armstrong 1979). The spectrum of the high-speed streams from coronal holes is steeper than Kolmogorov decay, which is attributed to dissipation at scales above 100 km (e.g., Manoharan et al.

P.K. Manoharan (✉)
Radio Astronomy Centre, National Centre for Radio Astrophysics,
Tata Institute of Fundamental Research, Udhagamandalam (Ooty), India

S.S. Hasan and R.J. Rutten (eds.), *Magnetic Coupling between the Interior and Atmosphere of the Sun*, Astrophysics and Space Science Proceedings,
DOI 10.1007/978-3-642-02859-5_27,

1994, 2000). There is considerable interest to understand the radial change of the fluctuations due to both waves and turbulence in the solar wind acceleration region. In this study, spectral features are analyzed over a range of distances from the Sun using interplanetary scintillation measurements made with the Ooty Radio Telescope at 327 MHz (Swarup et al. 1971).

2 Interplanetary Scintillation

Interplanetary scintillation (IPS) is the variability of distant compact radio source (e.g., a quasar or a radio galaxy) caused by microturbulence in the solar wind of spatial scales 10–1000 km (e.g., Manoharan et al. 1994). Scintillation measurements normally refer to the instantaneous departure of intensity ($\delta I(t)$) from the mean intensity of the source ($\langle I \rangle$), i.e., $\delta I(t) = I(t) - \langle I \rangle$. As the irregularities are convected by the solar wind, the statistical fluctuations of $\delta I(t)$ can be used to estimate the speed and turbulence spectrum of the solar wind, integrated along the line of sight to the radio source. However, for a given line of sight, the spectrum of scintillation drops rapidly with distance from the Sun, $C_{\mathrm{N}}^2(R) \sim R^{-4}$, and the scattering is therefore concentrated where the line-of-sight is closest to the Sun. The shape of the turbulence spectrum can be inferred from the temporal IPS spectrum, obtained by taking the Fourier transformation of intensity time series. The rms intensity variation $\langle \delta I(t)^2 \rangle^{1/2}$ is the integral of the power spectrum. The scintillation index, m, is estimated by

$$m^2 = \frac{1}{\langle I \rangle^2} \int_0^{f_{\mathrm{c}}} P(f)\,\mathrm{d}f, \tag{1}$$

where f_{c} is the cutoff frequency of the spectrum at which the scintillation equals the noise level. The systematic radial variation of $C_{\mathrm{N}}^2(R)$ can be obtained from the index vs. distance ($m - R$) plots as in Fig. 1. These smoothed plots represent average scintillations observed over several years for two well-known radio quasars (Manoharan 2008).

At given heliocentric distance, a compact source scintillates more than an extended one, because Fresnel filtering plays a key role in producing the intensity fluctuations and the scintillation is heavily attenuated by a large angular size $\Theta \geq \sqrt{\lambda/Z}$, where λ is the wavelength of observation and Z is the distance to the scattering screen. The observations reported in this study have been made with the Ooty Radio Telescope (ORT), which operates at $\lambda = 0.92$ m. In the case of near-Sun IPS measurements, the scattering medium is located at about 1 AU and therefore sources having angular size $\Theta > 500$ milliarcsec do not scintillate.

Figure 1 shows that as the Sun is approached, the scintillation increases to a peak value at a distance of $R \approx 40\,\mathrm{R}_\odot$, and then decreases for further closer solar offsets (e.g., Manoharan 1993), where 1 solar radius is $R_\odot = 6.96 \times 10^5$ km. The peak or transition distance, $R \approx 40\,\mathrm{R}_\odot$, is the characteristic of IPS measurement at $\lambda = 0.92$ m. It is a function of observing wavelength and moves close

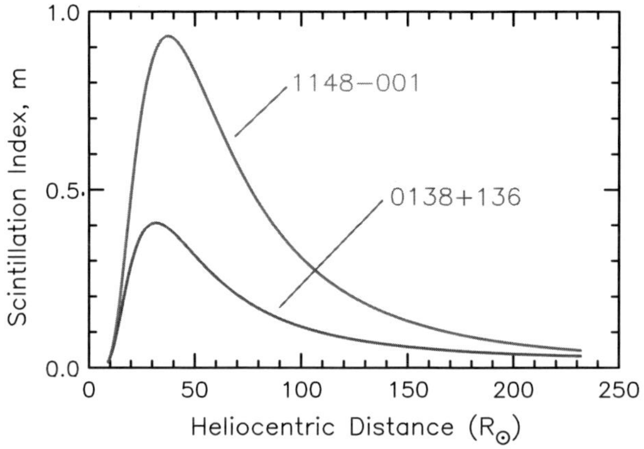

Fig. 1 Average scintillation index against heliocentric distance observed at 327 MHz. Radio quasars 1148-001 and 0138+136 have equivalent angular diameters of about 15 and 50 milliarcsec, respectively

to the Sun with decreasing wavelength (e.g., Coles 1978, Kolosov et al. 1982). At $R < 40\,\mathrm{R}_\odot$, the scintillation index of an ideal point source saturates at $m \approx 1$. In the weak-scattering region with $R \geq 40\,\mathrm{R}_\odot$, the decline in scintillation corresponds to a fall of turbulence approximately as $1/R^2$. The shape of the temporal spectrum of the intensity fluctuations, $P(f)$, is linearly related to the spatial turbulence spectrum, $\Phi_{\mathrm{N_e}}(q)$, which can be obtained by fitting the measured scintillation spectrum (Manoharan & Ananthakrishnan 1990, Manoharan et al. 1994). The turbulence spectrum is given by the power-law form $\Phi_{\mathrm{N_e}}(q) \sim q^{-\alpha}$, with the power index α varying between 3 and 3.8, depending on the solar-wind source location on the Sun. The spatial spectrum also includes Gaussian cutoffs at high spatial wavenumber q, set by the source-size visibility and the inner (dissipation) scale of the solar wind turbulence.

3 Scintillation Spectrum Near the Sun

In the strong-scattering region with $R < 40\,\mathrm{R}_\odot$, the turbulence is too intense and the relationship between the scintillation index and $C_{\mathrm{N}}^2(R)$ takes a complex form. The decrease in scintillation is due to the smearing of scintillation caused by the angular size of the radio source. In a physical sense, many independent random secondary sources exist inside the Fresnel zone, reduce the coherence, and cause reduction of the intensity fluctuations. However, the strong-scintillation spectrum is wider than the weak case, and contains information on a large range of turbulence scales. The present study reports a systematic analysis of intensity scintillation measurements of compact radio sources ($\Theta \approx 50$ milliarcsec) made with the ORT at 327 MHz. It allows to probe the solar wind from about 10–250 R$_\odot$. The closest solar offset is

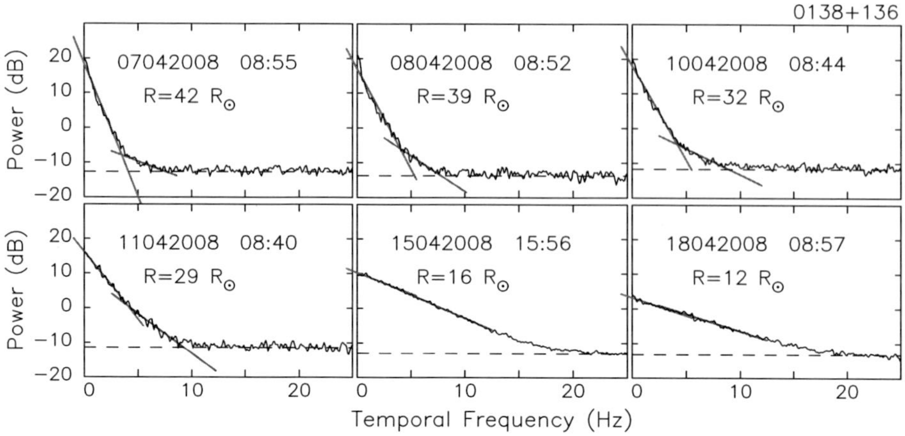

Fig. 2 Sample temporal power spectra of 0138+136 on log-linear scale, showing spectral shape variations with distance from the Sun. The date and time of observation and the heliocentric distance (R) are specified. These observations have been made at the eastern limb of the Sun so that the source approaches the Sun with increasing day number

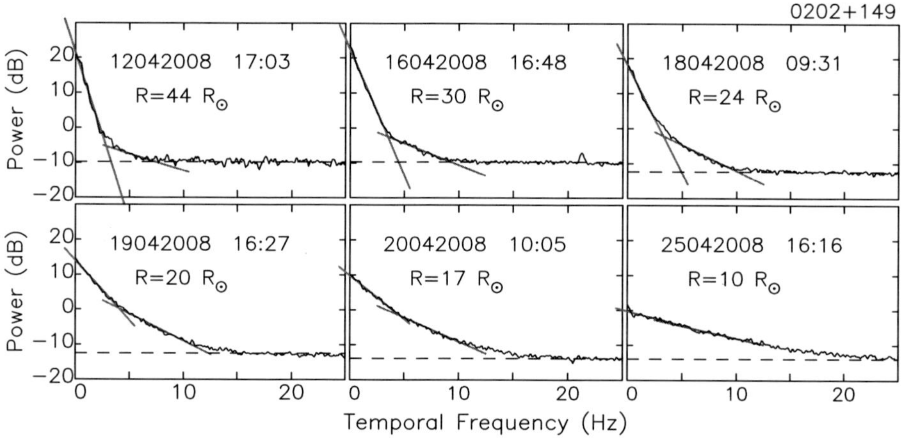

Fig. 3 Same as Fig. 2 for radio source 0202+149

mainly limited by the ORT beam width. Figures 2 and 3 display temporal scintillation spectra of radio quasars 0138+136 and 0202+149, observed at different solar offsets during April 2008. The sampling rate, 50 Hz, employed in the present study in principle extends the temporal frequency range of the spectrum to 25 Hz, which allows to infer the statistics of even small-scale turbulence. For example, for a typical value of the solar wind speed V, the spectrum can cover spatial wavenumbers in the range $0.002 < q = (2\pi f/V) < 0.2\ \mathrm{km}^{-1}$, corresponding to scales in the range 5–500 km.

Nearer to the Sun the spectrum broadens, suggesting systematic increase in turbulence associated with small-scale irregularity structures (<100 km). The flattening of the spectrum at $R < 40\,R_\odot$ indicates addition of small-scale turbulence. The remarkable change is that the high-frequency part of the spectrum gradually extends into the low-frequency part at distances closer than the transition point ($R < 40\,R_\odot$). The diminishing of spectral power at scales close to the Fresnel radius suggests the possibility of dominant effect of Fresnel filter, which can smear the scintillation. At $R > 40\,R_\odot$, the low-frequency part of the spectrum gradually steepens and merges with the slope of the density turbulence spectrum at scales smaller than the Fresnel radius. When a large number of spectra on a given radio source, observed on consecutive days over a period of 45 days, are displayed in movie mode, this gives a direct visualization, making the above results immediately apparent.

4 Radial Evolution of Small- and Large-Scale Turbulence

Figure 4 shows the power of turbulence associated with the low- and high-frequency portions of the spectrum at different solar offsets for radio sources 0138+136 and 0202+149. These plots illustrate the attenuation and enhancement of the scintillations, respectively, for the large-scale (>100 km) and the small-scale (<100 km) spectral regions and their radial variations. For most of the temporal spectra, the slope change from high- to low-frequency part is apparent, and whenever the spectrum monotonously increases towards the low frequency part, the half-value of the cutoff frequency (i.e., $f_c/2$) is considered to mark the separation of the scintillation between the low- and high-frequency parts.

It is obvious that the turbulence density associated with the low-frequency part is dominant at all heliocentric distances and that it closely follows the shape of the overall scintillation index vs. distance curve (Fig. 1). Manoharan (1993) has shown that the scintillation variation at $R > 40\,R_\odot$ is of power-law form, with $m \sim R^{-\beta}$ and $\beta = 1.7 \pm 0.2$. When the integration is accounted for, the scattering power changes as $C_N^2(R) \sim R^{-(2\beta+1)} = R^{-4.4\pm0.4}$. The scintillation in the low-frequency part of the spectrum is consistent with the above radial evolution. However, in the high-frequency part, the scintillation increases with decreasing solar offsets and tends to merge with the above portion. In the distance range $R = 15$–$100\,R_\odot$, the scintillation due to the high-frequency part follows the power-law $m_{\mathrm{high\_freq}} \sim R^{-b}$. Both sources show similar slopes $b \approx 2.0$ and 2.3. However, the average radial trend is much steeper, with $C^2_{N_{\mathrm{high\_freq}}}(R) \sim R^{-(2b+1)} = R^{-5.3}$, than the density turbulence slope $\sim R^{-4}$. The turbulence associated with small-scale fluctuations ($\leq$50 km) in the solar wind acceleration region steeply increases towards the Sun.

The strong scintillation spectra of selected radio sources observed at Ooty have also been compared with same-day observations at higher observing frequencies, for which the measurements fall in the weak scintillation regime. For example, IPS measurements with the Giant Metrewave Radio Telescope (GMRT) at 610 MHz

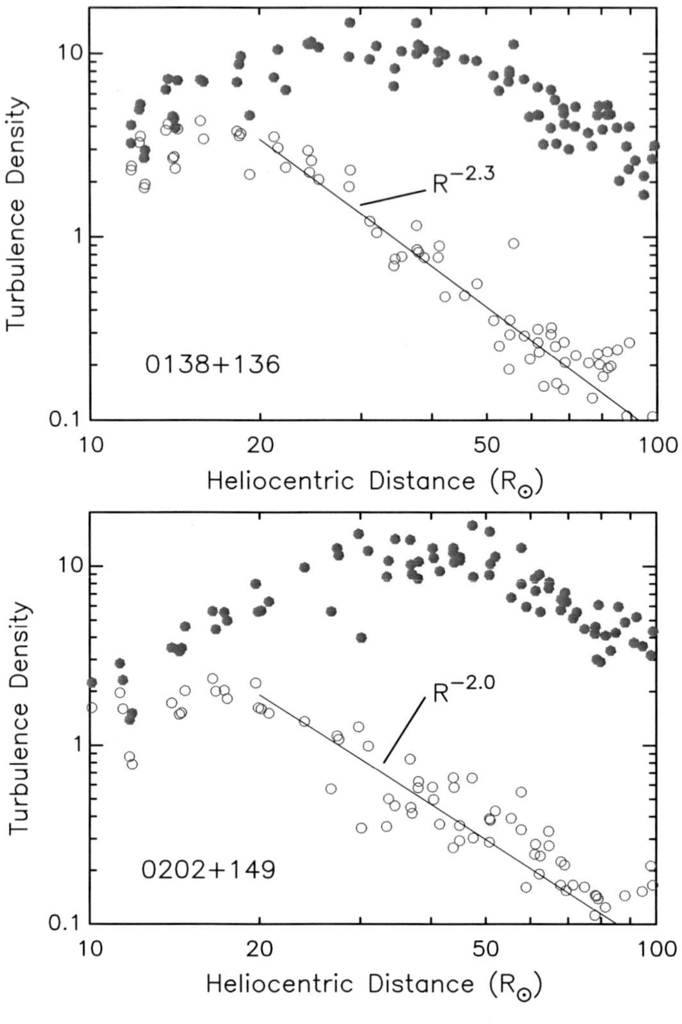

Fig. 4 Radial turbulence evolution in the low-frequency (*filled symbols*) and high-frequency portions (*open symbols*) of the spectrum

(Manoharan et al., in preparation) and the European Incoherent Scatter (EISCAT) system at 931 MHz (R.A. Fallows, private communication) reveal spectral features that are closely similar to that observed with the ORT at low and high temporal frequencies. The contributions of low and high spatial scales of the scintillation spectrum are not altered in the transition from strong to weak scintillation.

5 Discussion and Conclusion

Several IPS experiments have shown the turbulence spectrum to be $\Phi_{N_e} \sim q^{-\alpha}$, with the dissipative scale (i.e., inner scale or cutoff scale) size increasing linearly with distance as $l_i \approx (R/R_\odot)^{1\pm0.1}$ km at $R \leq 100\,R_\odot$ (Manoharan et al. 1987, Coles & Harmon 1989, Manoharan et al. 1994). Further, a flatter spectrum ($\alpha \leq 3$) and smaller dissipative scales ($l_i < 10$ km) have been observed in the near-Sun solar wind acceleration region ($R < 20\,R_\odot$) (Coles & Harmon 1989, Yakovlev et al. 1980, Yamauchi et al. 1998). The present result of flatter spectrum for the low-frequency part, which at larger distances merges with the density turbulence spectrum, is consistent with the earlier findings.

The effects of the angular structure of the radio source and of inner-scale turbulence dissipation are three-dimensionally Gaussian in shape, and tend to attenuate the high-frequency tail of the spectrum (Manoharan et al. 1987, Yamauchi et al. 1998). The inner-scale contribution is not significant at small solar offsets (it decreases and becomes small at regions close to the Sun). Furthermore, the effect of the angular size of the compact radio source ($\Theta \approx 50$ milliarcsec) is considerably small. However, the key point is that, in the near-Sun regions ($R < 40\,R_\odot$), a significant enhancement in scintillation power is measured at the high-frequency portion of the spectrum, well above the dissipation and source-size cutoff levels. To show scintillation above these cutoffs in the tail part of the spectrum, strong fluctuations are likely to be present, which are oriented in different directions than the radial flow of the solar wind. Therefore, the systematic and significant increase in power at the small-scale part of the spectrum suggests an active role of irregularities produced by magnetosonic waves in the solar wind, with multiple scale sizes and vector directions. The rapid radial change of the turbulence associated with the small-scale irregularities, $C^2_{N(\mathrm{high\_freq})}(R) \sim R^{-5.3}$, indicates that the dominant contribution is due to wave-generated turbulence in the solar wind acceleration region. Therefore, the overall near-Sun turbulence spectrum can be explained by the combined effects of the smeared density turbulence spectrum and the strong fluctuations generated by Alfvén waves at small scales ($\leq$50 km).

In summary, this preliminary analysis of the temporal spectrum of scintillations measured in the solar wind acceleration region provides evidence that, apart from density turbulence, small-scale fluctuations produced by magnetosonic waves plays a key role in shaping the spectrum. In comparison with the density turbulence, the effect of waves is significant but its importance decreases rather steeply with heliocentric distance. Its presence in the solar wind extends outside the acceleration region ($R > 20\,R_\odot$), although weaker in intensity. A more rigorous study of the small-scale microturbulence, its variation with the solar cycle and solar source regions will be reported in more detail elsewhere.

Acknowledgment The author thanks the observing/engineering team and research students of the Radio Astronomy Centre for help in performing the observations and the preliminary data reduction. This work is partially supported by the CAWSES–India Program, which is sponsored by the Indian Space Research Organization (ISRO).

References

Belcher, J. W., Davis, Jr., L. 1971, J. Geophys. Res., 76, 3534
Coleman, Jr., P. J. 1968, ApJ, 153, 371
Coles, W. A. 1978, Space Sci. Rev., 21, 411
Coles, W. A., Harmon, J. K. 1989, ApJ, 337, 1023
Harmon, J. K., Coles, W. A. 2005, J. Geophys. Res., 110, 3101
Higdon, J. C. 1986, ApJ, 309, 342
Kolosov, M. A., Yakovlev, O. I., Efimov, A. I., et al. 1982, Radio Sci., 17, 664
Manoharan, P. K. 1993, Solar Phys., 148, 153
Manoharan, P. K. 2008, The Solar Wind (Princeton Series in Astrophysics), 235–266
Manoharan, P. K., Ananthakrishnan, S. 1990, MNRAS, 244, 691
Manoharan, P. K., Ananthakrishnan, S., Pramesh Rao, A. 1987, in Sixth International Solar Wind Conference, eds. V. J. Pizzo, T. Holzer, & D. G. Sime, 55
Manoharan, P. K., Kojima, M., Gopalswamy, N., Kondo, T., Smith, Z. 2000, ApJ, 530, 1061
Manoharan, P. K., Kojima, M., Misawa, H. 1994, J. Geophys. Res., 99, 23411
Montgomery, D., Brown, M. R., Matthaeus, W. H. 1987, J. Geophys. Res., 92, 282
Swarup, G., Sarma, N. V. G., Joshi, M. N., et al. 1971, Nat, 230, 185
Tu, C.-Y., Marsch, E. 1995, Space Sci. Rev., 73, 1
Woo, R., Armstrong, J. W. 1979, J. Geophys. Res., 84, 7288
Yakovlev, O. I., Efimov, A. I., Razmanov, V. M., Shtrykov, V. K. 1980, Soviet Ast., 24, 454
Yamauchi, Y., Tokumaru, M., Kojima, M., Manoharan, P. K., Esser, R. 1998, J. Geophys. Res., 103, 6571

The Solar-Stellar Connection

J.H.M.M. Schmitt

Abstract The presence of strong magnetic fields on the solar surface has been known for more than 100 years, ever since Hale (1908) was the first to measure solar magnetic fields through the Zeeman effect. The coronal heating problem was established in the 30s and early 40s of the last century, when Grotrian and Edlén (see the discussion in Edlén (1945) on this issue) realized from the identification of so-called forbidden lines that the very outer layers of the Sun were much hotter than its photosphere. However, the connection between magnetic fields and coronal heating was not firmly made until the 70s, when the hundreds of high-resolution Skylab X-ray images of the Sun (Zombeck et al. 1978) demonstrated the extreme spatial inhomogeneity of its X-ray emission and the close association of X-ray activity with bipolar regions on its surface.

As far as stellar activity is concerned, the first systematic studies of stellar chromospheres were started in the 50s (Wilson 1963), mostly from intensity measurements of the Ca II H & K emission line cores. Such studies turned out to be extremely valuable because they allowed to establish the existence of stellar activity cycles similar to the solar cycle in a reasonably sized sample of stars (Baliunas et al. 1995). As Ca II H & K studies of stars suffer from a variety of selection effects regarding the spectral type and the rotation status of the investigated objects, truly unbiased surveys of stellar activity became possible with the advent of soft X-ray imaging. X-ray surveys of cool stars, carried out with the Einstein Observatory (Vaiana et al. 1981) and ROSAT (Schmitt 1997) telescopes, very clearly showed the universal character of the observed X-ray emission and – more generally – of magnetic activity throughout the cool part of the HR diagram, that is, among those stars with outer convection zones similar to the Sun.

Stellar X-ray emission provides a direct link from solar activity to stellar activity, and an interpretation of the stellar X-ray results without the Sun would have been virtually impossible. This is the essence of the so-called Solar-Stellar Connection. On the other hand, the activity observed nowadays from the Sun appears to be quite feeble compared to the activity – as measured through X-ray emission – from many other stars. While activity, again as measured through X-ray emission,

J.H.M.M. Schmitt (✉)
Hamburger Sternwarte, Universität Hamburg, Germany

S.S. Hasan and R.J. Rutten (eds.), *Magnetic Coupling between the Interior and Atmosphere of the Sun*, Astrophysics and Space Science Proceedings,
DOI 10.1007/978-3-642-02859-5_28,

is universal among cool stars, the Sun appears to be at the lower end of the observed activity scale, and only few (cool) stars output even less X-ray emission than the Sun. The challenge is to demonstrate that concepts to describe and understand solar phenomena can also be applied to stellar phenomena, taking place in a vastly different parameter space.

1 Introduction

The Sun is the very nearest and hence a very special star: the Sun can observationally be scrutinized in a detail that is not possible for other stars. Why should we then bother dealing with other stars at all? This question can be easily answered: solar physicists can only observe the Sun, but they cannot change any of its physical characteristics. We cannot make the Sun larger or smaller, more massive or less massive, we cannot make it spin faster or more slowly, and we cannot make the Sun younger and see what it looked like in its youth. Yet the phenomena we describe as "stellar activity" do depend on fundamental stellar parameters such as mass, age, and rotation rate. Thus, by studying other stars with other physical parameters, we eventually hope to improve our understanding of stellar and solar activity. This is precisely the concept of the solar-stellar connection. There is no general consensus or generally applied definition of solar and stellar activity. Usually one associates Sun spots, plage, flares, spicules, and related phenomena with magnetic activity on the Sun and similar definitions apply for stars. Linsky (1985) defines solar-like (activity) phenomena as "non-radiative in character, of fundamentally magnetic origin, and almost certainly due to a magnetic dynamo operating in or at the base of a convection zone."

Linsky's (1985) definition is very useful because it provides a recipe for identifying activity through searching for evidence of non-radiative heating and showing its magnetic nature. Direct measurements of magnetic fields on other stars are possible but rather difficult. Direct measurements of coronal magnetic fields are very difficult even for the Sun and impossible for stars at present. However, it is straightforward to search for the heating effects associated with magnetic activity. Such evidence for non-radiative heating can be obtained by observations of the heated thermal plasma in the UV or X-ray domain or by observations of nonthermal emission from highly energetic particles often accompanying and possibly intimately linked with the heating process(es). As nonradiative heating is – usually – confined in space and time, evidence for time variability or spatial structure is also good evidence for nonradiative heating. X-ray emission is generally considered to be a key indicator of "magnetic activity." Quiescent emission requires substantial heating up to X-ray temperatures and the detection of flaring emission from a star is a direct indicator of nonradiative heating.

The paradigm of stellar activity – as originally formulated by Rosner (1980) – is shown in Fig. 1. Linsky (1985) points out the importance of magnetic fields and their (presumable) dynamo origin. Rotation and turbulence, which is obviously present

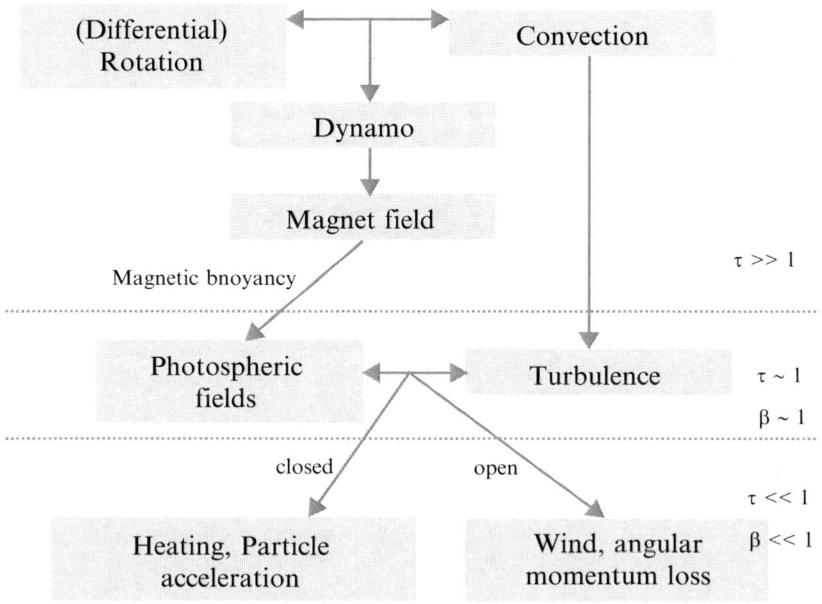

Fig. 1 The solar-stellar activity paradigm

in a stellar convection zone, are thought to be the key ingredients for dynamo action, which is held responsible for the generation of magnetic fields in the stellar interior. Magnetic buoyancy leads to the eruption of the magnetic fields and their emergence at the stellar surface, where the fields and any associated physical processes can be directly observed (optical depth $\tau < 1$). Observations show the magnetic topology to be very important. Strong X-ray emission is found only on closed field lines, filled with plasma and appropriately heated in regions of small plasma β; also particle generation can occur on very short time scales. Open field lines are the sources of the solar (and by analogy stellar) winds, which lead to angular momentum losses of the system. To what extent is the scheme shown in Fig. 1 of generic astrophysical interest? To what extent is the Sun typical for the activity on other stars? This is precisely the theme of the solar-stellar connection.

2 Sunspot Cycle and Total Solar Irradiance

Among the most prominent signatures of solar activity is the 11-year sunspot cycle (discovered by the amateur astronomer H. Schwabe in 1843). Eleven-year cycles show up in essentially all solar activity indicators and we now know that this cycle should be viewed as a 22-year magnetic cycle with systematic polarity reversals every 11 years. While the sunspot cycle is easy to detect in spatially resolved images

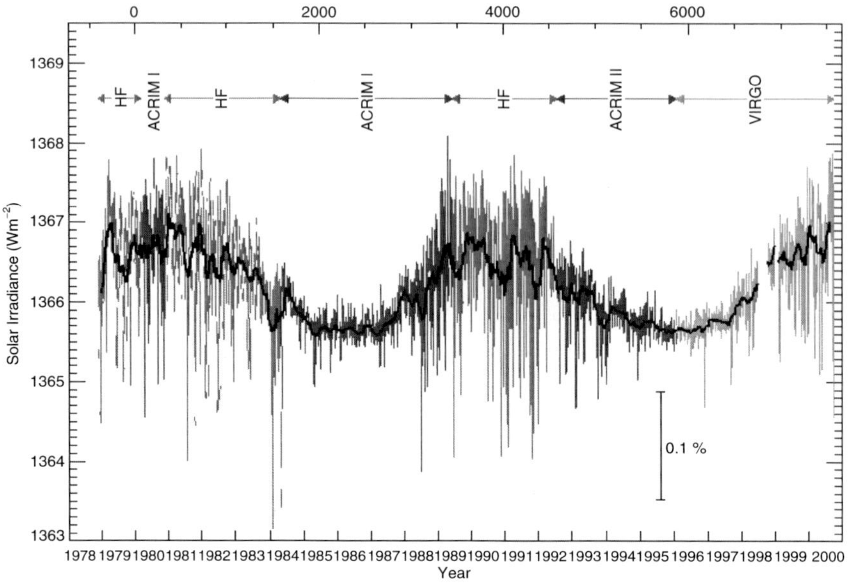

Fig. 2 Total solar irradiance with a variety of satellites. Image from http://www.pmodwrc.ch

and a "sunspot light curve," reaching back to the times of Galileo, is available, the cycle is actually rather difficult to detect in total flux (in the solar context called "irradiance") measurements, that is, those measurements that are at our disposal in a stellar context. Starting with Abbott and Langley in the 1880s, solar physicists have – unsuccessfully – attempted for almost a century to measure variations in the solar constant, until space borne high-precision measurements showed the solar "constant" to be actually variable. A current synopsis of three decades of solar irradiance measurements is shown in Fig. 2. The observed level of variations is about 0.15% peak-to-peak variability. Also, the cyclic nature of the variability is not easily disentangled from "variability noise." Further note that the variability dispersion is larger at times of maximum and the passages of sunspots show up in drops of total flux to levels otherwise found only under minimum conditions. Photometric variations of stars have of course been known for a long time, but in a stellar context measurements with an accuracy of the solar measurements have only recently become possible from space.

3 The X-Ray Sun

The X-ray properties of the Sun serve as a starting point of our discussion of X-ray properties of stars below. A whole armada of satellites has observed and is observing the corona of the Sun and thanks to modern communication technology; the

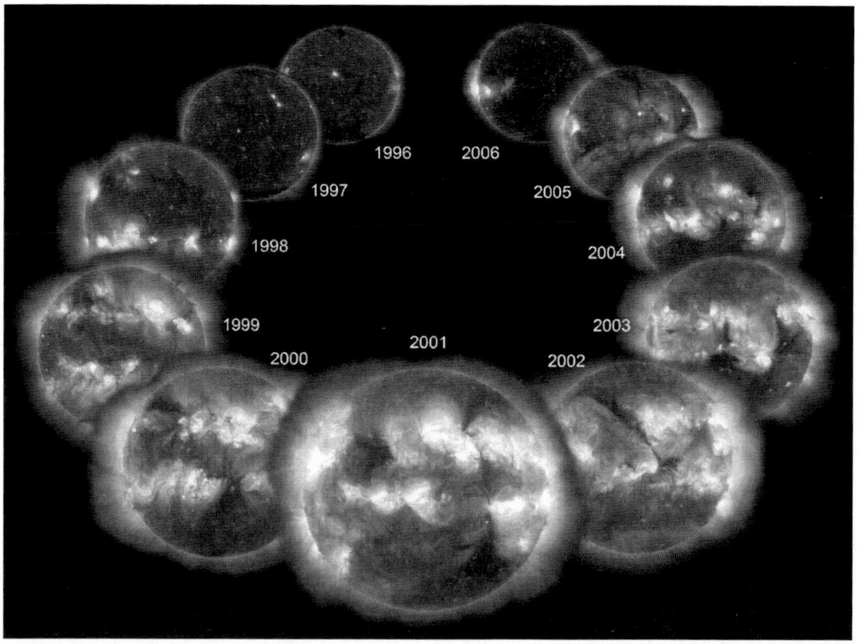

Fig. 3 Variation of the solar X-ray morphology as seen by SOHO. Image from http://sohowww.nascom.nasa.gov

resulting images are accessible to the general scientific and non-scientific public through the internet with only a few hours delay. The whole solar (X-ray) cycle has been followed for example by the EIT telescope onboard SOHO: at XUV and soft X-ray wavelengths one observes a very distinct morphological change of the appearance of the solar corona from solar maximum to solar minimum (Fig. 3). At solar maximum the appearance of the solar corona is dominated by so-called active regions, and at minimum few or none of those regions are on the solar disk and the X-ray output is produced by more diffuse, much fainter emission regions. This morphological change is accompanied by a substantial change in X-ray flux, which sensitively depends on the spectral band used. While at hard X-ray wavelengths the peak to peak variability exceeds two orders of magnitude, at softer wavelengths, where all of the stellar X-ray imaging is carried out, one observes a change on the order of one order of magnitude during a typical solar cycle. At any, rate the solar cycle is readily apparent in the Sun's total X-ray output.

Obviously, the solar corona appears far from being spherically symmetric, rather it shows an extremely high degree of spatial inhomogeneity and small-scale structure with vast brightness differences ($>10^4$) between adjacent features. Plasma confinement in numerous loop-like structures is readily apparent. Coronae of stars appear as point-like sources to all our X-ray telescopes, and the solar example shown in Fig. 3 leads us to expect lots of substructure below the instrumental resolution in our stellar X-ray data.

4 The Ca II H & K Sun

The variation of the solar emission in the cores of the calcium resonance lines and their association with magnetic fields has been known for a long time. Also in the stellar context a large body of measurements of the emission strengths of the cores of the Ca II H & K lines exists in the literature and various data archives; this method to measure activity was pioneered by O.C. Wilson in the 50s of the last century (Wilson 1963) and is the "classic" method to determine the solar cycle from spatially unresolved, "stellar" data through monitoring of the so-called Ca S-index, which measures the Ca II emission line strength relative to the nearby (pseudo-) continuum. In Fig. 4, I demonstrate that this method works well for the well known solar cycle, which is clearly visible in these data and correlates extremely well with the solar cycle seen in solar X-ray (Fig. 3) and solar irradiance data (Fig. 2). Using this Ca S-index, Baliunas et al. (1995) could demonstrate the presence of cycles in a reasonably sized sample of solar-like stars. At the same time one must realize that Ca II H & K studies of stars are plagued by a variety of selection effects regarding the spectral type and in particular the rotation status of the investigated objects; extreme care must be taken to correct for such biases (Schröder et al. 2009). Truly unbiased surveys of the activity of solar-like stars were possible with the advent of soft X-ray imaging, and the X-ray surveys of cool stars carried out with the Einstein Observatory (Vaiana et al. 1981) and ROSAT (Schmitt 1997).

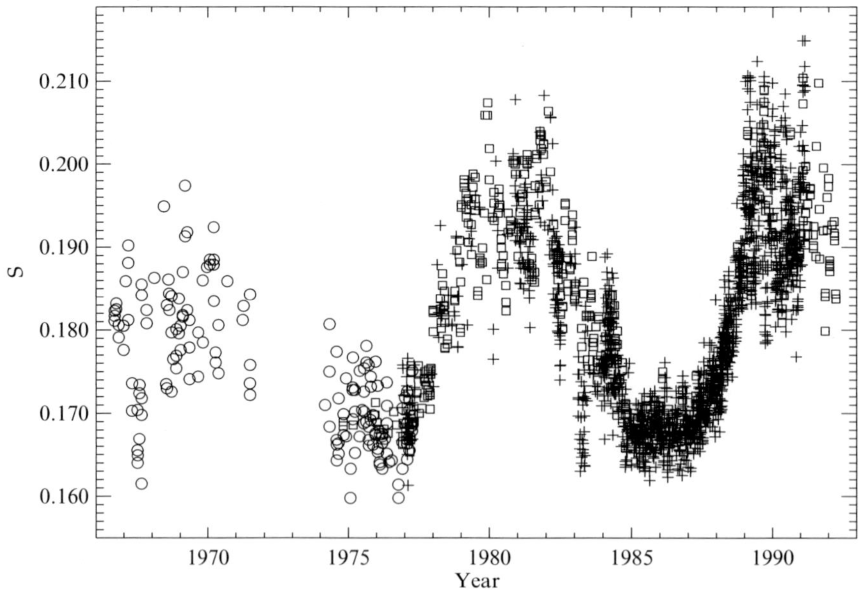

Fig. 4 Time series of the full-disk solar S-index. The 11-year solar cycle is clearly detectable

5 Which Stars are X-Ray Emitters?

Within the framework of the ROSAT All-Sky Survey (RASS), it was possible to carry out a sensitive and unbiased survey of X-ray emission from all types of stars. Limited stellar surveys had previously been carried out with the Einstein Observatory (Vaiana et al. 1981), and many of the key results of stellar X-ray astronomy had already been obtained with such data. The big, major surprise of the first data obtained with the Einstein Observatory was the fact that X-ray emission was produced by many different types of stars at levels vastly exceeding the X-ray output of the Sun by many orders of magnitude. If stars emitted X-rays at the level observed from the Sun, only stars in the immediate neighborhood of the Sun would be detectable. The characteristic value for the sensitivity limit of the RASS is a limiting X-ray flux of $\approx 2 \times 10^{-13}$ erg cm^{-2} sec^{-1}, which implies that X-ray emission at solar-like levels of $L_x \approx 2 \times 10^{27}$ erg sec^{-1} can be detected only out to distances of 10 pc, and consequently only a few hundred stars would have showed up as X-ray sources.

In actual fact one detects X-ray emission from thousands of stars. If one considers only the brightest stars contained in the Bright Star Catalog (BSC) with around 10,000 entries of stars brighter than a visual magnitude of 6.5 mag, and the nearest stars as contained in the Gliese catalog containing more than 3,000 stars within 25 pc around the Sun, one detects X-ray emission from thousands of stars as demonstrated in Fig. 5. The BSC is a magnitude-limited catalog and its composition in terms of

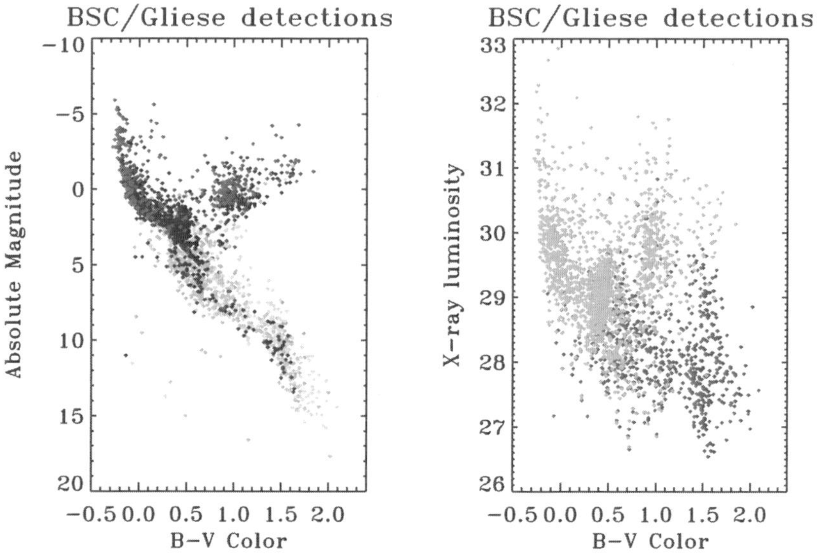

Fig. 5 *Left*: Color-magnitude diagram of RASS-detected BSC and/or Gliese catalog stars, from Hünsch et al. (1998, 1999). For main-sequence stars, the $B - V$ color is a measure of spectral type. The color coding refers to X-ray luminosity: *yellow* denotes $L_x < 10^{28}$ erg s^{-1}, *green* $L_x = 10^{28} - 10^{29}$ erg s^{-1}, *blue* $L_x = 10^{29} - 10^{30}$ erg s^{-1}, and *red* $L_x > 10^{30}$ erg s^{-1}. *Right*: X-ray luminosity for RASS-detected BSC and Gliese stars vs. $B-V$ color (color illustration are available in the on-line version)

spectral type is biased towards intrinsically bright stars, and consequently contains many stars of spectral type A and F as well as giant stars, with a deficit of intrinsically faint but nearby G, K, and M dwarfs, while the Gliese catalog, on the other hand, is a volume-limited catalog. By its very construction, the latter catalog is composed mostly of late-type dwarf stars of spectral type K and M, while its content of earlier type stars overlaps with the BSC.

In the lefthand panel of Fig. 5, I specifically plot a color-magnitude diagram of all RASS detected stars contained in the BSC and/or Gliese catalogs. As is apparent, all types of stars commonly placed in the color-magnitude diagram – with the exception of white dwarfs – are found to be X-ray emitters. For most stars shown in the lefthand panel of Fig. 5, trigonometric distances are known so that reliable X-ray luminosities can be computed. The righthand panel of Fig. 5 shows X-ray luminosity vs. $B-V$ color, which is used as an indicator of the stars' effective temperature, for the BSC stars (green) and Gliese stars (red). A huge spread of X-ray luminosities of up to four orders of magnitude from stars with given $B-V$ color is apparent. As most of the stars shown in Fig. 5 are main-sequence stars, it is also clear that the "fractional" X-ray luminosity, that is, the ratio $L_{\rm x}/L_{\rm bol}$, also varies over many orders of magnitude from star to star. An understanding of the cause of these variations is one of the central themes of stellar X-ray astronomy.

It is instructive to consider the ROSAT X-ray detection rates of main sequence stars listed in the BSC vs. $B-V$ color. As is apparent from Fig. 6, the detection rate is very large (essentially 100%) for stars of spectral type O, and I mention

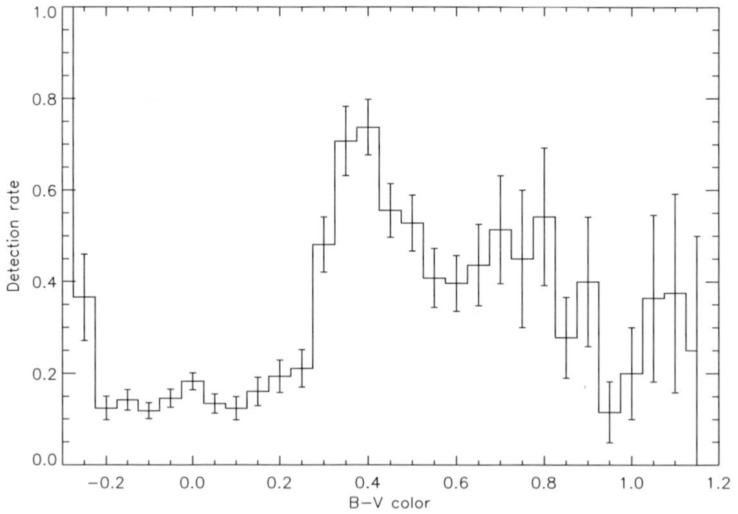

Fig. 6 Detection rate for main sequence stars contained in the BSC catalog as a function of $B-V$ color. Note the completeness of the detections among the early O-type stars and the large increase in detection rate at $B-V \approx 0.3$. For large $B-V$ the BSC catalog contains too few entries for meaningful discussion. In general, the detection rates for M-type dwarfs are high, while for M-type giants they are very small

in passing that these large detection rates are attributed to the omnipresent winds of these stars. Among the B- and A-type stars, the detection rate decreases to between 10 and 20%, and in fact the interpretation of the X-ray emission of these presumably fully radiative A-type stars, which, according to the activity paradigm should show no activity whatsoever, is somewhat controversial; it is clear that binarity plays a major role and in many, if not all, cases the observed X-ray emission from A-type stars can actually be attributed to low-mass companions. At the spectral type F near $B - V \approx 0.3$, the detection rate suddenly jumps to about 70%, whereupon it decreases with error bars becoming larger and larger towards redder and redder colors. It is extremely suggestive to associate this sudden jump in detection rate with the "onset of convection," which is known to occur at that spectral type (Schmitt et al. 1985; Schröder and Schmitt 2007). Thus, somehow, the occurrence of X-ray emission seems to be linked to the interior property of a star. This finding coupled with the rotation-activity connection provides strong support for a picture that views magnetic activity as universal for all stars with outer convection zones (and rotation). The activity observed on the Sun would just be a special manifestation of the occurrence of magnetic activity in a cool star that happens to be located very close to us.

6 Universality of Stellar X-Ray Emission

How universal is then the occurrence of X-ray emission among solar-like stars, that is, stars with outer convection zones? The diagrams shown in Fig. 5 were derived from X-ray flux limited surveys, which sample the parameter space of the highest X-ray luminosities very well. To explore the full range of X-ray luminosities, one needs to consider volume-limited surveys, which sample all known objects within a given distance range. Such studies have also been carried out using data from the RASS and the ROSAT pointing program (Schmitt and Liefke 2004). The current observational situation from the X-ray point of view is summarized in Fig. 7 (taken from Schmitt and Liefke 2004), which shows mean X-ray surface flux F_X as a measure of stellar activity vs. $B - V$ for volume-limited samples of stars with outer convection zones. As is clear from Fig. 7, X-ray emission and hence magnetic activity is universal for solar-like late-type stars, that is, stars with outer convection zones, in the sense that X-ray emission is detected from basically all observed stars.

How complete are now the detections in the NEXXUS data base in detail? Defining as nearby F/G-stars those with absolute magnitudes M_V in the range $3 \leq M_V \leq 5.80$, K stars those with $5.80 < M_V \leq 8.50$, and as M stars those fainter than $M_V = 8.50$, one finds the following results: Among 69 F/G stars within a distance of 14 pc around the Sun, only seven remain undetected, that is, detection rate within the volume out to 14 pc is therefore 94%, and all stars within 12 pc have been detected. Out of 51 K stars within 12 pc, only two stars have not been detected and hence the detection rate is 96%, while out of 65 M stars within 6 pc, six stars have not been detected. Most of the nondetected M-type stars are brown dwarfs or

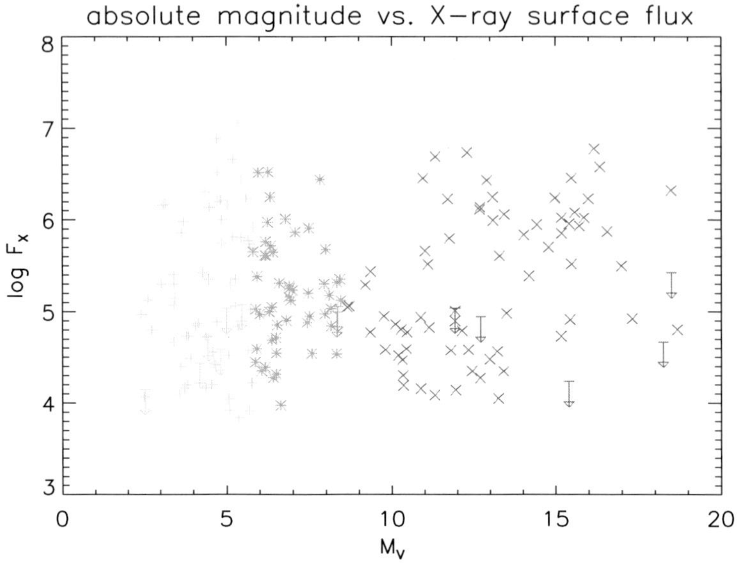

Fig. 7 X-ray surface flux F_X vs. $B - V$ color for nearby main sequence stars (from Schmitt and Liefke 2004)

very low-mass stars; the only nearby M star not detected in a short PSPC pointing is GJ 1002 and all stars within 4 pc have been detected. The conclusion from this exercise obviously is that the detection rate among the nearest stars is very large, and, in fact there is no reason to expect that these stars will not be detected once even more sensitive X-ray observations are available. In other words, the formation of X-ray emitting coronae appears to be universal for late-type main sequence stars. Thus, the presence of soft X-ray emission implies the presence of hot plasma with temperatures of at least 1–2 MK, and consequently some sort of coronal heating *must* take place in *all* stars with outer convection zones, be they little or large, old or young. It needs to be emphasized in this context that, first, all the upper limits in Fig. 7 are derived from ROSAT all-sky survey observations, while pointing observations usually result in detections, and, second, no systematic programs to study volume-limited samples of nearby stars have been carried out in the meantime with XMM-Newton or Chandra, so Fig. 7 is still state of the art.

7 Connection of X-Ray Emission with Other Stellar Parameters

Given the observed spread of X-ray luminosity for stars of given spectral type, a natural question to ask is which other stellar parameters except those determining the position of a star in the HR diagram determine the X-ray output level. For late-type main-sequence stars, the X-ray emission scales with rotation velocity (Pallavicini et al. 1981) and in fairness to optical astronomers one should acknowledge that this rotation-activity connection had already been recognized much earlier (Skumanich 1972). This relation is believed to trace directly back to the stellar dynamo. On theoretical grounds, the most adequate measure for the dynamo efficiency should be the Rossby number R_0, defined as the ratio between the rotation period and the convective turnover time. However, in observational studies R_0 often is replaced by the rotation period because the convection time scale is difficult to access observationally.

The current observational situation is summarized in Fig. 8, which shows rotation-activity relations for field dwarf stars and stars from various young open clusters such Hyades, Pleiades, α Per, IC 2391, IC 2602 (Pizzolato et al. 2003). As stellar rotation is thought to be braked by magnetic winds, the younger stars in clusters rotate much faster than the field stars, and they cover the left part of the diagram. Stars rotating with periods shorter than about 5 days are in the so-called saturation limit of $L_{\rm x}/L_{\rm bol} \sim 10^{-3}$, with no obvious dependence on rotation rate. If one computes for a solar-like star the mean X-ray surface flux corresponding to $L_{\rm x} \approx 10^{-3}\,L_{\rm bol}$, one finds $F_{\rm x} = 6.5 \times 10^7$ erg cm^{-2} s^{-1}, which agrees quite well with the maximally observed surface flux values (Fig. 7), if one assumes filling factors at ~10% level.

For slowly rotating stars (periods longer than ≈5 days), the observed X-ray luminosity scales inversely with the period, albeit with an unexplained scatter of at least one order of magnitude. Possible explanations for this scatter include longterm variability. Many of the data points used in Fig. 8 are derived from snapshot X-ray exposures, and may not be representative for the "mean" X-ray activity. Little information exists on the long-term variability of stellar coronae and, in particular, on the issue if stellar coronal activity cycles exist as observed for the Sun. The solar X-ray

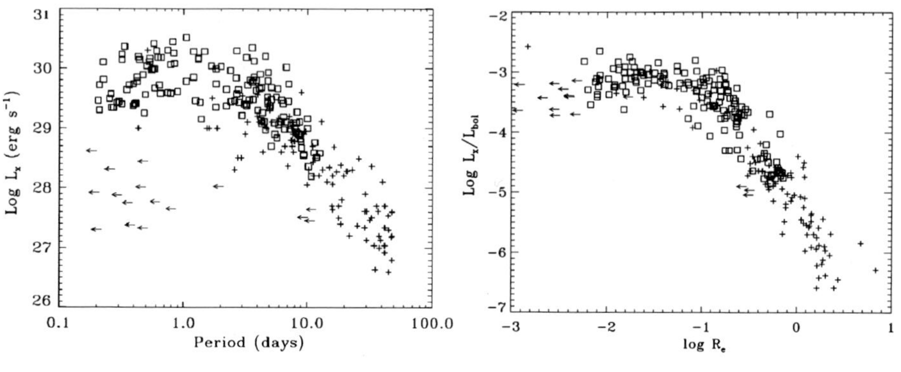

Fig. 8 Rotation-activity relation for a sample of about 250 field stars (*crosses*) and members of young open clusters (*squares*). *Left*: X-ray luminosity vs. rotation period. *Right*: fractional X-ray luminosity vs. Rossby number. *Leftward arrows* indicate field stars with periods derived from measurements of the rotation velocity $v_{\rm rot} \sin i$, yielding only upper limits to the rotation period because of the generally unknown inclination angle. Figures from Pizzolato et al. (2003)

emission varies – depending on the X-ray band considered – at least by an order of magnitude during a cycle, while the rotation rate stays clearly constant. Therefore, cycles or other kinds of long-term variability would provide a natural explanation for the observed scatter.

8 Coronal Structure

Based on the example of the solar corona, one also expects stellar coronae to exhibit a high degree of spatial structuring. Clearly, the angular resolution of current X-ray telescopes is insufficient to angularly resolve stellar coronae; therefore, one has to take recourse to eclipsing binaries to study the spatial structure of stellar coronae. At X-ray wavelengths those systems are especially interesting that consist of an X-ray bright and X-ray dark component. As most late B-type and A-type stars tend to be X-ray dark, eclipsing binaries containing an B/A-type star and a late-type star are particularly promising. One such case is the totally eclipsing binary α CrB, consisting of a A0 primary and a G5V type secondary. A total X-ray eclipse at the time of optical secondary minimum (i.e., with the A-type star in front) was detected with ROSAT (Schmitt and Kürster 1993). As demonstrated in the lefthand panel of Fig. 9 (taken from Güdel et al. (2003)), with XMM-Newton eclipse ingress and eclipse egress can be observed continously (rather than with gaps as with ROSAT), and it is the detailed shape of eclipse ingress and ingress that carries the information on the spatial structuring of the underlying G star corona. In this fashion, Güdel et al. (2003) could derive the coronal image of α CrB displayed in Fig. 9 (right hand panel); it clearly demonstrates that the corona of α CrB is far from being spatially homogeneous as is the case for the solar corona.

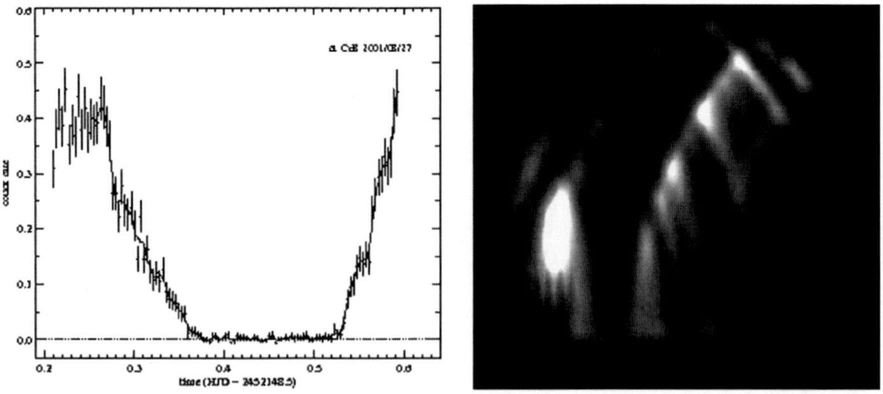

Fig. 9 *Left*: XMM-Newton light curve of the eclipsing binary α CrB. *Right*: reconstructed surface X-ray intensity distribution (from Güdel et al. (2003))

Similar observation was reported by Schmitt et al. (2003), who used XMM-Newton to observe a flare on the eclipsing binary Algol. The Algol system consists of an X-ray dark B8V primary and an X-ray bright K giant, both stars orbit each other with a period of 2.8 d. The rotation of the K giant is synchronized, and the star is therefore a rapid rotator and correspondingly active producing frequent flares. During the observations reported by Schmitt et al. (2003), an actually smaller flare occurred and, fortuitously, the flare occurred during optical secondary minimum, that is, with the X-ray dark primary in front of the X-ray bright secondary. The relevant data is shown in Fig. 10. The recorded X-ray light curve in various energy bands is shown in the lefthand panel of Fig. 10; as is typical for stellar flares, the spectral hardness increases during the flare. However, during the flare a sudden drop and later increase is clearly apparent. A natural interpretation of this light curve is of course an eclipse of the flaring plasma by the primary, and finally, by assuming an exponential light curve decay (in order to rectify the light curve) an image of the flaring plasma can be generated (righthand panel in Fig. 10, taken from Schmitt et al. (2003)), which shows that the flaring region was quite compact and located very close to the stellar limb. Figure 10 clearly shows that small regions of the stellar surface can be responsible for a major fraction of a star's total X-ray output.

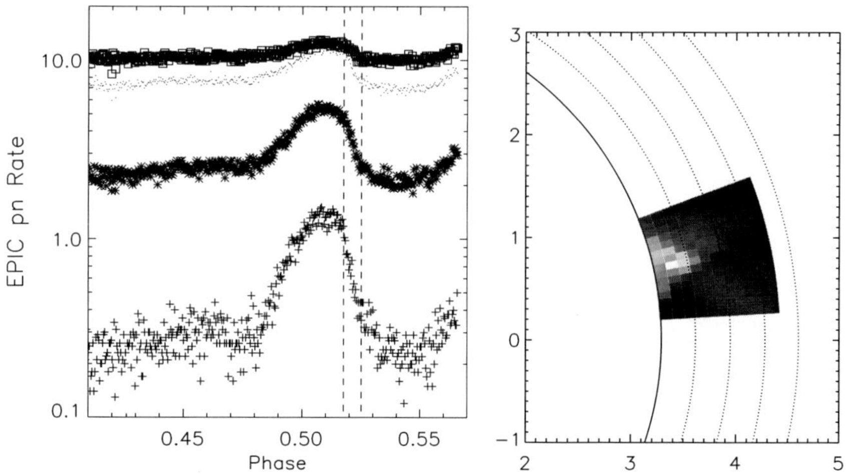

Fig. 10 *Left*: XMM-Newton light curves of Algol in the energy bands below 1 keV (*squares*), between 1–2 keV (*dots*), 2–5 keV (*asteriks*), and 5–10 keV (*pluses*). *Right*: reconstructed spatial distribution of the flare region on Algol B; the *circle* represents the limb of the K-type star. The units are expressed in solar radius; the *dashed circles* indicate heights in steps of 0.1 stellar radius. The system is shown at $\phi = 0.5556$ when the flare longitude of $\lambda = 70°$ is viewed exactly at the stellar limb. From Schmitt et al. 2003

9 Conclusions

I hope that this article demonstrates the enormous progress made in the field of solar-stellar connection research in the last decades. The field has matured and become an integral part of astrophysics. As is the case in many of its sub-disciplines, a multi-wavelength approach is required for a thorough understanding of the subject. The magnetic activity observed on the Sun manifests itself in many different forms in other stars, and compared to the top of the activity scale observed on the stars, our Sun – fortuitously – is among the less active stars. It is a challenge to demonstrate that physical models developed in the solar context can be adapted to interpret the stellar observations. The physical processes studied in solar and stellar activity, such as magnetic field generation through dynamo processes, plasma heating, particle acceleration, generation of magnetized winds etc., are generic processes of interest and relevance to astrophysics at large.

Acknowledgment We thank the conference organizers in Bangalore for putting together a very good meeting and our hosts from the IIA in Bangalore for taking such excellent care of us.

References

Baliunas, S. L., Donahue, R. A., Soon, W. H., et al. 1995, ApJ, 438, 269
Edlén, B. 1945, MNRAS, 105, 323
Güdel, M., Arzner, K., Audard, M., Mewe, R. 2003, A&A, 403, 155
Hale, G. E. 1908, ApJ, 28, 315
Hünsch, M., Schmitt, J. H. M. M., Voges, W. 1998, A&AS, 132, 155
Hünsch, M., Schmitt, J. H. M. M., Sterzik, M. F., Voges, W. 1999, A&AS, 135, 319
Linsky, J. L. 1985, Solar Phys., 100, 333
Pallavicini, R., Golub, L., Rosner, R., et al. 1981, ApJ, 248, 279
Pizzolato, N., Maggio, A., Micela, G., Sciortino, S., Ventura, P. 2003, A&A, 397, 147
Rosner, R. 1980, SAO Special Report, 389, 79
Schmitt, J. H. M. M. 1997, A&A, 318, 215
Schmitt, J. H. M. M., Kürster, M. 1993, Science, 262, 215
Schmitt, J. H. M. M., Liefke, C. 2004, A&A, 417, 651
Schmitt, J. H. M. M., Golub, L., Harnden, Jr., F. R., et al. 1985, ApJ, 290, 307
Schmitt, J. H. M. M., Ness, J.-U., Franco, G. 2003, A&A, 412, 849
Schröder, C., Schmitt, J. H. M. M. 2007, A&A, 475, 677
Schröder, C., Reiners, A., Schmitt, J. H. M. M. 2009, A&A, 493, 1099
Skumanich, A. 1972, ApJ, 171, 565
Vaiana, G. S., Cassinelli, J. P., Fabbiano, G., et al. 1981, ApJ, 245, 163
Wilson, O. C. 1963, ApJ, 138, 832
Zombeck, M. V., Vaiana, G. S., Haggerty, R., et al. 1978, ApJS, 38, 69

Summary and Perspective

N.O. Weiss

1 A Double Centenary

This meeting celebrated a double anniversary. A 100 years ago, George Ellery Hale had a hunch: he had noticed the vortical structure in Hα above sunspots, and he interpreted this as evidence for a vortex flow, which (he supposed) carried electric charges round a spot, and so provided an azimuthal electric current that would generate a magnetic field like that in a solenoid (Hale 1908a). His hunch was actually wrong since (as we now know) the solar plasma is electrically neutral – but when he looked for spectroscopic evidence of a magnetic field, using the recently discovered Zeeman effect, he found that there were kilogauss magnetic fields in sunspots (Hale 1908b). This was the first demonstration that magnetic fields were present outside the earth.

Meanwhile, John Evershed arrived at Kodaikanal. He already knew Hale and had stopped at Mt. Wilson on his way to India. Once installed, Evershed used an improved spectrograph to search for Doppler shifts corresponding to Hale's vortex motion – but instead he found radial outflows in the penumbrae of the spots (Evershed 1909a,b; 1910).[1] The twin discoveries, of magnetic fields and of the Evershed effect, initiated the modern era of solar physics. Since then, observations have always been ahead of theory.

The structure of the solar atmosphere is dominated by the magnetic fields that emerge into the photosphere, giving rise to chromospheric Ca II emission and to X-ray emission from a hot corona. These fields are also responsible both for driving a vigorous solar wind and for the coronal mass ejections that give rise to "space weather" in the heliosphere. Yet almost all our observations are confined to the outermost $10^{-10} M_{\odot}$, while the solar interior (where these fields are generated) can only be probed by the p-modes of helioseismology and by neutrinos.

N.O. Weiss (✉)
Department of Applied Mathematics and Theoretical Physics,
University of Cambridge, UK

[1] Both the Doppler shift and the magnetic doubling of spectral lines had actually been noticed long before, but had not been interpreted (Thomas and Weiss 2008). Evershed (1910) did also detect a slight spiral motion that was antisymmetric about the solar equator.

S.S. Hasan and R.J. Rutten (eds.), *Magnetic Coupling between the Interior and Atmosphere of the Sun*, Astrophysics and Space Science Proceedings,
DOI 10.1007/978-3-642-02859-5_29,

The papers at this meeting covered all aspects of solar magnetism, starting in the Sun's interior and moving outwards to discuss first small-scale fields (and how they can be measured), then sunspots themselves as well as the quiet Sun, before moving upwards into the atmosphere and the corona, with the violence of coronal mass ejections and of flares. Finally, solar activity was related to activity in other stars. The most striking impression given by these papers is of the enormous wealth of new observations that have been generated within the last 15 years, both from space – with Yohkoh, SoHO, and most excitingly with Hinode – and also from the ground – most notably with the Swedish Solar Telescope (SST) and the upgraded Dunn Solar Telescope (DST). Alongside these observations there have been new theoretical developments, and here the crucial availability of ever more powerful computers has made it possible to provide increasingly realistic simulations of convection and magnetoconvection at the solar surface.

In what follows, I shall try to convey what struck me as the most exciting issues that arose. I shall not attempt to assign credit for every item, nor will it be possible to cover every topic that was mentioned. I apologize therefore at the start to any authors who may feel that their own work has been inadequately represented or ignored.

2 The Solar Interior

Helioseismology has transformed our knowledge of the Sun's interior. The radial variations of the sound speed and the superadiabatic gradient are now well established and we know the depth of the convection zone with remarkable precision. Even more remarkable is the determination of the Sun's internal rotation from splitting of p-mode frequencies. We know that differential rotation persists throughout the convection zone, while the radiative zone rotates almost uniformly. Between the two lies the slender tachocline, which (it is generally agreed) plays a key role in the dynamo process that generates the cyclic fields responsible for solar activity. However, the internal structure of the tachocline is not yet properly understood, nor is it clear why it is so narrowly confined. It has become apparent, though, that a purely hydrodynamic description is inadequate, and that magnetic fields are significant within the tachocline, while its downward spreading is inhibited by fields within the radiative zone.

This situation calls for large-scale numerical computation – but any such simulation immediately becomes a major project. Over the years, the group at Boulder has developed anelastic codes that represent the hydrodynamics of the convection zone, including both the pattern of convection and that of differential rotation (but not the tachocline). These models have since been extended to include magnetic fields, in the hope of reproducing the cyclic behavior that is observed. Small-scale dynamo action is readily found but so far only nonreversing large-scale fields have appeared. This progress is encouraging, and we can look forward to the development of ever more realistic models in the future. Of course, it is easier to try to separate the different issues – differential rotation, baroclinic gradients, the slender tachocline, cyclic magnetic activity, the magnetohydrodynamic boundary layer at the top of the radiative zone – that are involved. The real worry is that we shall only be able to understand these individual features in the context of the fully global problem in which they are all involved together.

Owing to all these difficulties, much attention has been focused on mean field dynamo models, where the key ingredients are the known differential rotation, which draws out poloidal fields to form toroidal fields (the ω-effect) and the parametrized process that generates poloidal fields from toroidal fields (the α-effect). The latter can be ascribed to helical turbulence in the convection zone, or to the effect of the Coriolis force on instabilities driven by magnetic buoyancy. Meridional flows and the effects of flux expulsion and flux pumping have also to be considered. It is generally agreed that the ω-effect operates at or near the tachocline, producing strong toroidal fields around the interface between the convective and radiative zones, samples of which rise upwards to emerge as active regions. In interface dynamo models, the α-effect is concentrated near the tachocline, while in flux-transport models the α-effect is located near the surface, where poloidal fields spread owing to the combined effects of supergranular diffusion and meridional circulation. With plausible values of the turbulent diffusivity in the convection zone, these fields diffuse fairly rapidly to its base but, if the diffusivity is assumed to be much smaller, the new poloidal flux is carried on a slow conveyor belt to the bottom of the convection zone. For all these variant models, parameters can be tuned so as to produce convincing kinematic and (by introducing some plausible quenching mechanism) nonlinear models of the solar cycle. The zonal shear flows, or "torsional oscillations," which are the clearest manifestation of nonlinear behavior, precede the appearance of sunspots and propagate upwards through the convection zone at low latitudes.

There is an obvious demand for predictions of future activity cycles, as space weather affects satellites and missions into space. Early predictions of the new Cycle 24 varied widely. Extreme claims were made for a dynamo model that predicted a much more active cycle than the previous Cycle 23; other models, relying on measures of poloidal fields at the end of Cycle 23, predicted that its successor would be much weaker. The sluggish start of the new cycle makes the latter predictions seem more likely.[2] On a longer timescale, the Sun has been abnormally active for the past 80 years, and comparison with previous grand maxima over the past 10,000 years (using proxy records of cosmogenic isotope abundances) suggests that the current grand maximum is unlikely to last much longer (Abreu et al. 2008).

3 Small-Scale Fields and Polarimetry

The past decade has seen enormous advances not only in observations of fine-scale granular and subgranular structures with a resolution of 0.1 arcsec or less but also in accurate numerical modeling of small-scale magnetoconvection. Ambitious

[2] See http://solarscience.msfc.nasa.gov/images/ssn_predict_l.gif for the latest news.

simulations now incorporate ionization effects as well as radiative transfer, and yield photospheric images that can scarcely be distinguished from what is actually observed. The difference is that computations provide information about the vertical structure of the convection pattern, showing for instance how the $\tau_{500} = 1$ surface is corrugated and elevated over rising plumes. Simulations that extend to depths of 16 Mm indicate a self-similar pattern of convection, with vigorous downdrafts whose separations increase with increasing depth. Although large-scale fields tend to be pumped downwards, small-scale dynamo action is unimpeded and indeed is prevalent at all levels. Detailed models of granular convection show strong vertical fields in the intergranular lanes, connected by more diffuse horizontal fields that lie above the bright granules. Any area-based weighting emphasizes these weaker horizontal fields.

Advances in spectropolarimetry have made it possible to measure the 3D vector magnetic field and also to estimate the strength of weak, tangled fields within the granules. The "second solar spectrum" in linear polarization results from quantum scattering by atoms and molecules in the solar atmosphere, and the Hanle effect leads to depolarization (reduction of Stokes Q). This yields an average value of 130 G for the tangled field, though such fields are, understandably, locally stronger in intergranular lanes than within granules. With near-continuous, high-resolution polarimetry from space on Hinode, the evolution of fine-scale vector fields is now being measured with exquisite precision (de Wijn et al. 2009). Probability distribution functions for both vertical and horizontal fields can then be obtained, and these agree with theoretical predictions, suggesting also that small-scale fields are maintained by local dynamo action. Ground-based telescopes are also active at various locations across the world. The new CRISP spectropolarimeter on the SST resolves magnetic fields down to 0.15 arcsec and confirms the presence of up and down flows in and around granules; elsewhere, fields of 25–35 G have been measured within spicules. Meanwhile, a new stereoscopic magnetograph is being developed with the ATST in mind, and, within India, there are plans for high-resolution spectropolarimetry on the National Large Solar Telescope in Ladakh, while a new 0.5 m solar telescope is being installed at Udaipur Observatory.

4 Sunspots

Sunspots were naturally a major topic at this meeting. "What is a sunspot?" was the question posed in the Vainu Bappu Lecture, and the assembled experts tried to meet this challenge. Here again there have been great advances in our understanding of fine structure, especially that of the penumbra, derived both from high-resolution observations (most notably with the SST, and now with Hinode) and from recent detailed simulations. The famous G-band image in Fig. 1 shows the current state of the art.

Helioseismology has established that there is a more-or-less coherent flux tube extending downwards for at least 10 Mm below the surface of a sunspot, but there

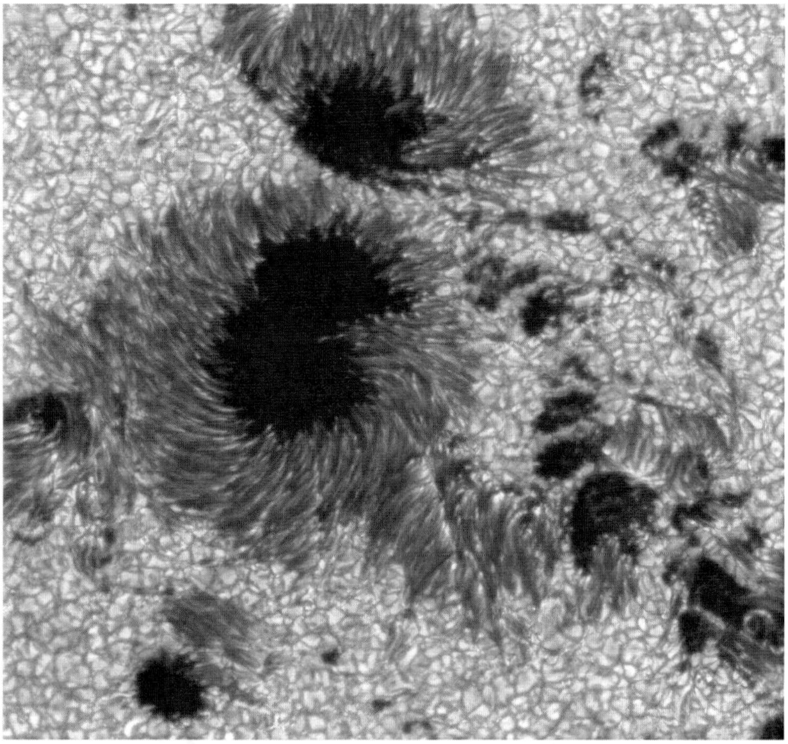

Fig. 1 High-resolution G-band image showing sunspots and pores in an active region, as well as the surrounding granulation, with bright points. Obtained with the Swedish solar telescope (courtesy of G. B. Scharmer and the Royal Swedish Academy of Sciences)

has been a long-standing controversy over the nature of this flux tube. Is it an inhomogeneous monolith or a tight cluster of individual flux tubes, separated by field-free plasma that is connected to the field-free plasma outside the spot? Unfortunately, current inversion techniques for magneto-acoustic waves do not yet allow helioseismology to probe the internal structure of the flux tube with sufficient precision to distinguish between monolith and cluster models, though there have been clear advances in studying wave propagation and mode transformations within spots and relating the results to umbral and penumbral oscillations. However, the agreement between high-resolution observations of bright umbral dots (e.g., Riethmüller et al. 2008) and detailed numerical simulations provides clear support for the monolithic picture: convection within the flux tube produces rising and expanding plumes with fields that are locally reduced, and these plumes correspond to umbral dots. The computations show dark streaks across the bright dots and even these have been detected by Hinode and the DST.

In all spot models, the field's inclination to the vertical increases with increasing radius, and this inclination is responsible for the filamentary structure of the penumbra. The transition from isolated plumes in the umbra to filamentary plumes in the penumbra is dramatically demonstrated in a recent numerical model of a circular sunspot (Rempel and Schüssler, in preparation). In the inner penumbra, the bright filaments contain thin dark cores with radial outflows and these features are reproduced in two independent numerical models with cartesian geometry. The overall pattern apparently corresponds to a form of roll-type convection and the dark cores are absorption features caused by density enhancements associated with buoyancy braking above the radially elongated rising plumes (as in umbral dots). Below the visible surface, the laterally expanding plumes lead to a reduction in the field strength; this local decrease in $|\mathbf{B}|$ bears some resemblance to the field-free regions in "gappy" penumbral models, though there is no connection to the field-free external plasma. Bright grains appear within the bright filaments, migrating inwards and into the umbra. There is also a pattern of oblique striations that apparently corresponds to a two-dimensional projection of three-dimensional structures. The motion of this pattern towards the umbra as a traveling wave is a characteristic feature of convection in a tilted magnetic field.

The outer penumbra is more complicated, owing to the large differences in inclination between the magnetic fields in bright and dark filaments and the associated Evershed outflow (which naturally received attention at this meeting).[3] The intricate magnetic structure – variously described as an "interlocking-comb," an "uncombed," or (better) an "interleaved" magnetic field – in which the vertical component of the field in dark filaments reverses sign towards the edge of the spot, is hard to explain unless the field lines are pumped downwards by the turbulent convection outside the spot. These arched field lines carry the Evershed flow, which appears to be a siphon flow, driven by a drop in gas pressure between the inner and outer footpoints. Surrounding the spot is an annular moat cell, with a radial outflow at the photosphere that is confirmed by f-mode helioseismic measurements. Numerical models indicate that this outflow is not just superficial but extends to an appreciable depth, though there must be a compensating inflow lower down that would help to preserve the spot. The reversed flow near the surface that has been reported from p-mode measurements may be an artefact resulting from an inadequate treatment of magneto-acoustic modes.

5 The Quiet Sun

Intergranular fields with kilogauss strengths are found throughout the quiet Sun (see Sect. 3). Both polarities are present, canceling each other out, at low and moderate latitudes. In polar regions, however, there are persistent patches with the same polarity (opposite at opposite poles). These unipolar patches have lifetimes up to 15 hr and contribute a significant net magnetic flux; their evolution is related to the

[3] See also Thomas and Weiss (2008).

ambient subphotospheric flow and they are associated with weaker horizontal fields. The field lines fan out above these patches to form a chimney up which Alfvén and magneto-acoustic waves can propagate. There is evidence for Alfvén waves traveling both upwards and downwards, and being partially reflected in the transition region.

The open field lines that form the polar coronal holes all emerge from these patches, giving rise to the fast solar wind. UV observations from SoHO (SUMER) and Hinode (EIS) provide spectroscopic information, which confirms that the fast solar wind emanates from these funnel-shaped magnetic fields, though there is also some evidence that the wind prefers interplume lanes. Hydrodynamic codes are now being developed to explore the extent to which these waves might heat the corona (in competition with the effects of nanoflares).

6 Coupling to the Atmosphere and Corona

The magnetic flux in active regions emerges into the photosphere from below, appearing initially as a series of bipoles that progressively reconnect to form a single arched feature in the atmosphere. The global process of flux emergence has been extensively modeled in the thin flux tube approximation, combining the effects of magnetic buoyancy, tension along the field and draining down it, as well as the Coriolis effects that are responsible for Joy's Law. One feature of these calculations is the need for the field to be twisted prior to eruption. The twisted fields transport magnetic helicity into the corona, with associated spin; the resulting structure shows some scatter about a pattern that is antisymmetric about the equator.

Within the corona there are loops that can be observed in EUV (if they are warm and overdense) or soft X-rays (if hot and underdense), as well as by low-frequency radio observations of plasma emission. As the footpoints of the loops get tangled by convective motions in the photosphere, some process – whether steady or impulsive – dissipates magnetic energy to heat the corona. Models involving steady heating seem to be becoming less plausible, leaving rare but very strong nanoflares as the most likely source. Recent results from Hinode show signs of very hot plasma that cools very rapidly, providing support for the concept of multistranded nanoflare storms. On the other hand, models involving heating by waves and flux tube instabilities are still being pursued.

Heating the chromosphere and transition region requires much more energy. The latest Hinode observations reveal fine dynamic fibrils and spicules with heights of about 500 km and velocities of 50–150 $\mathrm{km\,s^{-1}}$, together with high-speed upflows at the feet of coronal loops. Recently developed 3D codes, describing subphotospheric flux emergence into a pre-existing field, can describe the development of loops and spicules.

Moving further out, there are continuing radio observations of waves and turbulence in the solar wind, as revealed by interplanetary scintillation, which indicate the presence of Alfvén and ion cyclotron waves at distances of 50 $R_\odot$. In situ measurements by the SWIM mass spectrometer on the recent Indian lunar mission Chandrayaan-1 will be measuring the solar wind flux on and near the surface of the Moon.

7 Eruptive Phenomena: CMEs and Flares

CMEs have been extensively studied, first with LASCO on SoHO and now with SECCHI on STEREO. Most CMEs originate from active regions, in which case their incidence follows the solar activity cycle, rising from $0.3\,\mathrm{d}^{-1}$ at sunspot minimum to $4\,\mathrm{d}^{-1}$ at sunspot maximum. Other CMEs emerge from quiescent filament regions and there is an enhanced rate from polar filaments around sunspot maximum, when the polar fields are reversing, giving rise to a bimodal statistical distribution. Theoretical models of the rapid energy release assume that this energy is derived from the free magnetic energy associated with electric currents flowing in the corona, though magnetic helicity may also provide a constraint. The first stage is a rapid loss of equilibrium, associated with a changing field topology, and this is followed by rapid reconnection. The latter process has been examined in considerable detail: small motions of footpoints can lead to magnetic structures with complex topologies. The roles of null points, separatrix surfaces, and their intersections at separators can be distinguished: in three dimensions, reconnection is primarily at separators.

On a more global scale, the whole process from flux emergence to eruption can be followed in two- and three-dimensional numerical models, while the associated energetics can be obtained from observations. Once launched into the heliosphere, the eruption can travel as a magnetic cloud, with speeds of up to $2{,}800\,\mathrm{km\,s}^{-1}$ (corresponding to a total kinetic energy of 4×10^{33} erg). The nature of the cloud depends on the orientation of its axial and azimuthal fields, which in turn affects the development of geomagnetic storms if the cloud impinges on the Earth's magnetosphere.

CMEs and flares are clearly related, and they call for a comprehensive theoretical description, which can be extended down to the tiniest nanoflares that may well be responsible for heating the corona. Flares are triggered by emerging magnetic flux, which leads to the formation of a current sheet. The standard picture is of reconnection at such sheets, with the formation of magnetic islands and the ejection of plasmoids (found both in simulations and in laboratory experiments). New observations from Hinode demand further theories to explain chromospheric jets and penumbral microjets, as well as nanoflares and spicules.

8 The Solar-Stellar Connection

The most obvious sign of magnetic activity in distant stars is the presence of a hot corona, with associated emission of X-rays. The ROSAT mission established that all types of star emit X-rays; on the main sequence, strong coronal emission sets in at the transition from A to F type stars, with the appearance of a hydrogen convection zone, and continues through to include fully convective (M5 or later) stars; there is also evidence for the formation of coronae by accretion on to classical T Tauri stars. The Sun itself is a relatively weak X-ray source but stellar flares are common on more active stars, ranging from fully convective late-M dwarfs to massive post-main sequence giants.

Ca II emission provides a proxy measure of magnetic activity on solar-type stars, and many nearby stars have been studied. Magnetic activity is greatest on the most rapidly rotating stars, which are typically either newly arrived on the main-sequence or else close binaries, with spin locked tidally to orbital rotation. Isolated stars often arrive on the main sequence rotating close to their break-up rate, and gradually spin down owing to magnetic braking. The Sun is a middle-aged slow rotator; about a dozen other stars are known to exhibit cyclic activity and they all have similar rotation rates. Fast rotators show much more irregular variability.

The most active stars are typically rapidly rotating main sequence dwarfs (e.g., BY Draconis) or giant stars in close binary systems that have evolved off the main sequence (e.g., RS Canum Venaticorum). Evidence for large starspots came originally from photometric variability but has now been confirmed by Doppler imaging and, more recently, by Zeeman–Doppler imaging (see Thomas and Weiss 2008). A characteristic feature of these stars is the presence of large polar spots, indicating a magnetic field that is quite unlike that of the Sun. Figure 2 shows one of the most striking examples of a polar spot, on an RS Can Ven star with a radius of $11.5\,R_{\odot}$. Sunspots, which only cover 0.1% of the solar surface, are puny by comparison.

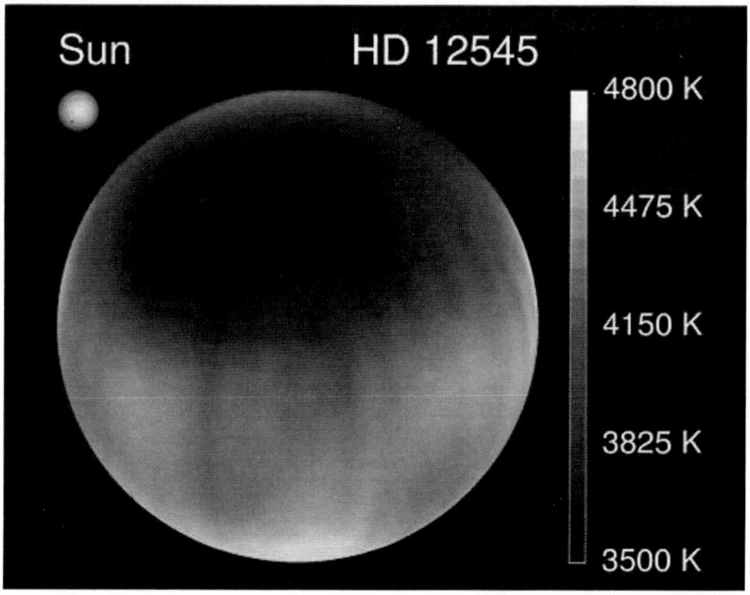

Fig. 2 Doppler image of the giant star HD12545 (XX Trianguli), with an enormous cool starspot near the pole. The sun is also shown to indicate the size of this polar spot (Courtesy of K. G. Strassmeier)

9 Epilogue

In this summary I have naturally focused on topics that most interested me, so that it has inevitably been selective. Like everyone else who was present in Bangalore, I was nevertheless impressed by the enormous range of topics that were considered, as well as by the vigor of current research on different aspects of solar magnetism. With so many new facilities, and so many talented young researchers entering the field, our subject has an exciting future ahead of it.

At this meeting we naturally looked back to Evershed's achievements at Kodaikanal 100 years ago. Those of us who were able to visit Kodaikanal afterwards saw that active research continues there today. Meanwhile, several new solar telescopes are coming into operation, culminating, we hope, with the ATST in 6 years' time, while the Solar Dynamics Observatory will soon be launched to join Hinode and STEREO in space. Within India, we offer our best wishes for the Space Coronagraph, and also for the 2 m National Solar Telescope in Ladakh. I was lucky enough to be able to visit its probable site at Pangong Lake a few months before this meeting, and it filled me with enthusiasm.

Finally, on behalf of everyone who came, I thank the Director of the Indian Institute of Astrophysics, Professor S.S. Hasan, and all his staff for bringing us to Bangalore and arranging so stimulating and enjoyable a meeting.

Acknowledgement I want to add my personal thanks to Siraj Hasan, Dipanku Banerjee, R. Ramesh, and everyone at IIA, as well as to those participants at this meeting whose contributions have formed the basis of my summary.

References

Abreu, J. A., Beer, J., Tobias, S. M., Weiss, N. O. 2008, Geophys. Res. Lett., 35, L20109
de Wijn, A. G., Stenflo, J. O., Solanki, S. K., Tsuneta, S. 2009, Space Sci. Rev., 144, 275
Evershed, J. 1909a, Kodaikanal Obs. Bull. 15, 63
Evershed, J. 1909b, MNRAS, 69, 454
Evershed, J. 1910, MNRAS, 70, 217
Hale, G. E. 1908a, ApJ, 28, 100
Hale, G. E. 1908b, ApJ, 28, 315
Riethmüller, T. L., Solanki, S. K., Zakharov, V., Gandorfer, A. 2008, A&A, 492, 233
Thomas, J. H., Weiss, N. O. 2008, Sunspots and Starspots, Cambridge University Press

Part III
Poster Presentations

Evidence for Return Meridional Flows in the Convection Zone

K.R. Sivaraman, H. Sivaraman, S.S. Gupta, and R.F. Howard

Abstract We report velocities of meridional motions derived from the latitude drifts of spot groups measured on photographic images of the Sun in the Kodaikanal observatory archives. They serve to measure different meridional flows for spot groups anchored at different depths. Comparison of spot-group rotation rates with the rotation profile resulting from helioseismic inversions yields estimates of the anchoring depths. The measured latitudinal drifts correspond to meridional flows at these depths. They show evidence of return meridional flows as required in a flux-transport dynamo.

1 Introduction

The flux transport dynamo model (the only dynamo model that is widely accepted among many solar dynamo models) envisages advective transport of magnetic flux by global meridional circulation. Flux from decayed active regions from the sunspot belts is transported to the poles by surface meridional motions aided by diffusion. The model invokes a return meridional flow, directed equatorward, in the deep interior in order to obey mass conservation. While the meridional flow, the key ingredient, has long been observed on the solar surface (poleward with velocities about 15–20 $\mathrm{m\,s^{-1}}$), the assumed equatorward return flow at greater depths, which together with the surface flow provides the conveyer belt mechanism for the flux transport, has not been detected observationally so far (Dikpati 2005 and references

K.R. Sivaraman (✉)
Indian Institute of Astrophysics, Bangalore, India

H. Sivaraman
853 Berryessa Court, Livermore, CA 94551, USA

S.S. Gupta
Indian Institute of Astrophysics, Kodaikanal, India

R.F. Howard
National Solar Observatory, Tucson, Arizona, USA

S.S. Hasan and R.J. Rutten (eds.), *Magnetic Coupling between the Interior and Atmosphere of the Sun*, Astrophysics and Space Science Proceedings,
DOI 10.1007/978-3-642-02859-5_30,

therein). In this paper, we show that the velocity of meridional motions at different depths in the convection zone can be inferred by monitoring latitude drifts of spot groups and sorting them into different area classes. Earlier studies have shown that spots have a rotation rate faster than the rotation rate of the surface plasma (Wilcox and Howard 1970), and that small spots with areas <5 microhemispheres (μ) rotate at faster rates than large spots with areas $>15\,\mu$ (Howard et al. 1984; Gupta et al. 1999). These findings have been explained by the hypothesis that the rotation rates measured from spots reflect the rotation rates of those plasma layers in the solar interior where the footpoints of the magnetic loops producing spots are anchored (Foukal 1972; Gilman and Foukal 1979). This hypothesis was strengthened by studies of meridional motions (Ribes et al. 1985) and by rotation rate determinations (Ribes et al. 1993) using spots as tracers. With the arrival of measurements of rotation rates vs. depth from helioseismology, it became possible to estimate the depths where the footpoints of the magnetic loops of spots of each area class (or of different lifetimes) are anchored by matching the spot rotation rates with those in the interior for the respective latitudes (Javaraiah and Gokhale 1997; Sivaraman et al. 2003).

2 Measurements

We here report velocities of meridional motions that are derived from our measurements of latitude drifts of spot groups using the daily photographic images of the Sun from the Kodaikanal Observatory (Indian Institute of Astrophysics) archives for the period 1906–1987. The plates were scanned with 0.2 arcsec sampling resolution; for plates made under good seeing conditions, speeds as low as a few meters per second can be measured with good accuracy because of the large number of spots available in the data bank.

We have divided the spot groups into three area classes, namely spots with umbral areas $<5\,\mu$); between 5 and $10\,\mu$, and those $>10\,\mu$. We determined the latitude drift of every spot group within each area class, and then obtained the latitude drift in degrees per day by dividing the drift by the elapsed time (expressed in days) between successive observations. These latitude drifts were converted from degrees per day to meters per second.

In the next step, we divided the latitude belts on both hemispheres (i.e., $\pm 40°$) into 5° strip zones. To reduce the noise, we averaged the meridional velocity in meters per second over all spot groups belonging to each area class per 5° latitude zone.

3 Results

Figure 1 shows the latitude drift (in deg/day) against latitude. Negative drifts represent equatorward motion and positive drifts poleward motion. In the central latitude zone $+35°$ to $-35°$, the meridional flow is equatorward with a mean velocity of

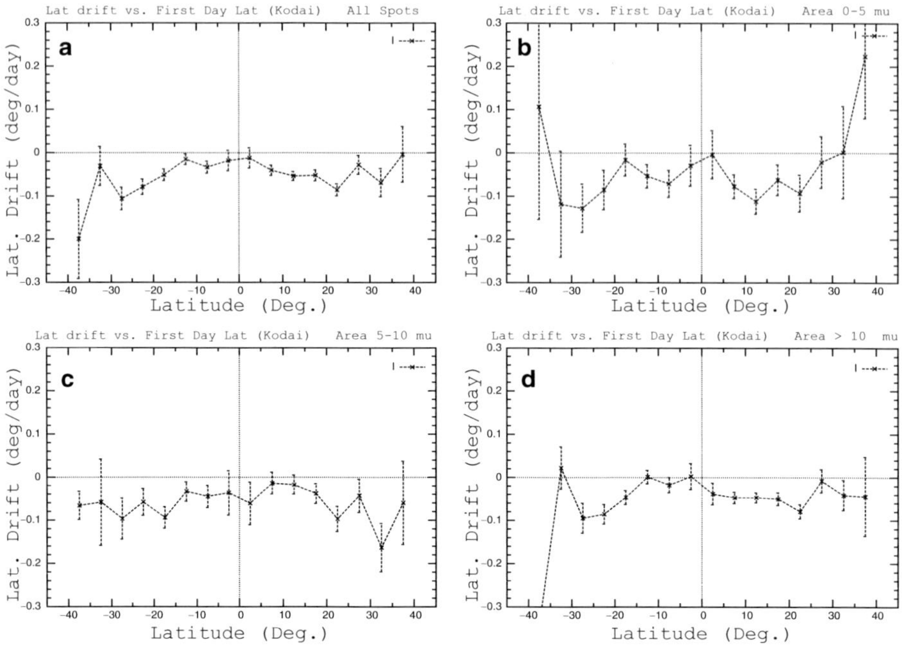

Fig. 1 Meridional motions in deg/day averaged over all the spot groups within 5° latitude bins against latitude, measured from Kodaikanal heliograms for the period 1906–1987. The latitude of each spot group on the first day of its life defines its 5° latitude bin. Negative drifts indicate equatorward motion in each hemisphere. *First panel*: all spot groups. *Second panel*: spot groups with area <5 μm. *Third panel*: spot groups with area 5–10 μm. *Fourth panel*: spot groups with area 10–15 μm

about 5 $\mathrm{m\,s^{-1}}$ when spot groups of all areas are combined in the flow velocity determination. The small spot groups (area 0–5 μ) show an equatorward flow with velocity $7 \pm 4\,\mathrm{m\,s^{-1}}$ in latitude zone ±30° and poleward flow beyond, while the flow is equatorward across the entire sunspot belt +40° to −40° for spot groups of the two other classes (5–10 μ and >10 μ), with velocities of 6 ± 3 and $5 \pm 2\,\mathrm{m\,s^{-1}}$, respectively.

By matching the average rotation rates of the spot groups in each class with a profile of the rotation rate against depth at the respective latitudes that has been derived from helioseismic inversions (Komm 2004), we find the anchor depths of the flux tubes corresponding to the three area classes of sunspot groups. These are $r/R_0 = 0.87$, 0.83, and 0.75, respectively, expressed in fractional solar radius, for spot groups with area 0–5, 5–10, and >10 μ, respectively.

We conclude that the latitudinal flow velocities that we have determined for the three classes of spot groups represent those of the meridional flows at these different depths in the convection zone, respectively. The equatorward components in Fig. 1 therefore sample return meridional flows.

References

Dikpati, M. 2005, Adv. Space Res., 35, 322
Foukal, P. V. 1972, ApJ, 173, 439
Gilman, P. A., Foukal, P. V. 1979, ApJ, 229, 1179
Gupta, S. S., Sivaraman, K. R., Howard, R. F. 1999, Solar Phys., 188, 225
Howard, R. F., Gilman, P. A., Gilman, P. I. 1984, ApJ, 283, 373
Javaraiah, J., Gokhale, M. H. 1997, A&A, 327, 795
Komm, R.W. 2004, private communication
Ribes, E., Mein, P., Mangeney, A. 1985, Nature, 318, 170
Nesme-Ribes, E., Ferreira, E. N., Mein, P. 1993, A&A, 274, 563
Sivaraman, K. R., Sivaraman, H., Gupta, S. S., Howard, R. F. 2003, Solar Phys., 214, 65
Wilcox, J. M., Howard, R. F., 1970, Solar Phys., 13, 251

Long-Term Variations in Meridional Flows

J. Javaraiah

Abstract Using sunspot group data from Greenwich and from the Solar Optical Observation Network during 1879–2008, we find that the mean meridional motion of the observed spot groups varies considerably on a time scale of about 10–20 years. During the current cycle 23, which is anomalous in the sense that the cycle pair 22 and 23 violated the Gnevyshev and Ohl rule, the mean meridional speed of the spot groups is substantially higher than it was during the last 10–11 cycles. We also find a roughly 18-year periodicity in the mean meridional motion of spot groups in the northern hemisphere, and a roughly 25-year periodicity in the north–south asymmetry of the mean meridional motion.

1 Introduction

Studying temporal variations in solar meridional flows is important for understanding the underlying mechanism of the solar cycle (e.g., Ulrich and Boyden 2005). Besides Doppler velocity measurements, observations of the longitudinal and latitudinal drifts of magnetic tracers, particularly sunspots, are traditionally used as proxies of solar fluid motions to study the solar rotational and meridional flows. Sunspot data are available for over a century, but the derived flow rates depend on the method of selection of the spots or spot groups (e.g., Javaraiah and Gokhale 1997; Javaraiah 1999; Sivaraman et al. 2010). Recently, Ruzdjak et al. (2005) found that evolutionary effects on the measured spot-group positions are small when estimating the mean meridional motion of the spot groups.

Here I use the largest available reliable set of sunspot group data by combining the Greenwich and the Solar Optical Observation Network (SOON) sunspot group data during the period 1879–2008 (taken from http://solarscience.msfc.nasa.gov/greenwch.shtml) and attempt to determine variations in the solar meridional flow that are of the order of the length of a solar cycle.

J. Javaraiah (✉)
Indian Institute of Astrophysics, Bangalore, India

S.S. Hasan and R.J. Rutten (eds.), *Magnetic Coupling between the Interior and Atmosphere of the Sun*, Astrophysics and Space Science Proceedings,
DOI 10.1007/978-3-642-02859-5_31,

2 Data Analysis and Results

The data reduction is the same as described in Javaraiah and Ulrich (2006). We determined the running-mean meridional motion (daily rate of the latitudinal drift) of the sunspot groups over 3-year moving time intervals (MTIs) through 1-year shifts, i.e., 1879–1881, 1880–1882, ..., 2006–2008, during the period 1979–2008. Figure 1 shows the variations of the mean meridional motion of the spot groups in the northern and the southern hemispheres and the corresponding north–south difference. We have used the 3-year MTIs because the statistics is too poor to plot yearly data. We determined periodicities in these mean motions by standard Fast Fourier transform (FFT) and Morlet wavelet power spectral analysis.

We find the following:

1. The mean meridional motion of sunspot groups varies considerably. The maximum amplitude of the variation of the mean motion of the spot groups in each hemisphere is about 5 m s^{-1}. The amplitude is found to be slightly large in a 10° latitude interval. There is also a suggestion that the patterns of the variations in

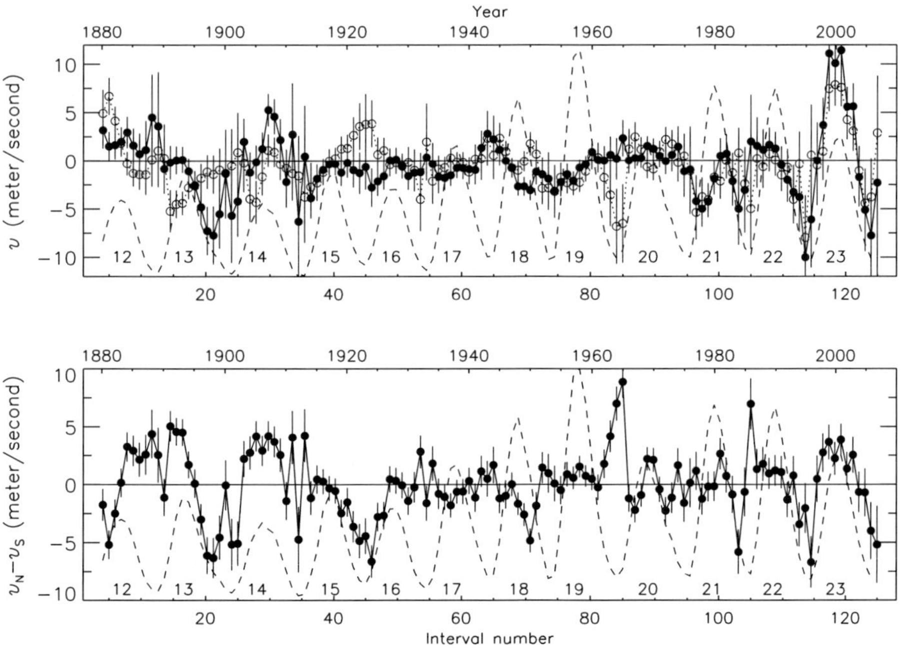

Fig. 1 Variations in mean meridional motion (a positive value represents the northbound motion) of the sunspot groups during the period 1879–2008 for the northern hemisphere (*solid curve*) and the southern hemisphere (*dotted curve*) and the corresponding north–south difference (v_N–v_S, *lower panel*), determined by binning the data into 3-year MTIs. In both panels the dashed curve represents the variation (smoothed by taking 3-year running average) in sunspot activity during 1879–2007. The Waldmeier cycle number is specified near each maximum epoch

the mean motion are significantly different during the odd- and even-numbered cycles. The hemispheric difference (north–south asymmetry) of the mean motion varies also with an amplitude of about $5\,\mathrm{m\,s^{-1}}$.
2. The variation in the mean motion of the spot groups during the current cycle 23 is the most pronounced in 100 years. During the maximum of this cycle, the mean motion is relatively strong (about $10\,\mathrm{m\,s^{-1}}$) and has northward direction in both the northern and the southern hemispheres.
3. In the FFT spectrum of the mean motion of the spot groups of the northern hemisphere, a broad peak at frequency $f \approx 1/18$ year^{-1} is found to be significant at >99% confidence level, while in the corresponding spectrum of the southern hemisphere no peak, including the one at $f \approx 1/18$ year^{-1}, is found to be significant at the 99% confidence level. The FFT spectrum of the north–south difference shows a well-defined and highly significant peak at $f \approx 1/25$ year^{-1}. Similar periodicities are also found in the corresponding Morlet wavelet spectra, but the available data are inadequate to determine precise frequencies and their time variations.

3 Discussion

According to the well-known Gnevyshev and Ohl (G-O) rule, an odd cycle is stronger than the immediately preceding even cycle (Gnevyshev and Ohl 1948). The cycle pair 22, 23 violated the G-O rule. A reason may be the difference in temporal behavior of the mean meridional flow velocities between cycles 22 and 23. In cycle 23, the activity was slightly larger in the southern hemisphere than in the northern hemisphere. The strong northbound meridional flow may have caused cancelation of much northern-hemisphere flux by equator-crossing flux from the southern hemisphere.

References

Gnevyshev, M. N., Ohl, A. I. 1948, AZh, 25, 18
Javaraiah, J. 1999, Solar Phys., 189, 289
Javaraiah, J., Gokhale, M. H. 1997, A&A, 327, 795
Javaraiah, J., Ulrich, R. K. 2006, Solar Phys., 237, 245
Ruzdjak, D., Brajsa, R., Sudar, D., Wöhl, H. 2005, Solar Phys., 229, 35
Sivaraman, K. R., Sivaraman, H., Gupta, S. S., Howard, R. F. 2010, In: Magnetic Coupling Between the Interior and the Atmosphere of the Sun, S. S. Hasan, R. J. Rutten (eds.), Astrophys. Space Sci. Procs., Springer, Heidelberg, these proceedings, 319
Ulrich, R. K., Boyden, J. E. 2005, ApJ, 620, L123

Low-Degree High-Frequency *p* and *g* Modes in the Solar Core

S. Mathur, R.A. García, and A. Eff-Darwich

Abstract Solar gravity (*g*) modes propagate within the radiative part of the solar interior and are highly sensitive to the physical conditions of the solar core. They would represent the best tool to infer the structure and the dynamics of the radiative interior, in particular the core, if they were properly detected and characterized. Although individual rotational splittings for *g* modes have not yet been calculated, we have to understand the effect of these modes, and also low-degree high-frequency *p* modes, on the inversion of the solar rotation rate between 0.1 and $0.2\,R_{\odot}$. In this work, we follow the methodology developed in Mathur et al. [2008, A&A, 484, 517] and García et al. [2008a, A.N., 329, 476], adding *g* modes and low-degree high-frequency *p* modes to artificial inversion data sets, in order to study how they convey information on the solar core rotation.

1 Introduction

Helioseismology has improved our understanding of the dynamical processes occurring within the solar interior (Thompson et al. 2003, and references therein), in particular the convective zone and the radiative zone down to $0.3\,R_{\odot}$ (Couvidat et al. 2003; García et al. 2004). However, the solar deep interior is still poorly constrained and the possible effect of the rotation rate in these regions on the solar structure distribution (e.g., sound speed and/or density) remains uncertain (Mathis and Zahn 2004, 2005; Turck-Chièze et al. 2008). These uncertainties should be added to the recently found discrepancies between helioseismic observations and theoretical modeling (Antia and Basu 2005; Turck-Chièze et al. 2004a; Mathur et al. 2007;

S. Mathur (✉)
Indian Institute of Astrophysics, Bangalore, India

R.A. García
Laboratoire AIM, CEA/DSM-CNRS-Univ. Paris Diderot, Gif-sur-Yvette, France

A. Eff-Darwich
Departamento de Edafología y Geología, Univ. de La Laguna, Tenerife, Spain
and
Instituto de Astrofísica de Canarias, Tenerife, Spain

S.S. Hasan and R.J. Rutten (eds.), *Magnetic Coupling between the Interior and Atmosphere of the Sun*, Astrophysics and Space Science Proceedings,
DOI 10.1007/978-3-642-02859-5_32,

Zaatri et al. 2007) when the new surface chemical abundances of Asplund et al. (2005) are used to calculate the structural model of the solar interior. In this context, gravity modes would add strong observational constraints about the structure and dynamics of the solar core. A few individual candidates have been detected (Turck-Chièze et al. 2004b) and also global asymptotic properties (García et al. 2007, 2008a) that were used by García et al. (2008b) to infer the rotation rate of the solar core and to constrain various physical quantities.

An artificial rotation rate (blue curve in Fig. 1) was used to calculate the frequency splittings for *p* and *g* modes. The shape of this profile (note that between 0.15 and 0.2 $R_\odot$, the rate is three times that in the rest of the radiative zone) was chosen to study the effect of the different modes used in the inversion sets (Table 1). The inversions were carried out through the Regularized Least Squared methodology (Eff-Darwich and Pérez Hernández 1997), the mode sets came from the artificial rotational rate, and the data uncertainties and noise were calculated from observations (García et al. 2008c; Korzennik 2005). We know that the inner turning points of low-degree high-frequency *p* modes lay around 0.1–0.2 $R_\odot$ (García et al. 2008c), while those of *g* modes are confined in a small region below 0.2 $R_\odot$ (Mathur et al. 2008). A priori, these modes would significantly improve the inversion results in the solar core, but it is necessary to understand the effect of observational uncertainties and the number of modes to set the accuracy needed from peak fitting techniques to properly characterize the core.

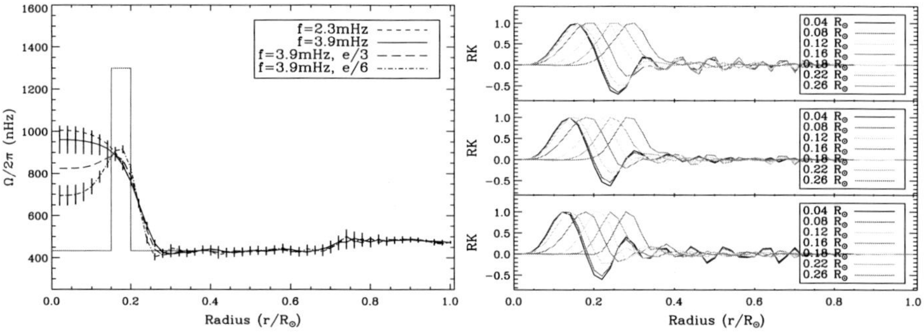

Fig. 1 *Left*: Rotation profiles (equatorial) reconstructed from different sets of *p* modes. *Right*: Resolution kernels of sets 1 (top), 3 (middle), and 4 (bottom)

Table 1 Artificial data sets, with ν (mHz) the mode frequency and ϵ its uncertainty

Data set	*g* modes	*p* modes $\ell = 1, 3$	ϵ for $\ell = 1, 3$	*p* modes $\ell > 3$
Set 1	–	$1 \leq \nu \leq 2.3$	ϵ	$1 \leq \nu \leq 3.9$
Set 2	–	$1 \leq \nu \leq 3.9$	ϵ	$1 \leq \nu \leq 3.9$
Set 3	–	$1 \leq \nu \leq 3.9$	$\epsilon/3$	$1 \leq \nu \leq 3.9$
Set 4	–	$1 \leq \nu \leq 3.9$	$\epsilon/6$	$1 \leq \nu \leq 3.9$
Set 5	$\ell = 2, n = -3$	$1 \leq \nu \leq 3.9$	ϵ	$1 \leq \nu \leq 3.9$
Set 6	$\ell = 1 - 2, n = -1$ to -10	$1 \leq \nu \leq 3.9$	ϵ	$1 \leq \nu \leq 3.9$

2 The Effect of Low-Degree High-Frequency *p* Modes

When low-degree high-frequency *p* modes are added to the inversion data sets (1–4), the increase in the rotation rate below $0.2\,R_\odot$ is detected in all cases but the maximum rate is not recovered (Fig. 1). Only for data sets 3 and 4, with the error bars significantly lowered, is the decrease of the rotation rate in the solar core detected, although the actual values are not recovered. The resolution kernels show that the deepest point sensed by the inversion shifts from 0.15 down to $0.12\,R_\odot$, explaining the peak sensitivity of these sets to the region between 0.1 and $0.15\,R_\odot$.

3 The Effect of *g* Modes in the Inversions

The effect of adding *g* modes on the inversions was analyzed adding 1 mode (set 5) and 20 modes (set 6) *g* to set 2 (Fig. 2). In both cases, the uncertainties in the *g* modes were 75 nHz, although two more sets (set 5+ and 6+) with the uncertainties reduced to 10 nHz were also computed. The inversion of set 5 is comparable to that of set 3 where no *g* modes were used. This result reflects the fact that individual *g* modes should be detected with accuracies significantly lower than 75 nHz. For sets 6, 5+, and 6+, the rotation rate recovered in the core is similar to the actual artificial rate. Hence, the detection of a few very accurate *g* modes will give similar information as a larger set of less accurate data. The resolution kernels show that the deepest point at which the inversion is sensitive shifts from 0.15 down to $0.03\,R_\odot$. However, the sensitivity is still quite low between 0.05 and $0.15\,R_\odot$ (set 6 in Fig. 2) when only *g* modes are used in the inversion. With a lower error bar, we probe several points between 0.03 and $0.15\,R_\odot$. Moreover, the resolution kernels are quite broad, explaining that we do not retrieve the rotation rate between 0.15 and $0.2\,R_\odot$ (average on the width of these kernels). To study this region, it would be necessary to significantly reduce the observational uncertainties for low degree and high frequency *p* modes. Finally, resolution kernels for sets 6, 5+, and 6+ exhibit less wiggles above $\approx 0.3\,R_\odot$: they are less polluted by the rotation of the outer layers.

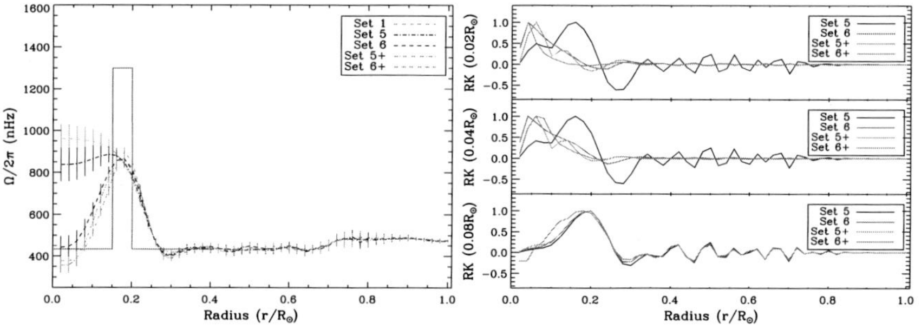

Fig. 2 *Left*: Same as Fig. 1 with 1 and 20 *g* modes ($\epsilon = 75$ and 10 nHz). *Right*: Resolution kernels for sets 5, 6, 5+, and 6+ at 0.02, 0.04, and $0.08\,R_\odot$

4 Conclusion

Gravity modes are necessary to improve our inferences on the rotational rate at the solar core. However, these modes should reach a certain level of accuracy, namely uncertainties for the frequency splittings significantly lower than 75 nHz, to assure that at least one *g* mode will be sufficient to greatly improve our sensitivity to the dynamics of the core. Although gravity modes will certainly improve inversion results below $0.1\,R_\odot$, it would be necessary to properly characterize low-degree and high-frequency *p* modes to resolve the region between 0.1 and $0.2\,R_\odot$.

References

Antia, H. M., Basu, S. 2005, ApJ, 620, L129
Asplund, M., Grevesse, N., Sauval, A. J. 2005, In: Cosm. Abund. as Rec. of Stel. Evol. and Nucleosynth., T. G. Barnes III, F. N. Bash (eds.), ASP Conf. Ser., vol. 336, p. 25
Couvidat, S., García, R. A., Turck-Chièze, S., et al. 2003, ApJ, 597, L77
Eff-Darwich, A. Pérez Hernández, F. 1997, A&AS, 125, 391
García, R. A., Corbard, T., Chaplin, W. J., et al. 2004, Solar Phys., 220, 269
García, R. A., Turck-Chièze, S., Jiménez-Reyes, S. J., et al. 2007, Science, 316, 1591
García, R. A., Jiménez, A., Mathur, S., et al. 2008a, A.N., 329, 476
García, R. A., Mathur, S., Ballot, J. 2008b, Solar Phys., 251, 135
García, R. A., Mathur, S., Ballot, J., et al. 2008c, Solar Phys., 251, 119
Korzennik, S. G. 2005, ApJ, 626, 585
Mathis, S., Zahn, J.-P. 2004, A&A, 425, 229
Mathis, S., Zahn, J.-P. 2005, A&A, 440, 653
Mathur, S., Turck-Chièze, S., Couvidat, S., García, R. A. 2007, ApJ, 668, 594
Mathur, S., Eff-Darwich, A., García, R. A., Turck-Chièze, S. 2008, A&A, 484, 517
Thompson, M. J., Christensen-Dalsgaard, J., Miesch, M. S., Toomre, J. 2003, ARA&A, 41, 599
Turck-Chièze, S., Couvidat, S., Piau, L., et al. 2004a, Phys. Rev. Lett., 93, 211102
Turck-Chièze, S., García, R. A., Couvidat, S., et al. 2004b, ApJ, 604, 455
Turck-Chièze, S., Phi Nghiem, P. A., Mathis, S. 2008, JPhCS, 118, 2030
Zaatri, A., Provost, J., Berthomieu, G., Morel, P., Corbard, T. 2007, A&A, 469, 1145

Interior and Exterior Clues of Solar Activity

S. Turck-Chièze, A.S. Brun, V. Duez, R.A. García, S. Mathis, L. Piau, D. Salabert, P.L. Pallé, S.J. Jiménez-Reyes, S. Mathur, R. Simoniello, and J.M. Robillot

Abstract Two research paths are described to obtain better understanding of the origin of global solar activity. First, observations with a multichannel resonant spectrometer may reveal the dynamics of the solar core, the tachocline, and the temporal evolution of activity between the photosphere and chromosphere. Such new observations will deliver constraints for 3D simulations of solar activity. Second, we examine the ab-initio introduction of a non-force-free field expressed in spherical harmonics into the solar structure equations and estimate its impact on the inner and subsurface layers, its time evolution, and its role in angular momentum transport.

1 Introduction

Understanding the great extrema of activity during the last millennium requires insight on solar dynamics down to the solar core on long time scales.

The GOLF/SoHO instrument (Gabriel et al. 1997) shows its promises to measure low-degree oscillation spectra of both the acoustic and gravity solar modes using the variability of Doppler velocity measurement of the sodium line. These contain indirect clues of solar magnetism within their complementary resonant cavities. However, the granulation noise and problematic instrument polarizers stayed a limitation to reach the core dynamics. We therefore decided to build a successor in proposing a new concept (Turck-Chièze et al. 2006). The multichannel spectrometer GOLF-NG was studied first with CNES support during 1999–2000 following the

S. Turck-Chièze (✉), A.S. Brun, V. Duez, R.A. García, S. Mathis, and L. Piau
Laboratoire AIM, CEA/DSM-CNRS-Univ. Paris Diderot, Gif-sur-Yvette, France

D. Salabert, P.L. Pallé, and S.J. Jiménez-Reyes
Instituto de Astrofísica de Canarias, Tenerife, Spain

S. Mathur
Indian Institute of Astrophysics, Bangalore, India

R. Simoniello
Physikalisch-Meteorogisches Observatorium Davos, Switzerland

J.M. Robillot
Observatoire de Bordeaux, France

S.S. Hasan and R.J. Rutten (eds.), *Magnetic Coupling between the Interior and Atmosphere of the Sun*, Astrophysics and Space Science Proceedings,
DOI 10.1007/978-3-642-02859-5_33, © Springer-Verlag Berlin Heidelberg 2010

concepts of GOLF (Gabriel et al. 1995) and the RM5 spectrometer (Ambroz et al. 1995), and was then built by CEA within a collaboration with IAC.

In addition, at the same time we decided to develop dynamical models of the Sun to understand the impact of rotation and its induced magnetic field on the radiative zone and the surface observables (Couvidat et al. 2003; Duez et al. submitted; Turck-Chièze et al. 2009).

2 The Dynamics of the Radiative Zone

Studies on the origin of solar activity are generally focused on the external 2% of the solar mass $M_{\odot}$, which constitutes the convective zone, but they have not yet fully explained the solar cycle variability. Little attention has been given to the radiative interior. SoHO allowed important progress on the dynamics of this zone through the high-quality Doppler velocity measurements with GOLF and MDI, outperforming the intensity measurements of VIRGO. GOLF succeeded in answering the different questions for which it was designed (Turck-Chièze et al. 2008):

- It put strong constraints on solar core thermodynamics, down to 6% $R_{\odot}$, and definitively establish the different neutrino fluxes
- It contributed to better understanding the subsurface layers during the solar cycle (Jiménez-Reyes et al. 2004, 2007), which sustains local and varying magnetic field
- It showed its capability to detect gravity-mode signatures (Turck-Chièze et al. 2004; García et al. 2007) although the central rotation remains uncertain (Mathur et al. 2008). These two studies are compatible only if the solar core turns quickly and if the axis of the core rotation is oblique to the solid rotation of the rest of the radiative zone.

The multichannel resonant GOLF-NG spectrometer detailed below is designed to improve on this performance and to enable solar-interior studies of waves produced by convective plumes or by volumetric stochastic excitation. The possible presence of deep-seated magnetic fields and the temporal variability of the crucial core and tachocline regions that strongly constrain the gravity-mode cavities will be studied with this instrument.

3 The Multichannel GOLF-NG Spectrometer

The GOLF-NG concept could be proposed as part of a payload devoted to gain a complete MHD vision of the Sun, while waiting for a more ambitious perspective as a formation-flying mission, such as DynaMICCS/HIRISE suggested to ESA in the Cosmic Vision plan 2016–2025 (Turck-Chièze et al. 2009) for space weather and space climate studies. In addition, multiple GOLF-NG instruments may be placed in a global ground-based network. Table 1 shows the performances of the prototype and of the final instrument.

GOLF-NG is a resonant scattering spectrophotometer that measures full-disk sunlight at 15 wavelengths in the Na I D_1 line (Turck-Chièze et al. 2006), instead of only four (in practice 2) for GOLF. GOLF-NG will alternately observe the blue and red wings every second, sampling different heights of the atmosphere. It benefits from the atmospheric variability of the granulation patterns to decrease the impact of this dominant phenomenon in the Fourier transform of the Doppler shift measurements around and below 1 mHz, leading to better detection of gravity and acoustic modes. It will also register atmospheric dynamics throughout the solar cycle. Thus, GOLF-NG will enable the development of a complete dynamical solar model producing insights on the core rotation and magnetic field, the variability of the tachocline through gravity-mode characteristics, and the interplay of convection, rotation, and magnetic field throughout and below the photosphere.

We decided to first construct a prototype (GOLF-NG phase I) to verify our ability to build an instrument that measures motions simultaneously at different heights in the solar atmosphere at 1-s cadence. The laboratory performance of this demonstrator is described in Turck-Chièze et al. (2008); the first analyses of the 2008 summer campaign in Tenerife are described in Salabert et al. (2008), including comparison with simultaneous MARK1 (the Tenerife instrument of the BiSON network) and GOLF observations.

Figure 1 shows residual daily velocities obtained with the GOLF-NG prototype. The 5-min oscillation is clearly visible. Multichannel Fourier power spectra compiled from 76 continuous days of observation by integrating the signals over 20 s are shown in Fig. 2. They sample different heights in the solar atmosphere, between 200 and 800 km. The potentiality of the multichannel technique is shown in the acoustic-mode area through increase of sensitivity with height as predicted by theory. Note,

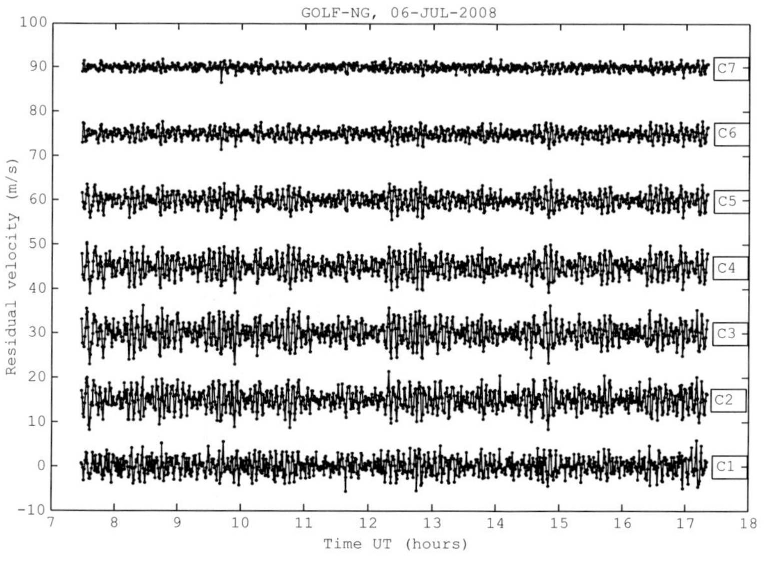

Fig. 1 Eight residual velocity signals (m s^{-1}) showing 5-min oscillations observed on 6 July 2008 with GOLF-NG at the Observatorio del Teide

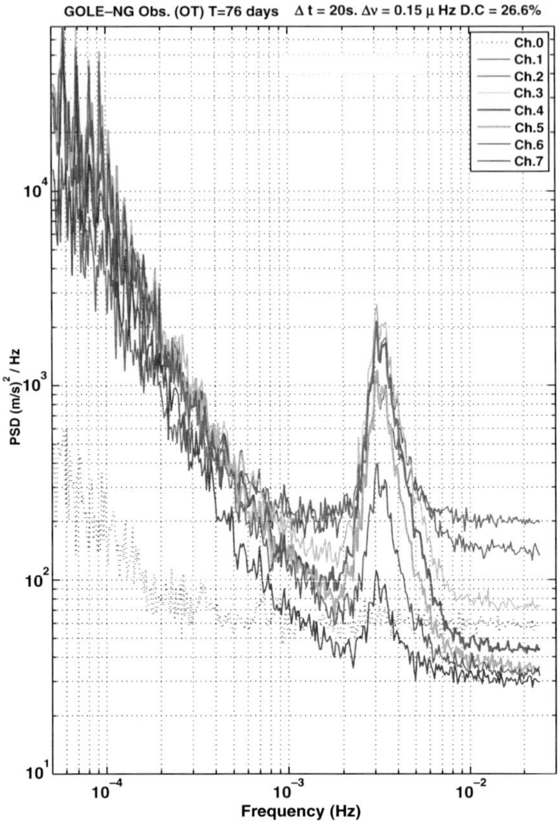

Fig. 2 Multichannel power spectra from the GOLF-NG prototype

Table 1 Main characteristics of the multichannel resonant spectrometer

Characteristics	Phase I: prototype	Phase II: final instrument
Na I D_1 line only	Yes	Yes
8 channels: $0 < B < 12\,\mathrm{kG}$	Yes	Yes
Continuum measurement	No	Yes
Liquid-crystal polariser	Yes	Yes
Counting rate for mid line	$5 \times 10^7\,\mathrm{s}^{-1}$	$> 1 \times 10^8\,\mathrm{s}^{-1}$
Instrumental uncertainty	Detector electronic noise	Statistical noise
Objective	Demonstrator	$0.3\,\mathrm{mm\,s}^{-1}$ in 10 days for $\nu > 5\,\mu\mathrm{Hz}$

in particular, the different behavior between different channels around 1 mHz, which is extremely promising for the extraction of low-amplitude acoustic modes in this region. The common background increase below 5×10^{-4} Hz is mainly due to photodiode matrix noise (Turck-Chièze et al. 2008). The detection performance must be improved, together with the thermal stability, in the planned phase-II development to obtain the objective specified in Table 1.

4 Building Secular Solar Dynamical Models

In parallel to 3D simulations of the radiative zone (Brun and Zahn 2006), we develop new solar models including internal rotation (Turck-Chièze et al. in preparation) and the impact of a large-scale magnetic field in the radiative zone and below the surface (Duez et al. submitted). These models will provide a physical framework to study momentum transport due to dynamical processes (Mathis and Zahn 2005), in particular, to partly explain the flatness of the rotation profile in the radiative zone above $0.4\,R_\odot$. Duez and Mathis (submitted) derive axymmetric magnetic configurations, which verify the magnetohydrostatic barotropic equilibrium as possible initial conditions for the internal fossil field, using the Grad–Shafranov equation. The field is non-force-free and contains both poloidal and toroidal components. An analysis of the impact on the solar structure is studied in Duez et al. (submitted). Figure 3

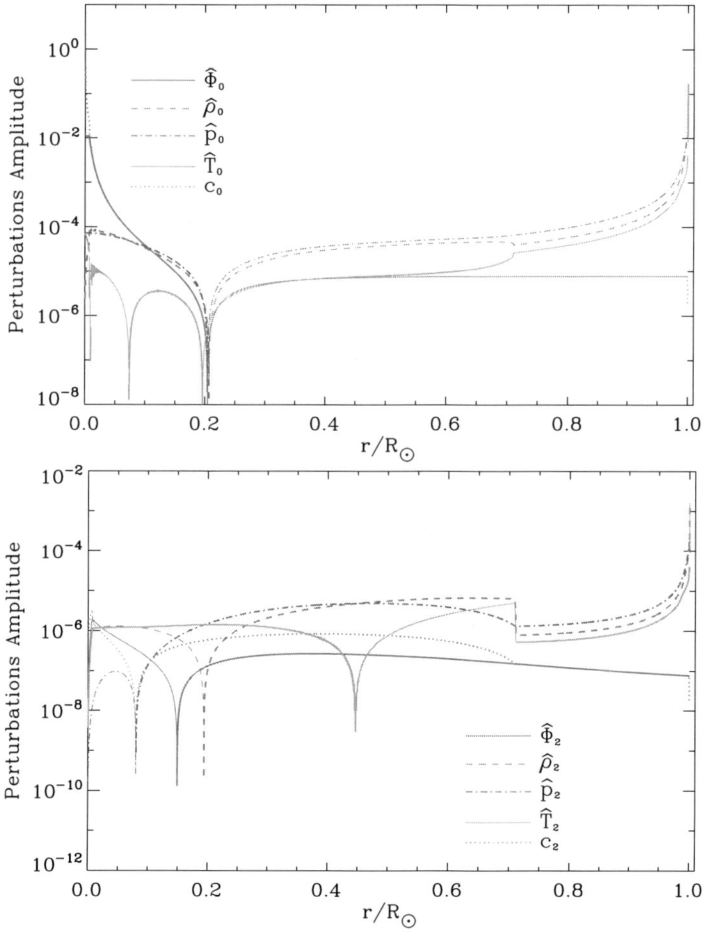

Fig. 3 $l = 0$ and $l = 2$ mode perturbations of the main variables for a deep field of 2.2 MG. From Duez et al. (2009)

shows that a deep field of 2.2 MG produces some effects just below the surface and that a detailed knowledge of this region could put constraints on such a deep field. Nevertheless, a more quantitative estimate will be obtained when all these equations will be introduced into the stellar codes. These theoretical predictions (potentially with different field configurations) will be confronted to observables of PICARD (Thuillier et al. 2006) and of GOLF-NG: shape deformation and gravity mode properties. In parallel to this theoretical effort, we need to initiate other theoretical efforts on the kind of modes that we may detect with GOLF-NG by using 3D models of the atmosphere. One needs also to estimate the theoretical visibility of internal and atmospheric gravity modes with height in the atmosphere (see Belkacem et al. 2009).

Acknowledgment We strongly thank the technical and instrumental support received from the CEA team and different IAC services.

References

Ambroz, P., Robillot, J. M., Bocchia, R. 1995, In: Helioseismology, ESA Special Publication, vol. 376, p. 233
Belkacem, K., Samadi, R., Goupil, M. J., et al. 2009, A&A, 494, 191
Brun, A. S., Zahn, J.-P. 2006, A&A, 457, 665
Couvidat, S., Turck-Chièze, S., Kosovichev, A. G. 2003, ApJ, 599, 1434
Duez, V., Mathis, S., ApJ, submitted
Duez, V., Mathis, S., Brun, A.S., Turck-Chièze, S., ApJ, submitted
Gabriel, A. H., Grec, G., Charra, J., et al. 1995, Solar Phys., 162, 61
Gabriel, A. H., Charra, J., Grec, G., et al. 1997, Solar Phys., 175, 207
García, R. A., Turck-Chièze, S., Jiménez-Reyes, S. J., et al. 2007, Science, 316, 1591
Jiménez-Reyes, S. J., Chaplin, W. J., Elsworth, Y., García, R. A. 2004, ApJ, 604, 969
Jiménez-Reyes, S. J., Chaplin, W. J., Elsworth, Y., et al. 2007, ApJ, 654, 1135
Mathis, S., Zahn, J.-P. 2005, A&A, 440, 653
Mathur, S., Eff-Darwich, A., García, R. A., Turck-Chièze, S. 2008, A&A, 484, 517
Salabert, D., Turck-Chieze, S., Barriere, J. C., et al. 2008, ArXiv e-prints
Thuillier, G., Dewitte, S., Schmutz, W. 2006, In: 36th COSPAR Sci' Assembly, p. 170
Turck-Chièze, S., García, R. A., Couvidat, S., et al. 2004, ApJ, 604, 455
Turck-Chièze, S., Carton, P.-H., Ballot, J., et al. 2006, Adv. Space Res., 38, 1812
Turck-Chièze, S., Carton, P. H., Mathur, S., et al. 2008, Astron. Nachr., 329, 521
Turck-Chièze S. et al. 2009, Exp. Astron., 23, 1017, Cosmic vision special issue
Turck-Chièze S., Palacios, A., Nghiem, P., ApJ, in preparation

Do Active Regions Modify Oscillation Frequencies?

S.C. Tripathy, K. Jain, and F. Hill

Abstract We investigate the variation of high-degree mode frequencies as a local response to the active regions in two different phases of the solar activity cycle. We find that the correlation between frequency shifts and the surface magnetic activity measured locally are significantly different during the two activity periods.

1 Introduction

The oscillation frequencies are known to vary in phase with the solar activity cycle. In most of the earlier studies (see Jain and Bhatnagar 2003, and references therein), the variation between the frequency shifts and activity, as measured by different proxies, demonstrated a linear relation. But a detailed analysis using the improved and continuously measured eigen-frequencies over solar cycle 23 indicates complex relationships: a strong correlation during the rising and declining phases and a significantly lower correlation during the minimum phase (Jain et al. 2009). In addition, there is no consensus as to the solar origin of these changes. There is some indication that the variation of the high-degree mode frequencies are spatially as well as temporally associated with active regions (Hindman et al. 2000; Howe et al. 2008). In this context, we investigate the variation of high-degree mode frequencies as a local response to active regions.

2 Analysis and Results

We use the ring-diagram technique (Hill 1988; Corbard et al. 2003) to calculate the high-degree mode frequencies and examine the frequency shifts during two Carrington Rotation (CR) periods of the solar cycle 23: one during the descending phase covering the period 21 October to 19 November 2003 (CR 2009) and the other during the low activity period of 20 June to 18 July 2007 (CR 2058). We analyze

S.C. Tripathy (✉), K. Jain, and F. Hill
GONG Program, National Solar Observatory, Tucson, USA

S.S. Hasan and R.J. Rutten (eds.), *Magnetic Coupling between the Interior and Atmosphere of the Sun*, Astrophysics and Space Science Proceedings,
DOI 10.1007/978-3-642-02859-5_34,

a set of 189 individual regions on the solar disk, which is commonly referred to as a dense-pack mosaic. Each region is about 15° × 15° in heliographic latitude and longitude and is tracked for a period of 1,664 min. The centers of the regions are separated by 7.5° in latitude and longitude and extend roughly to 52.5° from disk center. By repeating the analysis over the entire dense-pack mosaic, the frequencies are determined as a function of ring-day (1,664 min) and position on the solar disk. For each wavenumber and mode order, the frequency shifts ($\delta\nu$) are computed relative to the spatial average obtained from the 189 tiles. The strength of the magnetic field associated with each tile is also estimated by calculating a Magnetic Activity Index (MAI), which represents the average over all the pixels in a given tile. This is obtained from the magnetograms mapped and tracked in the same way as the Dopplergrams.

The frequency shifts corresponding to the p_3 ($n = 3$) mode with harmonic degree $\ell = 242$ are shown in Fig. 1. It reveals that the frequencies can vary as much as 20 μHz within an active region having high magnetic-field strength compared to the quiet Sun. In the same figure, we also plot the corresponding MAI (dotted line). A good agreement between $\delta\nu$ and MAI, particularly for large shifts, is clearly visible.

Figures 2 and 3 display images of the frequency shifts averaged over all modes (upper panels) as a function of the position of the solar disk for three days during CR 2009 and CR 2058, respectively. The lower panels in each figure correspond to the coeval MAI. For each of the images, each pixel corresponds to a single tile in the dense-pack mosaic. Comparing the top and bottom panels, it is clearly seen

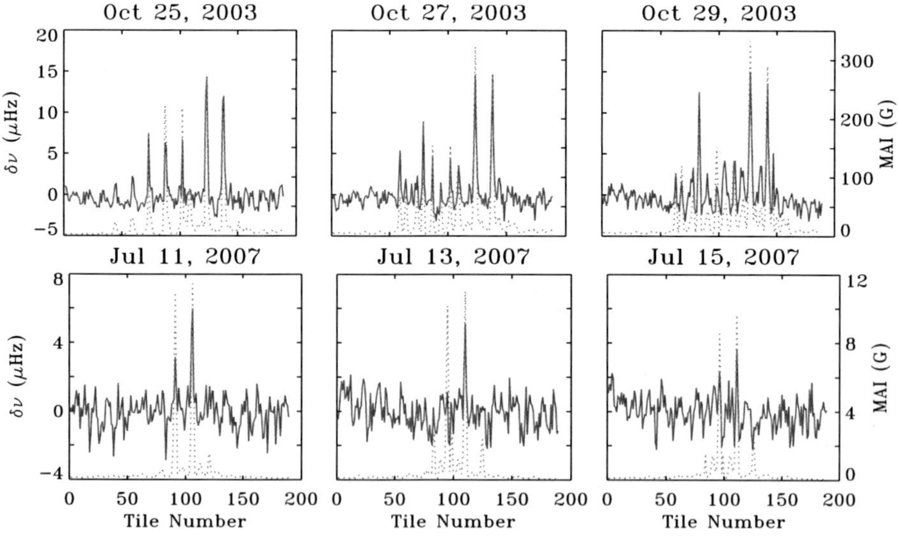

Fig. 1 The frequency shifts (*solid line*) for a single multiplet ($n = 3, \ell = 242$) for 189 dense-pack regions. The *upper* and *lower* panels represent CR 2009 and CR 2058, respectively. The *dotted line* denotes the MAI and is calculated using MDI 96-min magnetograms for CR 2009 and GONG 1-min magnetograms for CR 2058. The date on the top of the figures indicates the beginning date of the ring-day analysis

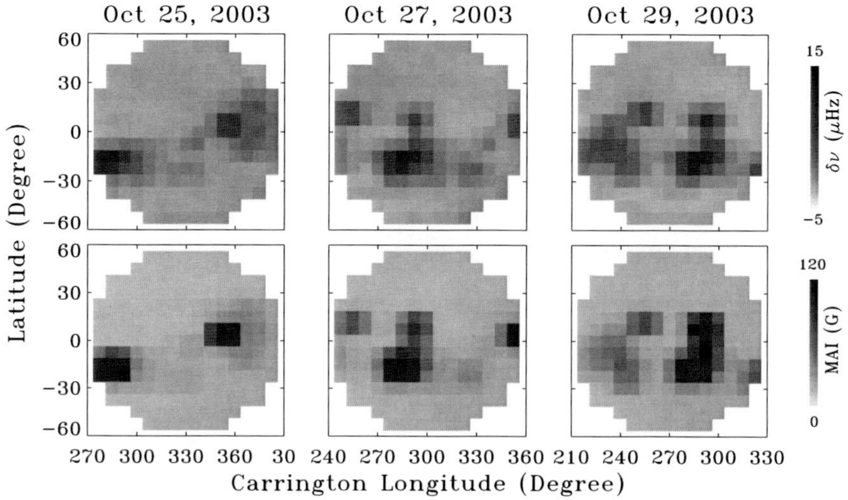

Fig. 2 Average frequency shifts of 189 dense pack tiles (*top panel*) and coeval MAI (*bottom panel*) for three days during CR 2009. The dates are indicated on the top of the panels

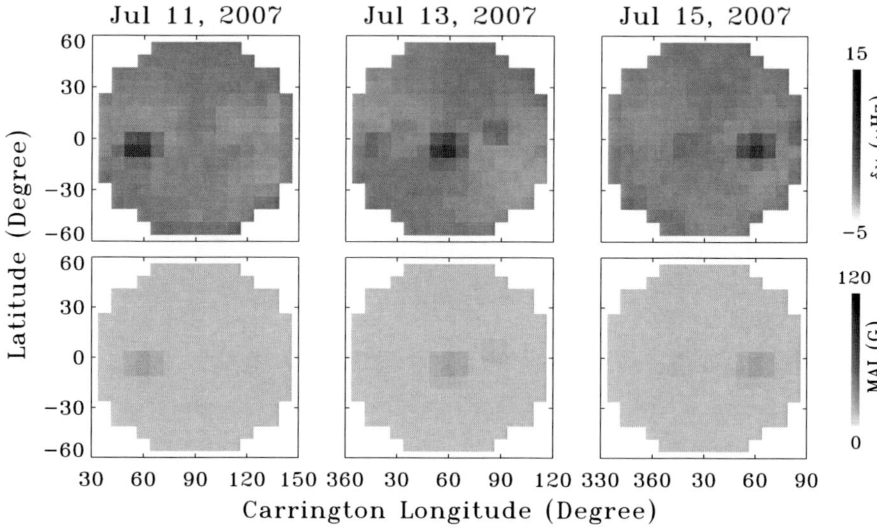

Fig. 3 Same as Fig. 2 but for low-activity period corresponding to CR 2058. The dates are indicated on the top of the panels

that the tiles with large frequency shifts match very well with the tiles at the same location having a large magnetic field. Thus, areas of high frequency shifts appear as active regions and viceversa. This confirms the findings of Hindman et al. (2000) that local frequency shift acts as a tracer of magnetic activity. However, during the low-activity period (Figure 3), where both the shifts and MAI are small, we note a few tiles with positive frequency shifts that have no counterparts in the MAI image implying a weaker agreement. To estimate how well we can associate the locations of active regions as locations of frequency shifts, we calculate the Pearson's correlation coefficient (r_p) between the shifts and MAI for each of the three ring-days. These are found to be 0.91, 0.93, and 0.88 and 0.74, 0.77, and 0.85, for CR 2009 and 2058, respectively, and confirms that the correlation between shifts and the surface magnetic activity during the two activity periods are significantly different. This result is consistent with the recent findings inferred from global modes (Jain et al. 2009).

Thus the argument that the solar-cycle variations in the global mode frequencies are due to global averaging of the local effect of active regions (Hindman et al. 2001) is only partially supported by our analysis. We believe that the weak component of the magnetic field, for example, ubiquitous horizontal field or turbulent field, must be taken into account to fully explain the frequency shifts, particularly during the minimal-activity phase of the solar cycle.

Acknowledgment We thank John Leibacher for a critical reading of the manuscript. This research was supported in part by NASA grants NNG 05HL41I and NNG 08EI54I. This work utilizes data obtained by the Global Oscillation Network Group program, managed by the National Solar Observatory, which is operated by AURA, Inc. under a cooperative agreement with the National Science Foundation. The data were acquired by instruments operated by the Big Bear Solar Observatory, High Altitude Observatory, Learmonth Solar Observatory, Udaipur Solar Observatory, Instituto de Astrofísica de Canarias, and Cerro Tololo Interamerican Observatory. This work also utilizes 96-min magnetograms from SOI/MDI on board *Solar and Heliospheric Observatory* (SOHO). SOHO is a project of international cooperation between ESA and NASA.

References

Corbard, T., Toner, C., Hill, F., Hanna, K. D., Haber, D. A., Hindman, B. W., Bogart, R. S. 2003, In: ESA SP-517, Local and Global Helioseismology: The Present and Future, H. Sawaya-Lacoste (ed.), ESA SP vol. 517, p. 255

Hill, F. 1988, ApJ, 333, 996

Hindman, B., Haber, D., Toomre, J., Bogart, R. 2000, Solar Phys., 192, 363

Hindman, B. W., Haber, D. A., Toomre, J., Bogart, R. S. 2001, In: Helio- and Asteroseismology at the Dawn of the Millennium, A. Wilson (ed.), ESA SP vol. 464, p. 143

Howe, R., Haber, D. A., Hindman, B. W., Komm, R., Hill, F., Gonzalez Hernandez, I. 2008, In: Subsurface and Atmospheric Influences on Solar Activity, R. Howe, R. W. Komm, K. S. Balasubramaniam, G. J. D. Petrie (eds.), ASP Conf. Ser., vol. 383, p. 305

Jain, K., Bhatnagar, A. 2003, Solar Phys., 213, 257

Jain, K., Tripathy, S. C., Hill, F. 2009, ApJ, 695, 1567

Deep-Focus Diagnostics of Sunspot Structure

H. Moradi and S.M. Hanasoge

Abstract In sequel to Moradi et al. [2009, ApJ, 690, L72], we employ two established numerical forward models (a 3D ideal MHD solver and MHD ray theory) in conjunction with time–distance helioseismology to probe the lateral extent of wave-speed perturbations produced in regions of strong, near-surface magnetic fields. We continue our comparisons of forward modeling approaches by extending our previous surface-focused travel-time measurements with a common midpoint deep-focusing scheme that avoids the use of oscillation signals within the sunspot region. The idea is to also test MHD ray theory for possible application in future inverse methods.

1 Introduction

In Moradi et al. (2009), we used two recently developed numerical MHD forward models in conjunction with surface-focused (i.e., center-to-annulus) time–distance measurements to produce numerical models of travel-time inhomogeneities in a simulated sunspot atmosphere. The resulting artificial travel-time perturbation profiles clearly demonstrated the overwhelming influence that MHD physics, as well as phase-speed and frequency filtering, have on local helioseismic measurements in the vicinity of sunspots.

However, there are numerous caveats associated with surface-focused time–distance measurements that use oscillation signals within the sunspot region, as the use of such oscillation signals is now known to be the primary source of most surface

H. Moradi (✉)
School of Mathematical Sciences, Monash University, Australia
and
Visiting Scientist: Indian Institute of Astrophysics, Bangalore, India

S.M. Hanasoge
Max-Planck-Institut für Sonnensystemforschung, Katlenburg-Lindau, Germany
and
Visiting Scientist: Indian Institute of Astrophysics, Bangalore, India

S.S. Hasan and R.J. Rutten (eds.), *Magnetic Coupling between the Interior and Atmosphere of the Sun*, Astrophysics and Space Science Proceedings,
DOI 10.1007/978-3-642-02859-5_35,

effects in sunspot seismology. These surface effects can be essentially categorized into two groups. The first revolves around the degree to which observations made within the sunspot region are contaminated by magnetic effects (e.g., Braun 1997; Lindsey and Braun 2005; Schunker et al. 2005; Braun and Birch 2006; Couvidat and Rajaguru 2007; Moradi et al. 2009), while the second concerns the degree to which atmospheric temperature stratification in and around regions may affect the absorption line used to make measurements of the Doppler velocity (e.g., Rajaguru et al. 2006, 2007).

There have been attempts in the past to circumvent such problems by adopting a time–distance measurement geometry known as "deep-focusing," which avoids the use of data from the central area of the sunspot by only cross-correlating the oscillation signal of waves that have a first-skip distance larger than the diameter of the sunspot (e.g., Duvall 1995; Braun 1997; Zhao and Kosovichev 2006; Rajaguru 2008). In this analysis, we follow up on the comparative study presented in Moradi et al. (2009) by using our two established forward models, in conjunction with a deep-focusing scheme known as the "common midpoint" (CMP) method to probe the sub-surface dynamics of our artificial sunspot.

2 The Flux Tube and Forward Models

The background stratification of our model atmosphere is given by an adiabatically stable, truncated polytrope (Bogdan et al. 1996), smoothly connected to an isothermal atmosphere. The truncated polytrope is described by: index $m = 2.15$, reference pressure $p_0 = 1.21 \times 10^5\,\mathrm{g\,cm^{-1}\,s^{-2}}$, and reference density $\rho_0 = 2.78 \times 10^{-7}\,\mathrm{g\,cm^{-3}}$. The flux tube (peak field strength of 3 kG) is modeled by an axisymmetric magnetic field geometry based on the Schlüter and Temesváry (1958) self-similar solution. The derived MHS sunspot model achieves a consistent sound-speed decrease (see Fig. 1), with a peak reduction of about 45% at the surface ($z = 0$) and less than 1% at $z = -2$ Mm, while the one-layered wave-speed enhancement is also confined to the near-surface layers, approaching 180% at the surface and around less than 0.5% at $z = -2$ Mm.

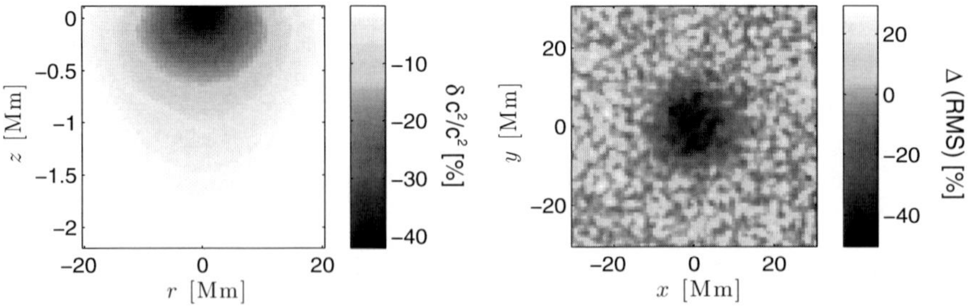

Fig. 1 Some properties of the model sunspot atmosphere. *Lefthand panel*: the near-surface thermal/sound-speed perturbation profile shown as a function of sound-speed squared. *Righthand panel*: a Doppler power map normalized to the quiet Sun

The two forward models presented in Moradi et al. (2009) are again used for our analysis. The first forward model integrates the linearized ideal MHD wave equations according to the recipe of Hanasoge (2008), where waves are excited via a precomputed deterministic source function that acts on the vertical momentum equation. To simulate the suppression of granulation related wave sources in a sunspot (e.g., Hanasoge et al. 2008), the source activity is muted in a circular region of 10 Mm radius. The simulations produce artificial line-of-sight (Doppler) velocity data cubes, extracted at a height of 200 km above the photosphere, effectually mimicking Michelson Doppler Imager (MDI) Dopplergrams. The data cubes have dimensions of $200 \times 200\,\mathrm{Mm}^2 \times 512\,\mathrm{min}$, with a cadence of 1 min and a spatial resolution of 0.718 Mm. Figure 1 depicts a normalized power map derived from the simulated Doppler velocity measurements.

The second forward model employs the MHD ray tracer of Moradi and Cally (2008), where 2D ray propagation is modeled through solving the governing equations of the ray paths derived using the zero-order eikonal approximation and the magneto-acoustic dispersion relation. It should be noted that neither forward model accounts for the presence of sub-surface flows.

3 Common Midpoint Deep-Focusing

The CMP method is often utilized in geophysics applications such as multichannel seismic acquisition (Shearer 1999). It measures the travel time at the point on the surface halfway between the source and the receiver (see Fig. 2). Cross-correlating numerous source–receiver pairs in this manner results in the method being mostly sensitive to a small region in the deep interior surrounding the lower turning point of the ray. A reworking of this method has been applied to helioseismic observations

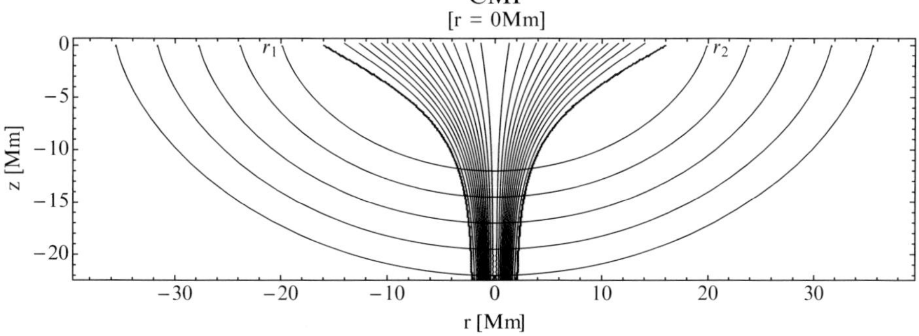

Fig. 2 An illustration of the CMP deep-focus geometry indicating the range of rays used for this study. The CMP method measures the travel time at the point on the surface located at the half-way point between a source ($\mathbf{r_1}$) and receiver ($\mathbf{r_2}$). For the above rays, the CMP is located on the central axis of the spot ($r = 0\,\mathrm{Mm}$)

by Duvall (2003), and has the obvious advantage of allowing one to study the wave-speed structure directly beneath sunspots without using the oscillation signals inside the perturbed region.

Our method for measuring time–distance deep-focus travel times is somewhat similar to the approach undertaken by Braun (1997) and Duvall (2003). First, the annulus-to-annulus cross-covariances (e.g., between oscillation signals located between two points on the solar surface, a source at $\mathbf{r}_1$ and a receiver at $\mathbf{r}_2$, as illustrated in Fig. 2) are derived by dividing each annulus ($\boldsymbol{\Delta} = |\mathbf{r}_2 - \mathbf{r}_1|$) into two semi-annuli and cross-correlating the average signals in these two semi-annuli. Then, to further increase the signal-to-noise ratio (SNR), we average the cross-covariances over three distances, respectively, slightly smaller than, and larger than, $\boldsymbol{\Delta}$. In the end, the five (mean) distances chosen (42.95, 49.15, 55.35, 61.65, and 68 Mm, respectively) are large enough to ensure that we sample only waves with a first-skip distance greater than the diameter of the sunspot at the surface (about 40 Mm).

Because of the oscillation signal at any location being a superposition of a large number of waves of different travel distances, the cross-covariances are very noisy and need to be phase-speed filtered first in the Fourier domain, using a Gaussian filter for each travel distance. The application of appropriate phase-speed filters isolates waves that travel desired skip distances, meaning that even though we average over semi-annuli, the primary contribution to the cross covariances is from these waves. In addition to the phase-speed filters, we also apply an f-mode filter that removes the f-mode ridge completely (as it is of no interest to us in this analysis), and we also apply Gaussian frequency filters centered at $\omega = 3.5, 4.0$, and 5.0 mHz with $\delta\omega = 0.5$ mHz band-widths to study frequency dependencies of travel times (e.g., Braun and Birch 2008; Moradi et al. 2009). To extract the required travel times, the cross-covariances are fitted by two Gabor wavelets (Kosovichev and Duvall 1997): one for the positive times and one for the negative times.

Even after significant filtering and averaging, the extracted CMP travel times are still inundated with noise. This is certainly an ever-present complication in local helioseismology, as there is a common expectation (with all local helioseismic methods and inversions) of worsening noise and resolution with depth. Realization noise associated with stochastic excitation of acoustic waves can significantly impair our ability to analyze the true nature of travel-time shifts on the surface (and by extension, also affect our interpretation of sub-surface structure). But, as we have full control over the wave excitation mechanism and source function, we have the luxury of being able to apply realization noise subtraction to improve the SNR and obtain statistically significant travel-time shifts from the deep-focus measurements. This is accomplished in the same manner as Hanasoge et al. (2007), that is, by performing two separate simulations, one with the perturbation (i.e., the sunspot simulation) and another without (i.e., the quiet simulation). We then subtract the travel times of the quiet data from its perturbed counterpart (see Fig. 3), allowing us to achieve an excellent SNR.

Finally, to compare theory with simulations, we also estimate deep-focusing time shifts using the MHD ray tracer of Moradi and Cally (2008). The single-skip magneto-acoustic rays are propagated from the inner (lower) turning point of their trajectories at a prescribed frequency (see e.g., Fig. 2). These rays do not undergo

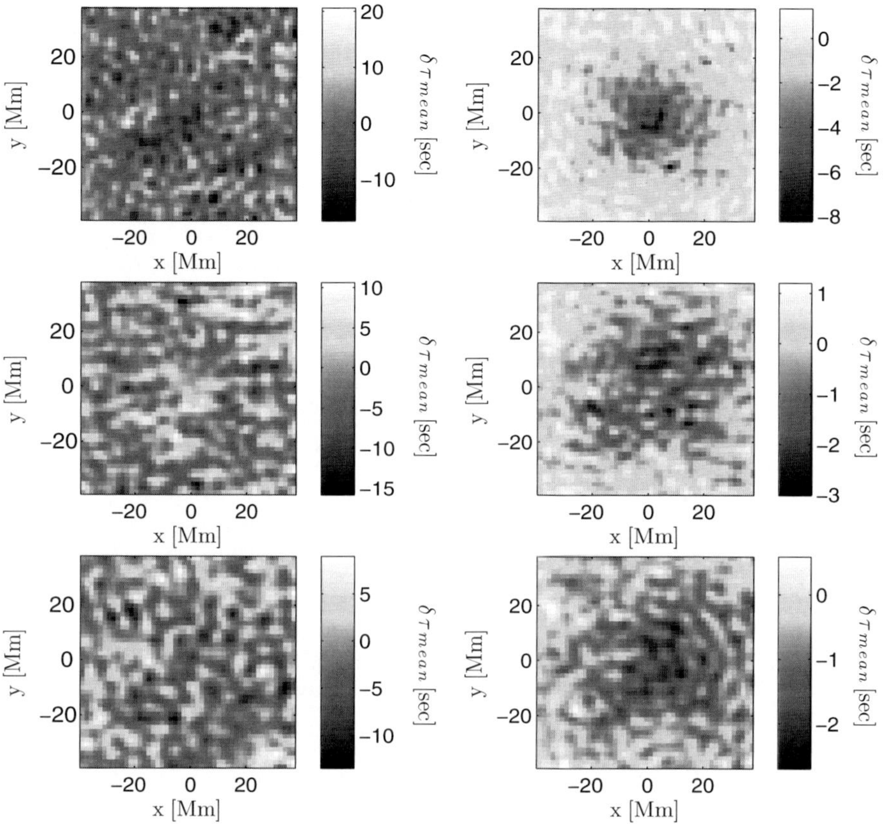

Fig. 3 Examples of (phase-speed filtered) CMP mean travel-time perturbation ($\delta\tau_{\rm mean}$) maps for $\mathbf{\Delta} = 42.95$ (*top*), $\mathbf{\Delta} = 49.15$ (*middle*), and $\mathbf{\Delta} = 61.65$ Mm (*bottom*). *Lefthand panels*: before realization noise subtraction. *Righthand panels*: after subtraction. A frequency filter centered at 5.0 mHz has been applied to the data

any additional filtering as the required range of horizontal skip distances are simply obtained by altering the depth at which the rays are initiated from. The resulting mean (phase) travel-time shifts ($\delta\tau_{\rm mean}$) derived from both forward models are presented in Figs. 3 and 4.

4 Results and Discussion

A number of travel time maps derived from the time–distance analysis, both before and after noise subtraction, are presented in Fig. 3. The impact of realization noise subtraction is self-evident in these figures as it is only after removing the background noise that we are able to detect statistically significant travel-time shifts. The umbral

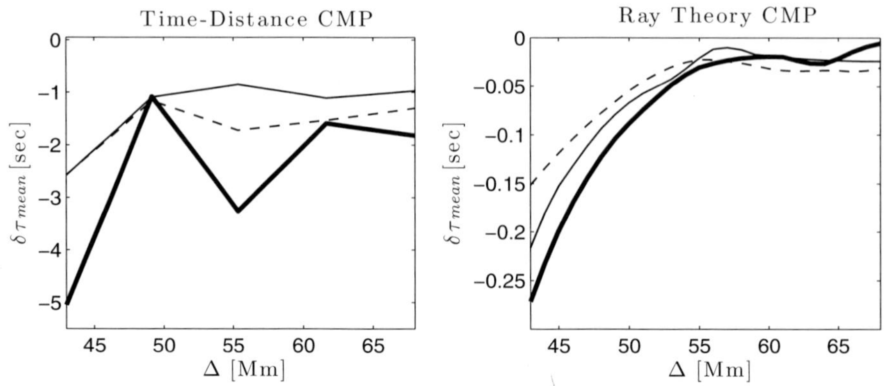

Fig. 4 Observed CMP travel-time shifts as a function of wave/ray travel distance (**Δ**). *Lefthand panel*: umbral averages of the CMP time shifts derived from time–distance analysis of the simulated data. *Righthand panel*: ray theory CMP travel-time shifts derived from rays propagated at various depths and with a CMP at $r = 0$ Mm. *Light sold lines* are indicative of frequency filtering centered at 3.5 mHz, *dashed lines* indicate 4.0 mHz, and *bold solid lines* indicate 5.0 mHz

averages of these time shifts are shown in Fig. 4. The $\delta\tau_{\rm mean}$ values range from a few seconds at 3.5 and 4.0 mHz to about 5 s at 5.0 mHz. However, even though the size of the measured time shifts are significant, there is no clear frequency dependence associated with them. As we only use waves outside of the perturbed region, surface effects can be effectively ruled out as the cause of the time shifts.

It is worth noting that linear inversions of surface-focused travel time maps of actual observations have suggested a two-layered wave-speed structure below sunspots – a wave-speed decrease of 10–15% down to a depth of 3–4 Mm, followed by a wave-speed enhancement, reportedly detected down to depths of 17–25 Mm below the surface (Kosovichev et al. 2000; Couvidat et al. 2006). However, with our forward model clearly prescribing both a shallow wave- and sound-speed perturbation profile (Fig. 1), it is hard to fathom that the time–distance $\delta\tau_{\rm mean}$ we are observing can be associated with some kind of anomalous deep sub-surface perturbation. In fact, both the sound-speed decrease and wave-speed enhancement at such depths registers at less than one-tenth of 1% and the value of the plasma β is in the range of $7–18 \times 10^3$ – in all likelihood not significant enough to produce a 3–5 s travel-time perturbation. To try and identify the root cause of these apparent travel-time shifts, it is useful to compare the time–distance CMP measurements with those derived from MHD ray theory in Fig. 4.

The ray theory CMP $\delta\tau_{\rm mean}$ clearly appears to be significantly smaller at all frequencies, with all observed time shifts registering at less than half a second. Certainly, these time shifts are more in line with our expectations, given the absence of any significant deep sound/wave-speed perturbation. But, we must bear in mind the differences between the two forward models before drawing our conclusions. With regards to helioseismic travel times, Bogdan (1997) has emphasized that they are not only sensitive to the local velocity field along the ray path, but also to conditions in the surrounding medium – a clear consequence of wave effects. As such, wave-like behavior needs to be considered when interpreting travel times – something which ray theory does not clearly account for, resulting in possible underestimation of deep-focus travel times.

On the other hand, we must also consider the effects of phase-speed filtering (which is absent in the ray theory calculations) on the time–distance measurements. If we look closely at the time–distance $\delta\tau_{\text{mean}}$ maps in Fig. 3, we notice that they are somewhat smeared in appearance, with the central sunspot region becoming increasingly sprawled-out across the map as we increase $\boldsymbol{\Delta}$. This behavior is most likely a consequence of both the phase-speed filtering (i.e, the size of the central frequency filter, the filter width, etc., see Couvidat and Birch 2006) and the averaging scheme applied to the cross-correlations – both of which are a necessity to improve the SNR in time–distance calculations. These effects, combined with the delocalized nature of the CMP measurements, may also introduce spurious travel-time shifts. However, further testing and control simulations are required to confirm this.

5 Conclusion

At present, it suffices to conclude that we do not have a definitive diagnosis with regards to the differences between the deep-focus time shifts produced by the two forward models. We may well be applying ray theory to regimes where it is seriously limited. Then, the same may also be said about local time–distance analysis. Whatever is the case, these preliminary results do provide us with the motivation to conduct further time–distance studies using the CMP method.

The direct, and also the indirect, effects of phase-speed filtering on deep-focus measurements, derived from both simulations and real data, also warrant more detailed examination, because any artifact produced by the filtering process is likely to be even more pronounced for phase-speed filtered MDI data for which we do not yet have the luxury of noise subtraction. We hope to address these issues in the near future with similar comparison studies.

Acknowledgment The authors express their gratitude to the Indian Institute of Astrophysics for their support and warm hospitality during their stay in Bangalore, which made this work possible. The authors also thank Charlie Lindsey and Paul Cally for insightful discussions regarding various aspects of this work. HM also acknowledges travel support provided by the Astronomical Society of Australia.

References

Bogdan, T. J. 1997, ApJ, 477, 475
Bogdan, T. J., Hindman, B. W., Cally, P. S., Charbonneau, P. 1996, ApJ, 465, 406
Braun, D. C. 1997, ApJ, 487, 447
Braun, D. C., Birch, A. C. 2006, ApJ, 647, L187
Braun, D. C., Birch, A. C. 2008, Solar Phys., 251, 267

Couvidat, S., Birch, A. C. 2006, Solar Phys., 237, 229
Couvidat, S., Birch, A. C., Kosovichev, A. G. 2006, ApJ, 640, 516
Couvidat, S., Rajaguru, S. P. 2007, ApJ, 661, 558
Duvall, Jr., T. L. 1995, In: GONG 1994. Helio- and Astro-Seismology from the Earth and Space, R. K. Ulrich, E. J. Rhodes, Jr., W. Dappen (eds.), ASP Conf. Ser., vol. 76, p. 465
Duvall, Jr., T. L. 2003, In: GONG+ 2002. Local and Global Helioseismology: The Present and Future, H. Sawaya-Lacoste (ed.), ESA-SP, vol. 517, p. 259
Hanasoge, S. M. 2008, ApJ, 680, 1457
Hanasoge, S. M., Couvidat, S., Rajaguru, S. P., Birch, A. C. 2008, MNRAS, 1291
Hanasoge, S. M., Duvall, Jr., T. L., Couvidat, S. 2007, ApJ, 664, 1234
Kosovichev, A. G., Duvall, T. L. 1997, In: SCORe'96: Solar Convection and Oscillations and Their Relationship, ASSL, vol. 225, p. 241
Kosovichev, A. G., Duvall, T. L. . J., Scherrer, P. H. 2000, Solar Phys., 192, 159
Lindsey, C. Braun, D. C. 2005, ApJ, 620, 1107
Moradi, H., Cally, P. S. 2008, Solar Phys., 251, 309
Moradi, H., Hanasoge, S. M., Cally, P. S. 2009, ApJ, 690, L72
Rajaguru, S. P. 2008, ArXiv e-prints, 802
Rajaguru, S. P., Birch, A. C., Duvall, Jr., T. L., Thompson, M. J., Zhao, J. 2006, ApJ, 646, 543
Rajaguru, S. P., Sankarasubramanian, K., Wachter, R., Scherrer, P. H. 2007, ApJ, 654, L175
Schlüter, A., Temesváry, S. 1958, In: Electromagnetic Phenomena in Cosmical Physics, B. Lehnert (ed.), IAU Symp., vol. 6, p. 263
Schunker, H., Braun, D. C., Cally, P. S., Lindsey, C. 2005, ApJ, 621, L149
Shearer, P. 1999, Introduction to Seismology, Cambridge University Press
Zhao, J., Kosovichev, A. G. 2006, ApJ, 643, 1317

Are Polar Faculae Generated by a Local Dynamo?

K.R. Sivaraman, H.M. Antia, and S.M. Chitre

Abstract Polar faculae (PF) are bright, small-scale structures measuring a few seconds of arc, populating the polar zones at latitudes >50°. They possess magnetic fields ranging from 150 to 1,700 Gauss and largely constitute the polar magnetic fields. Where and how their fields are generated in the solar interior remain open questions. Using measurements of PF rotation rates, we show that their anchor depths probably lie in subsurface layers at radius $r/R_{\odot} = 0.94$–1.00. If so, the PF fields are possibly generated by a local dynamo in a subsurface shear layer extending to $r/R_{\odot} > 0.94$.

1 Introduction

Polar faculae are bright, small-scale structures, a few seconds of arc in diameter, seen near the north and south poles of the Sun in white light images as well as in images taken in chromospheric lines (e.g., the Ca II K line). The numbers of PF occupying the polar zones vary cyclically and are 180° out of phase with the sunspot cycle (Waldmeier 1955, 1962; Sheeley 1964; Makarov and Sivaraman 1989). The PF appear at heliographic latitudes above 50° soon after every polar field reversal. The zones occupied by the PF then progressively expand and eventually fill up the high-latitude regions, reaching maximum density during sunspot minimum. With increasing activity, the zones of PF appearance shrink until they finally disappear at the polar field reversal around sunspot maximum. This sequence repeats at every sunspot cycle (Makarov and Sivaraman 1989, Fig. 4). In magnetograms, the PF appear in the form of "flux knots," bright or dark depending on

K.R. Sivaraman (✉)
Indian Institute of Astrophysics, Bangalore, India

H.M. Antia
Tata Institute of Fundamental Research, Mumbai, India

S.M. Chitre
Centre for Basic Sciences, University of Mumbai, Mumbai, India

S.S. Hasan and R.J. Rutten (eds.), *Magnetic Coupling between the Interior and Atmosphere of the Sun*, Astrophysics and Space Science Proceedings,
DOI 10.1007/978-3-642-02859-5_36,

their magnetic polarity. The magnetic fields of the PF appear to constitute polar magnetic fields (Sheeley 1991, 2008; Makarov and Makarova 1998; Varsik et al. 1999; Benevolenskaya 2004; Okunev et al. 2005). Based on spectropolarimetric measurements, Homann et al. (1997) showed that PF posses field strengths in the kilo-Gauss range. These authors estimated the magnetic flux per PF to be about 7×10^{19} Mx, while Varsik et al. (1999) estimated the flux to be about 10^{19} Mx from calibrated low-resolution magnetograms. Subsequent studies by Okunev (2001) and Okunev and Kneer (2004), using high-resolution observations, confirmed that PF posses kilo-Gauss fields. Recent high-resolution observations from the solar optical telescope on board HINODE also confirm this flux range (Tsuneta et al. 2008). Okunev and Kneer (2004) also detected many small-scale weak magnetic areas surrounding the PF with kilo-Gauss fields. Depending upon the field strength and the extent of these weak field areas surrounding the PF, the PF are seen to possess magnetic fields of strengths ranging from 150 to 1,700 G. The PF also display a high degree of dynamic behavior in terms of fast evolutionary changes (Okunev and Kneer 2004). Although enough observations on the evolutionary patterns of the PF and the magnetic fields associated with them are available, the outstanding question has been where and how the magnetic fields of PF are generated in the solar interior. Here we match the PF rotation rate observations of Varsik et al. (1999) with rotation rates in the interior from helioseismic data and show that the anchor depths of PF fluxtubes are likely to be located in the depth range $r/R_\odot = 0.94$–1.00. This result suggests that PF fields are generated by a local dynamo operating in a subsurface shear layer extending about 40,000 km below the photosphere.

2 Results

The suggestion that the observed surface rotation rates of sunspots mimic the rotation rates of the plasma layers at those depths in the solar interior where the foot points of the spot fluxtubes are anchored was put forward by Gilman and Foukal (1979) to explain the differences in rotation rate between small and large spots. This was validated using the Greenwich sunspot data by Javaraiah and Gokhale (1997) and more convincingly by Sivaraman et al. (2003), who estimated these anchor depths by matching the observed sunspot rotation rates with rotation profiles from helioseismic studies. Recently, this method has been applied to estimate the anchor depths of fluxtubes of magnetic features with field strengths about 600 G (Zhao et al. 2004). If this idea also holds for PF, then by matching observed PF rotation rates with helioseismic rotation profiles, it should be possible to estimate anchor depths of the PF flux loops.

PF rotation rates have been measured by feature tracking individual elements with lifetimes of >7 h. Varsik et al. (1999) used high-resolution magnetogram sequences of polar regions (>50°in latitude) obtained in 1989 and 1995 at the Big Bear Solar Observatory. Recently, Benevolenskaya (2007) utilized full-disk longitudinal field images from the Michelson Doppler Imager (MDI) onboard the Solar

Heliospheric Observatory (SOHO). Plots of rotation rate vs. latitude for the range 55–80° (Fig. 6 of Varsik et al. 1999 and Fig. 3 of Benevolenskaya 2007) show that the rotation slows down with increasing latitude. We have read off PF rotation rates for latitude 55–80° at every 5° latitude from the least-square fit in Fig. 6 of Varsik et al. (1999) and present these values in Table 1.

To obtain estimates of anchor depths of the flux loops of the PF, we have projected the rotation rates of PF at the corresponding latitudes from Table 1 on a plot of the rotation rate vs. depth in the solar interior in terms of fractional solar radius $r/R_{\odot}$, derived from the GONG data by Antia et al. (2008). We have represented the intercepts of the projected rotation rates of PF on these profiles by short horizontal bars on each of the GONG profiles in Fig. 1. We then read off the depths on the $r/R_{\odot}$ axis corresponding to the intercepts on the profiles at 55°, 60°,

Table 1 Rotation rates of polar faculae and the corresponding depths

Latitude (°)	Rotation rate (nHz)	$r/R_{\odot}$ (from Fig. 1)
55	392	0.94
60	376	0.97
65	358	1.00
70	342	0.99
75	325	0.98
80	312	0.98

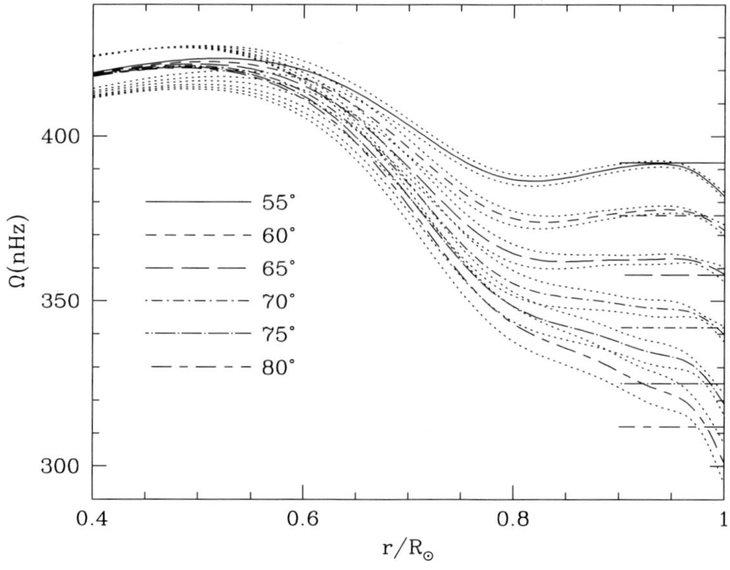

Fig. 1 Internal rotation rate vs. radius from GONG data (Antia et al. 2008) at different latitudes specified in the figure. The short horizontal lines represent the projection of the rotation rates of polar faculae from Table 1 on the GONG profiles at the respective latitudes. The depths corresponding to these intercepts (read off the x-axis) are given in the last column of Table 1

65°, 70°, 75°, and 80° latitudes. These depths lie in the range $r/R_{\odot} = 0.94$–1.00 (Table 1). This suggests that the anchor depths of the PF flux tubes lie in the subsurface layers in the depth range of $r/R_{\odot} = 0.94$–1.00 and magnetic fields of the polar faculae are possibly generated by a local dynamo operating in the subsurface shear layer at these depths. This possibility was discussed earlier, based on evidence both from observations and from modeling, by Cattaneo (1999), Dikpati et al. (2002), Benevolenskaya (2004), and Okunev et al. (2005).

Our observational study leads to the conclusion that the magnetic fields of polar faculae are likely generated by a local dynamo in a subsurface shear layer at $r/R_{\odot} > 0.94$.

References

Antia. H. M., Basu, S., Chitre. S. M. 2008, ApJ, 681, 680
Benevolenskaya, E. E. 2004, A&A, 428, L5
Benevolenskaya, E. E. 2007, Astron. Nachrichten, 328, 1016
Cattaneo, F. 1999, ApJ, 515, L39
Dikpati, M., Corbard. T. Thompson, M. J., Gilman, P. A. 2002, ApJ, 575, L41
Gilman, P. A., Foukal, P. V. 1979, ApJ, 229, 1179
Homann, T., Kneer, F., Makarov, V. I. 1997, Solar Phys., 175, 81
Javaraiah, J., Gokhale, M. H. 1997, A&A, 327, 795
Makarov, V. I., Makarova, V. V. 1998, In: Synoptic Solar Physics, K. S. Balasubramaniam, J. W. Harvey, D. M. Rabin (eds.), ASP Conf. Ser., 140, 347
Makarov, V. I., Sivaraman, K. R. 1989, Solar Phys., 123, 367
Okunev, O. V. 2001, Astron. Nachrichten, 322, 379
Okunev, O. V., Kneer, F. 2004, A&A, 425, 321
Okunev, O. V., Dominguez Cerdena, L., Puschmann, K. G., Kneer, F., Sanchez Almeida, J. 2005, Astron. Nachrichten, 326, 205
Sheeley, N. R., Jr. 1964, ApJ, 140, 731
Sheeley, N. R., Jr. 1991, ApJ, 374, 386
Sheeley, N. R., Jr. 2008 ApJ, 680, 1553
Sivaraman, K. R., Sivaraman, H., Gupta, S. S., Howard, R. F. 2003, Solar Phys., 214, 65
Tsuneta, S., Ichimoto, K., Katsukawa, Y., et al. 2008, ApJ, 688, 1374
Varsik, J. R., Wilson, P. R., Li, Y. 1999, Solar Phys., 184, 223
Waldmeier, M. 1955, ZAp, 38, 37
Waldmeier, M. 1962, ZAp, 54, 260
Zhao, J., Kosovichev, A. G., Duvall, T. L., Jr. 2004, ApJ, 607, L135

The Hanle Effect as Diagnostic Tool for Turbulent Magnetic Fields

L.S. Anusha, M. Sampoorna, H. Frisch, and K.N. Nagendra

Abstract The Hanle effect is calculated for a random magnetic field characterized by a finite correlation length and a probability density function of the magnetic field vector. It is shown that linear polarization is essentially independent of the magnetic field correlation length, but strongly depends on the distribution of the field strength.

1 Introduction

The Hanle effect provides a powerful diagnostic for the detection of weak turbulent solar magnetic fields (Stenflo 1982). Micro-turbulence, a magnetic field with an isotropic distribution and a single-valued field strength, is usually assumed in the interpretation of the spectropolarimetric data. Because of very high magnetic Reynolds number, the energy spectrum of the solar magnetic field covers a very wide range of scales. In the present paper, we study the Hanle effect due to a random magnetic field with a finite correlation length (comparable to a typical photon mean free path) and algebraic spectrum. An iterative method of solution of the ALI (Approximate Lambda Iteration) type is used to calculate the mean Stokes parameters and examine their dependence on the correlation length of the magnetic field. We also investigate the sensitivity of the polarization to the shape of magnetic field PDF (Probability Density Function).

L.S. Anusha (✉)
Indian Institute of Astrophysics, Bangalore, India
and
Laboratoire Cassiopée, Université de Nice and Observatoire de la Côte d'Azur, France

M. Sampoorna
Indian Institute of Astrophysics, Bangalore, India

H. Frisch
Laboratoire Cassiopée, Université de Nice and Observatoire de la Côte d'Azur, France

K.N. Nagendra
Indian Institute of Astrophysics, Bangalore, India

S.S. Hasan and R.J. Rutten (eds.), *Magnetic Coupling between the Interior and Atmosphere of the Sun*, Astrophysics and Space Science Proceedings,
DOI 10.1007/978-3-642-02859-5_37,

2 Transfer Equations

The random magnetic field vector $\boldsymbol{B}$ is represented by a Kubo–Anderson process (KAP). It is a Markov process, stationary, discontinuous, piecewise constant. A KAP is characterized by a correlation length $1/\nu$ (where ν is the number of jumping points per unit optical depth) and a probability density function $P(\boldsymbol{B})$ (Frisch 2006). The choice of this process allows one to write a transfer equation for a mean radiation field, still conditioned by the value of the vector magnetic field $\boldsymbol{B}$ (Frisch 2006; see also Frisch et al. 2009).

For the six-component vector $\mathcal{I}$, constructed with the six irreducible components I_Q^K describing linear polarization (Frisch 2007), this transfer equation may be written as

$$\mu \frac{\partial \mathcal{I}(\tau, x, \mu|\boldsymbol{B})}{\partial \tau} = \varphi(x)\,[\mathcal{I}(\tau, x, \mu|\boldsymbol{B}) - \mathcal{S}(\tau|\boldsymbol{B})] + \nu\Big[\mathcal{I}(\tau, x, \mu|\boldsymbol{B}) - \int \mathcal{I}(\tau, x, \mu|\boldsymbol{B}')P(\boldsymbol{B}')\,\mathrm{d}^3\boldsymbol{B}'\Big]. \quad (1)$$

This equation holds for a plane–parallel atmosphere and complete frequency redistribution. The mean conditional source vector $\mathcal{S}(\tau|\boldsymbol{B})$ is defined by

$$\mathcal{S}(\tau|\boldsymbol{B}) = \mathcal{G}(\tau) + \hat{N}(\boldsymbol{B}) \int_{-1}^{+1} \frac{\mathrm{d}\mu'}{2} \int_{-\infty}^{+\infty} \mathrm{d}x'\varphi(x')\hat{\Psi}(\mu')\mathcal{I}(\tau, x', \mu'|\boldsymbol{B}). \quad (2)$$

Here, $\mathcal{G}(\tau)$ is a given primary source term, $\hat{N}(\boldsymbol{B})$ describes the Hanle effect, and $\hat{\Psi}(\mu)$ the resonance scattering. Other notations are standard. The source vector $\mathcal{S}(\tau|\boldsymbol{B})$ satisfies an integral equation, which is solved by a Polarized ALI method (Nagendra et al. 1998) extended to random magnetic fields (Frisch et al. 2009). The mean field $\langle\mathcal{I}\rangle(\tau, x, \mu)$ is solution of a standard transfer equation in which the source term is given by the average of $\mathcal{S}(\tau|\boldsymbol{B})$ over the magnetic field PDF.

3 Results

3.1 Dependence on the Correlation Length

For intermediate values of the line optical thickness T (10–100), some sensitivity to the correlation length can be observed (Fig. 1a). For optically thin lines ($T \ll 1$) and optically thick ones ($T \geq 10^3$), the polarization is essentially independent of the magnetic field correlation length.

The very weak dependence of polarization, on the magnetic field correlation length, indicates that the polarization is created locally and can thus be estimated by a single scattering approximation. For optically thick lines, Stokes Q may then be approximated by

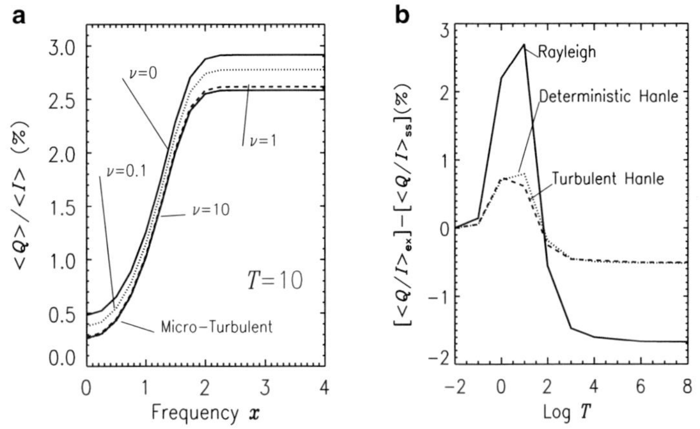

Fig. 1 (**a**) Correlation length effects on $\langle Q\rangle/\langle I\rangle$, for an exponential field strength distribution (see Table 1 1), with an isotropic angular distribution. (**b**) Difference between the single-scattering (ss) and exact (ex) solutions as a function of T, for $\tau = 0$, $x = 0$, and $\mu = 0.05$. The micro-turbulent results (*dashed curve*) are computed using the same model as for panel (**a**)

$$Q(0, x, \mu) \simeq -\frac{3}{2\sqrt{2}}(1-\mu^2)N^2_{00}(\boldsymbol{B})\bar{J}^2_0\left(\frac{\mu}{\varphi(x)}\right). \tag{3}$$

The function $\bar{J}^2_0\,(\mu/\varphi(x))$ is a measure of the anisotropy of the radiation field, and $N^2_{00}(\boldsymbol{B})$ is the element of the magnetic kernel coupling Stokes I to Stokes Q (or more precisely I^0_0 to I^2_0). Figure 1b shows the difference between the exact value of Q/I and the single scattering solution for several choices of the line thickness T and magnetic field. The single scattering approximation appears fairly reliable, in particular for optically thin and thick lines, for both turbulent Hanle and deterministic Hanle scattering problems.

3.2 Dependence on the Magnetic Field Vector PDF

The ratio $\langle Q\rangle/\langle I\rangle$ is calculated in the micro-turbulent limit for PDFs of the form

$$P(\boldsymbol{B})\,\mathrm{d}^3\boldsymbol{B} = P_{\mathrm{S}}(B)P_{\mathrm{A}}(\theta_{\mathrm{B}}, \chi_{\mathrm{B}})\,\mathrm{d}B\,\mathrm{d}\theta_{\mathrm{B}}\,\frac{\mathrm{d}\chi_{\mathrm{B}}}{4\pi}. \tag{4}$$

Here, B is the random field strength and (θ_{B},χ_{B}) are the polar angles of the field. The surface value of $\langle Q\rangle/\langle I\rangle$ is shown in Fig. 2b for an isotropic magnetic field ($P_{\mathrm{A}}(\theta_{\mathrm{B}}, \chi_{\mathrm{B}}) = \sin\theta_{\mathrm{B}}$), and different choices of the field strength distribution $P_{\mathrm{S}}(B)$ (see Table 1 and Fig. 2a). They were inspired by PDFs proposed in the literature (see e.g., Vögler et al. 2005; Stenflo and Holzreuter 2003; Sánchez Almeida 2007; Sampoorna et al. 2008). We observe that the polarization strongly depends on the choice of $P_{\mathrm{S}}(B)$. It grows when the probability of having very weak fields increases. A similar conclusion is suggested in Trujillo Bueno et al. (2004).

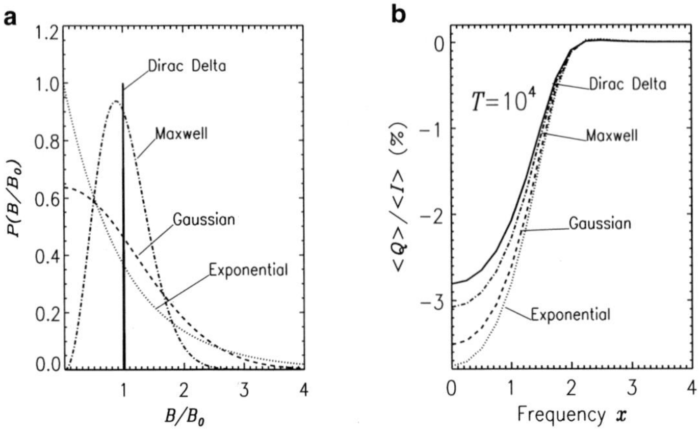

Fig. 2 (**a**) Various $P(B/B_0)$ as a function of (B/B_0). (**b**) The corresponding emergent $\langle Q\rangle/\langle I\rangle$ profiles for $\mu = 0.05$

Table 1 Field strength PDFs with mean field B_0 used in this paper[a]

$P_D(B) = \delta(B - B_0)$	$P_M(B) = \frac{32}{\pi^2 B_0}(B/B_0)^2 \exp\left[-\frac{4}{\pi}(B/B_0)^2\right]$
$P_G(B) = \frac{2}{\pi B_0}\exp\left[-\frac{1}{\pi}(B/B_0)^2\right]$	$P_E(B) = \frac{1}{B_0}\exp(-B/B_0)$

[a]D, Dirac delta; G, Gaussian; M, Maxwell; E, exponential

4 Conclusions

The choice of the magnetic field strength distribution strongly affects the mean magnetic field strength deduced from the Hanle effect analysis. The sensitivity to the correlation length is in general weak, and in most cases of astrophysical interest, micro-turbulence can be safely assumed. The sensitivity to the correlation length is related to the number of scattering events that are needed to create the emergent polarization. In the presence of random magnetic fields, the single scattering approximation is applicable, especially for lines with small and large optical depths. For Rayleigh scattering, contributions from higher orders of scattering become important so that the single scattering approximation cannot be used except for optically thin lines.

References

Frisch, H. 2006, A&A, 446, 403
Frisch, H. 2007, A&A, 476, 665
Frisch, H., Anusha, L. S., Sampoorna, M., Nagendra, K. N., A&A, accepted
Nagendra, K. N., Frisch, H., Faurobert-Scholl, M. 1998, A&A, 332, 610
Sampoorna, M., Nagendra, K. N., Frisch, H., Stenflo, J. O. 2008, A&A, 485, 275

Sánchez Almeida, J. 2007, ApJ, 657, 1150
Stenflo, J. O. 1982, Solar Phys., 80, 209
Stenflo, J. O., Holzreuter, R. 2003, Astron. Nachrichten, 324, 397
Trujillo Bueno, J., Shchukina, N., Asensio Ramos, A. 2004, Nature, 430, 326
Vögler, A., Shelyag, S., Schüssler, M., et al. 2005, A&A, 429, 335

Phase III of the USO Solar Vector Magnetograph

S. Gosain and P. Venkatakrishnan

Abstract The solar vector magnetograph (SVM) is a modern imaging spectropolarimeter installed at Udaipur Solar Observatory (USO). Earlier phases saw the development of the instrument using off-the-shelf components with in-house software development. Subsequently, improvements were done in the opto-mechanical design of the sub-systems and the telescope tracking system. The third phase of the instrument development saw three major improvements: (1) installation of a web-camera-based telescope guiding system, developed in-house, (2) high-cadence spectropolarimetry using liquid-crystal variable retarders and a fast CCD camera, and (3) inclusion of the Na I D_1 line for chromospheric observations, in addition to the regularly used photospheric Fe I 6302 Å line.

1 Introduction

The solar vector magnetograph (SVM) is a filter-based imaging Stokes polarimeter. It takes monochromatic pictures of active regions using a tunable, narrow pass-band Fabry-Perot (FP) filter with a full width at half maximum (FWHM) of about 120 mÅ at 6,300 Å. Stokes spectra are constructed for every pixel in the field of view by scanning the FP filter. Polarization modulation was done with quarter waveplates during phases I and II, but for phase III liquid crystal variable retarders (LCVRs) are planned. The details of the instrument design parameters, hardware, and software subsystems can be found in Gosain et al. (2004, 2006, 2008). The main features of the instrument are (1) straight and symmetric optical design, (2) direct pointing towards the Sun with no oblique reflections in the entire optical path and therefore minimal instrumental polarization, and (3) dual-beam polarization analyzer to minimize seeing effects.

S. Gosain (✉) and P. Venkatakrishnan
Udaipur Solar Observatory, Udaipur, India

S.S. Hasan and R.J. Rutten (eds.), *Magnetic Coupling between the Interior and Atmosphere of the Sun*, Astrophysics and Space Science Proceedings,
DOI 10.1007/978-3-642-02859-5_38,

2 Recent Developments

Webcam-based Guider System. A low-cost webcam-based guider system was developed for SVM. The telescope mount is equatorial equipped with an MKS-4000 telescope control system. The guider-scope produces a full-disk solar image on the webcam. Interactive software was developed in-house to grab images from the webcam and compute RA and DEC drifts due to telescope tracking errors. The errors are corrected by sending TTL pulses to the telescope drive. The residual rms guiding error is about $1''$.

Comparison with HINODE/SP Observations. Comparison of spectropolarimetric observations of a sunspot penumbra was carried out between HINODE SOT/SP and SVM. In spite of dilution of the polarization signals in the SVM observations due to the combined effects of lower spatial and spectral resolution, leakage from adjacent Fabry–Perot transmission channels and the smearing due to seeing Milne-Eddington inversions of the two data-sets with the MELANIE inversion code (Socas-Navarro 2001) give similar values for the magnetic parameters (Table 1). These initial results indicate that full-Stokes inversion of SVM observations can yield reliable magnetic field parameters. A more detailed comparison of the two instruments for a larger field-of-view is being carried out and will be reported in a forthcoming paper.

3 Ongoing Developments

LCVR-based Polarization Modulator. LCVR modulators have been procured from Meadowlark Optics and integrated in the SVM optical scheme. The calibration of the retardation voltages for different wavelengths and temperatures is underway.

Automated Polarimeter Calibration Unit. The calibration of the quarter-waveplate polarimeter of phases I and II was done manually following the POLIS scheme (Beck et al. 2005). Currently this scheme is being automated for regular calibration.

Fast CCD Camera and Na I D_1 Line. Our earlier camera, an Apogee6E, had only a slow framerate of 1/s. To take advantage of the faster LCVR modulators, a new CCD camera "Sensicam" from PCO was procured and will be integrated into the SVM during phase III. This will also make the spectral scanning faster. Furthermore, chromospheric vector-field measurement is planned using the Na I D_1 line. The filters have been procured and calibration of LCVRs for this line is underway.

Table 1 Comparison of inversion results

Parameter	SVM Udaipur	HINODE/SOT-SP
B (Gauss)	$1{,}302 \pm 200$	$1{,}277 \pm 120$
Inclination (degree)	126 ± 12	120 ± 8
Straylight	0.38 ± 0.13	0.0 ± 0.0

4 Conclusions

The comparison with HINODE/SP observations shows that reliable magnetic field parameters can be derived with the SVM. The upgrade to higher cadence will make the observations very valuable for studies of magnetic field evolution in relation to flares, of line profile changes during flares, of high-frequency magnetic oscillations and waves in active regions, and of the evolution of magnetic energy and helicity in regions with rapid flux emergence.

References

Beck, C., Schmidt, W., Kentischer, T., Elmore, D. 2005, A&A, 437, 1159
Gosain, S., Venkatakrishnan, P., Venugopalan, K. 2004, Exp. Astron., 18, 31
Gosain, S., Venkatakrishnan, P., Venugopalan, K. 2006, J. Astrophys. Astron., 27, 285
Gosain, S., Tiwari, S., Joshi, J., Venkatakrishnan, P. 2008, J. Astrophys. Astron., 29, 107
Socas-Navarro, H. 2001, In: Advanced Solar Polarimetry – Theory, Observation, and Instrumentation, M. Sigwarth (ed.), ASP Conf. Ser., vol. 236, p. 487

Revisit of the Classic Wilson Effect

S.P. Bagare

Abstract We have examined the Wilson effect in 580 epochs of sunspots observed at the Kodaikanal Observatory. Only unipolar spots that are well isolated from opposite polarity spots or pores (49%) exhibit the classic Wilson effect. The larger the isolation, the stronger is the effect. Unipolar spots also show a decrement of the effect that seems to be related to age. However, bipolar spots (45%) do not display the Wilson effect, while spots associated with a mixed-polarity magnetic configuration (6%) show an inverse effect. We are now studying dynamic responses of spot penumbrae to their ambient magnetic configuration.

1 Introduction

The classic Wilson effect is a geometric phenomenon observed in single isolated sunspots. As the spot approaches the limb, the width of the penumbra on the disk-center side decreases more rapidly than the width on the limb-ward side. This effect is ascribed to a saucer-shaped depression of the spot due to increased transparency of the sunspot atmosphere owing to its low temperature and gas pressure (Thomas and Weiss 2008). Summarizing earlier observations, Gokhale and Zwaan (1972) found the depression to be 400–800 km. However, most studies, especially those that involved observations of large numbers of sunspots (Chistyakov 1962; Collados et al. 1987), reported a very significant percentage of cases with no Wilson effect or with an inverse Wilson effect. These results have generally been ignored as being attributable to nonuniformity in the shape of penumbrae or various difficulties in the actual measurement of the effect (McIntosh 1981).

There have also been many reports on the presence of asymmetries in the type and extent of the observed Wilson effect between the eastern and the western hemispheres. The presence of the inverse Wilson effect and these east-west asymmetries were ascribed to a tilt of the spot axis, or to intrinsic elongation, or to shortening of

S.P. Bagare (✉)
Indian Institute of Astrophysics, Bangalore, India

S.S. Hasan and R.J. Rutten (eds.), *Magnetic Coupling between the Interior and Atmosphere of the Sun*, Astrophysics and Space Science Proceedings, DOI 10.1007/978-3-642-02859-5_39,

the eastern or western portions of penumbrae. Later studies involving geometry and stability considerations have shown that these arguments are untenable (Meyer et al. 1977; Hejna and Solovev 1985).

Bagare and Gupta (1998) examined 20 sunspots and concluded that the associated magnetic configuration appears to play a key role in the occurrence or absence of the classic Wilson effect. We examine this aspect in detail here.

2 Observations and Measurements

We used full-disk photoheliograms with 20-cm solar diameter that were obtained at the Kodaikanal Observatory during the years 1978–1982. Figure 1 illustrates such a full-disk Kodaikanal photoheliogram. Careful measurement of penumbral widths was done at 580 epochs of 253 sunspots located at a distance of 40°–80° from the central meridian to estimate the Wilson effect parameter f for each. The parameter f is the ratio of the penumbral width on the limb-ward side to that on the disk-center

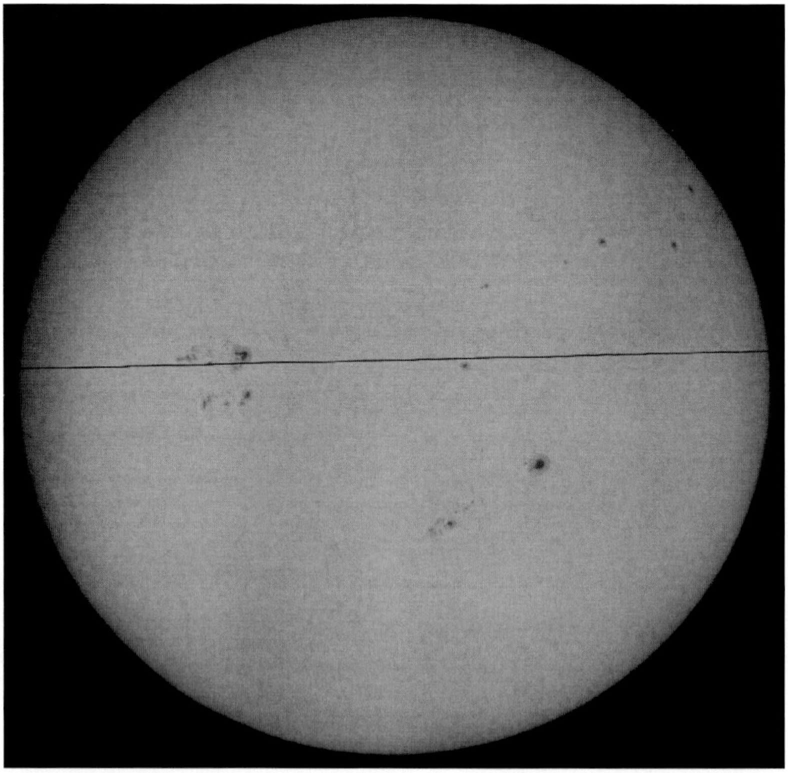

Fig. 1 Full-disk photoheliogram taken on 3 March 1982 at the Kodaikanal Observatory. The line specifies geocentric east–west as coordinate reference

side (see Bray and Loughhead 1964). Sunspots with well-defined penumbrae were selected for the study, irrespective of whether it was a single spot, or accompanied by other spots or pores, or a member of a complex group. Spots with light bridges, split umbrae, or broken or complex penumbrae were not included. The measurements were made on a high-accuracy digitizing pad. Zeeman spectra at $\lambda = 630.3$ nm, obtained with the main spectrograph at the Solar Tower Telescope at Kodaikanal, were used to evaluate the longitudinal component of the magnetic field for some of the studied sunspots. These estimations agree closely with those published in the monthly Solar Data Bulletins (SDB) of the Academy of Sciences, Leningrad, which provide daily sketches and magnetic field strengths of all sunspots and strong pores. These were used to evaluate the magnetic configurations associated with the sunspots.

3 Results and Conclusion

The classic Wilson effect is displayed in 49% of our samples but with a wide range of f-values (1.2–3.2). In 45% of the cases they showed no Wilson effect (with f-values in the range 0.8–1.2, which is the range for most of the sampled sunspots when they passed around the central meridian). The remaining 6% of the spots exhibited significant inverse Wilson effect (f-values in the range of 0.4–0.8).

We examined the spots and these measurements, taking into account the magnetic configurations associated with the spots using the magnetic classification suggested by Bray and Loughhead (1964). We found the following to hold very convincingly:

- Unipolar sunspots of magnetic class α, α p, or α f (the α-type) invariably display the classic Wilson effect. The more isolated the spot, the larger the effect.
- Spots of magnetic class β, β p, or β f do not display the Wilson effect. The presence of an opposite polarity sunspot or a strong pore in a neighboring group seems to kill the Wilson effect.
- Spots of class $\beta\gamma$ or γ display the inverse Wilson effect. The presence of mixed polarity pores is invariably associated with the occurrence of the inverse effect. The larger the number of pores and the mixing of polarities, the larger is the inverse effect.

In Fig. 2a–c, we illustrate the above three categories of sunspots. It appears that sunspots respond dynamically, and in observable manner, to the ambient magnetic

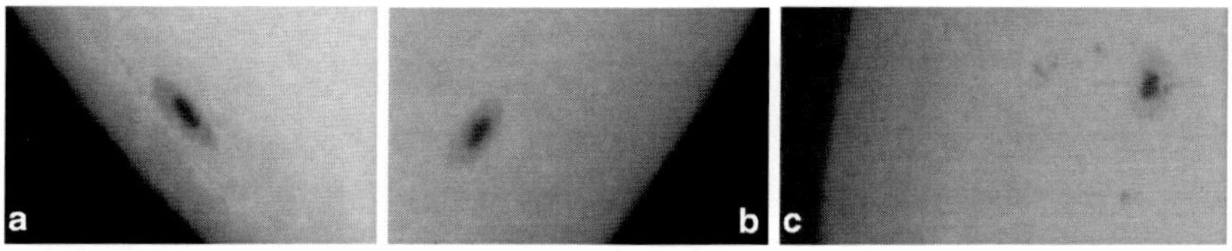

Fig. 2 Illustration of three different categories of sunspots: α-type, β-type, and γ-type, in panels **a**, **b**, and **c**, respectively

field configuration. We have examined many cases of change-over in magnetic class; the spots then promptly display the corresponding changes in their f-value. In addition, we find significant correlation between the f-value of the sunspot and the heliographic distance to the nearest opposite-polarity sunspot or pore, even up to distances of 40°.

We searched for east-west asymmetries and found that only in the case of α-type spots, there is a clear decrement of f between the eastern limb and the western limb. We defined a decrement factor and studied this in a significant number of unipolar cases. The decrement appears to be associated with the aging of the sunspot, as suggested by Bray and Loughhead (1964).

In their attempts to explain the absence of Wilson effect, or the presence of an inverse effect, some investigators have suggested that "the spot behaves as if the umbra were a cusp." However, our study of the geometry of the Wilson effect does not support this picture. Also, theoretical considerations require the presence of an umbral depression for the sunspot to be stable. In the light of these issues, we are carefully re-examining our data seeking clues on what seem to be dynamic responses of the penumbrae to the ambient magnetic configurations. We aim to publish these results shortly.

Acknowledgment The Evershed effect centenary conference provided an excellent stimulus through outstanding presentations and critical but constructive discussions.

References

Bagare, S. P., Gupta, S. S. 1998, Bull. Astron. Soc. India, 26, 197

Bray, R. J., Loughhead, R. E. 1964, In: Sunspots, Int. Astrophys. Series, Chapman and Hall, London

Chistyakov, V. F. 1962, Soviet Ast., 5, 471

Collados, M., del Toro Iniesta, J. C. , Vazquez, M. 1987, Solar Phys., 112, 281

Gokhale, M. H., Zwaan, C. 1972, Solar Phys., 26, 52

Hejna, L., Solovev, A. A. 1985, Bull. Astron. Inst. Czechosl., 36, 183

McIntosh, P. S. 1981, In: The Physics of Sunspots, L. E. Cram, J. H. Thomas (eds.), Sacramento Peak Observatory

Meyer, F., Schmidt, H. U., Weiss, N. O. 1977, MNRAS, 179, 741

Thomas, J. H. Weiss, N. O. 2008, In: Sunspots and Starspots, Cambridge Astrophysics Series 46, Cambridge University Press

The Waldmeier Effect in Sunspot Cycles

B.B. Karak and A.R. Choudhuri

Abstract We discuss two aspects of the Waldmeier Effect, namely (1) the rise times of sunspot cycles are anti-correlated to their strengths (WE1) and (2) the rates of rise of the cycles are correlated to their strengths (WE2). To avoid overlapping effect and the lack of single peaks in some cycles, we carefully define rise times to study WE1. From analysis of four different data sets, we conclude that WE1 exists in all, although the significance levels are poor in some cases due to insufficient data. We also find strong evidence for WE2 in all the data sets. We study this effect theoretically by introducing suitable stochastic fluctuations in our regular solar dynamo model. Some preliminary results are presented.

1 Introduction

The Waldmeier effect is an important feature of solar cycles. However, with "Waldmeier effect" two different measures are meant by different authors, causing confusion in the literature. The first is the observation that the rise times of sunspot cycles are anti-correlated to their strengths (i.e., the stronger cycles have shorter rise times). The second is that the rates of rise of the cycles are correlated to their strengths (i.e., the stronger cycles rise faster). Let us refer to these somewhat different correlations as WE1 and WE2.

Hathaway et al. (2002) found evidence for WE1 in both the Zürich sunspot numbers and the group sunspot numbers, while Dikpati et al. (2008) claim that this effect does not exist in sunspot area data. We believe that this discrepancy was caused mainly by not defining the rise time carefully. A new sunspot cycle sometimes begins before the end of the previous cycle, making it difficult to ascertain exactly when the cycle began. Some cycles have plateau-like maxima with multiple peaks, so that it is difficult to say when the rising phase ended. We analyze the observational data by carefully defining the rise time and the rate of rise. A theoretical study to understand aspects of the observational data is planned in addition.

B.B. Karak (✉) and A.R. Choudhuri
Department of Physics, Indian Institute of Science, Bangalore, India

S.S. Hasan and R.J. Rutten (eds.), *Magnetic Coupling between the Interior and Atmosphere of the Sun*, Astrophysics and Space Science Proceedings,
DOI 10.1007/978-3-642-02859-5_40,

2 Observational Study

We have studied four different data sets: (1) Wolf sunspot numbers (cycles 12–23), (2) group sunspot numbers (cycles 12–23), (3) sunspot area data (cycles 12–23), and (4) 10.7 cm radio flux (available only for the last 5 cycles). All data sets have been smoothed by a Gaussian filter with a FWHM of 1 year. For a cycle with amplitude P, we take the rise time to be the time during which the activity level changes from $0.2P$ to $0.8P$. The rise time defined in this way has a good anti-correlation with the cycle amplitude for all the data sets, the correlation coefficients and the significance levels for the four data types being (1) −0.50 and 90.16% for sunspot numbers; (2) −0.42 and 82.12% for group sunspots; (3) −0.31 and 67.3% for sunspot areas; and (4) −0.33 and 41% for 10.7 cm radio flux. The results for sunspot numbers and sunspot areas are shown in panels (a) and (b) of Fig. 1. These results are very sensitive to the averaging bin size. If we average the data with a FWHM of 2 year instead of 1 year, we obtain the following correlation coefficients and significance levels for the four data sets: (1) −0.63 and 97.28%; (2) −0.60 and 96.13; (3) −0.36 and 75.38%; and (4) −0.67 and 78.34. If we calculate the rise time differently by taking the end of rise phase to be the time when the cycle reaches a strength of $0.7P$ or $0.9P$ (rather than $0.8P$), then the correlation coefficients also change, having a value of about −0.7 in one case comparable to what Hathaway et al. (2002) reported.

We also study the second Waldmeier effect WE2 in all four data sets. We calculate the rate of rise by determining the slope between two points at a separation of 1 year, with the first point 1 year after the sunspot minimum. We find strong correlation between the rates of rise and the amplitudes of the sunspot cycles.

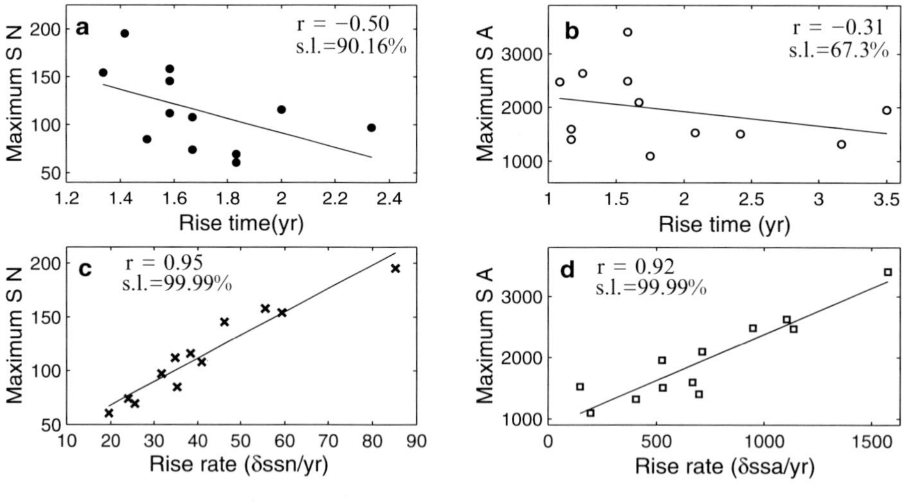

Fig. 1 Observational evidences for WE1 (*upper row*) and WE2 (*lower row*). The upper row shows scatter diagrams plotting the peak values of (**a**) sunspot number and (**b**) sunspot area against rise times. The lower row shows scatter diagrams plotting the peak values of (**c**) sunspot number and (**d**) sunspot area against rise rates. The quantities r and s.l. are the correlation coefficient and its significance level

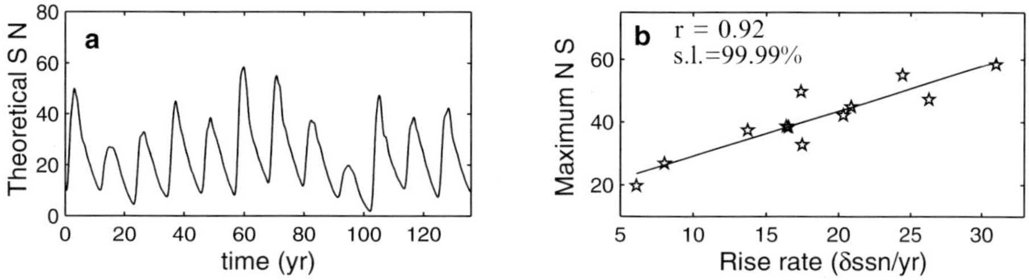

Fig. 2 (**a**) Theoretical sunspot numbers vs. time. (**b**) Scatter diagram plotting peak sunspot numbers against rise rates

Results for sunspot number and sunspot area are shown in panels (c) and (d) of Fig. 1. Cameron and Schussler (2008) have computed the rise rate slightly differently and obtained almost similar results.

We conclude that there is evidence for both WE1 and WE2 in different kinds of data sets.

3 Preliminary Theoretical Results

We plan to carry out a theoretical study based on our flux transport dynamo model (Nandy and Choudhuri 2002; Chatterjee et al. 2004) to explain the Waldmeier effect. We believe that the fluctuations in the Babcock–Leighton process of poloidal field generation is the main source of irregularities in solar cycles (Choudhuri et al. 2007; Jiang et al. 2007). Variations in meridional circulation are likely to introduce additional irregularities. We are investigating whether the cycle irregularities produced theoretically in this manner show both effects WE1 and WE2. Preliminary results of WE2 are shown in Fig. 2.

References

Cameron, R., Schüssler, M. 2008 ApJ, 685, 1291
Chatterjee, P., Nandy, D., Choudhuri, A. R. 2004, A&A, 427, 1019
Choudhuri, A. R., Chatterjee, P., Jiang, J. 2007, Phys. Rev. Lett., 98, 131103
Dikpati, M., Gilman, P. A., de Toma, G. 2008, ApJ, 673, L99
Hathaway, D. H., Wilson, R. M., Reichmann, E. J. 2002, Solar Phys., 211, 357
Jiang, J., Chatterjee, P., Choudhuri, A. R. 2007, MNRAS, 381, 1527
Nandy, D., Choudhuri, A. R. 2002, Sci, 296, 1671

Flare-Driven Acoustic Modes in the Sun

B. Kumar and P. Venkatakrishnan

Abstract We have analyzed disk-integrated velocity signals at the solar surface obtained from full-disk Doppler images as observed with MDI (Michelson and Doppler Imager) on board SoHO (Solar and Heliospheric Observatory) to study the effect of solar flares on acoustic velocity oscillations. It is seen that during a flare these oscillations are enhanced significantly in the higher frequency band, beyond the acoustic cut-off-frequency in the solar photosphere of about 5 mHz, while there is feeble to no enhancement of these oscillations in the 5-min (3.3 mHz) band. Enhancement in the high-frequency component of the acoustic spectrum of the Sun during major flares has also been reported recently in disk-integrated intensity observations of the Sun.

1 Introduction

Solar flares are known to release a large amount of energy. Wolff (1972) suggested that large solar flares can stimulate free modes of oscillation of the entire Sun. The assumed mechanism is that flares would cause a thermal expansion that would act as a mechanical impulse by causing a wave of compression to move sub-sonically into the solar interior. Several researchers have tried to study the effect of flares on the acoustic velocity oscillations of the Sun since the discovery of the 5-min oscillations of the solar surface in the 1960s.

The progress in this field was relatively slow in the beginning, but it escalated more recently with the advent of continuous data from instruments such as MDI (Michelson and Doppler Imager) on board SoHO (Solar and Heliospheric Observatory) and GONG (Global Oscillation Network Group). First of all, Haber et al. (1988) reported an average increase in the power of intermediate degree modes after a flare. However, Braun and Duvall (1990) could not detect acoustic-wave excitation from an X-class flare. Using the SoHO/MDI Dopplergrams, Kosovichev and Zharkova (1998) reported the first detection of "solar quakes" inside the Sun,

B. Kumar (✉) and P. Venkatakrishnan
Udaipur Solar Observatory, Physical Research Laboratory, Udaipur, India

S.S. Hasan and R.J. Rutten (eds.), *Magnetic Coupling between the Interior and Atmosphere of the Sun*, Astrophysics and Space Science Proceedings,
DOI 10.1007/978-3-642-02859-5_41,

observed during the X2.6 flare of 9 July 1996. Following this result, Donea et al. (1999) found an acoustic source associated with a flare using seismic images produced with helioseismic holography techniques. They found considerable flare-induced acoustic power in the 3.5 mHz band. The flare also induced power in the 6 mHz band. Application of ring-diagram analysis to several flare-producing active regions showed that the power of the global p-modes appears to be larger compared with the power in non-flaring regions of similar magnetic field strength (Ambastha et al. 2003).

Additionally, Donea and Lindsey (2005) have reported emission of seismic waves from large solar flares using helioseismic holography. Some of the large solar flares have been observed to produce enhanced high frequency acoustic velocity oscillations in localized parts of active regions (Kumar and Ravindra 2006). Further, Kumar (2007) reported large downflows associated with major flares appearing at these localized positions. Such large flows appeared during the rise time of the soft X-ray flux from the Sun observed with GOES. One such event is illustrated here in Fig. 1.

Venkatakrishnan et al. (2008) have shown that there was co-spatial evolution of seismic sources and H-alpha flare kernels during the major flare of 28 October 2003. Recently, Karoff and Kjeldsen (2008) have found that the correlation between X-ray flare intensity and the energy in the acoustic spectrum of disk-integrated

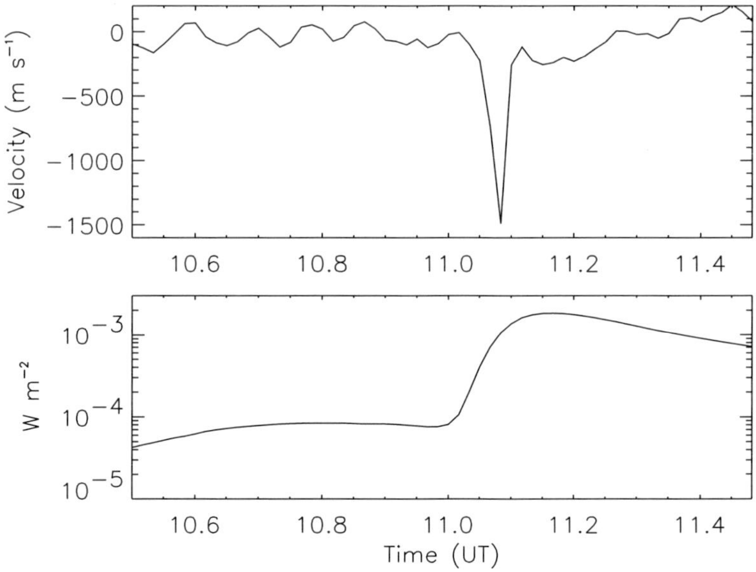

Fig. 1 The *upper panel* shows the velocity flows in localized positions of active region NOAA 10486 for the time before, during, and after the flare on 28 October 2003. The *lower panel* is the soft X-ray flux detected by the GOES satellite. It is evident that large downflow is triggered in the active region during the rise time of the soft X-ray flux emitted during the flare. Taken from Kumar (2007)

intensity oscillations (as observed with VIRGO onboard SoHO) is stronger for high-frequency waves than for the well known 5-min oscillations. In this study, we have searched for similar effects in disk-integrated velocity signals from the solar surface for the major solar flare (X17.6/4B) of 28 October 2003, using the full-disk Dopplergrams obtained with SoHO/MDI. This study will be extended to other large solar flares of cycle 23.

2 Data and Analysis

In regular observing mode, SoHO/MDI obtains full-disk Doppler images of the solar photosphere at a cadence of one per minute and with a spatial sampling rate of 2 arcsec per pixel. We have used a sequence of MDI Dopplergrams obtained on 28 October 2003 during 10:00–12:00 UT, examining the nature of the global solar velocity oscillations in quiet and in flaring conditions. To replicate the disk-integrated intensity observations by SoHO/VIRGO, we have integrated the velocity signals from the SoHO/MDI full-disk Dopplergrams. These Doppler images are collapsed into single velocity value, excluding the noisy pixels along the solar limb. This process is applied to the time series of Doppler images from an hour before the flare and then for an hour spanning the flare. It is believed that by collapsing the full-disk Doppler images, the acoustic modes with maximum $l = 0,1,2,3$ remain while the modes higher than these are averaged out. Thus, the collapsed velocity value should be representative of the global acoustic modes.

The temporal evolution of the disk-integrated velocity signals during the 2 h are shown in the upper panel of Fig. 2. They have been Fourier-transformed to estimate the power spectra shown in the lower panel of Fig. 2. The velocity oscillations are enhanced significantly during the flare in the higher frequency band (5–6.5 mHz), while there is weak or no enhancement in the lower frequency band (2–4 mHz). Thus, the disk-integrated velocity power spectrum also illustrates the presence of high frequency oscillations in addition to the dominant 5-min oscillations of the Sun, even in non-flaring conditions.

3 Discussion and Conclusions

The existence of high-frequency acoustic waves was first discovered in high-degree observations at the Big Bear Solar Observatory (Libbrecht and Kaufman 1988, Libbrecht 1988) and later in GOLF low-degree disk-integrated observations (Garcia et al. 1998). Recently, they have also been seen in BiSON disk-integrated radial-velocity data (Chaplin et al. 2003) and VIRGO intensity data (Jiménez et al. 2005). The presence of high frequency power obtained from disk-integrated SoHO/MDI full-disk Dopplergrams is in good agreement with the above findings using disk integrated intensity/velocity observations. The enhancement of high frequency power

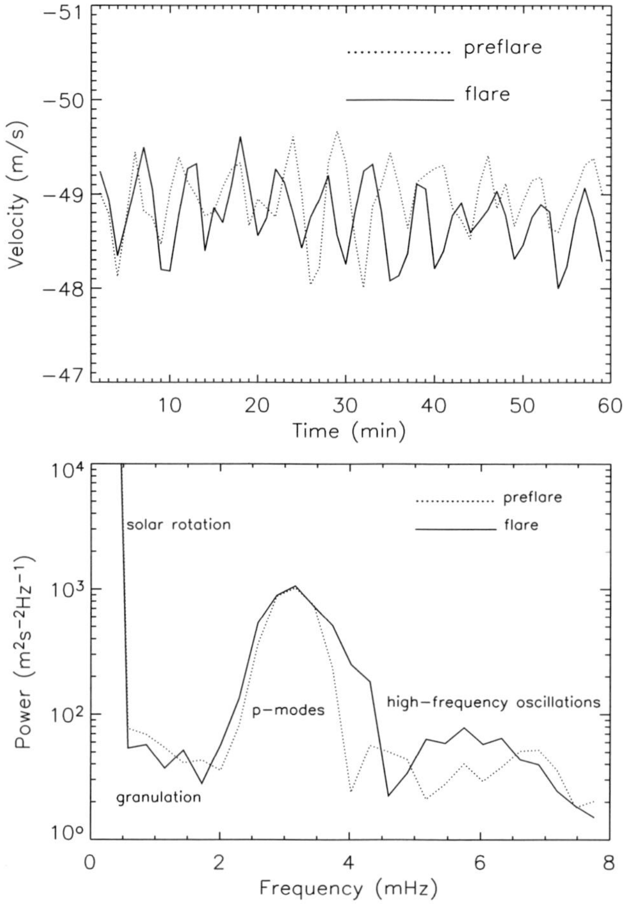

Fig. 2 The *upper panel* shows the temporal evolution of disk-integrated velocity measurements from SoHO/MDI during the hour before the flare (*dotted curve*) and for the hour spanning the flare (*solid curve*) for the flare of 28 October 2003. The *bottom panel* shows the smoothed corresponding power spectra

in these velocity oscillations during the major solar flare (X17.6/4B) of 28 October 2003 is comparable with the flare related enhancements reported by Karoff and Kjeldsen (2008) in disk-integrated intensity oscillations as observed with VIRGO.

Two models have been proposed to account for these low-degree high-frequency oscillations with frequencies higher than the atmospheric acoustic cut-off frequency. The first model that was originally proposed by Kumar and Lu (1991) explains the high-frequency waves as an interference phenomenon between ingoing and outgoing waves from a localized source just beneath the photosphere. The other model, which was originally proposed by Balmforth and Gough (1990), suggests that the high-frequency waves are partly reflected by the sudden change in temperature at the transition region between the chromosphere and the corona, and not at the photosphere as is the case for ordinary p-modes. The latter model is supported by observations of partial wave reflection at the transition region using time–distance helioseismology (Jefferies et al. 1997). In either case, the amount of energy that is stored in the high-frequency waves is extremely low compared to the amount of energy stored in the acoustic oscillations with frequencies below the atmospheric cut-off frequency.

These observations open area of study concerning the excitation of global high-frequency oscillations by local tremors due to major solar flares. We are in the process of analyzing more such events to establish flare-related changes in global solar oscillations.

Acknowledgment The use of data from SoHO/MDI and GOES is gratefully acknowledged. SoHO is a joint mission under cooperative agreement between ESA and NASA. We are thankful to Douglas Gough, Robertus Erdelyi, and Christophe Karoff for useful discussions.

References

Ambastha, A., Basu, S., Antia, H. M. 2003, Solar Phys., 218, 151
Balmforth, N. J., Gough, D. O. 1990, Solar Phys., 128, 161
Braun, D. C., Duvall, Jr., T. L. 1990, Solar Phys., 129, 83
Chaplin, W. J., Elsworth, Y., Isaak, G. R., et al. 2003, In: GONG+ 2002. Local and Global Helioseismology: The Present and Future, H. Sawaya-Lacoste (ed.), ESA-SP, 517, 247
Donea, A.-C., Braun, D. C., Lindsey, C. 1999, ApJ, 513, L143
Donea, A.-C., Lindsey, C. 2005, ApJ, 630, 1168
Garcia, R. A., Palle, P. L., Turck-Chieze, S., et al. 1998, ApJ, 504, L51+
Haber, D. A., Toomre, J., Hill, F. 1988, In: Advances in Helio- and Asteroseismology, J. Christensen-Dalsgaard, S. Frandsen (eds.), IAU Symposium, vol. 123, p. 59
Jefferies, S. M., Osaki, Y., Shibahashi, H., et al. 1997, ApJ, 485, L49
Jiménez, A., Jiménez-Reyes, S. J., García, R. A. 2005, ApJ, 623, 1215
Karoff, C., Kjeldsen, H. 2008, ApJ, 678, L73
Kosovichev, A. G., Zharkova, V. V. 1998, Nat, 393, 317
Kumar, B. 2007, PhD thesis, Udaipur, India.
Kumar, B., Ravindra, B. 2006, J. Astrophys. Astron., 27, 425
Kumar, P., Lu, E. 1991, ApJ, 375, L35
Libbrecht, K. G. 1988, ApJ, 334, 510
Libbrecht, K. G., Kaufman, J. M. 1988, ApJ, 324, 1172
Venkatakrishnan, P., Kumar, B., Uddin, W. 2008, MNRAS, 387, L69
Wolff, C. L. 1972, ApJ, 176, 833

Dynamics of Active Regions Revealed by Tracking of Doppler Features

M. Švanda, M. Sobotka, M. Klvaňa, and V. Bumba

Abstract We investigate the large-scale horizontal dynamics of active regions in the 23rd solar cycle. The large-scale horizontal velocity fields were measured applying the local correlation tracking (LCT) algorithm to the processed high-cadence full-resolution full-disc MDI Dopplergrams. We performed the selection of NOAA active regions in the available dataset and followed their individual evolution in time. The statistical study of this sample gives us a unique opportunity to study the dynamics of active regions at various stages of their evolution. In few cases, we found behavior that is consistent with the dynamical disconnection of sunspots from the magnetic roots.

1 Introduction

The dynamics of active regions in the solar photosphere and in the close sub-photospheric layers is closely related to the solar dynamo process. The magnetic flux emerges from beneath the surface and forms the sunspots and other active phenomena. At some point, the magnetic field starts to disperse, the sunspots disappear, and the active region dies, forming surge-like structures of the trailing polarity expanding to the solar poles, where they contribute to the solar field reversals.

The measurements of the dynamical behavior of active regions can bring new insights in what is going on with the magnetic field under the surface. In particular, some ideas supported by the theory (Fan et al. 1994; Schüssler and Rempel 2005) suggested that at some point, bipolar magnetic regions may disconnect from their magnetic roots and form an isolated island-like feature. The mechanism is based on the buoyant upflow of plasma along the field lines. Such flows arise in the upper

M. Švanda (✉)
Ondřejov Observatory, Academy of Sciences, Ondřejov, Czech Republic
and
Astronomical Institute, Charles University, Prague, Czech Republic

M. Sobotka, M. Klvaňa, and V. Bumba
Ondřejov Observatory, Academy of Sciences, Ondřejov, Czech Republic

S.S. Hasan and R.J. Rutten (eds.), *Magnetic Coupling between the Interior and Atmosphere of the Sun*, Astrophysics and Space Science Proceedings,
DOI 10.1007/978-3-642-02859-5_42,

part of a rising flux loop during the final phases of its buoyant ascent towards the surface. The combination of the pressure build-up by the upflow and the cooling of the upper layers of an emerged flux tube by radiative losses at the surface leads to a progressive weakening of the magnetic field in several mega-meter depth. When the field strength has become sufficiently low, convective motions ablate the flux tube into thin, passively advected flux fragments, thus providing a dynamical disconnection of the emerged part from its parent magnetic structure. This instant should be observed as the change in the dynamical regime, because the floating island does not reflect the deep dynamics anymore. In this contribution, we investigate this issue in a more detail.

2 The Data and Method

In recent papers (e.g., Švanda et al. 2006), we introduced a method to measure the large-scale dynamics in the solar photosphere. It is based on tracking supergranular structures in full-disk Dopplergrams. Its application allows us to compute 24-h averaged horizontal flow fields with a resolution of 60″ at a noise level of $15\,\mathrm{m\,s^{-1}}$. In the magnetized regions, this method makes it possible to measure the apparent motion of supergranular-scale magnetic features.

In our set of 1,004 flow maps, we identified 564 active regions, of which 522 were observed in more than one flow map and 194 emerged less than 60° from the central meridian. From this sub-set we selected 69 bipolar regions that survived at least 4 days within central meridian distance less than 60°. This drastic reduction of the sample was necessary to avoid any possible bias.

For each active region in the sample, we computed the large-scale flow field in the area, the average rotation speed, the magnetic field area (with threshold of 100 G), and the mean magnetic field intensities in both leading and trailing polarities, sampled per 12 h. We investigated the change in the dynamics that may be connected to the effect of disconnection from the magnetic roots. The expected change should display sudden deceleration because the radial gradient of the mean rotation is negative in nearby subphotospheric layers.

3 Results and Conclusions

We found that 75% of the active regions in the selected sample show the characteristic change in dynamics. These active regions display a sudden decrease in their proper rotation rate, typically within 3 days after their emergence. Dynamical disconnection from their magnetic roots is a natural explanation for this behavior. Our study suggests that this phenomenon is quite common among bipolar active regions.

We observe a systematic phase shift between the time of the dynamical regime change and the time of the maximum in the area. We interpret this as the time during

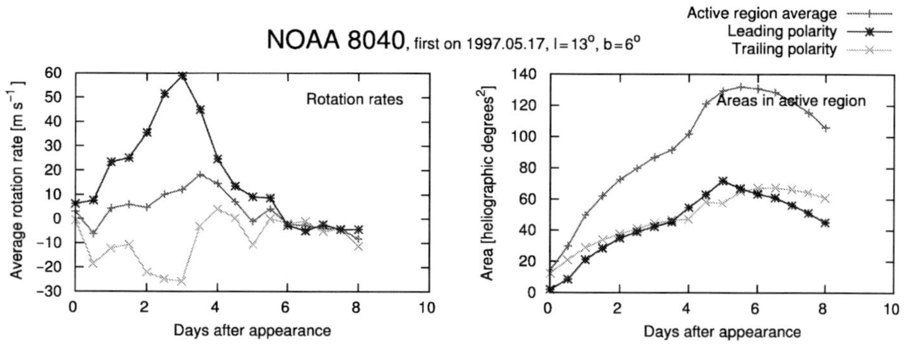

Fig. 1 The evolution of the mean rotation speed with respect to the long-term background, active region area, and the mean field in active region NOAA 8040. This is an example of the active region displaying the signature of dynamical disconnection

which the magnetic island rises from its parent magnetic structure to the surface after the disconnection. After this takes place, no more magnetic field is fed into the region and the region starts to diminish.

The systematic shift is 1.67 ± 0.72 days. Assuming a 1-kG fluxtube emerging at the Alfvén speed, the theoretical shift is 180 days if the fluxtube emerges from the bottom of the convection zone and 7 days when emerging from a subsurface shear layer at $0.95\,R_\odot$. The measured lag corresponds to the depth 17 ± 4 Mm. The Alfvén speed computation is based on model S of Christensen-Dalsgaard et al. (1996). The actual rise speed of an expanding fluxtube should not differ from the Alfvén speed in order of magnitude. The computation shows that the disconnection cannot occur deeper than some 40 Mm for a 4-kG fluxtube to reproduce the observed behavior. Although these numbers represent only a rough estimate, our results show that the disconnection takes place just a few Mega-meter below the photosphere.

We conclude that dynamical disconnection is an important part of the life of bipolar active regions. Our results favor a sub-surface shear layer as the location of origin for surface magnetic activity.

Acknowledgment This work was supported by the Grant Agency of Academy of Sciences of the Czech Republic under grant No. IAA30030808 and by research project No. MSM0021620860 of the Ministry of Education of the Czech Republic.

References

Christensen-Dalsgaard, J., Dappen, W., Ajukov, S. V., et al. 1996, Science, 272, 1286
Fan, Y., Fisher, G. H., McClymont, A. N. 1994, ApJ, 436, 907
Schüssler, M., Rempel, M. 2005, A&A, 441, 337
Švanda, M., Klvaňa, M., Sobotka, M. 2006, A&A, 458, 301

The Growth of a Primitive Penumbra

P. Sreejith, A. Tritschler, and K. Sankarasubramanian

Abstract We report on the penumbral formation in active region NOAA 10837. The penumbra observed on continuum intensity images grew from a quiet-Sun area to a primitive penumbra and then to a fully developed penumbra over about 5 h. The growth indicates nonlinear development with time.

1 Introduction

Sunspots are manifestations of strong magnetic fields that emerge in the solar photosphere (Bray and Loughhead 1964, Solanki 2003). Although sunspots are stable configurations when compared to the dynamical time scales of other features on the Sun, the observed umbral and penumbral fine-structures are very dynamic and subject to constant change and transformation on small spatial scales. Our understanding of these processes and the nature of the fine structure improved significantly in the last decade (Rimmele 2008, Brummell et al. 2008), but we still lack detailed knowledge about the key process of penumbral formation and decay. Observations of this process are very rare, and in many cases limited to imaging information only. It is known that sunspot penumbrae form relatively fast; observing them at the right time is a difficult task to accomplish. In this paper, we report on high spatial resolution observations of the transformation of a primitive penumbra into a penumbral segment in the follower spot of active region NOAA 10837. Continuum images with a time cadence of 1 min were used to see the growth of this penumbrae.

P. Sreejith (✉)
ISRO Satellite Centre, Bangalore, India
and
University of Calicut, Kerala, India

A. Tritschler
National Solar Observatory, Sunspot, NM, USA

K. Sankarasubramanian
ISRO Satellite Centre, Bangalore, India

S.S. Hasan and R.J. Rutten (eds.), *Magnetic Coupling between the Interior and Atmosphere of the Sun*, Astrophysics and Space Science Proceedings,
DOI 10.1007/978-3-642-02859-5_43,

2 Observations

The observations were performed on 22 December 2005 at the 0.76 m Dunn Solar Telescope (DST) in Sunspot, New Mexico, USA. The observed active region was at S11 W41, with viewing angle $\mu = \cos\theta = 45.5°$. The high-order adaptive optics system at the DST (Rimmele et al. 2004) fed a seeing-corrected beam to a set of back-end instruments. The seeing during the observation was variable, very good occasionally but average most of the time. The main back-end instruments used for this observation were the Diffraction-Limited SpectroPolarimeter (DLSP; Sankarasubramanian et al. 2006) and the Universal Birefringent Filter (UBF). The UBF is a tunable Lyot filter with a pass-band that varies between 120 and 250 mÅ in the visible from 5,000 to 7,000 Å. The UBF was tuned to Fe I 5434 Å and Hα. The iron line is a magnetically insensitive line with Landé factor $g = 0$. We also recorded G-band and Ca II K images on large-format CCDs.

We observed this during about 5 h. Only good images were kept for further analysis, choosing these from the contrast in a quiet-Sun region. We cross-checked this selection manually.

The UBF images were corrected for dark and flat fields. Residual differential image motion due to seeing was corrected using a destretch algorithm (November and Simon 1988). The UBF filtergrams were then carefully aligned with each other and were scaled down to 0.09 arcsec pixel^{-1} in order to match the DLSP images. Here we present preliminary results from the UBF continuum images only.

3 Results and Conclusions

Figure 1 shows UBF continuum images at three different stages of the penumbral formation. Two penumbral forming areas are marked as P1 and P2. The time of observation, in UT, is specified at the top of each image. The intensity range is specified below the right-hand part.

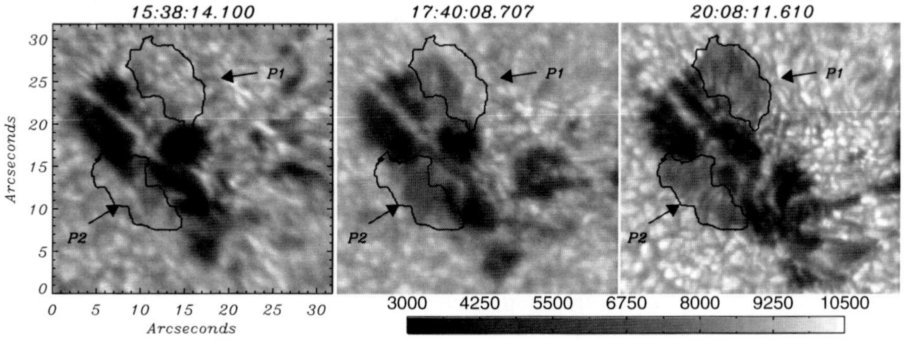

Fig. 1 UBF continuum images at three different stages of penumbral formation

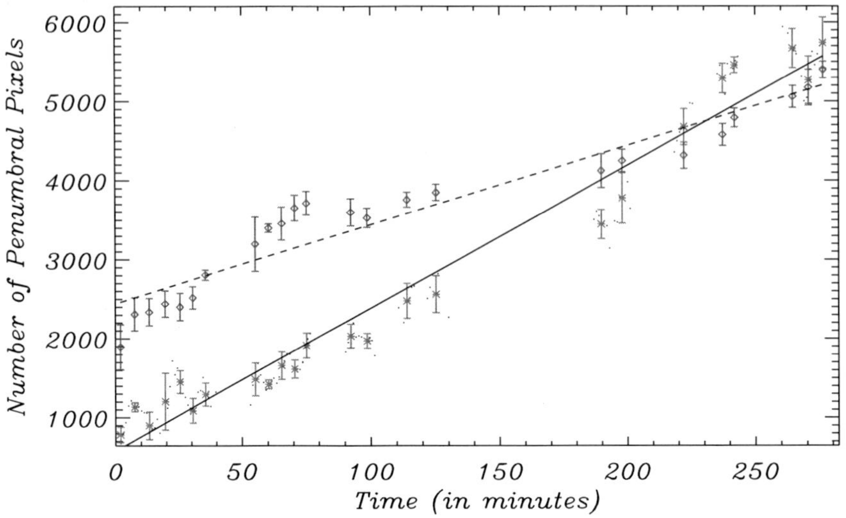

Fig. 2 Penumbral area in number of pixels against time. The lower measurements (*asterisks*) correspond to area P1, the upper measurements (*diamonds*) to area P2

Usually, histograms of the continuum intensity over a region with a mature sunspot show a triple-peaked distribution with the peaks corresponding to the umbra with $I = (0.2 - 0.3)\, I_{\mathrm{QS}}$, the penumbra with $I \approx 0.8\, I_{\mathrm{QS}}$, and quiet Sun with $I \equiv I_{QS}$ (e.g., Leka and Skumanich 1998). All pixels that contribute to the second histogram peak, that is, the intensity range $I = (0.7 - 0.9)\, I_{\mathrm{QS}}$, are defined as penumbral for this study. The two areas P1 and P2 were selected manually. The number of pixels classified as penumbra was evaluated as function of time and binned over 5-min bins to reduce the noise. These mean penumbral areas are plotted in Fig. 2 for areas P1 and P2 with linear fits.

Figure 2 shows that the actual measurements deviate from the straight line fits for P2 more than for P1. Higher-order polynomials fit the data better than the linear fits. This suggests that the growth rate of the penumbral area is nonlinear, as predicted earlier by, for example, Brummell et al. (2008). Although many theoretical models have been formulated to explain penumbral formation (Brummell et al. 2008, Spruit and Scharmer 2006, Schlichenmaier et al. 1998), none discusses the rate of penumbral evolution.

A detailed numerical simulation of penumbrae may provide more information on the exact nature of the growth behavior. It should also be pointed out that the effect of the difference in line-of-sight integration between these two penumbra (one samples the limb-side penumbra, the other the center-side penumbra) is not clear. The other physical parameters of this particular observation, especially vector field and velocity maps, will be inspected to obtain for more detailed information about the growth rate of this penumbra.

Acknowledgment This observations were done at National Solar Observatory (NSO), Sunspot, NM, USA. We thank the observers for all their help during the observations.

References

Bray, R. J., Loughhead, R. E. 1964, Sunspots, Chapman and Hall, London

Brummell, N. H., Tobias, S. M., Thomas, J. H., Weiss, N. O. 2008, ApJ, 686, 1454

Leka, K. D., Skumanich, A. 1998, ApJ, 507, 454

November, L. J., Simon, G. W. 1988, ApJ, 333, 427

Rimmele, T. 2008, ApJ, 672, 684

Rimmele, T. R., Richards, K., Hegwer, S., et al. 2004, In: SPIE Conf. Ser., S. Fineschi, M. A. Gummin (eds.), SPIE, vol. 5171, p. 179

Sankarasubramanian, K., Lites, B., Gullixson, C., et al. 2006, in Fourth Solar Polarization Workshop, eds. R. Casini & B. W. Lites, ASP Conf. Ser., 358, 201

Schlichenmaier, R., Jahn, K., Schmidt, H. U. 1998, A&A, 337, 897

Solanki, S. K. 2003, A&A Rev., 11, 153

Spruit, H. C., Scharmer, G. B. 2006, A&A, 447, 343

Evershed Flow Velocities During 100 Years

K.M. Hiremath

Abstract An extensive literature survey indicates that the peak velocities of the Evershed flows in sunspots at the photospheric level are independent of the strength of the solar activity cycle. This result is discussed in the context of the existing models for the cause of the Evershed flow.

1 Introduction

Nearly 100 years ago, John Evershed discovered at the Kodaikanal Observatory from spectroscopic observations that there is a stationary radial outflow of matter from the umbra–penumbra boundary, with a peak velocity of about 2 km s^{-1} and abruptly ending at the penumbral boundary. Many observations after Evershed's discovery have confirmed this phenomenon unambiguously. This discovery is called the Evershed effect (e.g., Schröter 1967; Sivaraman 1984; Solanki 2003; Thomas and Weiss 2004; Ichimoto 2010; Thomas 2010) and remains one of the unsolved problem in solar physics.

The sun is a variable star with sunspot activity varying on a time scale of 11 years; when one considers the polarity of the sunspots, the sun experiences a 22-year magnetic cycle. Observations show that the magnetic field strength of sunspots varies with the solar activity level (Penn and Livingston 2006). As the spot structure and dynamics mainly depend upon the strength of the magnetic field, one would expect that these physical variables, especially the speed of the Evershed flow, must be a function of solar activity. We test this expectation with an extensive compilation of Evershed flow measurements dating back to its discovery.

K.M. Hiremath (✉)
Indian Institute of Astrophysics, Bangalore, India

S.S. Hasan and R.J. Rutten (eds.), *Magnetic Coupling between the Interior and Atmosphere of the Sun*, Astrophysics and Space Science Proceedings, DOI 10.1007/978-3-642-02859-5_44, © Springer-Verlag Berlin Heidelberg 2010

2 Data and Results

Figure 1 shows peak velocities of the photospheric Evershed flow against the sunspot number for the year of observation, compiled from many studies[1].

Contrary to expectation, the figure shows no indication that the Evershed peak velocity depends on the level of solar activity.

3 Discussion and Conclusion

Obviously, these peak velocities are an inhomogeneous collection. They sample different epochs in spot life span and different spot latitudes. Assuming that these factors play no role, the existing theories for the origin of the Evershed flow are examined for consistency with the result in Fig. 1. Except for the siphon flow model (Montesinos and Thomas 1997), most of the models predict that the velocity of the Evershed flow depends on the level of solar activity (Maltby and Eriksen 1967;

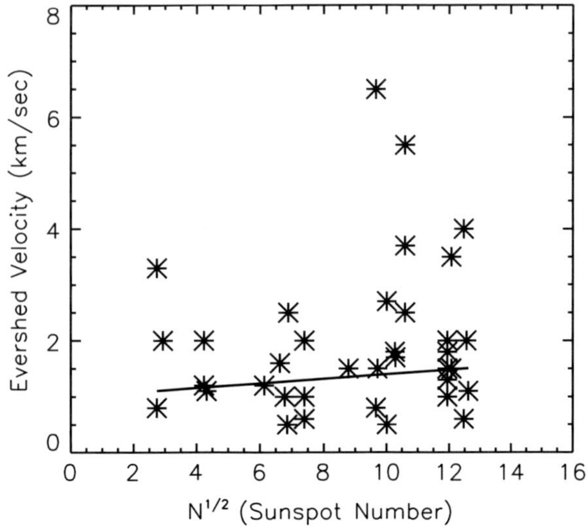

Fig. 1 The variation of the peak velocity of the photospheric Evershed flow with solar activity. The line is a least-squares fit

[1] Evershed (1909a); St. John (1913); Julius (1925); Maltby (1960, 1964); Servajean (1961); Holmes (1961, 1963); Bhatnagar (1964, 1967); Schröter (1965); Maltby and Eriksen (1967); Adler (1968); Mamadazimov (1972); Lamb (1975); Dialetis et al. (1985); Wiehr et al. (1986); Ichimoto (1987); Alissandrakis et al. (1988); Balthasar (1990); Dere et al. (1990); Adam and Petford (1991); Boerner and Kneer (1992); Wiehr and Degenhardt (1992); Solanki et al. (1993, 1994); Degenhardt and Wiehr (1994); Shine et al. (1994a,b); Rimmele (1994, 1995); Stanchfield et al. (1997); Bai et al. (1998); Rüedi et al. (1998, 1999); Hirzberger and Kneer (2001); Bellot Rubio et al. (2003); Deng et al. (2007); Ichimoto et al. (2007).

Chitre 1968; Gokhale and Hiremath 1986; Heinemann et al. 2007; Scharmer et al. 2008). Our model (Gokhale and Hiremath 1986), which stresses the importance of the overall equilibrium of the sunspot flux tube, implying a net upflow near the umbra–penumbra boundary and a weak downflow in the umbra, is consistent with recent high-resolution observations (Stanchfield et al. 1997; Hirzberger and Kneer 2001; Deng et al. 2007).

In conclusion, our current understanding of the physics of the Evershed effect is similar to Evershed's (1909b) own conclusion a hundred years ago: *"It must be admitted that the discovery of this radial movement, while descriptive of current theories, cannot be said to bring enlightenment as to the origin of this movement, yet no other interpretation of the observed line displacements seems possible."*

Acknowledgment The author is thankful to R. J. Rutten for suggestions that improved this manuscript and to M. H. Gokhale for useful discussions.

References

Adam, M. G., Petford, A. D. 1991, Solar Phys., 135, 319
Adler, S. M. 1968, Astron. J., 73, 53
Alissandrakis, C. E., Dialetis, D., Mein, P., Schmieder, B., Simon, G. 1988, A&A, 201, 339
Bai, T., Scherrer, P. H., Bogart, R. S. 1998, In: Structure and Dynamics of the Interior of the Sun and Sun-like Stars, S. Korzennik (ed.), ESA-SP, vol. 418, p. 607
Balthasar, H. 1990, Solar Phys., 125, 31
Bhatnagar, A. 1964, The Evershed effect in sunspots, PhD Thesis, Agra Univ. India
Bhatnagar, A. 1967, Kodaikanal Obs. Bull., 180, A13
Bellot Rubio, L. R., Balthasar, H., Collados, M., Schlichenmaier, R. 2003, A&A, 403, L47
Boerner, P., Kneer, F. 1992, A&A, 259, 307
Chitre, S. M. 1968, Solar Phys., 4, 168
Degenhardt, D., Wiehr, E. 1994, A&A, 287, 620
Deng, N., Choudhary, D. P., Tritschler, A., et al. 2007, ApJ, 671, 1013
Dere, K. P., Schmieder, B., Alissandrakis, C. E. 1990, A&A, 233, 207
Dialetis, D., Mein, P., Alissandrakis, C. E. 1985, A&A, 147, 93
Evershed, J. 1909a, Kodaikanal Obs. Bull., 15, 63
Evershed, J. 1909b, Mem. Kodaikanal Obs., 1, part 1
Gokhale, M. H., Hiremath, K. M. 1986, Adv. Sp. Res., 6, 47
Heinemann, T., Nordlund, Å., Scharmer, G. B., Spruit, H. C. 2007, ApJ, 669, 1390
Hirzberger, J., Kneer, F. 2001, A&A, 378, 1078
Holmes, J. 1961, MNRAS, 122, 301
Holmes, J. 1963, MNRAS, 126, 155
Ichimoto, K. 1987, PASJ, 39, 329
Ichimoto, K., Suematsu, Y., Tsuneta, S., et al. 2007, Science, 318, 1597
Ichimoto, K. 2010, In: Magnetic Coupling Between the Interior and the Atmosphere of the Sun, S. S. Hasan, R. J. Rutten (eds.), Astrophys. Space Sci. Procs., Springer, Heidelberg, these proceedings
Julius, W. H. 1925, Bull. Astron. Inst. Netherlands, 2, 219
Lamb, S. A. 1975, MNRAS, 172, 205
Maltby, P. 1960, Ann. d'Astrophyss., 23, 983
Maltby, P. 1964, Astrophys. Norvegica, 8, 205
Maltby, P., Eriksen, G. 1967, Solar Phys., 2, 249

Mamadazimov, M. 1972, Solar Phys., 22, 129
Montesinos, B. Thomas, J. H. 1997, Nat, 390, 485
Penn, M. J., Livingston, W. 2006, ApJ, 649, L45
Rimmele, T. R. 1994, A&A, 290, 972
Rimmele, T. R. 1995, ApJ, 445, 511
Rüedi, I., Solanki, S. K., Keller, C. U. 1999, A&A, 348, L37
Rüedi, I., Solanki, S. K., Keller, C. U., Frutiger, C. 1998, A&A, 338, 1089
Scharmer, G. B., Nordlund, Å., Heinemann, T. 2008, ApJ, 677, L149
Schröter, E. H. 1965, ZfAp, 62, 256
Schröter, E. H. 1967, In: Solar Physics, J. N. Xanthakis (ed.), p. 325
Servajean, R. 1961, Ann. d'Astrophys., 24, 1
Shine, R. A., Title, A. M., Tarbell, T. D., et al. 1994a, In: Solar Surface Magnetism, R. J. Rutten, C. J. Schrijver (eds.), p. 197
Shine, R. A., Title, A. M., Tarbell, T. D., et al. 1994b, ApJ, 430, 413
Sivaraman, K. R. 1984, Kodaikanal Obs. Bull., 4, 11
Solanki, S. K. 2003, A&A Rev., 11, 153
Solanki, S. K., Montavon, C., Livingston, W. 1993, In: IAU Colloq. 141: The Magnetic and Velocity Fields of Solar Active Regions, H. Zirin, G. Ai, H. Wang (eds.), ASP Conf. Ser., vol. 46, p. 52
Solanki, S. K., Montavon, C. A. P., Livingston, W. 1994, A&A, 283, 221
St. John, C. E. 1913, ApJ, 37, 322
Stanchfield, II, D. C. H., Thomas, J. H., Lites, B. W. 1997, ApJ, 477, 485
Thomas, J. H., Weiss, N. O. 2004, ARA&A, 42, 517
Thomas, J. H. 2010, In: Magnetic Coupling between the Interior and the Atmosphere of the Sun, S. S. Hasan, R. J. Rutten (eds.), Astrophys. Space Sci. Procs., Springer, Heidelberg, these proceedings
Wiehr, E., Degenhardt, D. 1992, A&A, 259, 313
Wiehr, E., Knoelker, M., Grosser, H., Stellmacher, G. 1986, A&A, 155, 402

The Thermal Structure of Sunspots

K.M. Hiremath and B.B. Akshatha

Abstract We use SOHO/MDI continuum images to measure the average thermal structure of bipolar spots during their initial appearance. Although the average thermal structure of penumbrae is independent of sunspot size and life span, the thermal structure of umbrae depends on the life span. This suggests that penumbrae are a shallower phenomenon and that umbrae with different life spans have different anchoring depths in the convection zone.

1 Introduction

Sunspot intensities have been analyzed by Jensen et al. (1955); Bray and Loughhead (1964, and references therein), Ekmann (1974), Albregtsen and Maltby (1981), Brandt et al. (1990), Solanki (2003, and references therein), Vaquero et al. (2005), Penn and MacDonald (2007), Mathew et al. (2007), and Wesolowski et al. (2008) irrespective of the sunspot life spans. However, as spots with different life spans have different anchoring depths (Hiremath 2002; Hiremath and Lovely 2007 and references therein), observations of sunspot dynamics, magnetic field, and thermal structure during their initial appearance on the solar surface can be used to infer physical variables at depth in the convective envelope. Rather than rely on ad-hoc models like mixing length theory, as was done independently by Shibahashi et al. (1998), we aim to estimate the thermal structure of the entire convective envelope by using the thermal structure of sunspots during their initial appearance on the surface. To this purpose we analyze the thermal structure of sunspots for different sizes and life spans during their initial appearance.

K.M. Hiremath (✉)
Indian Institute of Astrophysics, Bangalore, India

B.B. Akshatha
Department of Physics, Mangalore University, Mangalore, India

S.S. Hasan and R.J. Rutten (eds.), *Magnetic Coupling between the Interior and Atmosphere of the Sun*, Astrophysics and Space Science Proceedings,
DOI 10.1007/978-3-642-02859-5_45,

2 Data and Analysis

We use full-disk, level-two calibrated, continuum intensity images from SOHO/MDI taken during 2000–2003. Their resolution is 4″and the noise 0.3% (Scherrer et al. 1995). Only non-recurrent sunspots that were born and vanished on the visible part of the solar disk are considered, leaving 160 spots on both hemispheres in the 0–15° latitude range. The reason for selecting this latitude range is that for these spots the anchoring depth during initial appearance can be estimated (Hiremath 2002). We use the same criteria for spot selection as in Hiremath and Lovely (2007). The life span of a spot group is defined as the number of days between the first and the last appearance on the same part of the solar disk. Solar Geophysical Data is used for the life-span estimation.

The average intensity flux with its standard error, the total intensity flux, and the number of pixels in the umbra and the whole spot were estimated. From these we computed the intensity of the penumbra and estimate temperatures assuming LTE and Stefan–Boltzmann's law.

3 Results and Conclusions

The majority of the sunspots belong to bipolar groups at their initial appearance on the surface. We therefore evaluated the average intensity and temperature separately for leading and following spots and then merged these data to obtain larger statistical significance. Figure 1 illustrates the variation of the average temperature of the umbra, the penumbra, and the whole spot at the time of spot appearance against their size and the spot life span.

The left-hand plot shows that the full-spot average temperature is independent of spot size. This might be due to mixing spots with different life spans, because

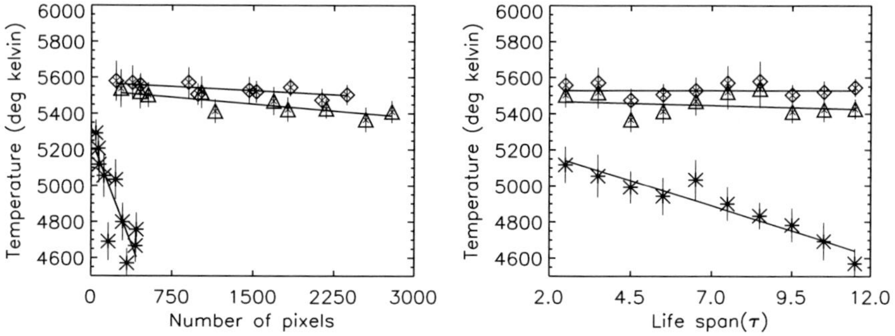

Fig. 1 *Left*: variation of umbral temperature (*crosses*), penumbral temperature (*diamonds*), and whole-spot temperature (*triangles*) against their area. *Right*: the same against spot life span. The lines are least-square fits

Hiremath (2002) and Hiremath and Lovely (2007) suggested that sunspot dynamics and fields vary with life spans due to different anchoring depth in the convective zone. However, the righthand plot also shows independence of the full-spot temperature averaged with spot life span.

On the other hand, the umbrae do show dependency, suggesting that they are anchored at different depths. The obvious interpretation is that penumbrae are shallower, probably because, closer to the surface where radiation losses dominate, the penumbral structure is heated uniformly and hence is independent of either life span or size. This interpretation is consistent with the theoretical study of Jahn and Schmidt (1994).

Acknowledgment The authors are thankful to R.J. Rutten for suggestions that improved this manuscript, and to P. Scherrer, T. Duvall, M. Christina Rabello-Soares, R. Wachter, and Y. Liu of Stanford University for useful discussions. SOHO, from which we used MDI continuum images, is a project of international cooperation between ESA and NASA.

References

Albregtsen, F., Maltby, P. 1981, Solar Phys., 71, 269
Brandt, P. N., Schmidt, W., Steinegger, M. 1990, Solar Phys., 129, 191
Bray, R. J., Loughhead, R. E. 1964, Sunspots, Chapman and Hall, London
Ekmann, G. 1974, Solar Phys., 38, 73
Hiremath, K. M. 2002, A&A, 386, 674
Hiremath, K. M. Lovely, M. R. 2007, ApJ, 667, 585
Jahn, K., Schmidt, H. U. 1994, A&A, 290, 295
Jensen, E., Nordø, J., Ringnes, T. S. 1955, Astrophys. Norvegica, 5, 167
Mathew, S. K., Martínez Pillet, V., Solanki, S. K., Krivova, N. A. 2007, A&A, 465, 291
Penn, M. J. MacDonald, R. K. D. 2007, ApJ, 662, L123
Scherrer, P. H., Bogart, R. S., Bush, R. I., et al. 1995, Solar Phys., 162, 129
Shibahashi, H., Hiremath, K. M., Takata, M. 1998, In: Structure and Dynamics of the Interior of the Sun and Sun-like Stars, S. Korzennik (ed.), ESA-SP, vol. 418, p. 537
Solanki, S. K. 2003, A&A Rev., 11, 153
Vaquero, J. M., Gordillo, A., Gallego, M. C., Sanchez-Bajo, F., Garcia, J. A. 2005, Obs., 125, 152
Wesolowski, M. J., Walton, S. R., Chapman, G. A. 2008, Solar Phys., 248, 141

Bright Points in G-Band and Ca II H Images from Hinode

C.L. Pradeep and R. Kariyappa

Abstract We analyze a time sequence simultaneous G-band and Ca II H images taken with Hinode/SOT on 14 April 2007 during 17:00–18:00 UT. In each sequence, we selected 20 bright points and derived their light curves. Power-spectrum analysis was performed to determine periodicities in these light curves, which are 2–5 min for the Gband bright points and 3–4 min for the Ca II H bright points, respectively. Comparison of the light curves indicates the presence of a phase difference, which suggests the occurrence of propagating waves that may be responsible for heating the chromosphere in Ca II H bright points.

1 Observations, Analysis, and Results

We use a 1-h (17:00–18:00 UT) time sequence of simultaneous G-band and Ca II H images obtained on 14 April 2007 using the Solar Optical Telescope (SOT) onboard the Hinode mission, covering a quiet region near the center of the solar disk. We selected 20 G-band bright points (GbBPs) and 20 Ca II H bright points (CaBPs) for analysis (Fig. 1).

We derived light curves for all BPs by summing the intensity values per bright point over square sub-image cutouts containing each selected GbBP or CaBP. These clearly show intensity oscillations. We performed power spectrum analysis to determine the corresponding periodicities. Figure 2 shows examples for GbBP1 and CaBP1. The GbBPs show periodicities in the 2–5 min range and the CaBPs in the 3–4 min range. The light curves of BP pairs show phase differences between the G-band and Ca II H. This suggests that the waves propagate from the photosphere to the chromosphere. Detailed Fourier phase-difference analysis is needed to determine the phase lag and to clarify the nature of the waves that may be responsible for the heating of the chromosphere at the sites of Ca II H bright points.

C.L. Pradeep (✉)
University of Hyderabad, Hyderabad, India

R. Kariyappa
Indian Institute of Astrophysics, Bangalore, India

S.S. Hasan and R.J. Rutten (eds.), *Magnetic Coupling between the Interior and Atmosphere of the Sun*, Astrophysics and Space Science Proceedings, DOI 10.1007/978-3-642-02859-5_46, © Springer-Verlag Berlin Heidelberg 2010

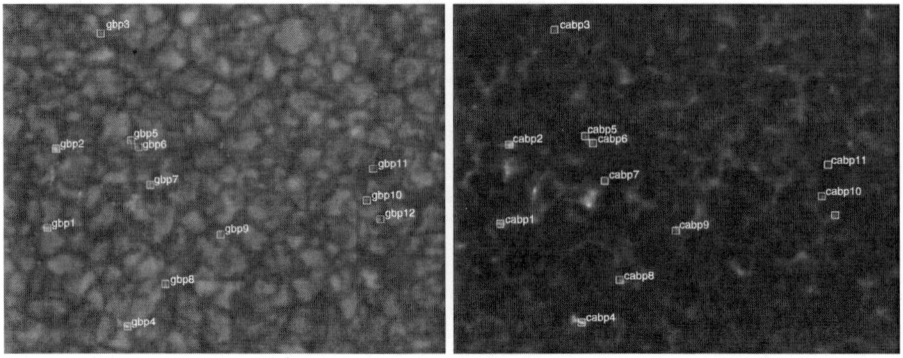

Fig. 1 Sample G-band and Ca II H images. The analyzed bright points are marked

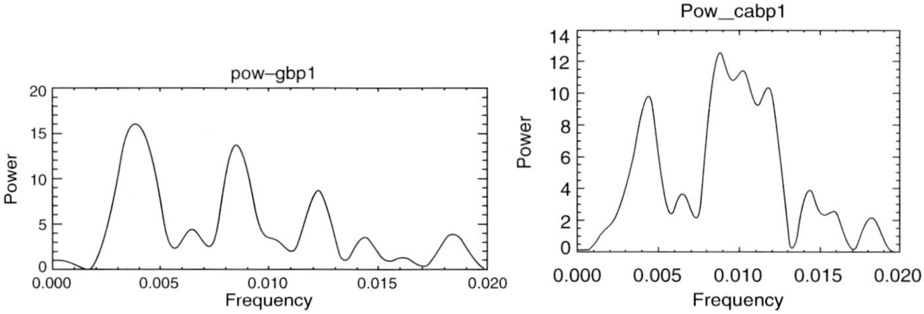

Fig. 2 Power spectra for GbBP1 and CaBP1. Frequency units: Hz

Acknowledgment Hinode is a Japanese mission developed and launched by ISAS/JAXA, with NAOJ as domestic partner and NASA (USA) and STFC (UK) as international partners. It is operated by these agencies in co-operation with ESA and NSC (Norway). We thank the conference organizers for a very good meeting and the editors for excellent instructions and detailed comments on this contribution. This research work was carried out at IIA when Pradeep was a Summer Project student during May–July 2008. He extend his gratitude to the Director of the IIA for financial support to attend the conference.

Waves in the Transition Region

E. Scullion, R. Erdélyi, and J.G. Doyle

Abstract Fundamental questions concerning coronal heating and the origin of the fast solar wind may be answered through greater understanding of fine structures in the lower solar atmosphere. We present results from a 2.5-D ideal-MHD simulation from a new numerical code, plus new supportive observational material. The simulation was run with the Sheffield Advanced Code (SAC, Shelyag et al. 2008, A&A, 486, 655) and is supported by multi-instrument evidence of a surface wave obtained from co-alignment of observations with Hinode/EIS and SoHO/MDI.

1 Introduction

Upward propagating magneto-acoustic waves (leaked *p*-modes), which contribute to the formation of spicules (De Pontieu et al. 2004), can puncture the transition region and become guided across the transition layer as a shock wave, resulting in a circular surface wave. Ubiquitious spicules form in leaked *p*-modes with photospheric velocities, which travel along steep magnetic field lines (Sterling 2000; Malins and Erdélyi 2007; Erdélyi et al. 2007).

2 Evidence

EIS He II 256.32 ($10^{4.7}$ K) images have been co-aligned with a SoHO MDI full-disk magnetogram. Upon emergence of a jet, a wave front forms and propagates

E. Scullion (✉)
Solar Physics and Space Plasma Atmospheric Research Centre, University of Sheffield, UK
and
Armagh Observatory, Northern Ireland

R. Erdelyi
Solar Physics and Space Plasma Atmospheric Research Centre, University of Sheffield, UK

J.G. Doyle
Armagh Observatory, Northern Ireland

S.S. Hasan and R.J. Rutten (eds.), *Magnetic Coupling between the Interior and Atmosphere of the Sun*, Astrophysics and Space Science Proceedings,
DOI 10.1007/978-3-642-02859-5_47,

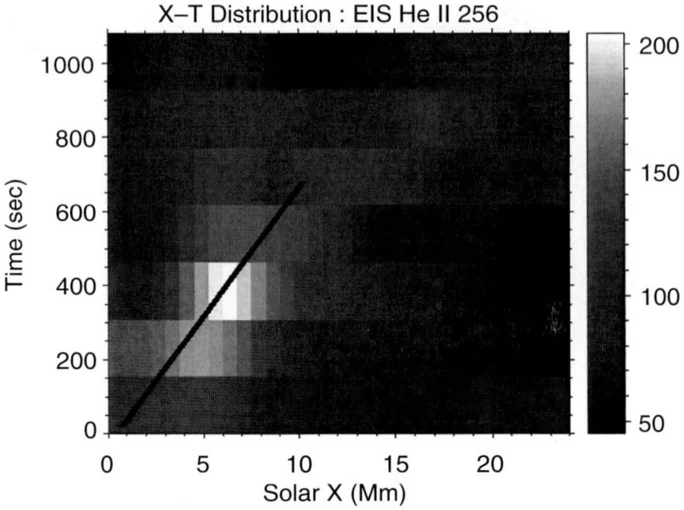

Fig. 1 $X - T$ plot showing an observed wavefront with a measurable speed, i.e., the gradient of the fitted black line

across the transition layer with a range of 20–30″ from the source of the jet. The evolving 40″ FOV image presents this ring-like wavefront. Crosscuts of the dataset in the region of the wavefront allowed us to produce a time–distance ($X - T$) plot to measure the propagation speed of this wavefront as being about 12 km s^{-1}, that is, the sound speed (see Fig. 1).

The numerical simulation recreates the solar atmosphere (with z = height from surface, x = Solar X) from the photosphere to the corona at a height of 8 Mm. An antiparallel magnetic field configuration is established (center) to enable the formation of a central current sheet in two dimensions (x, y = constant, z). The VAL-IIIc atmosphere (Vernazza et al. 1981) is initially driven along the z-direction at the base with a sinusoidal 300 s (5 min) oscillating source, which has periodic motion in the y-direction. These oscillations have been shown to power spicule formation from shocks, with some heating effects due to ion-neutral collisonal damping of MHD waves (James and Erdélyi 2002). At time step 143.88 s in the simulation, we see clear indications of a large jet along the current sheet, reaching a height of approximately 4 Mm above the solar surface.

Surface wave motions at the transition region have until now been unsimulated and unobserved. Upon reaching the transition region, p-modes can expand radially outwards across the transition layer, due to the steep pressure gradient that exists there. From the simulation we measure the separation of the peaks to be approximately 4.5 Mm behind and 168 s after the rise of the jet, deduced from the cross section of the V_x component of the wave propagation at z = 2.3 Mm. A 4.5-Mm ring diameter at 168 s is equivalent to 6″ observational spreading within just under 3 min. From Fig. 1, the observed wavefront reached about 4 Mm separation in just over 3 min. The simulated wave front propagates by approximately 2.25 Mm in 168 s, so the surface motion has a group velocity of 13 km s^{-1}. Clearly, the simulation and observation are in excellent agreement.

Wave-driven jets can reach a height of 4 Mm with rise velocities (V_z) of about 0.5 km s^{-1}. To simulate these effects, in the case of larger jets such as macrospicules which are observed to be an order of magnitude larger and faster than normal spicules, we will most likely require magnetic reconnection in the current sheet and so obtain much larger rise velocities. We propose that, given the high frequency of spicules and other jet formations in the solar atmosphere, a torrent of surface–wave interactions at the transition region contributes to the heating of the upper coronal atmosphere.

We intend to submit the details of this study soon to a peer-reviewed journal.

Acknowledgment We thank the conference organizers for a very good meeting. This work was carried out at the Armagh Observatory and SP2RC, University of Sheffield. RE acknowledges M. Kéray for patient encouragement. RE is also grateful to NSF, Hungary (OTKA, Ref. No. K67746) and the Science and Technology Facilities Council (STFC), UK for financial support.

References

De Pontieu, B., Erdélyi, R., James, S. P. 2004, Nat, 430, 536
Erdélyi, R., Malins, C., Tóth, G., de Pontieu, B. 2007, A&A, 467, 1299
James, S. P., Erdélyi, R. 2002, A&A, 393, L11
Malins, C., Erdélyi, R. 2007, Solar Phys., 246, 41
Shelyag, S., Fedun, V., Erdélyi, R. 2008, A&A, 486, 655
Sterling, A. C. 2000, Solar Phys., 196, 79
Vernazza, J. E., Avrett, E. H., Loeser, R. 1981, ApJS, 45, 635

Viscous Damping of Alfvén Surface Waves with Steady Flows

G.D. Rathinavelu, M. Sivaraman, and A.S. Narayanan

Abstract The dispersion relation of Alfvén Surface Waves propagating along the moving viscous plasma–vacuum interface has been derived and solved numerically. The plasma layer is below while the upper layer is assumed to be vacuum. The layer of plasma is assumed to be moving parallel to the direction of the background magnetic field. This interface supports two modes of viscous-damped Alfvén Surface Waves when the plasma is static. However, when the flow is introduced, new modes start evolving in addition to the existing modes and the damping due to viscosity increases.

1 Introduction

Magnetohydrodynamic (MHD) waves are believed to play an important role in the solar atmosphere. They are widely considered as possible source of heating of various parts of the solar atmosphere. Among the suggested wave dissipation mechanisms (viscosity, thermal conductivity, phase mixing, ohmic heating, resonance absorption, etc.), magnetic-field oriented heating has been strongly favored as the solar atmosphere is highly structured (Vaina & Rosner 1978). Hence, the heating mechanisms cannot solely depend on the plasma parameters but are also sensitive to the magnetic field geometry. Many researchers have studied the viscous dissipation of magnetosonic waves in the solar corona (Kalomeni et al. 2007), with thermal conductivity (Kumar & Kumar 2006; Kumar et al. 2006) and surface MHD wave damping due to viscosity in the solar chromosphere (Ruderman et al. 2000). Recently, Alfvén waves have been detected in the solar corona (Tomczyk et al. 2007), chromosphere (De Pontieu et al. 2007), prominences (Okamoto et al. 2007), and solar X-ray jets (Cirtain et al. 2007). Thus, the ubiquitous presence of Alfvén waves is now accepted by the community. This motivates a study of the

G.D. Rathinavelu (✉) and M. Sivaraman
Department of Physics, Gandhigram Rural University, Tamil Nadu, India

A.S. Narayanan
Indian Institute of Astrophysics, Bangalore, India

S.S. Hasan and R.J. Rutten (eds.), *Magnetic Coupling between the Interior and Atmosphere of the Sun*, Astrophysics and Space Science Proceedings,
DOI 10.1007/978-3-642-02859-5_48,

effect of flows on the viscous damping of Alfvén waves in a structured atmosphere, in particular on Alfvén surface waves (henceforth ASW). Such study may reveal interesting phenomena that may occur in the solar atmosphere. In this work, we study the viscous damping of ASW at a single interface, separating an incompressible viscous plasma and vacuum.

2 Dispersion Relation

In the MHD approximation, the linearized equations for an incompressible viscous plasma with uniform density, embedded in a uniform magnetic field, with a single interface can be simplified (assuming wave solutions) to yield the following dispersion relation:

$$[(x-R)^2-1]^2+[i\nu(x+3(x-R))-\beta^2][(x-R)^2-1]+[2\nu^2(x-R)(2x+R)](T-1)=0, \tag{1}$$

where $T = \tau/k = [1-(x^2-1)/V(x-R)]^{1/2}$, $\beta = B_{02}/B_{01}$, $x = \omega/kV_{A1}$, $\nu = \nu_1 k/\rho_{01}V_{A1}$, and $R = U/V_{A1}$ are the magnetic field ratio, normalized phase velocity, viscosity, and Alfvén Mach number, respectively. By algebraic simplification, the above relation can be written (after dividing by the common factor (x^2-1) which represents the bulk mode) as

$$(x-R)^6+C(x-R)^5+D(x-R)^4+E(x-R)^3+F(x-R)^2+G(x-R)+H=0, \tag{2}$$

where the coefficients are functions of ν, β, and R. We skip presenting them for the sake of brevity.

3 Discussion and Conclusion

The dispersion relation is solved numerically and delivers six roots of the form $x = x_r + ix_i$, with $0 < x_r, x_i < 0$. Figure 1 depicts the solution of (2) for interface parameter $\beta^2 = 0.02$ and normalized flow velocities $R = 0$ and $R = 1.5$ in terms of k_r, k_i being a function of ν. In the absence of flow ($R = 0$), we clearly observe two ASW modes. However, when $R > 0$, the second mode becomes evanescent above a critical value of ν, and a new mode appears at a higher value of ν (the second half of Fig. 1). It seems that the flow suppresses the existing modes and supports evolution of new ones.

Figure 2 exhibits the dispersive characteristics of the modes when β is changed to 0.2, with the same flow velocities as in Fig. 1. It exhibits the same dispersive properties as for the smaller value of β, except that the evanescence of the existing Alfvén modes and the appearance of new modes occur at higher ν values than in Fig. 1. However, when the flow is Alfvénic, a strange mode with $k_i >> k_r$ appears,

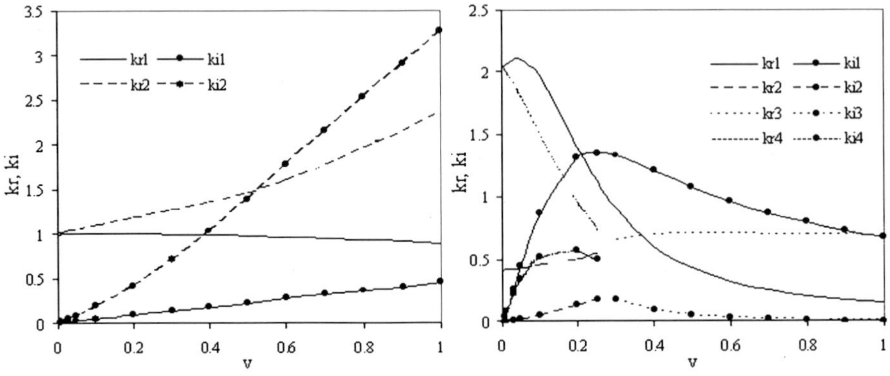

Fig. 1 k_r and k_i as a function of v with $\beta^2 = 0.02$ for (**a**) $R = 0$ and (**b**) $R = 1.5$

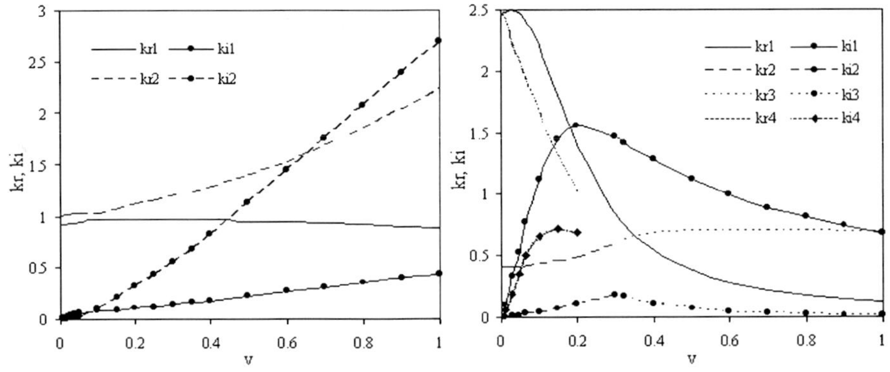

Fig. 2 k_r and k_i as a function of v with $\beta^2 = 0.2$ for (**a**) $R = 0$ and (**b**) $R = 1.5$

with lower and upper cutoffs at the lower v value, which was not observed for other flow velocities.

In conclusion, two ASW modes, modified by viscosity, are observed in the absence of a flow. The presence of a flow ($R > 0$) suppresses the wave numbers of the existing modes and supports the evolution of new modes.

Acknowledgment G. David Rathinavelu thank the organizers of the meeting for providing financial support for his participation.

References

Cirtain, J. W., Golub, L., Lundquist, L., et al. 2007, Science, 318, 1580
De Pontieu, B., McIntosh, S. W., Carlsson, M., et al. 2007, Science, 318, 1574
Kalomeni, B., Hurkal, D. O., Pekunlu, E. R., et al. 2001, Turk. J. Phys., 25, 195
Kumar, N., Kumar, P. 2006, Solar Phys., 236, 137

Kumar, N., Kumar, P., Singh, S. 2006, A&A, 453, 1067
Okamoto, T. J., Tsuneta, S., Berger, T. E., et al. 2007, Science, 318, 1577
Ruderman, M. S., Oliver, R., Erdelyi, R., et al. 2000, A&A, 354, 261
Tomczyk, S., McIntosh, S. W., Keil, S. L., et al. 2007, Science, 317, 1192
Vaina, G. S., Rosner, R. 1978, ARA&A, 16, 393

Statistical Detection of Propagating Waves in a Polar Coronal Hole

G.R. Gupta, E. O'Shea, D. Banerjee, M. Popescu, and J.G. Doyle

Abstract Waves are important in the heating of the solar corona and the acceleration of the solar wind. We have examined a long spectral time series sampling a southern coronal hole, observed on the 25 February 1997 using the SUMER spectrometer onboard SoHO. The observations used the spectra lines N IV 765 Å, formed in the transition region, and Ne VIII 770 Å, formed in the low corona. The spectra indicate the presence of compressional waves with periods of about 18 min, and also significant power at shorter periods. Using Fourier techniques, we measured the phase delays between the intensity as well as the velocity oscillations in the two lines as a function of frequency. From these measurements we derive the travel time of the propagating oscillations and so the propagation speeds of the waves producing the oscillations. As the measured propagation speeds are subsonic, we conclude that the observed waves are slow magneto-acoustic ones.

1 Introduction

Coronal holes are regions of cool and low density plasma that appear relatively dark at coronal temperatures. The predominantly unipolar magnetic field in coronal hole regions is thought to give rise to the fast solar wind. A number of studies, for example, Ofman et al. (1997, 2000); Banerjee et al. (2001); Popescu et al. (2005)

G.R. Gupta (✉)
Indian Institute of Astrophysics, Bangalore, India
and
Joint Astronomy Programme, Indian Institute of Science, Bangalore 560012, India

E. O'Shea
Armagh Observatory, College Hill, Armagh BT61 9DG, N. Ireland

D. Banerjee
Indian Institute of Astrophysics, Bangalore 560034, India

M. Popescu
Armagh Observatory, College Hill, Armagh BT61 9DG, N. Ireland

J.G. Doyle
Armagh Observatory, College Hill, Armagh BT61 9DG, N. Ireland

S.S. Hasan and R.J. Rutten (eds.), *Magnetic Coupling between the Interior and Atmosphere of the Sun*, Astrophysics and Space Science Proceedings,
DOI 10.1007/978-3-642-02859-5_49,

have measured oscillations in coronal holes in the polar off-limb regions of the Sun. All these studies point to the presence of compressional waves thought to be slow magnetoacoustic waves as found by Deforest and Gurman (1998); O'Shea et al. (2006, 2007). Recently, Gupta et al. (2009) have reported the detection of these waves in the disk part of the polar coronal hole (hereafter PCH). They also find a difference in nature of the compressional waves between bright (network) and dark (internetwork) regions in the PCH. In this contribution, we extend such analysis to another dataset. More detail is given in Gupta et al. (2009).

2 Observations and Data Analysis

The data used in this analysis were taken on 25 February 1997, during 00:00–13:59 UT with the $1 \times 300''$ slit of SUMER and an exposure time of 60 s in the N IV 765 Å and Ne VIII 770 Å lines in a southern PCH. Details of the data reduction are given in Gupta et al. (2009).

The chromosphere and transition region show enhanced-intensity network boundaries and darker internetwork cells. Presumably, the magnetic field is predominantly concentrated at the network boundaries and, within coronal holes, the footpoints of coronal funnels emanate from these network boundaries. As the observing duration of this dataset is very long, the locations of bright and dark pixels along the slit change with time. For this reason, we have analyzed the whole dataset pixel by pixel and timeframe by timeframe.

For example, for one given moment, we first determined the average intensity along the slit. All pixels having an intensity higher than 1.25 times this average intensity were chosen as bright pixels. If such pixels are bright for at least 60 min (or 60 timeframes), then these are considered to be a bright network location over that time interval. The bright pixel identification is done only for the low-temperature N IV line; the network pixels obtained from it are assumed to be the same in the higher-temperature Ne VIII line.

3 Results and Discussion

Figure 1 shows a representative example of the oscillations measured in a bright region of the PCH. We use wavelet analysis to provide information on the temporal signal variation (Torrence and Compo 1998). Further details on this wavelet analysis are found in Gupta et al. (2009); O'Shea et al. (2001) and references therein. Figure 1 shows oscillations of about 18 min periodicity in both lines at the same location. This suggests that these two layers are linked by a propagating wave passing from one layer to the other. To test this hypothesis and to ascertain the nature of the propagating waves, we measured phase delays in intensity and in Dopplershift between the two lines at each of the measurable pixels along the slit for a full frequency

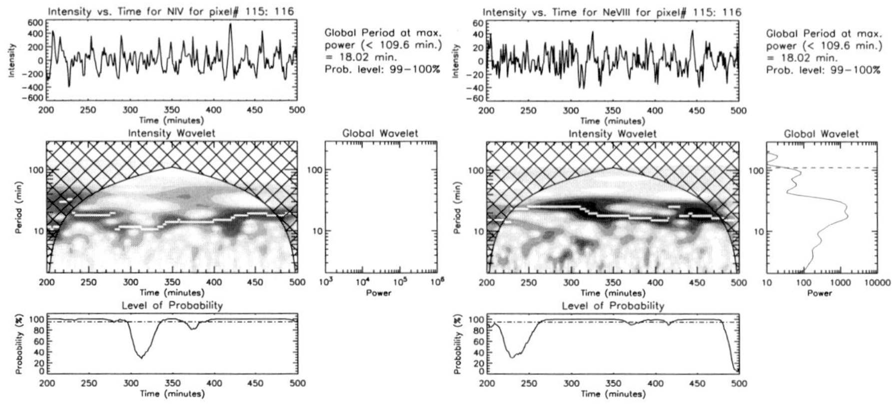

Fig. 1 Wavelet result for a bright location (pixels 115–116, corresponding to $Y \approx -915''$) in the N IV and Ne VIII intensities. In each set, the *top panels* show the relative (background trend removed) radiant flux, the *central panels* show the color-inverted wavelet power spectrum, the *bottom panels* show the variation of the probability estimate associated with the maximum power in the wavelet power spectrum (marked with *white lines*), and the *center-right panels* show the global (averaged over time) wavelet power spectrum. Above the global wavelet, the period measured from the maximum power from the global wavelet and the probability estimate are specified

range, following the method of Athay and White (1979), O'Shea et al. (2006), and Gupta et al. (2009), which deliver the phase delay $\Delta\phi = 2\pi f T$, where f is the frequency and T the time delay in seconds. From this definition, we expect that the phase difference varies linearly with f and will change by 360° over frequency intervals of $\Delta f = 1/T$, producing parallel lines in plots of $\Delta\phi$ vs. f at fixed frequency intervals ($\Delta f = 1/T$), corresponding to a fixed time delay T.

The method to identify the parallel lines in the phase plot (left-hand panel of Fig. 2) and their spacing and significance in the histogram (right-hand panel) is discussed in more detail in Gupta et al. (2009). In the bright location, we did not have good enough statistics to come to a reliable conclusion.

In the dark locations of the PCH (left-hand panel of Fig. 2), we measured a time delay of -2607 ± 491.2 s between the two lines, indicating upwardly propagating waves. Notice the many closely spaced, steeply sloped, parallel phase lines that correspond to this long time delay. In the right-hand panel of Fig. 2, these parallel phase lines correspond to significant peaks above 95% significance in the histogram. These peaks are equally spaced at the phase difference of 360° that would be expected for unimpeded propagating waves.

The measured time delays from this PCH may be used to estimate propagation speeds of the waves assumed to be causing the oscillations. To calculate the propagation speeds, one needs information on the formation height difference of the two lines. A value of 4,095 km has been calculated using the limb brightening technique as described in O'Shea et al. (2006); Gupta et al. (2009). Using this height difference and the measured time delay, we find a speed of $-1.6 \pm 0.3\,\mathrm{km\,s^{-1}}$ for the dark region, which is smaller than the sound speed at that height. Hence, the identified waves can be identified as slow magneto-acoustic waves. These waves are rather slow and may not carry enough energy flux for the acceleration of the solar wind.

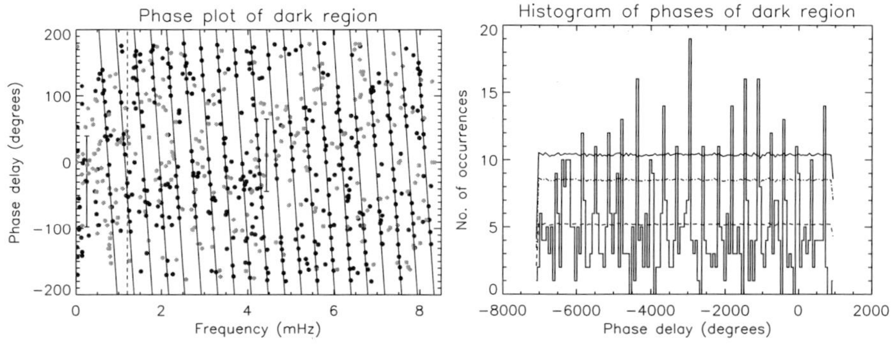

Fig. 2 *Left-hand panel*: Phase delays measured between oscillations in the spectroscopic line pair at dark locations. The phases from brightness oscillations are shown as *grey dots*, and those from Doppler velocities as *black dots*. Overplotted are parallel lines corresponding to fixed time delays. The vertical dashed line at 1.2 mHz indicates that phase values below this frequency may be affected by solar rotation. Representative errors on the phase measurements are indicated by error bars. *Right panel*: Histogram showing the distribution of the phase delay measurements as a function of frequency for the dark locations. The *horizontal dotted*, *dot-dashed*, and *long dashed lines* show the 68.3% (1σ), 90% (1.64σ), and 95% (2σ) confidence levels, evaluated from Monte-Carlo simulations with 5,000 permutations

References

Athay, R. G., White, O. R. 1979, ApJ, 229, 1147
Banerjee, D., O'Shea, E., Doyle, J. G., Goossens, M. 2001, A&A, 380, L39
Deforest, C. E., Gurman, J. B. 1998, ApJ, 501, L217
Gupta, G. R., O'Shea, E., Banerjee, D., Popescu, M., Doyle, J. G. 2009, A&A, 493, 251
Ofman, L., Romoli, M., Poletto, G., Noci, G., Kohl, J. L. 1997, ApJ, 491, L111
Ofman, L., Romoli, M., Poletto, G., Noci, G., Kohl, J. L. 2000, ApJ, 529, 592
O'Shea, E., Banerjee, D., Doyle, J. G. 2006, A&A, 452, 1059
O'Shea, E., Banerjee, D., Doyle, J. G. 2007, A&A, 463, 713
O'Shea, E., Banerjee, D., Doyle, J. G., Fleck, B., Murtagh, F. 2001, A&A, 368, 1095
Popescu, M. D., Banerjee, D., O'Shea, E., Doyle, J. G., Xia, L. D. 2005, A&A, 442, 1087
Torrence, C., Compo, G. P. 1998, Bull. American Meteo. Soc., 79, 61

Network Loop Oscillations with EIS/Hinode

A.K. Srivastava, D. Kuridze, T.V. Zaqarashvili, B.N. Dwivedi, and B. Rani

Abstract We analyze a time sequence of He II 256.32 Å images obtained with EIS/Hinode, sampling a small magnetic loop in magnetic network. Wavelet analysis indicates 11-min periodicity close to the loop apex. We interpret this oscillation as forcing through upward leakage by the fundamental acoustic eigenmode of the underlying field-free cavity. The observed loop length corresponds to the value predicted from this mechanism.

1 Introduction

Field-free cavities under bipolar magnetic canopies (Centeno et al. 2007) in the vicinity of magnetic network are likely to serve as resonators for fast magnetoacoustic waves (Kuridze et al. 2007). Srivastava et al. (2008) have studied the properties of the fundamental fast magnetoacoustic mode in brightened magnetic network. It leaks through the magnetic network into the upper solar atmosphere. Recently, Martinéz González et al. (2007) found evidence for low-lying loops in magnetic internetwork. In EIS/Hinode observations of bright magnetic network, we found a small loop located near the south pole. We search for magnetoacoustic oscillations in this loop through wavelet analysis.

2 Observations

The observations were acquired on 11 March 2007 during 19:04–19:54 UT in the study HPW005_QS_Slot_60m. The slot-center position was ($X = 118''$, $Y = -973''$), with a $40'' \times 512''$ field of view (Fig. 1). The data were binned

A.K. Srivastava (✉) and B. Rani
Aryabhatta Research Institute of Observational Sciences, Manora Peak, Nainital, India

D. Kuridze and T.V. Zaqarashvili
Abastumani Astrophysical Observatory, Tbilisi, Georgia

B.N. Dwivedi
Department of Applied Physics, Institute of Technology, Banaras Hindu University, Varanasi, India

S.S. Hasan and R.J. Rutten (eds.), *Magnetic Coupling between the Interior and Atmosphere of the Sun*, Astrophysics and Space Science Proceedings,
DOI 10.1007/978-3-642-02859-5_50,

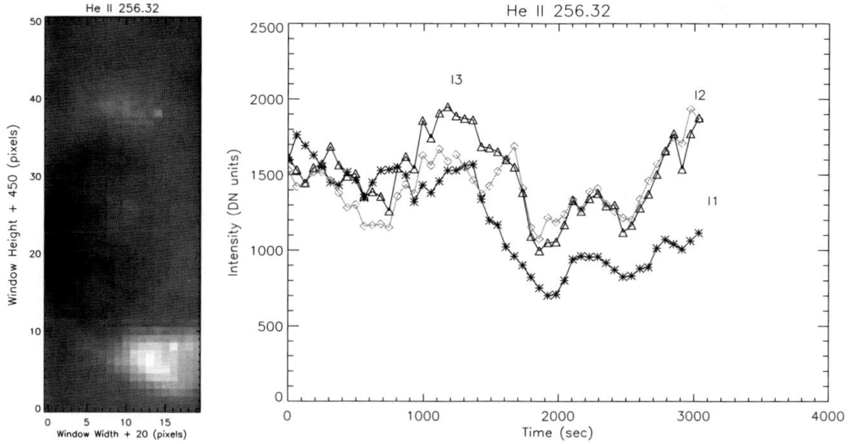

Fig. 1 *Left*: partial image in He II 256.32 Å showing bright magnetic network. *Right*: intensity vs. time for the loop apex (I2) and footpoints (I1 and I3). The similarity of the temporal intensity profiles indicates the presence of a closed system such as a small-scale loop in this bright magnetic network

over $1'' \times 1''$. The exposure time was about 60 s per spectral line, and the integration time per step of the time series was 60 s. The data reduction is described in Srivastava et al. (2008).

3 Magnetoacoustic Oscillations

Using standard wavelet analysis, we found intensity oscillations at the loop apex with periodicity about 11 min (Fig. 2). We propose that this oscillation results from eigen-mode oscillation leakage from a field-free cavity under the magnetic loop.

To test this suggestion we solve the ideal MHD equations in cylindrical geometry for such a cavity and an overlying magnetic loop, and merge the solutions at the interface to obtain an analytical dispersion relation (Kuridze et al. 2007, Srivastava et al. 2008). Figure 3 plots this dispersion relation for the fundamental harmonic ($m = 1$ in cylindrical coordinates). The observed periodicity is reproduced when the length of the cavity–loop interface is about 3,500 km. This estimate agrees well with the observed loop length.

We conclude that indeed the fundamental harmonic of the field-free cavity may force the observed oscillation in the small overlying loop.

Acknowledgment AKS thanks E. O'Shea for the Randomlet software. Hinode is a Japanese mission developed and launched by ISAS/JAXA, with NAOJ as domestic partner and NASA (USA) and STFC (UK) as international partners. It is operated by these agencies in co-operation with ESA and NSC (Norway).

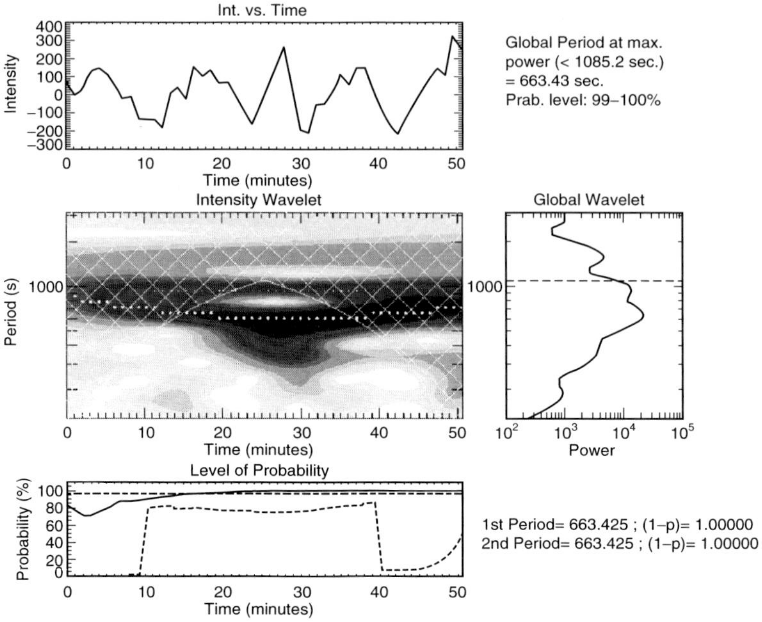

Fig. 2 Wavelet analysis for He II 256.32 Å, following Mathioudakis et al. (2006). The *top panel* shows the intensity variation, the *middle panel* the wavelet power spectrum, the *bottom panel* the probability

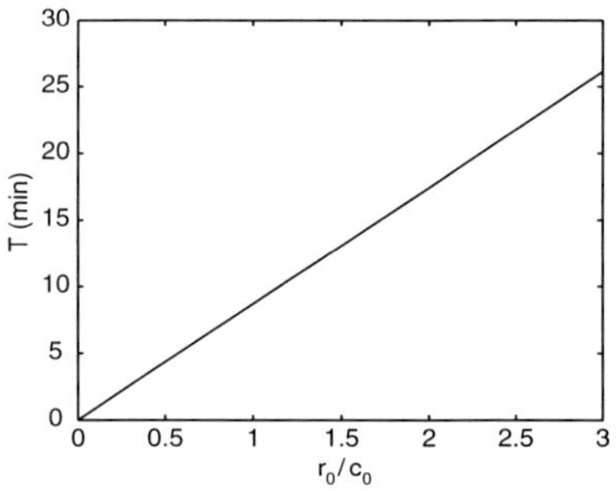

Fig. 3 The period of the $m = 1$ harmonic against the ratio of the cavity size r_0 and the sound speed c_0

References

Centeno, R., Socas-Navarro, H., Lites, B., et al. 2007, ApJ, 666, L137

Kuridze, D., Zaqarashvili, T.V., Shergelashvili, B. M., Poedts, S. 2007, Annales Geophysicae, 26, 2983

Martinéz González, M. J., Collados, M., Ruiz Cobo, B., Solanki, S. K. 2007, A&A, 469, L39

Mathioudakis, M., Bloomfield, D. S., Jess, D. B., Dhillon, V. S., Marsh, T. R. 2006, A&A, 456, 323

Srivastava, A. K., Kuridze, D., Zaqarashvili, T. V., Dwivedi, B. N. 2008, A&A, 481, L97

Dynamical Evolution of X-Ray Bright Points with Hinode/XRT

R. Kariyappa, B.A. Varghese, E.E. DeLuca, and A.A. van Ballegooijen

Abstract We analyzed a 7-h long time sequence of soft X-ray images obtained on 14 April 2007 from a quiet region using the X-Ray Telescope (XRT) onboard Hinode. The aim was to observe intensity oscillations in coronal XBPs of different brightness and to study differences, if any, in the periodicity of the intensity variations and the heating mechanism during their dynamical evolution. We have compared the XRT images with GONG magnetograms using Coronal Modeling Software (CMS), and found that some of the XBPs are located at magnetic bipoles. The coronal XBPs are highly dynamic and oscillatory in nature, showing a wide variety of time scales in their intensity variations.

1 Introduction

Coronal X-ray bright points (XBPs) were discovered using a soft X-ray telescope on a sounding rocket in the late 1960s (Vaiana et al. 1978). Their nature remained enigmatic. Later, using Skylab and Yohkoh/SXT X-ray images, XBPs were studied in detail (Golub et al. 1974; Longcope et al. 2001; Hara and Nakakubo-Morimoto 2003). The number of XBPs that is daily present on the visible hemisphere of the Sun varies from several hundred to a few thousand (Golub et al. 1974), with 800 on the entire solar surface at any given time (Zhang et al. 2001). The number of coronal bright points varies inversely with the solar activity cycle (Sattarov et al. 2002; Hara and Nakakubo-Morimoto 2003). The XBP diameters are about 10–20″ (Golub et al. 1974). Their lifetime ranges from a few hours to a few days (Zhang et al. 2001; Kariyappa and Varghese 2008).

In this contribution, we report the analysis of XBPs on soft X-ray images obtained from Hinode/XRT and on magnetograms from GONG. We briefly discuss the dynamical evolution of the XBPs in relation to the magnetic field.

R. Kariyappa (✉) and B.A. Varghese
Indian Institute of Astrophysics, Bangalore, India

E.E. DeLuca and A.A. van Ballegooijen
Harvard-Smithsonian Center for Astrophysics, Cambridge, USA

S.S. Hasan and R.J. Rutten (eds.), *Magnetic Coupling between the Interior and Atmosphere of the Sun*, Astrophysics and Space Science Proceedings,
DOI 10.1007/978-3-642-02859-5_51,

2 Results and Discussion

We use a 7-h (17:00–24:00 UT) time sequence of soft X-ray images obtained on 14 April 2007 with the X-Ray Telescope (XRT) onboard the Hinode mission, using the Ti_poly filter for a quiet region near the center of the solar disk. We selected 14 XBPs for analysis, marking them as XBP1, XBP2, ..., XBP14, and two background, very dark coronal comparison regions as XBP15 and XBP16. The XRT images have been calibrated using the SSW subroutine xrt_prep.pro (Kariyappa and Varghese 2008). We also obtained the full-disk magnetograms obtained with GONG during the XRT observing period. These magnetograms have been co-registered with the XRT images using the Coronal Modeling Software (CMS) developed by the fourth author.

The XBPs, defined as the sites where intense brightness enhancement is seen, are highly dynamic in nature. We derived light curves for the XBPs by summing their brightness over small square image cut-outs covering the selected XBPs.

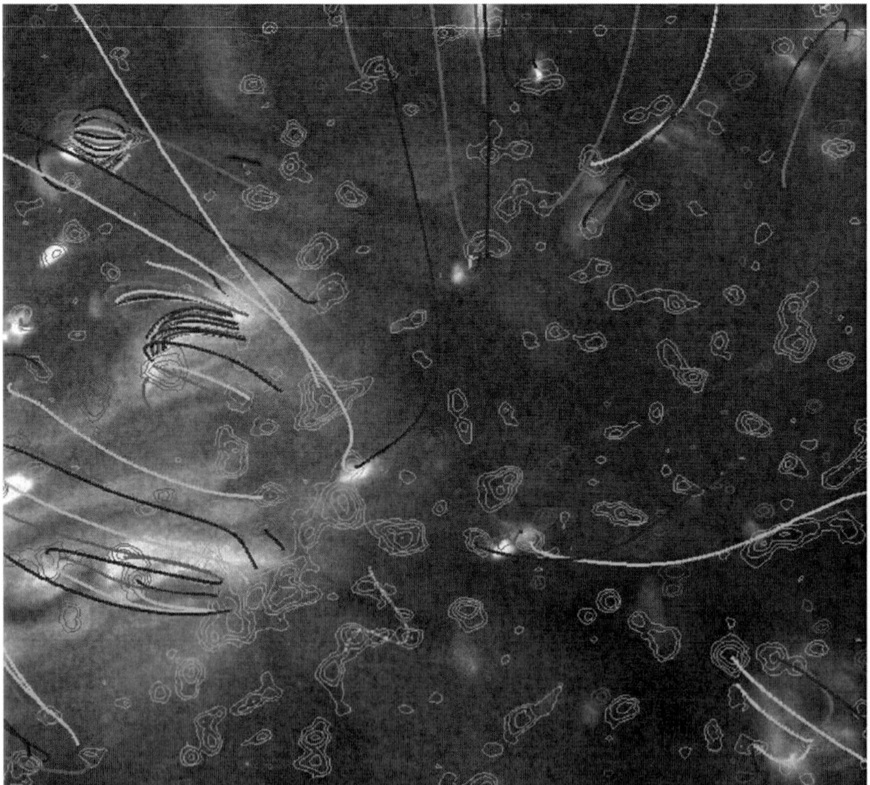

Fig. 1 GONG magnetogram overlayed on an XRT image using CMS modeling. The magnetic field lines are computed from a potential-field extrapolation of the magnetogram

These clearly show intensity oscillations. We performed power-spectrum analysis to determine the corresponding XPB periodicities. They vary from a few minutes to multiple hours.

The light curves of the 14 XBPs have been carefully examined. We found that the XBPs can be grouped into three classes depending on their brightness level. Class I shows very large brightness, Class II medium brightness, and Class III weak brightness during the dynamical evolution. It is evident from the XBP light curves and power spectra that the periodicities seem identical for the three classes (see Kariyappa and Varghese 2008), which suggests that the XBP heating mechanism is similar for all three.

In Fig. 1 we have overlayed an X-ray image with the corresponding GONG magnetogram through CMS modeling. Detailed comparisons of this type suggest that some XBPs lie in between positive (red) and negative (green) magnetic-element polarity, that is, these XBPs are located at magnetic bipoles. The existence of different classes of XBP with different brightness enhancement may be related to the different strengths of the magnetic field with which they have been associated. Detailed analysis of a larger number of XBPs together with magnetograms is needed and is in progress.

Acknowledgment Hinode is a Japanese mission developed and launched by ISAS/ JAXA, collaborating with NAOJ as a domestic partner, and NASA (USA) and STFC (UK) as international partners. Scientific operation of the Hinode mission is conducted by the Hinode science team organized at ISAS/JAXA. Support for the post-launch operation is provided by JAXA and NAOJ (Japan), STFC (U.K.), NASA (U.S.A.), ESA, and NSC (Norway). This work utilizes data obtained by GONG, managed by NSO, which is operated by AURA, Inc., under a cooperative agreement with NSF.

References

Golub, L., Krieger, A. S., Silk, J. K., Timothy, A. F., Vaiana, G. S. 1974, ApJL, 189, L93
Hara, H., Nakakubo-Morimoto, K. 2003, ApJ, 589, 1062
Kariyappa, R., Varghese, B. A. 2008, A&A, 485, 289
Longcope, D. W., Kankelberg, C. C., Nelson, J. L., Pevtsov, A. A. 2001, ApJ, 553, 429
Sattarov, I., Pevtsov, A. A., Hojaev, A. S., Sherdonov, C. T. 2002, ApJ, 564, 1042
Vaiana, G. S., Krieger, A. S., Van Speybroeck, L. P., Zehnfennig, T. 1970, Bull. Am. Phys. Soc., 15, 611
Zhang, J., Kundu, M., White, S. M. 2001, Sol. Phys., 88, 337

Helicity at Photospheric and Chromospheric Heights

S.K. Tiwari, P. Venkatakrishnan, and K. Sankarasubramanian

Abstract In the solar atmosphere, the twist parameter α has the same sign as magnetic helicity. It has been observed using photospheric vector magnetograms that negative/positive helicity is dominant in the northern/southern hemisphere of the Sun. Chromospheric features show dextral/sinistral dominance in the northern/southern hemisphere and sigmoids observed in X-rays also have a dominant sense of reverse-S/forward-S in the northern/southern hemisphere. It is of interest whether individual features have one-to-one correspondence in terms of helicity at different atmospheric heights. We use UBF Hα images from the Dunn Solar Telescope (DST) and other Hα data from Udaipur Solar Observatory and Big Bear Solar Observatory. Near-simultaneous vector magnetograms from the DST are used to establish one-to-one correspondence of helicity at photospheric and chromospheric heights. We plan to extend this investigation with more data including coronal intensities.

1 Introduction

Helicity is a physical quantity that measures the degree of linkage and twistedness in the field (Berger and Field 1984). It is derived from a volume integral over the scalar product of the magnetic field **B** and its vector potential **A**. Direct calculation of helicity is not possible due to the nonuniqueness of the vector potential **A** and the limited availability of data sampling different layers of the solar atmosphere. The force-free parameter α estimates one component of helicity, that is, twist, the other component being writhe, which cannot be derived from the available data. This α is a measure of degree of twist per unit axial length. It has the same sign as magnetic helicity (Pevtsov et al. 2008, Pevtsov 2008). It is now well known that negative/positive helicity dominates in the northern/southern hemisphere. For

S.K. Tiwari (✉) and P. Venkatakrishnan
Udaipur Solar Observatory, Physical Research Laboratory, Udaipur, India

K. Sankarasubramanian
Space Astron. & Instrument. Div., ISRO Satellite Center, Bangalore, India

S.S. Hasan and R.J. Rutten (eds.), *Magnetic Coupling between the Interior and Atmosphere of the Sun*, Astrophysics and Space Science Proceedings,
DOI 10.1007/978-3-642-02859-5_52,

active regions the hemispheric helicity rule holds in the photosphere, see Hagino and Sakurai (2005) and references therein. Similarly for the chromospheric and coronal helicity rules, see Bernasconi et al. (2005) and references therein, and Pevtsov et al. (2001) and references therein. The topology of chromospheric and coronal features decide the sign of the associated helicity. Chirality is the term used for the sign of the helicity in these features. Thus, helicity is a physical measure of chirality. The chirality of active region features shows correspondence with the sign of the helicity in the associated lower/upper atmospheric features. For example, the chirality of X-ray features with S (inverse-S) shapes are associated with sinistral (dextral) filaments (Martin 2003, Rust 2003). Chae (2000) reported for a few cases that active filaments showing dextral/sinistral chirality are related with negative/positive magnetic helicity. Pevtsov et al. (2001) demonstrated correspondence between photospheric and coronal chirality for a few active regions. However, this needs to be confirmed. We have reported similar helicity signs at photospheric, chromospheric, and coronal heights for a few active regions (Tiwari et al. 2008).

Comparison between magnetic helicity signs at different heights in the solar atmosphere may be a useful tool to predict solar eruptions leading to interplanetary events. Also, it may help to constrain modeling chromospheric and coronal features taking the photosphere as boundary condition. However, the data required to do this are not directly available and are often nonconclusive. Vector magnetic fields are not available as routinely as is necessary to derive photospheric twist values. Chromospheric Hα images may be available most of the time by combining data from different telescopes, but are not always conclusive due to lack of angular resolution. Analysis of coronal loop observations is required to determine coronal helicity signs, but these are also not available routinely. Above all, it is hard to find data taken simultaneously at different heights in the solar atmosphere. In this work we combine photospheric and chromospheric data from multiple solar observatories and telescopes. They were often not taken at precisely the same time. We therefore assume that the sign of the magnetic helicity does not change within a few hours.

2 Sign of Magnetic Helicity

The sign of helicity in the photosphere is usually found from the force-free parameter α, for example, $\alpha_{\rm best}$ (Pevtsov et al. 1995), averaged α, for example, $< \alpha_z > = < \mathbf{J_z}/\mathbf{B_z} >$ (Pevtsov et al. 1994) with current density $\mathbf{J_z} = \nabla \times \mathbf{B_z}$, where $\mathbf{B_z}$ is the vertical component of the magnetic field. Some authors have used the current helicity density $H_c = \mathbf{B_z} \cdot \mathbf{J_z}$ (Bao and Zhang 1998; Hagino and Sakurai 2005). A good correlation was found between $\alpha_{\rm best}$ and $\langle \alpha_z \rangle$ by Burnette et al. (2004) and Leka et al. (1996). The force-free parameter α has the same sign as magnetic helicity (Pevtsov et al. 2008). Also, the current helicity (which is not a conserved quantity like magnetic helicity) has the same sign as that of magnetic helicity (Seehafer 1990; Hagyard and Pevtsov 1999; Pevtsov 2008; Sokoloff et al. 2008). In this study, we use the sign of the global α parameter as the sign of magnetic helicity, giving the twist present in the active region.

Numerical measurement of the sign of the chromospheric magnetic helicity is not possible due to non-availability of vector magnetic field observations at these heights. However, the twist present in morphological intensity features were reported already long ago (Hale 1925; Richardson 1941) to tend to follow the hemispheric helicity rule, independent of the solar cycle. Later, many researchers studied the chirality of different chromospheric features such as filaments, fibrils, filament channels etc. (Martin 1998, 2003). We use the chirality of these chromospheric features, mostly whirls observed in Hα, to derive its association with the photospheric sign of magnetic helicity.

3 Data

Apart from a few data sets, most are obtained from different solar observatories and telescopes due to the unavailability of all required data from the same place. The vector magnetic field data were obtained from the Advanced Stokes Polarimeter (ASP, Elmore et al. 1992) as well as the Diffraction Limited Spectropolarimeter (DLSP, Sankarasubramanian et al. 2004, 2006), both mounted at the DST. Near-simultaneous Hα images from the Universal Bi-refringent Filter (UBF) at the DST are used whenever obtained along with the ASP and DLSP. For the vector field observations that do not have corresponding UBF data, Hα images from Udaipur Solar Observatory (USO) and Big Bear Solar Observatory (BBSO) were used. We made sure that in these cases the Hα images were collected within less than a day. Standard and well-established processing was done to derive vector fields. The procedure is described in the references given above.

4 Results and Discussion

Table 1 shows how the sign of helicity at the photospheric level and of the chirality in associated features at chromospheric heights are related with each other. Figure 1a, b clearly show that the Hα whirls follow the transverse magnetic field vectors measured at photospheric heights. The positive/negative helicity derived from the global twist in this sunspot in the photospheric vector data is directly associated with the sinistral/dextral chirality derived from the chromospheric Hα data.

In this preliminary analysis, we thus conclude that the sign of helicity (positive/negative) derived from global twist present around sunspots in the photosphere has one-to-one correspondence with the (sinistral/dextral) sense of chirality observed in the associated chromospheric data. We mostly use the chirality of chromospheric whirls to derive the chromospheric helicity sign. It is known (Martin 1998, 2003) that filaments, filament channels, etc. have the same sense of chirality as the whirls above the associated active regions. The chirality of filaments associated with an active region can therefore be used to determine the chromospheric sense of chirality when high resolution Hα data are not available.

Table 1 Sign of helicity at the photospheric and the chromospheric level

Observation date	AR NOAA number	Photospheric helicity	Chromospheric chirality
06 Feb 2007	NOAA 10941	Negative	Dextral
09 Jan 2007	NOAA 10935	Negative	Dextral
11 Dec 2006	NOAA 10930	Negative	Dextral
28 Dec 2005	NOAA 10841	Positive	Sinistral[a]
22 Dec 2005	NOAA 10838	Positive	Sinistral
27 Aug 2005	NOAA 10804	Negative	Dextral[a]
28 Aug 2005	NOAA 10803	Positive	Sinistral
26 Aug 2005	NOAA 10800	Positive	Sinistral
04 July 2005	NOAA 10781	Positive	Sinistral[a]
17 Apr 2005	NOAA 10752	Positive	Sinistral
09 Apr 2003	NOAA 10330	Positive	Sinistral[a]
03 Sep 2001	NOAA 09601	Positive	Sinistral[a]
01 Sep 2001	NOAA 09601	Positive	Sinistral[a]
30 Aug 2001	NOAA 09601	Positive	Sinistral[a]
01 Sep 2001	NOAA 09596	Positive	Sinistral
30 Aug 2001	NOAA 09596	Positive	Sinistral
30 Aug 2001	NOAA 09591	Negative	Dextral[a]
26 Aug 2001	NOAA 09590	Negative	Dextral
24 Aug 2001	NOAA 09590	Negative	Dextral
24 Aug 2001	NOAA 09585	Positive	Sinistral[a]

[a]Dominant sense

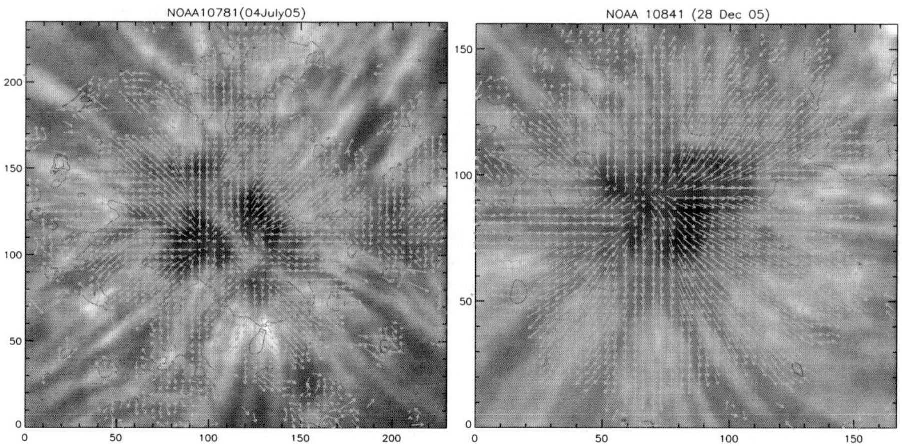

Fig. 1 Two examples of chromospheric Hα images of sunspots with the corresponding photospheric field measurements overlaid. The contours and vectors show longitudinal and transverse magnetic fields, respectively. The axis divisions are pixel numbers

Acknowledgment We thank the conference organizers for a very good meeting and the editors for excellent instructions.

References

Bao, S., Zhang, H. 1998, ApJ, 496, L43+
Berger, M. A., Field, G. B. 1984, J. Fluid Mech., 147, 133
Bernasconi, P. N., Rust, D. M., Hakim, D. 2005, Solar Phys., 228, 97
Burnette, A. B., Canfield, R. C., Pevtsov, A. A. 2004, ApJ, 606, 565
Chae, J. 2000, ApJ, 540, L115
Elmore, D. F., Lites, B. W., Tomczyk, S., et al. 1992, SPIE Conf. Series, 1746, 22
Hagino, M., Sakurai, T. 2005, PASJ, 57, 481
Hagyard, M. J., Pevtsov, A. A. 1999, Solar Phys., 189, 25
Hale, G. E. 1925, PASP, 37, 268
Leka, K. D., Canfield, R. C., McClymont, A. N., van Driel-Gesztelyi, L. 1996, ApJ, 462, 547
Martin, S. F. 1998, In: IAU Colloq. 167: New Perspectives on Solar Prominences, D. F. Webb, B. Schmieder, D. M. Rust (eds.), ASP Conf. Ser., vol. 150, p. 419
Martin, S. F. 2003, Adv. Space Res., 32, 1883
Pevtsov, A. A. 2008, J. Astrophys. Astron., 29, 49
Pevtsov, A. A., Canfield, R. C., Latushko, S. M. 2001, ApJ, 549, L261
Pevtsov, A. A., Canfield, R. C., Metcalf, T. R. 1994, ApJ, 425, L117
Pevtsov, A. A., Canfield, R. C., Metcalf, T. R. 1995, ApJ, 440, L109
Pevtsov, A. A., Canfield, R. C., Sakurai, T., Hagino, M. 2008, ApJ, 677, 719
Richardson, R. S. 1941, ApJ, 93, 24
Rust, D. M. 2003, Adv. Space Res., 32, 1895
Sankarasubramanian, K., Gullixson, C., Hegwer, S., et al. 2004, SPIE Conf. Series, 5171, 207
Sankarasubramanian, K., Lites, B., Gullixson, C., et al. 2006, In: Procs. 4th Solar Polarization Workshop, R. Casini, B. W. Lites (eds.), ASP Conf. Ser., vol. 358, p. 201
Seehafer, N. 1990, Solar Phys., 125, 219
Sokoloff, D., Zhang, H., Kuzanyan, K. M., et al. 2008, Solar Phys., 248, 17
Tiwari, S. K., Joshi, J., Gosain, S., Venkatakrishnan, P. 2008, In: Turbulence, Dynamos, Accretion Disks, Pulsars and Collective Plasma Processes, S. S. Hasan, R. T. Gangadhara, V. Krishan (eds.), Astrophys. & Space Sci. Procs., Springer, Heidelberg, p. 329

Evolution of Coronal Helicity in a Twisted Emerging Active Region

B. Ravindra and D.W. Longcope

Abstract Active-region magnetic fields are believed to be generated near the shear layer of the convection zone by dynamo processes. These magnetic fields are concentrated into fluxtubes, which rise, due to buoyancy, through the convection zone to appear in the form of bipoles at the photosphere. Thin-fluxtube simulations suggest that active regions require twist to emerge. All regions are observed to emerge with some twist; some of them show larger twist than others. A theoretical model [Longcope and Welsch 2000, ApJ, 545, 1089] predicts that an emerging fluxtube injects helicity into the corona for one or two days after its initial emergence through rotation of its footpoints driven by magnetic torque. There have been very few observational studies of helicity injection into the corona by emerging flux. This paper presents a study of helicity during the emergence of active region NOAA 8578. The time history of spin helicity injection, related to footpoint rotation, suggests that an Alfvén wave packet crossed the apex of the emerging fluxtube.

1 Introduction

Active regions often emerge as bipolar regions on the solar surface resembling Ω shaped fluxtubes. These newly emerged active regions appear as sunspots, concentrated with intense magnetic field strength. Longcope and Welsch (2000) predicted, through a simplified dynamical model, that a fraction of the current generated in the twisted emerging fluxtube will enter the corona. Any initial current mismatch between the photosphere and the corona results in a magnetic torque that drives rotation of the photospheric footpoints. This rotation is part of a torsional Alfvén wave propagating downward along the fluxtube and injecting helicity upward into the corona.

B. Ravindra (✉)
Indian Institute of Astrophysics, Bangalore, India

D.W. Longcope
Department of Physics, Montana State University, Bozeman, USA

S.S. Hasan and R.J. Rutten (eds.), *Magnetic Coupling between the Interior and Atmosphere of the Sun*, Astrophysics and Space Science Proceedings,
DOI 10.1007/978-3-642-02859-5_53,

By decomposing the photospheric helicity flux into spin and braiding components, it is possible to interpret an observed emergence in terms of the model of Longcope and Welsch (2000). In this paper we perform such a decomposition on high-resolution MDI magnetograms (0.6″ pixel size) taken at 1-min cadence during the emerging of active region NOAA 8578.

2 Results

Co-aligned portions of MDI magnetograms were corrected for magnetic field underestimation (Berger and Lites 2003) and then 5-min boxcar-averaged to reduce noise. The horizontal velocities were computed and partitioning of the magnetogram sequence was performed as in Longcope et al. (2007).

The emergence of AR 8578 is shown as a magnetogram time sequence in Fig. 1. The active region started to emerge at 07:00 UT on 08 June 1999 at a rapid rate of 10^{20} Mx h^{-1}. As the active region emerged, the bipoles moved away from each other at a rate of 292 m s^{-1}. The active region emerged with a tilt away from the East-West direction, and after 40 h of emergence aligned itself along the East-West direction as can be seen in Fig. 1.

The two largest regions resulting from the partition, P01 and N01, are shown in the left-hand panel of Fig. 2. The right-hand part of Fig. 2 shows the rotation rate of the regions P01 and N01, computed with the method of Longcope et al. (2007). As soon as the active region started emerging, region P01 rotated anticlockwise, but it gradually changed its direction of rotation to clockwise. The region then changed

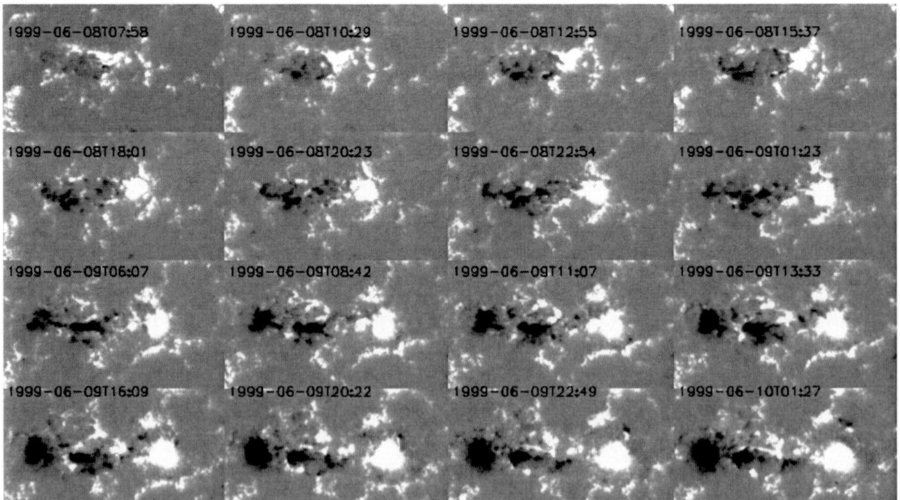

Fig. 1 Emergence of a bipole in active region NOAA 8578 on the solar surface close to disk center at latitude 19°. The date and time of the magnetogram snapshots are specified on each image

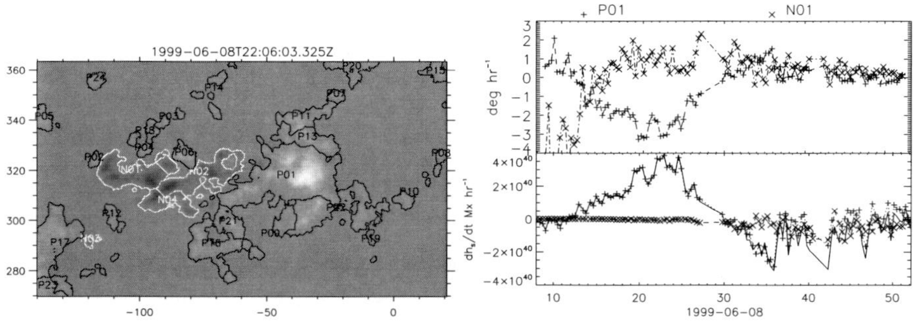

Fig. 2 *Left*: partitioning of the active region using the tessellation algorithm. P (*black contours*) denotes positive and N (*white contours*) denotes negative polarity. *Right*: rotation of regions P01 (+) and N01 (×) during emergence. The *top panel* shows the mean rotation rate, in degrees per hour. The *bottom panel* shows the spin helicity injection resulting from the rotation

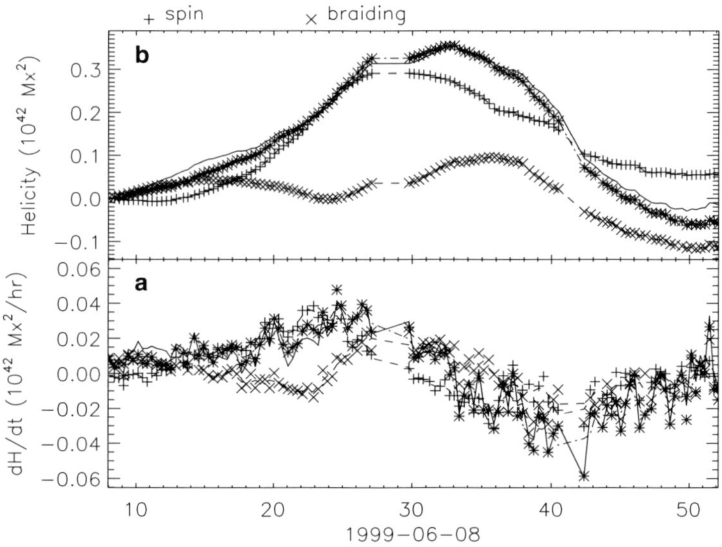

Fig. 3 (**a**) Various contributions to the helicity fluxes are plotted as a function of time: braiding (*multi*), spinning (*plus*), sum of braiding and spin (*asterisk*), and the total helicity flux (*solid curve*). (**b**) Time-integrated spin, braiding, and total helicity fluxes are plotted as a function of time

rotation direction again, after 20 h, returning to anticlockwise rotation. The rotation rate was about 2–3° h^{-1} in the beginning and ended at 0.5° h^{-1} after 30 h. Negative region N01 followed a mirror-image evolution, first rotating clockwise, then changing its direction of rotation, and then changing back to clockwise. In other words, P01 and N01 maintained rotation direction opposite to each other, even when they changed rotation direction twice.

Figure 3a shows the spin and braiding contributions to the helicity flux. It is clear that the braiding and spinning helicity fluxes had opposite sign for the first 15 h and that they interchanged sign later. The total injected helicity was positive (Fig. 3b) and the spin helicity contributed the most. It is evident from the lower right-hand panel of Fig. 2 that positive region P01 contributed the larger amount of spin helicity due to its larger magnetic flux. By the end of the observation, spinning helicity and braiding had contributed roughly equal amounts of helicity, of opposite signs, and the net injected helicity was almost zero.

It is striking that the reversal in spinning sense removes most of the helicity that was initially injected during the first day of emergence. The total helicity increases for about 20 h to roughly $4 \times 10^{41}\,\mathrm{Mx}^2$, mostly due to the clockwise rotation of region P01. Once this pole reversed, the spin helicity flux remained negative for roughly 20 h, causing the total to decrease back to almost zero.

3 Discussion

Active region NOAA 8578 followed the Hale–Nicholson polarity law after its emergence. The leading polarity rotated in one direction and the following polarity region rotated in the opposite direction. They both changed rotation direction but kept rotating opposite to one another. This rotation led to the injection of positive helicity during the first 25 h of emergence, followed by injection of negative helicity (or removal of positive helicity) over the next 20 h, canceling almost all the helicity that was initially injected.

This observation is consistent with the propagation of twist along one of the active region's submerged legs as a torsional Alfvén wave packet. The shape of the packet would explain the variation in rotation direction of one polarity. Its propagation through the corona would then cause the other polarity to rotate in the opposite sense. As the Alfvén speed is much larger in the corona than in the convection zone, the two footpoints should follow one another's evolution without perceptible delay, as observed. An isolated Alfvén wave packet would carry no net helicity, which is also consistent with our observations.

References

Berger, T. E., Lites, B. W. 2003, Solar Phys., 213, 213
Longcope, D. W., Ravindra, B., Barnes, G. 2007, ApJ, 668, 571
Longcope, D. W., Welsch, B. T. 2000, ApJ, 545, 1089

Power-Law Nanoflare Heating

L. Prasad and V.K. Joshi

Abstract Nanoflares are small impulsive events in the corona that dissipate magnetic energy in the range 5×10^{23} to 10^{26} ergs. We model their characteristics through assuming a power-law event distribution.

1 Introduction

The nanoflare hypothesis of Parker (1988) is that the corona contains a large assembly of high-temperature elemental magnetic filaments or loops, created together with the coronal magnetic field through randomly distributed impulsive heating events. They heat the loop plasma to 5×10^6 K (Shimizu 1995) and supposedly are one of the main agents generating the high coronal temperature (Yoshida and Tsuneta 1996; Watanabe 1995). Nanoflare coronal heating in terms of power-law distributions has been discussed by Dennis (1985). Kopp and Poletto (1993) extended this work to power-law indices $\alpha > 2$.

2 Power-Law Estimates

We express the total energy release by nanoflaring as the power law

$$\frac{\mathrm{d}N}{\mathrm{d}E} = A\,E^{-\alpha}, \tag{1}$$

with α the power law index, $\mathrm{d}N$ the number of small impulsive events between energy range E and $E + \mathrm{d}E$, and A a normalization factor (Hudson 1991; Kreplin 1997). The nanoflare rate over the whole solar surface is $R_{\mathrm{total}} = \int_{E_{\mathrm{min}}}^{E_{\mathrm{max}}} (\mathrm{d}N/\mathrm{d}E)\,\mathrm{d}E = A/(\alpha - 1)\,E_{\mathrm{min}}^{1-\alpha}$.

L. Prasad (✉) and V.K. Joshi
Kumaun University, Nainital, India

S.S. Hasan and R.J. Rutten (eds.), *Magnetic Coupling between the Interior and Atmosphere of the Sun*, Astrophysics and Space Science Proceedings,
DOI 10.1007/978-3-642-02859-5_54,

Table 1 Nanoflare rates per loop

Semilength (km)	Base area (cm^2)	Total number of events	Average rate (s^{-1})
1 000	6.3×10^{14}	9.8×10^{7}	0.004
2 000	2.5×10^{15}	2.6×10^{7}	0.017
5 000	1.6×10^{16}	3.9×10^{6}	0.105
10 000	6.3×10^{16}	9.8×10^{5}	0.419

The nanoflare rate per elementary flux loop R_l is $R_l = R_{total}/N_l$, with N_l the number of loops on the Sun or $R_l = (A/N_l)(\alpha - 1)\, E_{min}^{1-\alpha}$. Table 1 lists numerical values for different loop lengths. At smaller lengths more nanoflaring occurs. For $\alpha = 2$ the occurrence frequency is $f = f_0\, E^{-2}$, where f_0 is the cutoff. In that case the coronal heating is

$$R_c = f_0 \int_{E_{min}}^{E_{max}} E^{-2}\, E\, dE = f_0 = \ln \frac{E_{max}}{E_{min}}. \tag{2}$$

Setting $E_{min} = 5 \times 10^{23}$ erg and $E_{max} = 5 \times 10^{26}$ erg gives $R_c = 5.16 \times 10^4$ ergs cm^{-2} s^{-1}, close to the observed value 4.5×10^5 erg cm^{-2} s^{-1} reported by Krucker and Benz (1998).

Acknowledgment It is a pleasure to acknowledge the kind hospitality of IUCAA, Pune, and UGC, New Delhi, for partial support under Project No. F31-23/2005/(SR).

References

Dennis, B. R., 1985, Solar Phys., 100, 465
Hudson, H. S., 1991, Solar Phys., 133, 357
Kopp, R. A., Poletto, G., 1993, Solar Phys., 418, 496
Kreplin, R. W., Dere K. P., Horan D. M., Meekins J. F., 1997, In: The Solar Output and Its Variation, O.R. White (ed.), Colorado, p. 287
Krucker, S., Benz, A. O., 1998, Solar Phys., 501, L213
Parker, E. N. 1988, ApJ, 330, 474
Shimizu, T., 1995, PASJ, 47, 251
Yoshida, T., Tsuneta, S., 1996, ApJ, 459, 342
Watanabe, T., 1995, Solar Phys., 157, 169

Spectroscopic Diagnostics of Polar Coronal Plumes

K. Wilhelm, B.N. Dwivedi, and W. Curdt

Abstract Polar coronal plumes seen during solar eclipses can now be studied with space-borne telescopes and spectrometers. We briefly discuss such observations from space with a view to understanding their plasma characteristics. Using these observations, especially from SUMER/*SOHO*, but also from EUVI/*STEREO*, we deduce densities, temperatures, and abundance anomalies in plumes and inter-plume regions, and discuss their implications for better understanding of these structures in the Sun's atmosphere.

1 Introduction

Polar coronal plumes are ray-like structures aligned along open magnetic field lines in polar coronal holes. A total eclipse of the Sun shows these rays in white light, depicting the magnetic configuration of the Sun in a coronal hole. Many studies have been carried out to relate these rays to the coronal magnetic field inferred by current-free photospheric magnetic field extrapolation. The coronal plumes and the inter-plume regions seem to play a rôle in the acceleration mechanism of the fast solar wind. They have been extensively observed from space across the electromagnetic spectrum. Investigations have been made to unravel the appearance and disappearance of these plumes. The fact remains that we know little about them, probably because we have no direct knowledge of the coronal magnetic field. The identification of the sources that produce coronal plumes and their contribution to the fast solar wind is still a matter of investigation (DeForest et al. 1997; Wang et al. 1997; Wilhelm et al. 1998; Gabriel et al. 2003; Teriaca et al. 2003; Antonucci et al. 2004; Curdt et al. 2008). To understand the processes of plume formation, we need to know the physical conditions in plumes and the surrounding inter-plume

K. Wilhelm (✉) and W. Curdt
Max-Planck-Institut für Sonnensystemforschung, Katlenburg-Lindau, Germany

B.N. Dwivedi
Max-Planck-Institut für Sonnensystemforschung, Katlenburg-Lindau, Germany
and
Department of Applied Physics, Institute of Technology, BHU, Varanasi, India

S.S. Hasan and R.J. Rutten (eds.), *Magnetic Coupling between the Interior and Atmosphere of the Sun*, Astrophysics and Space Science Proceedings,
DOI 10.1007/978-3-642-02859-5_55,

environment, such as electron densities, n_e, and electron temperatures, T_e, the effective ion temperatures and non-thermal motions, the plume cross-section relative to the size of the coronal hole, and the plasma bulk speeds.

In this paper, we briefly discuss the observations of polar coronal plumes from space with a view for understanding their plasma characteristics. Using these measurements, especially from SUMER/*SOHO*, we deduce electron densities and temperatures as well as abundance anomalies in plumes. This will improve the understanding of these structures in the Sun's atmosphere, which are the subject of an International Team Study at ISSI, Bern.[1]

2 Spectroscopic Observations of Coronal Plumes

In the framework of a *Hinode/STEREO/SOHO* cooperation, observations of coronal plumes in a coronal hole were performed in April 2007, using spectrographs and imagers (cf., EUVI/*STEREO*) aboard these spacecraft (cf., Curdt et al. 2008). SUMER performed a scan in the southern coronal hole of the Sun from 7 April 2007 (01:01 UTC) to 8 April 2007 (12:19 UTC). Emission was observed from the O VI, Ne VIII, Mg VIII, Mg IX, Si VIII, Si IX, Al IX, and Na IX lines. The spectral lines were recorded almost simultaneously at each location. Contribution functions and the FIP (First Ionization Potential) values of the corresponding elements are shown in Fig. 1. Density-dependent Si VIII and temperature-dependent Mg IX line ratios

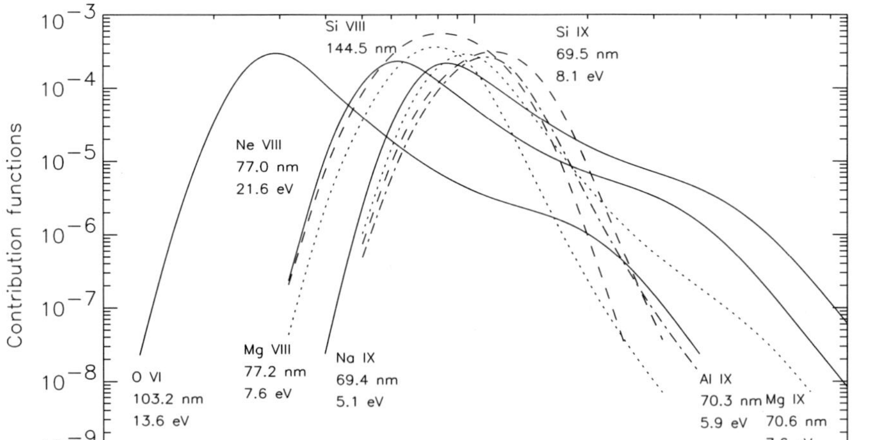

Fig. 1 Contribution functions of the observed lines and the FIP values of the corresponding elements, based on ionic fractions from Mazzotta et al. (1998). Neon and oxygen have high FIP values, while the other elements have low values <10 eV

[1] http://www.issibern.ch/teams/solarcoronal

were observed to produce n_e and T_e maps. Line-width studies allowed us to monitor the ion temperatures, which are much higher than the electron temperatures.

3 Results and Discussion

Figure 2 shows a large raster above the southern coronal hole, obtained in several VUV emission lines. All maps are noisy above a height of 150 Mm. The density and temperature maps are, therefore, averaged over larger height ranges along the line of sight. A detailed analysis of similar observations obtained in 2005 has shown that the plume density is about five times higher than that of the environment in this altitude range (Wilhelm 2006). The electron temperature, T_e, in plumes is lower than that in the interplume regions (cf., Wilhelm et al. 1998). This is confirmed by the Mg IX T_e-sensitive line pair in the present data. The insensitivity of the Si VIII ratio to scattered radiation is discussed by Wilhelm et al. (1998). It is caused by the lines being barely visible on the disk (Curdt et al. 2001) so that the stray-light is subtracted by the standard background correction for coronal observations.

Electron density and temperature maps are shown in Fig. 3a, b, respectively, in which gray represents cooler plasma conditions. All radiance maps (except for Al IX) show the plume structures. It is still to be investigated why the Al IX radiance map does not show plume structure. The line ratio Ne VIII/Mg VIII can monitor the abundance variations between high-FIP and low-FIP elements. However, the different temperatures in plume and inter-plume regions should be taken into account in view of the high-temperature tail of the lithium-like Ne^{7+}. The contribution functions of Ne VIII and Mg VIII overlap considerably in the temperature range just below 1 MK. The contribution functions of Ne VIII and Mg IX are more similar at

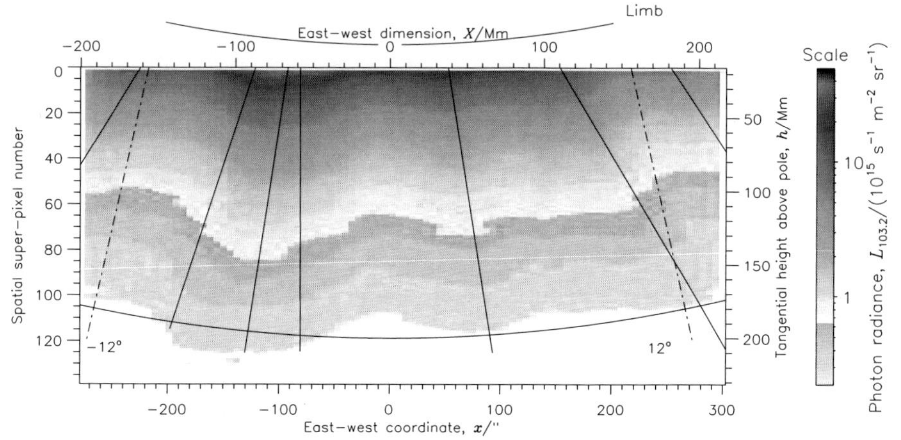

Fig. 2 A large raster above the southern coronal hole was obtained in several VUV emission lines. It took 36 h starting on 7 April 2007 at 01:01 UTC from West to East. The photon radiance of O VI $\lambda 1032$ is shown here. Radial *dashed-dotted lines* are shown at $\pm 12°$ off the pole

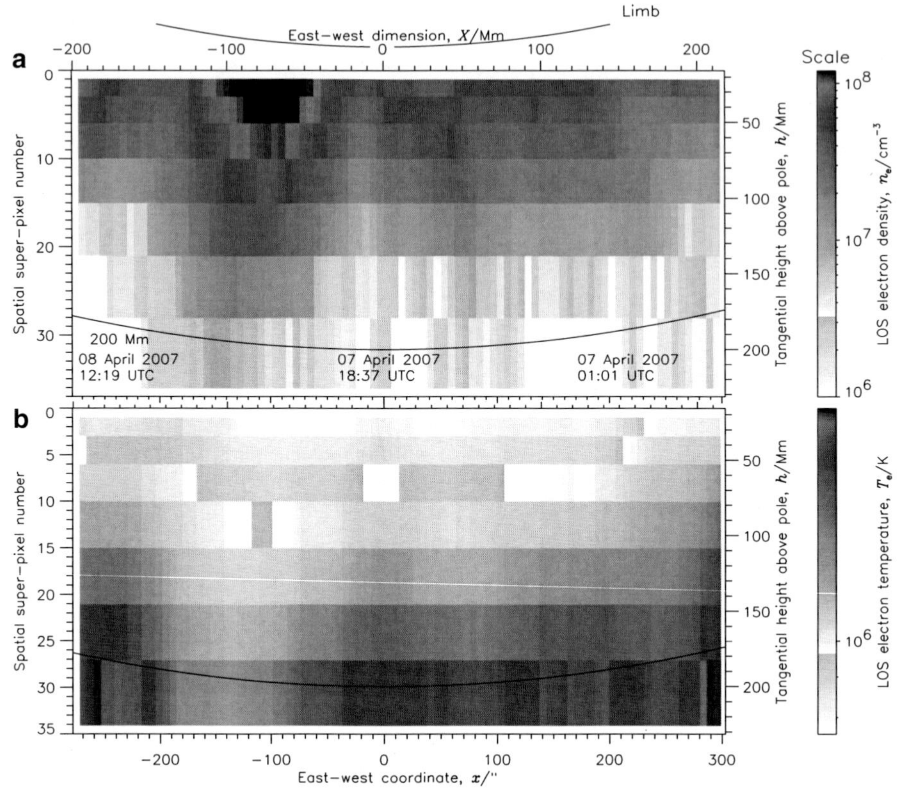

Fig. 3 (**a**) Electron density map; (**b**) Electron temperature map (*gray* represents cooler plasma conditions) (color illustration are available in the on-line version)

higher temperatures than that of Ne VIII and Mg VIII. And yet, the same signature is visible, which indicates that a temperature effect hardly plays a rôle. In particular, the Ne VIII/Na IX ratio shows that the abundance anomaly is real and not a temperature effect. The estimated FIP bias for Ne and Mg is 1.5–2. The search for Doppler shifts in the O VI line did not indicate any significant flows in plumes, although some of them were directed out of the plane of the sky (cf., Curdt et al. 2008). The super-radial expansion of plumes is evident in the O VI map.

Investigating lines of high- and low-FIP elements in the SUMER data set, we find the following:

- A temperature effect hardly plays a role, reconfirming the existence of abundance anomalies in plumes
- T_e is lower in plumes than in inter-plume regions in agreement with earlier findings
- Plumes have higher densities than their environment
- No significant flows could be detected in coronal plumes below 150 Mm.

Implications of deducing physical parameters, such as ion and electron temperatures, densities, abundance anomalies, outflow velocities in plumes and inter-plume regions are crucial to develop theoretical models of these features and of high-speed solar wind. This will also help in understanding the plume footpoint (e.g., XBPs), the relationship between processes at the footpoint and the plume characteristics, and to explore whether there is a relationship between plumes and the fast solar wind. We plan to expand this work and to communicate it as an article to the Astrophysical Journal.

References

Antonucci, E., Dodero, M. A., Giordano, et al. 2004, A&A, 416, 749
Curdt, W., Brekke, P., Feldman, U., et al. 2001, A&A, 375, 591
Curdt, W., Wilhelm, K., Feng, L., Kamio, S. 2008, A&A, 481, L61
DeForest, C. E., Hoeksema, J. T., Gurman, J. B., et al. 1997, Sol. Phys., 175, 393
Gabriel, A. H., Bely-Dubau, F., Lemaire, P. 2003, ApJ, 589, 623
Mazzotta, P., Mazzitelli, G., Colafrancesco, S., Vittorio, N. 1998, A&AS, 133, 403
Teriaca, L., Poletto, G., Romoli, M., Biesecker, D. A. 2003, ApJ, 588, 566
Wang, Y.-M., Sheeley, N. R., Jr., Dere, K. P., et al. 1997, ApJ, 484, L75
Wilhelm, K. 2006, A&A, 455, 697
Wilhelm, K., Marsch, E., Dwivedi, B. N., et al. 1998, ApJ, 500, 1023

A Flaring Polar Filament

C. Dumitrache

Abstract A special, huge polar filament producing multiple CMEs is analyzed. Around this double-S shaped filament a two-ribbon flare occurred which rose into a CME.

1 Introduction

This work is part of a series of papers concerning studies of a special sigmoid polar filament. It was observed on the solar disk between 11 and 19 August 2001. At first, two parallel filaments appeared, but on 13 August a double S-shape linked these two into a single huge feature (Dumitrache 2008). On 15 August, multiple CMEs were registered (Dumitrache submitted). A double-S polar filament rose above the polarity inversion line between the remnants of a complex of activity (a cluster of active regions that had developed during the previous solar rotation) and the unipolar magnetic field surrounding it. In the same rotation (CR 1979), another active region (AR9571) was located towards the South-West but too far away to influence the polar filament.

In this paper, we analyze an unusual event that occurred on 14 August 2001 in the polar filament: the occurrence of a spotless flare. It produced a CME registered by LASCO at UT 16:30 as an LDE (long decay event). Eruptive flares tend to lead to such long-decay events (MacCombie and Rust 1979) and to be more strongly associated with coronal mass ejections. They are cooler and fainter than active-region flaring events. A two-ribbon flare occurred in the main body of the complex double-S filament, representing only one event in the long and rich history of its evolution. Unfortunately, lack of data during the flare development inhibits detailed analysis of the flare onset. The two ribbons of the flare were situated on the filament sides, while the post-flare loops covered all of the filament channel in the direction where the filament had minimum curvature.

C. Dumitrache (✉)
Astronomical Institute of Romanian Academy, Bucharest, Romania

S.S. Hasan and R.J. Rutten (eds.), *Magnetic Coupling between the Interior and Atmosphere of the Sun*, Astrophysics and Space Science Proceedings,
DOI 10.1007/978-3-642-02859-5_56,

The Stanford whole-sun magnetograms indicate that large-scale magnetic reconnections started on 13 August 2001 (Dumitrache 2008). On 15 August, the double-S filament gave repeated CMEs in a squall while part of it disappeared temporary to reform during the next rotation. We suggest that these large-scale reconnections caused the filament destabilization and the explosive events.

2 Observations

The association of S-shaped soft X-ray features with eruptivity is well established. An X-ray C2.3 GOES type flare was registered on 14 August between UT 10:56 and 14:04, with its maximum at UT 12:42. The soft X-ray telescope of the Yohkoh mission observed this flare between UT 10:47 and 16:42 with a mass ejection starting at UT 12:14 (Fig. 1). A Hα image (Fig. 2) displays a two-ribbon flare with the filament in the middle of the ribbons, but actually the latter are the foot points of the bright post-flare arcade. The maximum of the flare was accompanied by a small wave visible in EIT images and more evident in soft X-rays. This wave stopped in the bright plage near the filament South end (marked with a circle). The flare onset was situated very close to this region. The flare development as seen in EIT observations is displayed in Fig. 3.

A halo CME was registered by SOHO at UT 16:01 at 618 km s^{-1} speed. On 14 August 2001, deceleration of the differential rotation of the filament occurred as estimated from the d'Azambuja law. Actually, the foots of the filament end approached each other so that the curve of the filament's tilt angle had an inflexion point. On

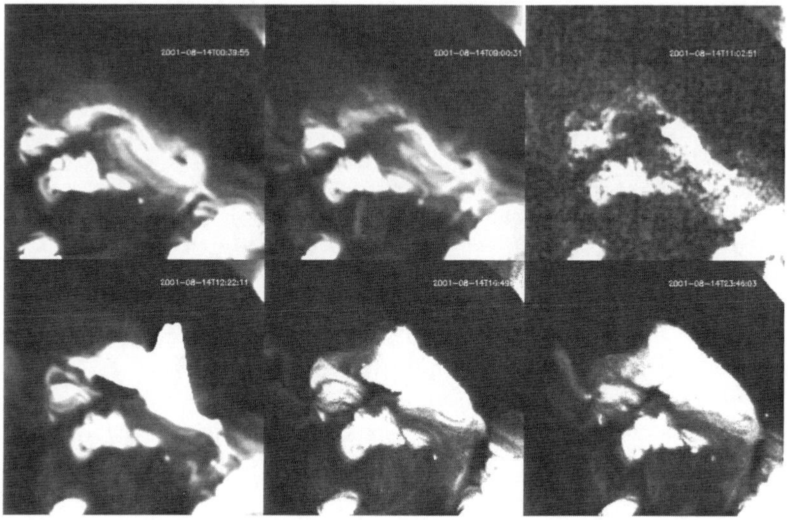

Fig. 1 Soft X-ray observations

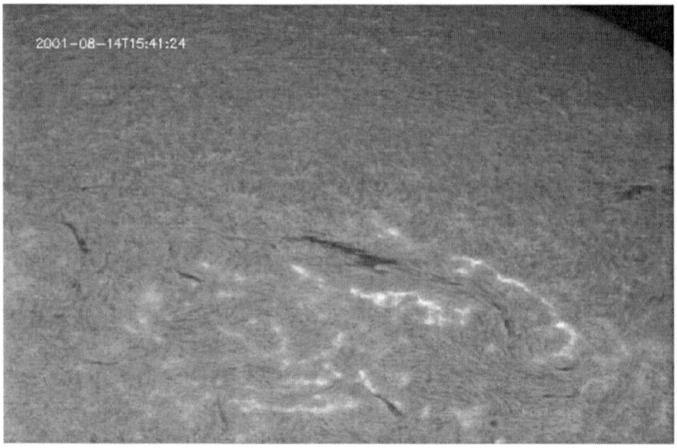

Fig. 2 Hα observations

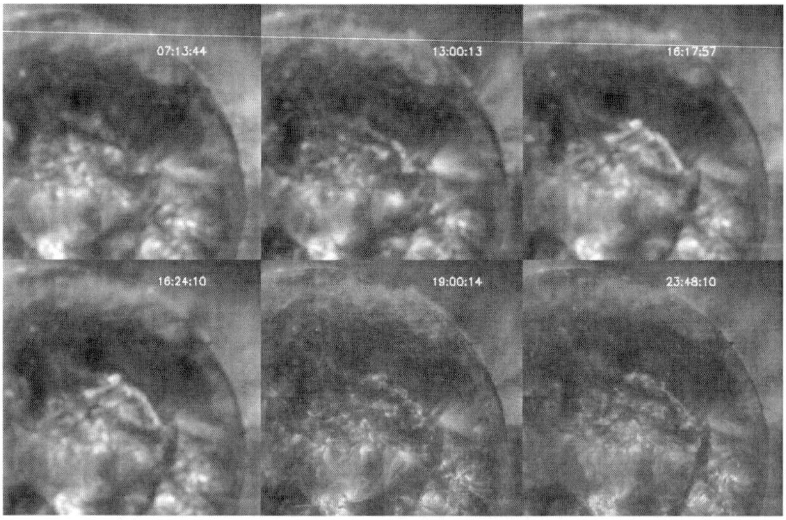

Fig. 3 EIT observations

15 August, the filament had three CMEs in a squall; on this date all diagnostics showed major changes. The estimated filament helicity decreased during the whole observing period.

We have extrapolated the 3D coronal magnetic field from the available MDI magnetograms in the filament channel zone using a Powell method to determine the local extreme (Lee 2002). The results suggest that at UT 07:45, before the flare onset, there were no loops above the filament channel, whereas afterwards post-flare loops are visible above the filament channel, connecting the six dipoles considered in the computations. These loops persisted many hours and were denser shortly after the flare end. Many magnetic reconnections occurred during the flare decay phase.

The flaring filament presented here was a long-lived feature. We intend to study many other aspects of its dynamics and development.

References

Dumitrache, C. 2008, In: Exploring the Solar System and the Universe, V. Mioc, C. Dumitrache, N. A. Popescu (eds.), AIPC, vol. 1043, p. 298

Dumitrache, C., In: Universal Heliophysical Processes, IAU Symp. 257, submitted

Lee, J. K. 2002, MA Thesis, University of Alabama

MacCombie, W. J., Rust, D. M. 1979, Solar Phys., 61, 69

Solar X-Ray Processes

P.C.V. Mallik, J.C. Brown, and A.L. MacKinnon

Abstract Past analyses of solar flares have ignored nonthermal recombination (NTR) emission as a means of producing Hard X-rays (HXRs) in the corona and chromosphere. However, Brown and Mallik (2008, A&A, 481, 507) have shown that NTR can be significant and even exceed nonthermal bremsstrahlung (NTB) emission for certain flare conditions that are quite common. For hot enough plasma ($T > 10$ MK), HXR emission of a few deka-keV has a large contribution from NTR onto highly ionized heavy elements, especially Fe. Consequently, including NTR has implications for the magnitude and the form of the inferred electron spectrum, $F(E)$, and hence for fast-electron density and energy budgets and for the acceleration mechanisms. We show under what circumstances NTR dominates in deka-keV HXR emission. It is important to note that at high temperatures, HXR emission from thermal electrons (recombination and bremsstrahlung) becomes important. However, NTR dominates over NTB without being swamped by thermal emission in the photon energy (ϵ) regime of 20–30 keV and temperature range of 10–25 MK (Fig. 1, left). By integrating the flux for all $\epsilon > 20$ keV, i.e., looking at the source luminosity function above 20 keV, we were able to show that by including NTR, the acceleration requirements are less demanding for every event, but to varying degrees based on temperature (T), spectral index (δ) and electron low-energy cut-off (E_c). Our key result is that, for $T > 10$ MK and $\delta \approx 5$, including NTR reduces the demand for nonthermal electrons by up to 85%. Our paper with these results will be submitted to ApJ Letters.

In related work on diagnosing ion acceleration and relativistic electron acceleration in flares, we have also been reconsidering the role of Inverse Compton (IC) scattering of photospheric photons. Gamma-ray observations clearly show the presence of 100 MeV electrons and positrons in the solar corona, by-products of GeV energy ions. In our work on IC scattering, we aim to show that IC can be an important tool in diagnosing neutron distributions near the Sun. This is because secondary positrons are produced in the same nuclear reaction processes as free neutrons. Korchak (1971, Solar Phys., 18, 284) showed that IC may be significant in low-density regions of the solar atmosphere, for example, the corona. Using the

P.C.V. Mallik (✉), J.C. Brown, and A.L. MacKinnon
Department of Physics and Astronomy, University of Glasgow, Scotland, UK

S.S. Hasan and R.J. Rutten (eds.), *Magnetic Coupling between the Interior and Atmosphere of the Sun*, Astrophysics and Space Science Proceedings,
DOI 10.1007/978-3-642-02859-5_57,

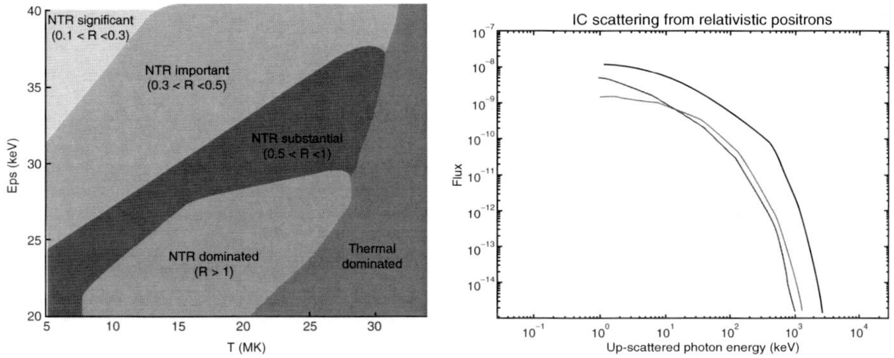

Fig. 1 *Left*: A shaded plot of the (ϵ, T) plane showing regimes of NTR importance for accelerated ratio, $f_c = 0.01$. For $f_c = 0.1$, the thermal-dominated boundary moves about 7 MK to the right. *Right*: Photon spectra at the Sun (photons per keV per second per source positron) from positrons produced by protons with a power-law energy distribution, i.e., E^{-3}. The darker (*black*) curve is for the flux from the limb while the lighter (*blue* and *red*) curves are for the flux near disk center (color illustration are available in the on-line version)

calculations of Moskalenko and Strong (2000, ApJ, 528, 357), we tailored the radiation field geometry to suit that of the solar surface. Our model shows how highly energetic electrons and positrons up-scatter photospheric UV or visible photons into the HXR regime. We calculated the IC spectrum using typical electron (power-law) and secondary positron (by-products of power-law proton) distributions observed near the Sun. The trends match our expectation that the most energetic photons are likely to come from the limb (Fig. 1, right). This is because of relativistic effects and the geometry, which is such that the photons produced by head-on collisions, that is, those with maximum energy, are those produced at the limb as viewed by the observer. Note that because of relativistic effects, the radiation is highly beamed, that is, we see the photons produced by the particles that are coming straight towards us (Jones 1968, Phys. Rev., 167, 1159). The spectra are also very hard, and hence if observed, would be unmistakable. If detectable by modern instruments, this would be a great tool to determine secondary positron distributions near the Sun, thereby giving us a handle on the neutron fluxes to be encountered by future inner heliosphere space missions. Our paper on this work will be submitted to A&A.

Acknowledgment This work was made possible by a UK STFC Rolling Grant and Dorothy Hodgkin's Scholarship. We thank the conference organizers of IIA, Bangalore, for a very well-organized and memorable meeting.

References

Brown, J. C., Mallik, P. C. V. 2008, A&A, 481, 507
Jones, F. C. 1968, Phys. Rev., 167, 1159
Korchak, A. A. 1971, Solar Phys., 18, 284
Moskalenko, I. V. and Strong, A. W. 2000, ApJ, 528, 357

Time-Varying Thermal Emission in Solar Flares

R. Jain, A.S. Rajpurohit, M. Aggarwal, R. Jamwal, and A. Awasthi

Abstract We study thermal emission in solar flares using high-resolution X-ray spectra obtained with the Si detector of the Solar X-ray Spectrometer (SOXS) mission onboard the GSAT-2 Indian spacecraft launched in 2003. We model the spectral-temporal evolution of the medium-hard X-ray flux in terms of an evolving multi-temperature plasma governed by thermal conduction cooling and find agreement with the observations. By measuring the DEM power-law index for five M-class flares, we find that the emission in the 6–20 keV energy range is dominated by temperatures 15–50 MK, while the power-law index of the thermal spectrum varies over 2.2–6.1. The mean value of the thermal conduction cooling time is 1,440 s; the temperature-dependent cooling time varies from 22 to 102 s.

1 Introduction

Aschwanden (2007) proposed three physical processes that lead to measurable time delays as a function of energy in solar flares: the time-of-flight dispersion of free-streaming electrons, the collisional trapping of electrons, and the Neupert effect (Hudson 1991; Dennis and Zarro 1993). The characteristic times for heating and cooling of the X-ray emitting plasma in solar flares are estimated from the time profile and temperature and the emission measure of the thermal X-ray burst and the over-all length scale of the flare-heated plasma at thermal X-ray maximum. The heating is assumed to be due to magnetic reconnection, the cooling to heat conduction, and radiation. As the conductive losses increase rapidly with temperature, several authors have concluded that conduction dominates radiation at least during the initial phase when $T \geq 10^7$ K (Culhane et al. 1970; Moore and Datlowe 1975). Culhane et al. (1970) treated the problem of flare-produced plasma cooling by examining three cooling processes: Coulomb collisions of hot

R. Jain (✉), A.S. Rajpurohit, M. Aggarwal, and A. Awasthi
Physical Research Laboratory, Navrangpura, Ahmedabad, India

R. Jamwal
Department of Physics, Barkatullah University, Bhopal, India

S.S. Hasan and R.J. Rutten (eds.), *Magnetic Coupling between the Interior and Atmosphere of the Sun*, Astrophysics and Space Science Proceedings,
DOI 10.1007/978-3-642-02859-5_58,

electrons with lower-temperature ions, conductive cooling, and radiative cooling. They concluded that the first process is unimportant because it is not applicable to solar flares (Somov 1978), and that radiative cooling, which varies as the square of the electron density, is ineffective below particle density $10^{11}\,\mathrm{cm}^{-3}$. Based on observed flare parameters, Antiochos and Sturrock (1978), Culhane et al. (1994) and Aschwanden and Alexander (2001) suggested that thermal conduction dominates in high-temperature soft X-ray loops, while radiative cooling dominates in the later cooling phase in post-flare loops, as observed at EUV wavelengths. However, during the flare rise phase, time profiles in different energy bands are observed to have different temporal characteristics, which is not understood quantitatively in terms of conductive cooling times.

In this paper we address this issue by considering multi-temperature plasma with temporal evolution. The bremsstrahlung spectrum $F(\varepsilon)$ of an isothermal plasma with temperature T as a function of photon energy $\varepsilon = h\upsilon$ is given by

$$F(\varepsilon) = F_0 \int \frac{\exp(-\varepsilon/k)}{T^{1/2}} \frac{\mathrm{dEM}(T)}{\mathrm{d}T}\,\mathrm{d}T, \tag{1}$$

with $F_0 = 8.1 \times 10^{-39}\,\mathrm{s}^{-1}\,\mathrm{cm}^{-2}\,\mathrm{keV}^{-1}$ and $\mathrm{dEM}(T)/\mathrm{d}T$ specifies the temperature sensitivity of the differential emission measure $\mathrm{DEM} = n_{\mathrm{e}}{}^2\,\mathrm{d}V$ in volume element $\mathrm{d}V$ with temperature T and electron density n_{e} (Aschwanden 2007). Taking $\mathrm{DEM} = 10^{49}\,\mathrm{cm}^{-3}$ and increasing the temperature from $T = 15$ to $T = 50\,\mathrm{MK}$ in steps of 5 MK, we obtain the flare spectra in the left-hand panel of Fig. 1. These multi-temperature bremsstrahlung spectra together represent multi-thermal flare plasma. Their sum (dotted curve) suggests the presence of high-temperature gas that is generally referred to as superhot in solar flares.

The DEM distribution is characterized by the power law $\mathrm{dEM}(T)/\mathrm{d}T \approx (T/T_0)^{-4.5}$, so that the photon spectrum may be written in power-law form as

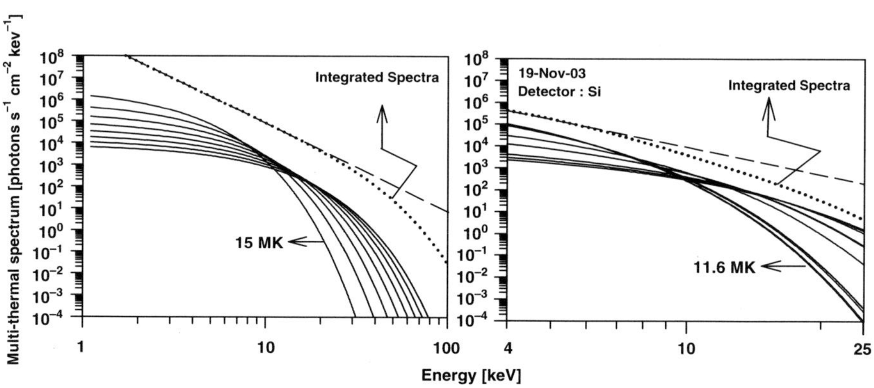

Fig. 1 *Left*: Theoretical bremsstrahlung spectra for temperatures $T = 15, 20, \ldots, 50\,\mathrm{MK}$ and $\mathrm{DEM} = 10^{49}\,\mathrm{cm}^{-3}$, in steps of 5 MK (*thin curves*) and their sum (*dotted*). The *dashed line* is the bremsstrahlung spectrum with power-law slope $\gamma = 4$. *Right*: Model for the bremsstrahlung spectrum of the 19 November 2003 flare with contributions from the peak temperatures and DEM distributions using multi-thermal power-law fitting

$F(\varepsilon) \propto \varepsilon^{-4.0}$. The sum (dotted curve) in the left-hand panel of Fig. 1 represents power-law bremsstrahlung with slope −4, which suggests that apparent super-hot plasma in a solar flare may actually represent the energy-dependent temporal evolution of multi-temperature plasma.

2 Observations, Analysis, Results

To test the above interpretation, we probe multi-temperature plasma in five M-class flares using full-disk X-ray emission observations with the Si detector of the Solar X-Ray Spectrometer (SOXS), which was launched onboard the GSAT-2 Indian spacecraft in 2003 (Jain et al. 2005) and provides unprecedentedly high spectral resolution (0.7 keV) in the 4–25 keV energy range. We exploit it here to model such data from five selected flares listed in Table 1. The data were taken from the SOXS website at http:www.prl.res.in/~soxs-data.

The SOXS X-ray spectra were reduced with package OSPEX in the SolarSoft distribution, accounting for the response of the Si detector and yielding peak plasma temperatures and DEM estimates through least-χ^2 fitting. The right-hand panel of Fig. 1 shows spectra for the 19 November 2003 flare as example, modeled by applying contributions of peak-temperature plasma measured at different times, with average measured DEM $= 0.5 \times 10^{49}$ cm^{-3} and $\delta = 5.5$ as listed in Table 1. The sum of all thermal contributions, represented by the dotted curve, may be expressed as $F \propto \varepsilon^{-3.6}$ or $\gamma = -3.6$.

The left-hand panel of Fig. 2 shows the temporal evolution of the 19 November 2003 flare, in terms of its intensity in 13 energy bands. This flare evolved in the beginning via slow plasma heating seen in soft X-rays (4–10 keV) and then through impulsive transport followed by a long-duration thermal component. This figure also reveals delay of the flare onset at higher energy while peaking earlier relative to the low-energy curves. For example, in the left-hand panel of Fig. 2, the flux in the highest energy band (20–25 keV) peaked at 03:59:14 UT, while it peaked in the next lower energy band (15–20 keV) at 03:59:30 UT, at a delay of 16 s. The right-hand panel of Fig. 2 shows these delays as function of energy, and demonstrates that the

Table 1 Parameters of the studied flares. The X-ray peak intensities are in units of 10^4 photon count cm^{-2} s^{-1} in the 7–10 keV band

Date	Peak Time	Peak Intens.	GOES Class	Loop Length (Mm)	n_e $\times 10^{10}$ (cm^{-3})	t_c (s)	Avg. DEM $\times 10^{49}$ (cm^{-3})	DEM index (δ)	Photon index (γ)
19-11-03	04:02	5.8	M1.7	38	8.8	77	0.5	5.5	3.6
07-01-04	04:00	38.0	M4.5	220	0.3	22	0.21	7.9	2.1
31-10-04	05:31	1.1	M2.3	20	50.0	102	6.6	1.1	6.1
25-04-05	05:36	47.5	M2.3	55	4.5	54	0.09	5.6	3.1
17-09-05	06:05	18.4	M9.8	110	1.9	42	0.06	8.3	4.2

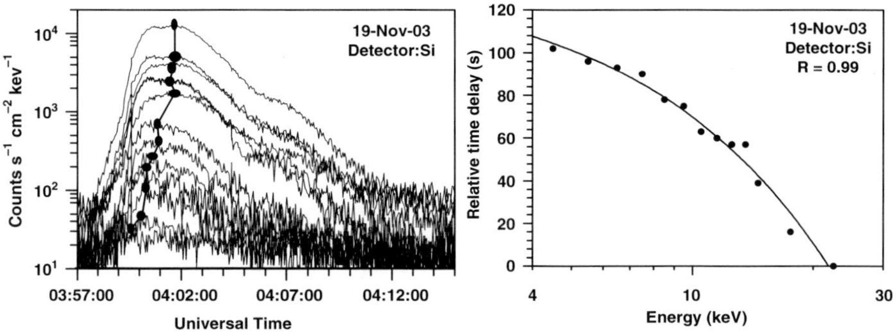

Fig. 2 *Left*: Temporal profiles in 13 energy bands (ϵ = 4–5, ..., 14–15, 15–20, and 20–25 keV) of the 19 November 2003 flare. Each peak time is shown by a *circle*. The curve connecting these circles illustrates the delay between successive profiles. *Right*: Relative time delay between two consecutive energy bands vs. energy for this flare

rise time to peak emission was longer at lower energy. We measured such behavior for all five flares. It also suggests the presence of multi-thermal plasma.

Multi-temperature components in flare plasma are due to heating and cooling processes during the flare evolution. An intimate relationship between the nonthermal and thermal time profiles was found long ago in the sense that the thermal emission often closely resembles the integral of the nonthermal emission, which is known as the Neupert effect. This is, however, strictly true only for the asymptotic limit of very long cooling times (t_c). The cooling time at a given energy band can be estimated from the decay time in a flare time profile. However, flare plasma cooling at initially high temperature (T $\geq$ 10 MK) is dominated by conduction cooling, observable in soft X-rays, while the later phase detected in EUV (T $\leq$ 2 MK) is dominated by radiative cooling. The conduction cooling time (t_c) is given by the ratio of the thermal energy to the conductive loss rate and can be expressed as power-law dependence on temperature

$$t_c(T) = t_{c0}\,(T/T_0)^{-\beta}, \tag{2}$$

where T is the flare peak temperature, T_0 = 11.6 MK, $\beta = 5/2$, and the thermal conduction cooling time t_{c0} is given by

$$t_{c0} = 344\,\left(L/10^9\right)^2\left(n_e/10^{11}\right), \tag{3}$$

where L is the half loop length and n_e the electron density. We estimated the length and volume of the flaring loops on corresponding Hα images recorded at Hiraiso Solar Observatory in Japan (Akioka 1998). The left-hand panel of Fig. 3 shows the Hα loop of the 9 November 2003 flare. Its length L is measured from position A to B. Comparison of the volume with the DEM estimate from the X-ray spectra yields the electron density n_e and hence the thermal conduction cooling time t_{c0}, which

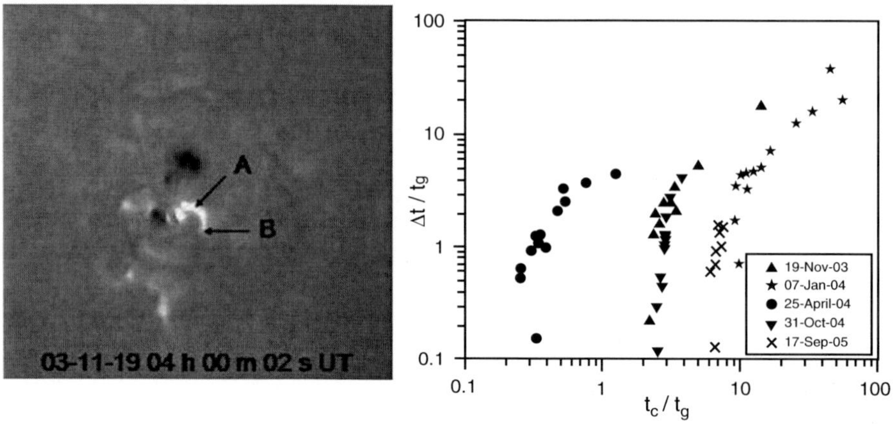

Fig. 3 *Left*: Hα filtergram taken on 19 November 2003 at 04:00:02 UT. *Right*: Time delay Δt obtained by convolving a Gaussian profile with halfwidth t_g with an e-folding time t_c for each flare. The time scales are normalized by t_g

is 1,088 s in this case, and through (2), the conductive cooling time t_c, which was 77 s for this flare. The five flares yielded t_{c0} values between 1,088 and 2,016 s and t_c values between 22 and $1,02$ s (Table 1).

The right-hand panel of Fig. 3 shows convolutions of the time profiles per energy band with Gaussians of width t_g. The convolution delays the peak (Δt). The cooling time t_c does not vary when the time delay $\Delta t \leqslant 1$, while for $\Delta t \geqslant 1$ the cooling time increases nearly logarithmically for each flare. This property corresponds to the Neupert effect, that is, the soft X-ray profile corresponds to the temporal integral of the hard X-ray emission. It is the asymptotic limit for very long cooling time $t_c \to \infty$.

3 Conclusion

Our observations provide direct evidence of multi-thermal plasma in all five flares and suggest that hard X-ray spectra should be fitted with multi-thermal spectra rather than isothermally. The relative delay measurements show that the rise-to-peak time to reach peak is longer for lower energy. This delay increase indicates also that flares are multi-thermal plasma. We also found that the cooling time t_c is constant for $\Delta t \leqslant 1$ and increases nearly logarithmically for $\Delta t \geqslant 1$, retrieving the Neupert effect in the asymptotic limit.

Acknowledgment We are grateful to M.J. Aschwanden and K.J. Shah for fruitful discussions during the development of our model and to R.J. Rutten for much editorial help.

References

Akioka, M. 1998, In: Synoptic Solar Physics, K. S. Balasubramaniam, J. Harvey, D. Rabin (eds.), ASP Conf. Ser., 140, 475
Antiochos, S. K., Sturrock, P. A. 1978, ApJ, 220, 1137
Aschwanden, M. J., Alexander, D. 2001, Solar Phys., 204, 91
Aschwanden, M. J. 2007, ApJ, 661, 1242
Culhane, J. L., Vesecky, J. F., Phillips, K. J. H. 1970, Solar Phys., 15, 394
Culhane, J. L., Phillips, A. T., Inda-Koide, M., et al. 1994, Solar Phys., 153, 307
Dennis, B. R., Zarro, D. M. 1993, Solar Phys., 146, 177
Hudson, H. S. 1991, BAAS, 23, 1064
Jain, R., Dave, H., Shah, A. B., et al. 2005, Solar Phys., 227, 89
Moore, R. L., Datlowe, D. W. 1975, Solar Phys., 43, 189
Somov, B. V. 1978, Solar Phys., 60, 315

Multi-Wavelength View of Flare Events on 20 November 2003

P. Kumar, P.K. Manoharan, and W. Uddin

Abstract We analyze two flare events that occurred in active region NOAA 501 on 20 November 2003. The Hα and magnetogram measurements show interaction between two filaments, which produced a slowly rising flare event, corresponding to two stages of magnetic reconnection. The relative clockwise rotation between the two sunspot systems caused filament destabilization. The cusp-shaped magnetic field in the main phase of the second flare and its evolution in correlation with ribbon separation provide evidence for the cause of the CME eruption. The propagation and orientation of the CME with respect to the ecliptic plane is illustrated by IPS images.

1 Flare Events on 20 November 2003

The period of October–November 2003 is well known for its extreme solar activity, corresponding to ARs 484, 486, and 488 (Uddin et al. 2006). The continuous emergence of magnetic flux in and around AR 501 (return of AR484, in the next rotation) also caused several intense flares. Here, we report two flare events (1N/M1.4 and 2B/M9.6) that occurred near N00 W05 on 20 November 2003. GOES soft X-ray measurements at 0.5–4 Å and 1–8 Å bands (Fig. 1) show a broad profile of gradual rise and decline for the event at 02:12 UT. However, the second event (peak at 07:47 UT) showed a quick rise in intensity and a gradual decrease. Before the start of the latter event, an impulsive flare (C3.8) was observed at 07:25 UT.

P. Kumar (✉) and W. Uddin
Aryabhatta Research Institute of Observational Sciences, Nainital, India

P.K. Manoharan
Radio Astronomy Centre (NCRA), Tata Institute of Fundamental Research, Udhagamandalam (Ooty), India

S.S. Hasan and R.J. Rutten (eds.), *Magnetic Coupling between the Interior and Atmosphere of the Sun*, Astrophysics and Space Science Proceedings,
DOI 10.1007/978-3-642-02859-5_59, © Springer-Verlag Berlin Heidelberg 2010

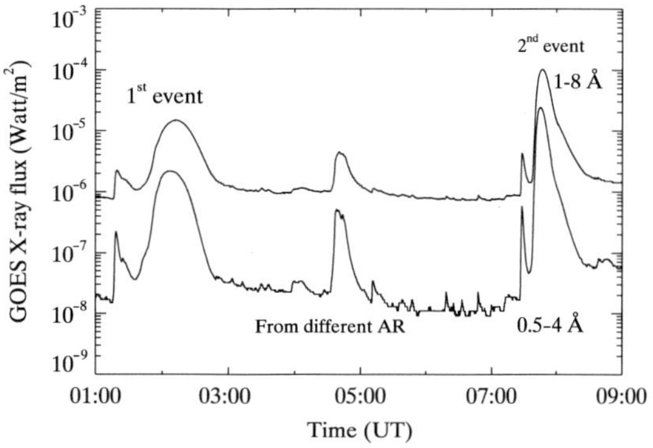

Fig. 1 GOES X-ray measurements in the 0.5–4 Å and 1–8 Å wavelength bands

2 Reorganization of Magnetic Structures

Hα images were recorded at ARIES, Nainital, with a cadence of about 15–20 s (details in Joshi et al. 2007). The high sampling rate allows us to follow the evolution of the complex twisted field structures in the chromosphere in relation to the magnetic configuration in the photosphere seen on MDI images (Fig. 2). Before the onset of the first flare, a system of two filaments goes through a gradual evolution (Fig. 2a–c, indicated by arrows) while they also approach each other at a speed of about 10 km s^{-1} (Fig. 3). The interaction between them triggers the energy release during 01:50–02:10 UT and leads to field merging seen in the Hα images. The change in filament orientation after the energy release indicates a relaxed state of the magnetic configuration.

During 02:40–07:30 UT, there was no significant activity at the flare site (Fig. 1). However, the energy buildup from 01:30 to 08:00 UT is revealed by the rotation of two sunspots of opposite polarity (the northern spot moving clockwise, the southern one anticlockwise) at the flare site (see arrows in the MDI images). During the onset of the second event at about 07:35 UT, the twisted filament system, with one footpoint attached to the sunspot group, goes through heavy destabilization and the filaments approach each other similarly as in the first event. The associated brightening, observed near the filament, suggests that reconnection occurred between twisted field lines and nearby small-scale fields. Furthermore, as the merging of the filaments progressed (Fig. 2d–h), the flare brightness rose to maximum intensity (Fig. 1). The Hα images also revealed restructuring of the system after the flare maximum.

A dark cusp-shaped structure formed after the flare maximum (Fig. 2f) and started to move upward. Its speed component in the plane of the disk was about 5–6 km s^{-1} during 07:45–07:50 UT (Fig. 3). At 07:51 UT, the cusp erupted and some part of it fell back. The interesting point is that the brightening observed along the flare ribbons and the rate of separation between the ribbons correlate well with the eruption rate of the cusp (Fig. 3). The high correlation (about 93%) suggests that the magnetic reconnection and the rise of the filament system played a prime role in the initiation of the coronal mass ejection (CME). After the flare, the filament system returned to its original, relaxed state.

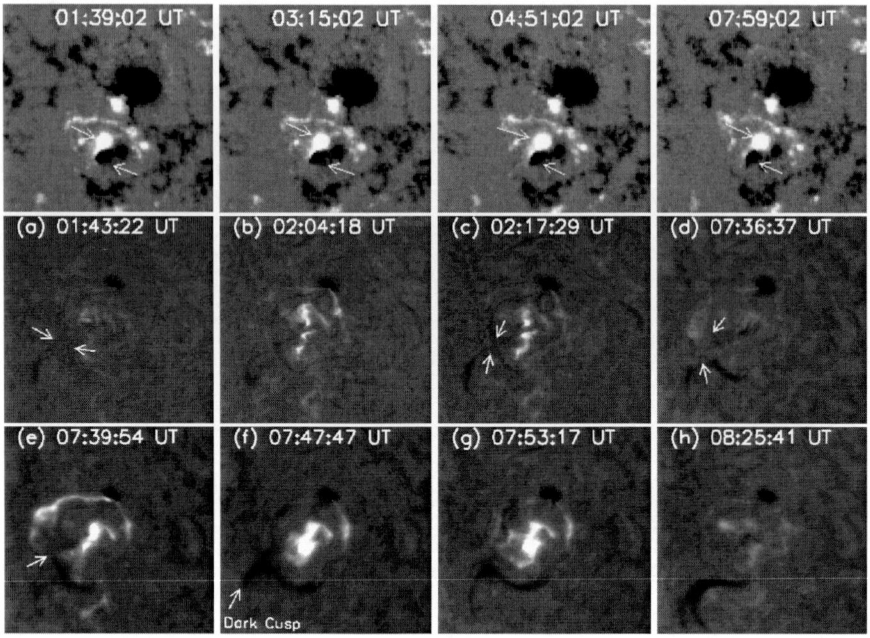

Fig. 2 MDI magnetograms (*top row*) and Hα images (*middle and bottom rows*) during the flare events

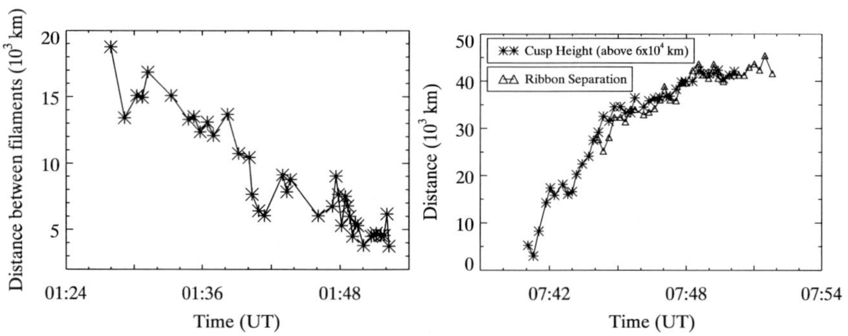

Fig. 3 *Left*: separation of the two filaments, showing that they approached each other. *Right*: cusp height variation and ribbon separation during the flare event

Figure 4 shows EIT and LASCO images taken at the time of the CME onset. The CME was tracked further out using the interplanetary scintillation (IPS) technique, which shows the geometry of the CME while crossing the Earth's magnetosphere.

3 Results and Discussion

This multi-wavelength study provides evidence that opposite rotation of opposite polarity regions plays a crucial role in building the magnetic energy required for

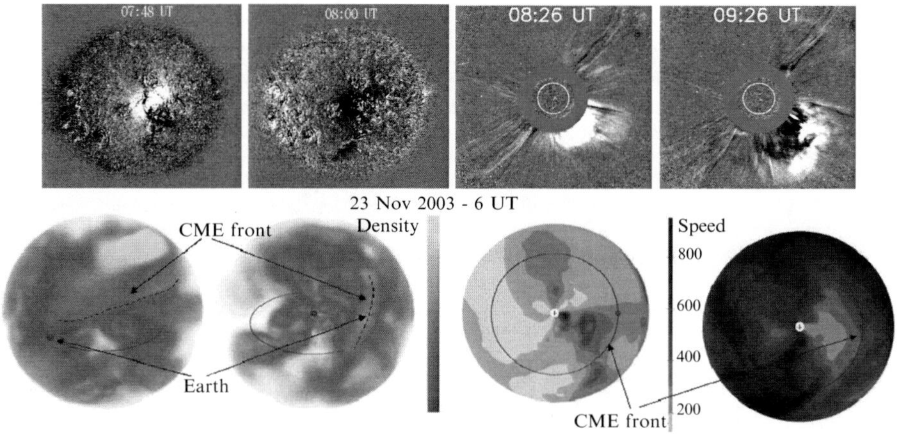

Fig. 4 *Top*: EIT difference images of the flare, showing the flare maximum at 7:48 UT and coronal dimming (*left two images*) and difference images of the CME from the C2 coronagraphs on SOHO/LASCO, showing the CME association with the second, M9.6 flare (*right two images*). *Bottom*: 3D view of the CME in Ooty IPS images, showing the orientation of the flux rope of about 70° with respect to the ecliptic plane

the flare process. Sunspot rotation is the primary driver of helicity production and injection into the corona (Tian et al. 2006, van Driel-Gesztelyi et al. 2002). Newly emerging flux plays a major role in the destabilization of filaments. The cusp shape suggests the formation of a magnetic null point in the high corona (Manoharan et al. 2003). The correlation between the separation of flare ribbons and the expansion of the cusp structure indicates that large-scale reconnection and particle acceleration occurred during the cusp eruption. The IPS technique shows that the flux rope is oriented about 70° with respect to the ecliptic plane. Therefore, in spite of a strong shock, the Earth-directed CME caused only a moderate storm (Dst −85 nT) at the Earth.

Acknowledgment SOHO (EIT, LASCO, and MDI images) is a project of international cooperation between ESA and NASA. PKM acknowledges the partial support for this study by CAWSES-India Program, which is sponsored by ISRO.

References

Joshi, B., Manoharan, P. K., Veronig, A. M., Pant, P., Pandey, K. 2007, Solar Phys., 242, 143
Manoharan, P. K., Kundu, M. R. 2003, ApJ, 592, 597
Tian, L., Alexander, D. 2006, Solar Phys., 233, 29
Uddin, W., Chandra, R., Ali, S. S. 2006, JAA, 27, 255
van Driel-Gesztelyi, L., Schmieder, B., Poedts, S. 2002, ESA SP-477, 47

Time-Delay Between Solar Soft X-Ray and EUV Flare Emissions

M.B. Dhanya and A. Bhardwaj

Abstract Temporal variations in soft X-rays (1–8 Å) and EUV (26–34 nm) flare emission are studied for the high-activity period October–November 2003. Although EUV emission is known to have impulsive and gradual peaks during flares, we find that these appear in different ways. Both the impulsive and the gradual EUV peaks were delayed with respect to the 1–8 Å soft X-ray peak. The delay of the 26–34 nm impulsive peak shows a linear relation with the logarithmic X-ray peak, while the gradual peak does not show such dependence. This result has implications on the flare emission mechanism and the standard flare model.

1 Introduction

Multiwavelength studies of solar flares have helped to explore the underlying physical mechanism (see Benz 2008 for a recent review). The flare emission is separated impulsive and gradual phases based on the temporal behavior of the hard X-rays, microwaves, EUV, and soft X-rays (Kane 1974). Several observations (e.g., Castelli and Richards 1971; Doschek 1972; Kelly and Rense 1972; Wood and Noyes 1972; Horan et al. 1983) have shown that the impulsive phase, which is characterized by short duration, is dominated by hard X-ray, impulsive EUV, and microwave emission, while the second component called the gradual phase is dominated by soft X-ray and gradual EUV emissions.

We have compared the time-intensity profiles of 1–8 Å soft X-ray and 26–34 nm EUV for C, M, and X-class flares during the October-November 2003 period, which was an unusually active period in the declining phase of the solar cycle 23. We have chosen those flares that exhibit distinct impulsive and gradual peaks in EUV and investigated the time delay of the impulsive and gradual EUV peaks with respect to the soft X-ray peak.

M.B. Dhanya (✉) and A. Bhardwaj
Planetary Science Branch, Space Physics Laboratory, Vikram Sarabhai Space Centre, Trivandrum, India

S.S. Hasan and R.J. Rutten (eds.), *Magnetic Coupling between the Interior and Atmosphere of the Sun*, Astrophysics and Space Science Proceedings, DOI 10.1007/978-3-642-02859-5_60,

The solar soft X-ray (1–8 Å) data is taken from the XRS onboard GOES with a time resolution of 1 min. The 26–34 nm EUV data is taken from the SEM instrument onboard SOHO with time resolution of 15 s.

2 Results and Discussion

The time-intensity profiles of soft X-ray and EUV radiation for the X17.2 flare on 28 October 2003 are shown in Fig. 1. It can be seen that a group of closely situated peaks constitutes the impulsive phase of emission. Considering the peak with highest peak flux, the impulsive peak leads the soft X-ray peak by about 5 min, whereas the gradual EUV peak lags the soft X-ray peak by about 6 min.

Of the 41 flares chosen for this study, consisting of 9 X-class, 10 M-class, and 22 C-class flares, 76% have impulsive peaks in 26–34 nm leading the soft X-ray peak with time delays in the range 2–13 min, whereas 24% has zero time delays. Figure 2 shows the time delays vs. soft X-ray peak flux in the 1–8 Å band on a logarithmic scale. This plot is fitted linearly. The time lead for the impulsive peak with respect to the soft X-ray peak shows a power law relation with the soft X-ray peak flux.

We have also investigated the time delay between the 26–34 nm gradual peak with respect to the soft X-ray peak. There were some flares for which the gradual phase of emission is seen in the EUV but the gradual peak position cannot be determined accurately. Thus the number of flares considered for the time delay of gradual peak got reduced to 5 X, 8 M, and 14 C-class flares. We found that the time delay does not show any dependence on the peaking of the X-ray flux.

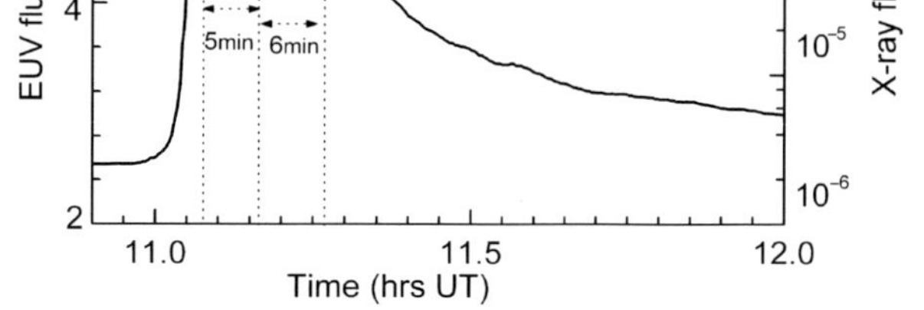

Fig. 1 Time profiles of X-ray flux in the wavelength band 1–8 Å as measured by the GOES satellite and the EUV flux in the wavelength band 26–34 nm as measured by SEM/SOHO for the X17.2 flare on 28 October 2003

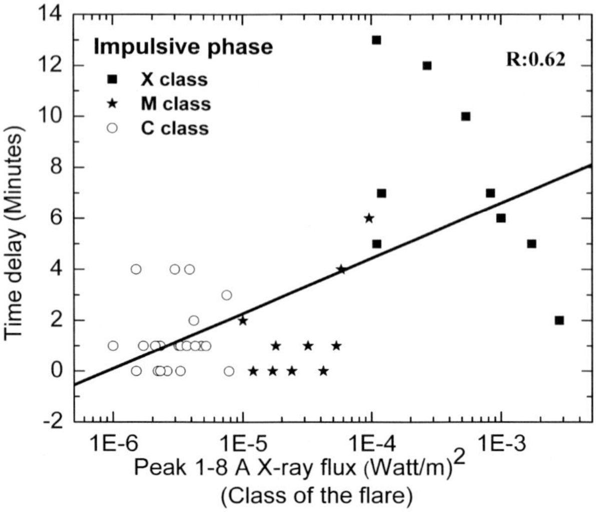

Fig. 2 Variation of time delay between the impulsive EUV peak and the soft X-ray peak with respect to the soft X-ray peak flux. The *solid line* is a linear fit with correlation coefficient $R = 0.62$. The time delay is measured as EUV peak–SXR peak. Read 1E-6 as 1×10^{-6}

References

Benz, A. O., 2008, Living Rev. Solar Phys., 5, 1
Castelli, J. P., Richards, D. W. 1971, J. Geophys. Res., 76, 8409
Doschek, G. A., 1972, Space Sci. Rev., 13, 765
Horan, D. M., Kreplin, R. W., Dere, K. P. 1983, Solar Phys., 85, 303
Kane, S. R., 1974, In: Coronal Disturbances, G. Newkirk, Jr. (ed.), IAU Symp. 57, 105
Kelly, P. T., Rense, W. A. 1974, Solar Phys., 26, 431
Wood, A. T. Jr, Noyes, R. W. 1972, Solar Phys., 24, 180

Major Surge Activity of Super-Active Region NOAA 10484

W. Uddin, P. Kumar, A.K. Srivastava, and R. Chandra

Abstract We observed two surges in Hα from the super-active region NOAA 10484. The first surge was associated with an SF/C4.3 class flare. The second one was a major surge associated with a SF/C3.9 flare. This surge was also observed with SOHO/EIT in 195 Å and NoRH in 17 GHz, and showed similar evolution in these wavelengths. The major surge had an ejective funnel-shaped spray structure with fast expansion in linear (about 1.2×10^5 km) and angular (about 65°) size during its maximum phase. The mass motion of the surge was along open magnetic field lines, with average velocity about 100 km s^{-1}. The de-twisting motion of the surge reveals relaxation of sheared and twisted magnetic flux. The SOHO/MDI magnetograms reveal that the surges occurred at the site of companion sunspots where positive flux emerged, converged, and canceled against surrounding field of opposite polarity. Our observations support magnetic reconnection models for the surges and jets.

1 Introduction

During October–November 2003, major solar activity originated from three super-active regions, namely NOAA AR 10484, 10486, and 10488. On 25 October, we observed two surges between 01:50 UT and 04:15 UT that originated from NOAA AR 10484. The first surge was small; the second one was very dynamic and explosive in nature. Using multi-wavelength data we present a morphological study of these surges in order to understand the physical processes behind their activity.

W. Uddin (✉), P. Kumar, and A.K. Srivastava
Aryabhatta Research Institute of Observational Sciences, Nainital, India

R. Chandra
Observatoire de Paris, LESIA, Meudon, France

S.S. Hasan and R.J. Rutten (eds.), *Magnetic Coupling between the Interior and Atmosphere of the Sun*, Astrophysics and Space Science Proceedings,
DOI 10.1007/978-3-642-02859-5_61,

2 Observations

Hα images of these events were obtained at Aryabhatta Research Institute of Observational Sciences (ARIES), Nainital, India, on 25 October 2003 using the 15-cm f/15 Coudé solar tower telescope equipped with a Bernard Halle Hα filter, at intervals of 15–20 s and with a pixel size of 1″. We also used data from SOHO/MDI (cadence 96 min, 1.98″ pixels, Scherrer et al. 1995), SOHO/EIT (cadence 12 min, 2.5″ pixels, Delaboudinière et al. 1995), and NoRH (cadence 10 s, 5″ pixels, Takano et al. 1997).

The Hα observations nicely show the dynamic evolution of the recurrent surge activities from 01:50 UT to 04:15 UT (Fig. 1). The surge activities occurred at the following satellite sunspot of the active region. First, a small surge was associated with a small (SF/C4.3) flare, which started at 01:55 UT, reached maximum at 01:57 UT, and ended at 01:59 UT. The arrows show flare 1 and surge 1 in the Hα images. Another subflare (SF/C2.6) then started at 02:59 UT, peaked at 03:00 UT, and ended at 03:07 UT without surge activity. At 03:32 UT, another eruptive subflare (SF/C3.9) started with the second, major, dynamic, and explosive surge. It reached maximum at 03:52 UT and continued up to 04:15 UT. The soft X-ray flux showed two flares during this main surge eruption. The temporal evolution of these 25 October 2003 flares from NOAA AR 10484 is presented in Fig. 2. The surge evolved with initially small velocity, but in the ascending phase its velocity grew and it showed funnel-like structures during its maximum phase at 03:59 UT, which

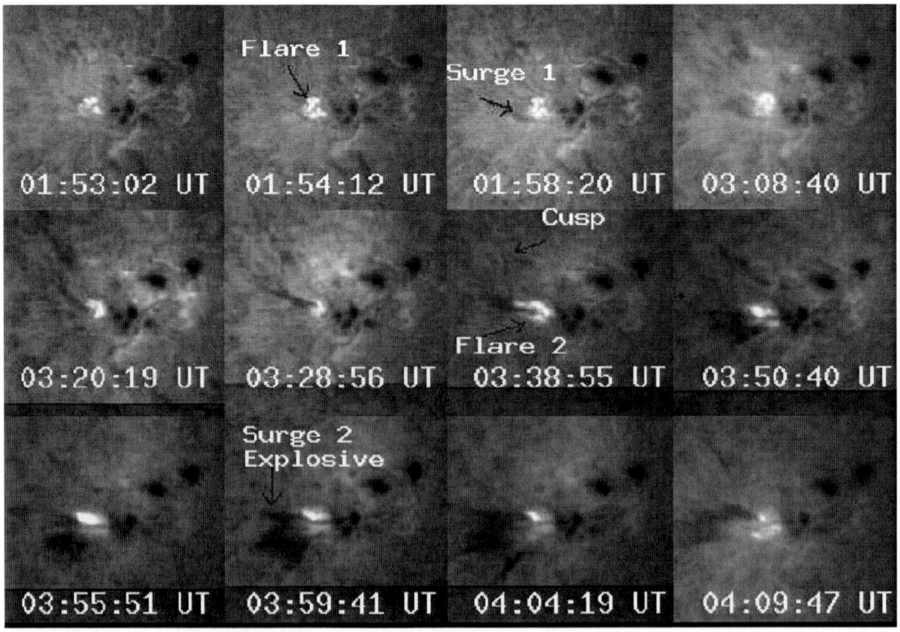

Fig. 1 Sample images of surge evolution in Hα. The field of view is 300″ × 300″

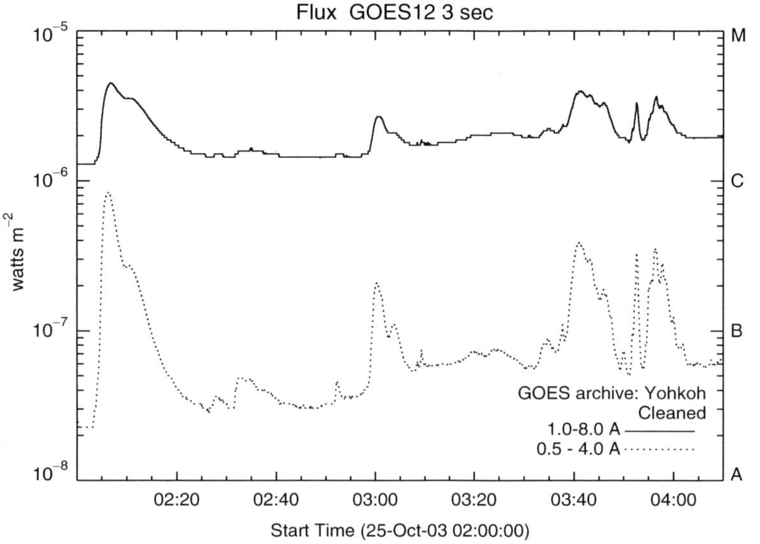

Fig. 2 GOES time profiles of the 25 October 2003 flares from NOAA AR 10484

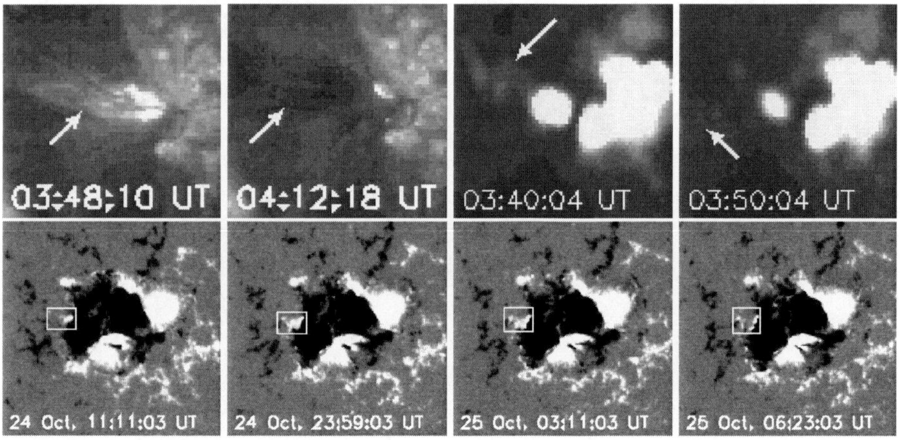

Fig. 3 *Top*: two EIT 195 Å images and two NoRH 17 GHz images showing the surge eruption. *Bottom*: SOHO/MDI magnetograms that show flux emergence and cancelation within the box. The field of view of each frame is $300'' \times 300''$

indicate spray-type behavior (the arrow indicates this explosive surge 2 in Fig. 1). The two-ribbon structure at the footpoint of the surge was also seen during the surge eruption. The Hα movie shows the change in orientation of the surge (from North-East to South-West) and also de-twisting motion was observed during its evolution.

The EIT 195 Å observations show similar morphology of the surge as in Hα: two-ribbon structure and mass motion at 03:48 UT (arrow in Fig. 3). The Nobeyama 17 GHz images also show the orientation change of the surge material as being similar to that in the Hα images at 03:40 UT (Fig. 3).

The MDI observations show positive flux emergence before the 5–6 h of surge activity nearby the satellite sunspot (marked by the box in Fig. 3). This flux emerged before the surge activity and disappeared after the event.

3 Results and Discussion

From our detailed investigation it is evident that the surges were associated with many C-class subflares, which indicates magnetic field annihilation at the site of surge activity. The MDI magnetograms reveal positive flux emergence and its cancelation by surrounding opposite polarity fields. The major surge rose upwards with an average velocity of about 100 km s^{-1}; its orientation change and de-twisting motion demonstrated relaxation of sheared and twisted magnetic field. The funnel-shaped structure of the surge is due to the material that follows the open magnetic field lines at the site (visible in the EIT images). The surge shows similar evolution in Hα (chromospheric temperature), EIT 195 Å (coronal temperature), and Nobeyama radio observations at 17 GHz (nonthermal coronal emission).

These multi-wavelength data indicate that the first reconnection took place between the newly emerged positive polarity sunspot and the pre-existing surrounding field. Subflaring then occurred, plasma was heated up to 10 MK, and transported towards open and closed field lines, which led to the formation of the two small flare loops visible in Hα and EIT, and the funnel-shaped surge structures. This scenario was earlier reported by Yokoyama and Shibata (1995) on the basis of numerical simulations. The flaring loops that formed nearby the footpoint of the surge and the type-III radio burst that was observed during this event are evidence favoring the magnetic-reconnection surge model (Shibata et al. 1994, Schmieder et al. 1995, Canfield et al. 1996). Our observational results support this model of magnetic reconnection for surges. These are only the preliminary results; we are planning to carry out more detailed study of these observations.

Acknowledgment We thank the conference organizers for a very good meeting and the editors for excellent instructions. R.C. thanks the CIFIPRA for his postdoc grant.

References

Canfield, R. C., Reardon, K. P., Leka, K. D., et al. 1996, ApJ, 464, 1016
Delaboudinière, J.-P., Artzner, G. E., Brunaud, J., et al. 1995, Solar Phys., 162, 291
Scherrer, P. H., Bogart, R. S., Bush, R. I., et al. 1995, Solar Phys., 162, 129
Schmieder, B., Shibata, K., van Driel-Gesztelyi, L., Freeland, S. 1995, Solar Phys., 156, 245
Shibata, K., Nitta, N., Strong, K. T., et al. 1994, ApJ, 431, L51
Takano, T., Nakajima, H., Enome, S., et al. 1997, In: Coronal Physics from Radio and Space Observations, G. Trottet (ed.), Lecture Notes in Physics, vol. 483, p. 183
Yokoyama, T., Shibata, K. 1995, Nat, 375, 42

Coronal Magnetic Field Estimation Using Type-II Radio Bursts

K.R. Subramanian, E. Ebenezer, and K.H. Raveesha

Abstract Coronal magnetic fields at two different heights were estimated using multiple type-II radio bursts observed on 23 January 2003. The strength of the magnetic field was estimated using 1–5 times the Newkirk density value. With these densities the magnetic field varied from 1.47 to 2.16 Gauss for the first type-II burst, and from 1.13 to 1.76 Gauss for the second one. Its strength was found to have power-law variation with height, with the index varying from −3 to −2 for densities 1–5 times Newkirk's value.

1 Observations, Analysis, Results

Meter-wavelength type-II radio bursts can be used to derive the strength of the coronal magnetic field by relating the speed of type-II radio bursts with the Alfvén velocity. In the so-called multiple type-II bursts, two type-II bursts occur in sequence, the second at lower frequency than the first (Subramanian and Ebenezer 2006).

Type-II emission originates in the vicinity of the transition region of shock waves. For fundamental/harmonic radiation, the properties of the shock across the shock front (up/down stream) are related (Vrsnak et al. 2004), with the density jump related to the instantaneous bandwidth and to the Alfvénic Mach number. The radial velocity can be derived from the drift rate in a type-II burst assuming a density model and radial propagation. The relative bandwidth and Alfvénic Mach number are related as

$$\triangle f/f = (f_u - f_l)/f_l = \sqrt{(N_2/N_1 - 1}, \qquad (1)$$

$$N_2/N_1 = 4M_{\mathrm{A}}^2/(3 + M_{\mathrm{A}}^2), \qquad (2)$$

K.R.Subramanian (✉) and E. Ebenezer
Indian Institute of Astrophysics, Bangalore, India

K.H. Raveesha
Department of Physics, CMRIT, Bangalore, India

S.S. Hasan and R.J. Rutten (eds.), *Magnetic Coupling between the Interior and Atmosphere of the Sun*, Astrophysics and Space Science Proceedings,
DOI 10.1007/978-3-642-02859-5_62,

where $M_A = V_r/V_A$ is the Alfvénic Mach number, with V_r the radial velocity of the type-II burst and V_A the Alfvén speed.

Our solar type-II burst data of 23 January 2003 were used to estimate the strength of the magnetic field in the solar corona at two different heights. Figure 1 shows the dynamic spectrum. The first burst started at 04:32 UT around 130 MHz, the second at 05:59 UT around 100 MHz, and so at lower frequency. The relative bandwidths were 0.29 ± 0.03 and 0.39 ± 0.04. The corresponding Alfvénic Mach numbers are 1.46 and 1.67. The drift rate was $0.10\,\mathrm{MHz\,s^{-1}}$ for the first burst and $0.07\,\mathrm{MHz\,s^{-1}}$ for the second. We used Newkirk's density model $N_e = M \times 4.2 \times 10^4 \times 10^{4.32}/\rho$, with ρ measured in solar radii, for the calculation of the burst speed assuming density enhancement factor $M = 1 - 5$. The mean frequencies of the two bursts were 102 and 74 MHz. Magnetic field strengths were estimated from the relation

$$V_A = 1.9 \times 10^4 \times B/f, \tag{3}$$

with B the field strength and f the mean frequency (Dulk and McLean 1978). The variation of the magnetic field with height is shown in Fig. 2. It can be written as a power law of the form $B(R) = \alpha R^{\gamma}$, with the index γ varying between -3 and -2 for density enhancement factors $M = 1 - 5$.

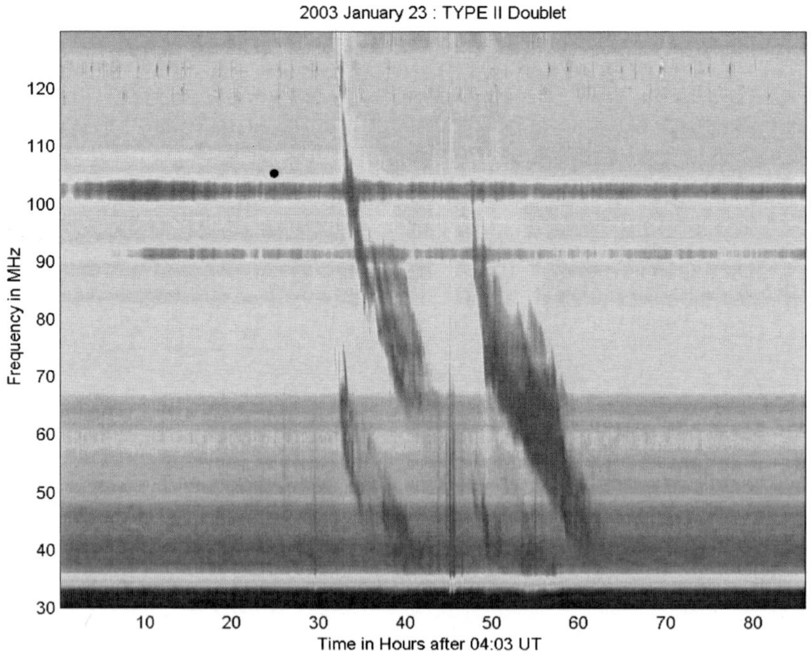

Fig. 1 Example of a multiple type-II radio burst observed with the Gauribidanur digital solar radio spectrograph. Two type-II bursts are seen in sequence: the first one starting at 04:32 UT, the second one at 04:49 UT. Both show fundamental/harmonic structures

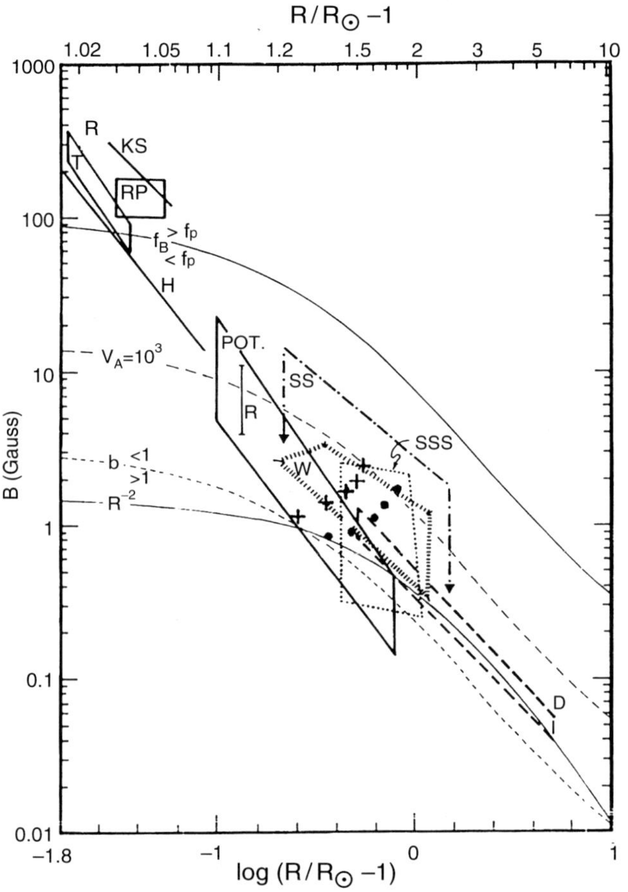

Fig. 2 Variation of the strength of the magnetic field as function of distance to the Sun. The estimated magnetic field values at 102 and 74 MHz are also shown. The *plus* and *filled circle* symbols specify the field strengths for density enhancement factors $M = 1-5$ at 102 and 74 MHz. The emission height and corresponding field strength increase with M at both frequencies

References

Dulk, G. A., MClean, D. J. 1974, Solar Phys., 57, 219
Vrsnak, B., Magdalenic, J., Zloblec, P. 2004, A&A, 413, 753
Subramanian, K. R., Ebenezer, E. 2006, A&A, 451, 683

Acceleration of CMEs Associated with Eruptive Prominences

A.D. Joshi and N. Srivastava

Abstract The association of coronal mass ejections (CMEs) with erupting prominences has been known for a long time. However, most studies focus on CMEs that cannot be observed close to the solar surface. We present a study of two CMEs that were associated with eruptive prominences, using data from the STEREO and SOHO space missions to study the CME accelerations. Our results confirm the scaling law proposed by Chen and Krall (2003).

1 Introduction

It has been well established (Webb et al. 1976; Munro et al. 1979; Gopalswamy et al. 2003; etc.) that eruptive prominences have large correlation with CMEs. It has also been reported (Zhang and Dere 2006, Zhang et al. 2004) that CMEs show rapid acceleration between 2 and 3 $R_\odot$, measured from the Sun's center. However, the factors that determine this height are not yet well understood.

In this paper, we examine CME acceleration for two CMEs that occurred on 9 April 2008 and 19 May 2007, respectively, and were observed with the coronagraphs onboard SOHO as well as the STEREO A and B spacecrafts.

2 Observations

The CME on 9 April 2008 first appeared in the STEREO A COR1 field of view (FOV) at 10:05 UT. It showed a classical three-part structure, and was seen near the south-west limb of the Sun. Data from STEREO A and B (COR1, COR2 and EUVI 304) as well as SOHO (LASCO C2, C3, and EIT 304) were used to obtain the projected height–time profiles. As the leading edge of this CME was visible only in the first few frames, a distinguishable feature in the bright central core of this CME

A.D. Joshi (✉) and N. Srivastava
Udaipur Solar Observatory, Udaipur, India

S.S. Hasan and R.J. Rutten (eds.), *Magnetic Coupling between the Interior and Atmosphere of the Sun*, Astrophysics and Space Science Proceedings,
DOI 10.1007/978-3-642-02859-5_63, © Springer-Verlag Berlin Heidelberg 2010

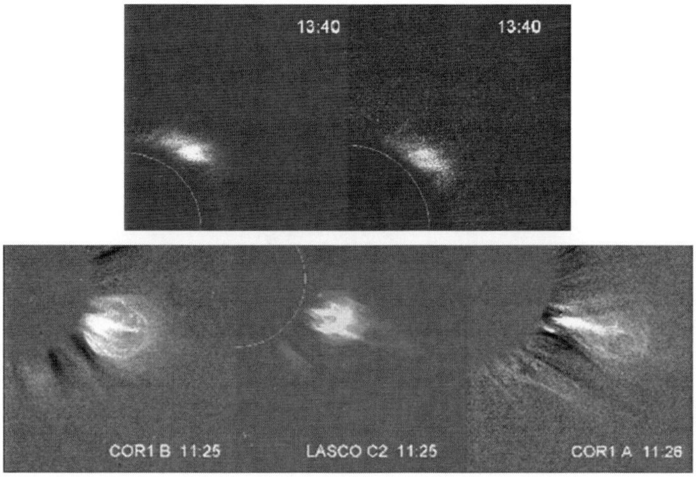

Fig. 1 *Upper images*: simultaneous frames of the CME event of 19 May 2007 from COR1 B and COR1 A. *Lower images*: simultaneous frames of the CME event of 9 April 2008 from COR1 B, LASCO C2, and COR1 A

was tracked in all the frames. This CME had a prominence very close to the solar limb associated with it.

The 19 May 2007 CME was a relatively faint CME seen close to the north-west limb. It first appeared in the STEREO A COR1 FOV at 13:25 UT. Its projected height was determined using the coronagraph images from SOHO and STEREO. This CME did not display a full three-part structure. The speed and acceleration of a sterak-like feature were estimated from polynomial fitting of its height–time curves.

Figure 2 shows the projected height, speed, and acceleration for the event on 19 May 2007. EUVI 304 images were used to measure the separation between the two footpoints of the prominences preceding the two CMEs.

3 Discussion and Summary

Based on a statistical study, Zhang and Dere (2006) reported that most CMEs exhibit bimodal acceleration profiles: a main acceleration phase occurring close to the Sun's surface and a residual acceleration phase occurring at a greater height. The profiles for the two events studied here show similar behavior. We also find that for the 19 May 2007 event, maximum acceleration in the main phase was greater than that for the 9 April 2008 event by almost an order of magnitude. The maximum heights and accelerations attained by the CMEs are given in Table 1, while Table 2 gives the speeds, measured in the plane of sky, of the CME leading edges and the central bright knots that correspond to the erupting prominences.

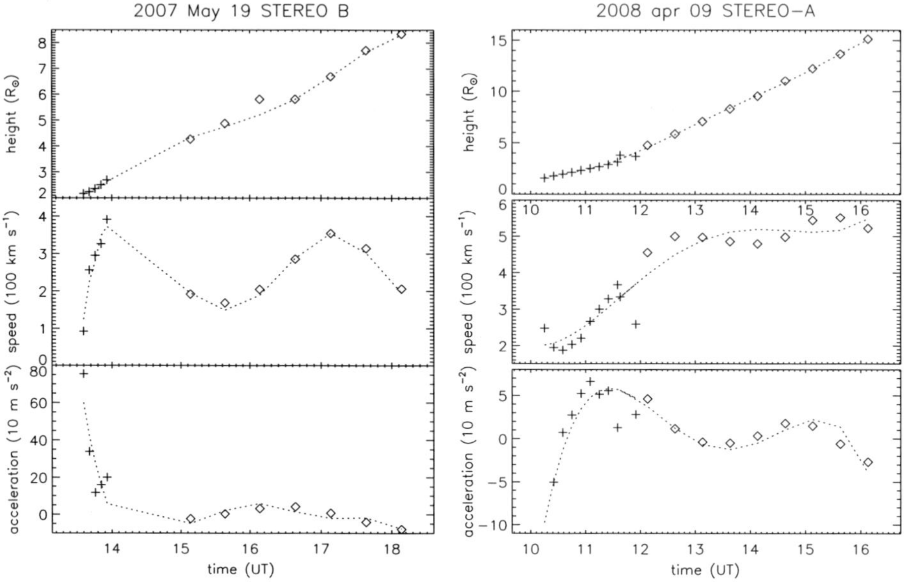

Fig. 2 Projected height, speed, and acceleration of the 19 May 2007 and 9 April 2008 CMEs as seen from STEREO B and STEREO A. COR1 data are denoted by *plus signs* and COR2 data by *diamonds*

Table 1 The average v_{avg} and maximum v_{max} speed and the maximum acceleration a_{max} recorded by each instrument is given along with the projected height and time corresponding to the instant at which a_{max} was achieved by the two CMEs

CME	Coronagraphs	v_{avg} (km s^{-1})	v_{max} (km s^{-1})	a_{max} (m s^{-2})	Height ($R_{\odot}$)	Time (UT)
2007	COR1 and COR2 A	273	346	429	1.70	13:25
May 19	COR1 and COR2 B	372	476	457	1.68	13:21
	LASCO C2 and C3	363	423	285	2.41	14:00
2008	COR1 and COR2 A	454	551	66	1.56	10:15
Apr 09	COR1 and COR2 B	385	444	113	1.55	10:25
	LASCO C2 and C3	388	548	131	2.60	11:06

Chen and Krall (2003) have proposed a scaling law for those CMEs that are associated with eruptive prominences. It states that the height at which CMEs attain maximum acceleration is greater than half the separation of footpoints of the associated prominence. The footpoint separation for the two filaments was found to be about 0.21 $R_{\odot}$, measured from the EUVI 304 images. The footpoint separations of the associated prominences compared with the height a_{max} are given in Table 1 and show that this scaling applied.

The acceleration of both CMEs occurred in two phases: the main phase and the residual phase (Fig. 2 and Table 1). This is in accordance with Zhang and Dere (2006), who found average accelerations of 331 and 0.9 m s^{-2} for the main and

Table 2 Plane-of-sky speeds in units of km s^{-1} for the leading edge (LE) and the associated prominences (Prom.) for 19 May 2007 and 9 April 2008 as measured from COR1 and COR2 images

	19 May 2007				09 Apr 2008			
	COR1		COR2		COR1		COR2	
	Prom.	LE	Prom.	LE	Prom.	LE	Prom.	LE
STEREO A	250	222	195	318	237	315	450	491
STEREO B	289	255	259	357	219	246	381	441

residual phase, respectively. Our study also agrees with Joshi and Srivastava (2007) who found for all but one event that prominences with footpoint separation greater than $0.20\,R_{\odot}$ have a CME associated with them.

4 Future Plan

Presently, we are engaged in obtaining the three-dimensional reconstructed heights of the 9 April 2008 CME, from which we will be able to determine the true CME velocity. In addition, we will carry out a similar reconstruction for EUVI 304 images of the filament on 19 May 2007, which happens to be close to disk center, in order to study filament motion in the early stages of eruption. We will also look for signatures of filament material from this event in interplanetary space using in situ data from various spacecrafts.

Acknowledgment The authors thank the SOHO/EIT and LASCO consortia for providing the data used in this analysis. We also acknowledge the STEREO/SECCHI consortium for providing the data.

References

Chen, J., Krall, J. 2003, JGRL (Sp. Phys.), 108, 1410
Gopalswamy, N., Shimojo, M., Lu, W., et al. 2003, ApJ, 586, 562
Joshi, V., Srivastava, N. 2007, Bull. Astron. Soc. India, 35, 447
Munro, R. H., Gosling, J. T., Hildner, E., et al. 1979, Solar Phys., 61, 201
Webb, D. F., Krieger, A. S., Rust, D. M. 1976, Solar Phys., 48, 159
Zhang, J., Dere, K. P. 2006, ApJ, 649, 1100
Zhang, J., Dere, K. P., Howard, R. A., Vourlidas, A. 2004, ApJ, 604, 420

Interplanetary Consequences of a Large CME

M. Lahkar, P.K. Manoharan, K. Mahalakshmi, K. Prabhu, G. Agalya, S. Shaheda Begum, and P. Revathi

Abstract We analyze a coronal mass ejection (CME) that resulted from an intense flare in active region AR486 on 4 November 2003. The CME propagation and speed are studied with interplanetary scintillation images, near-Earth space mission data, and Ulysses measurements. Together, these diverse diagnostics suggest that the internal magnetic energy of the CME determines its interplanetary consequences.

1 Introduction: Intense Flare and CME

An intense flare and associated coronal mass ejection (CME) occurred on 4 November 2003, during 19:50–20:10 UT in active region AR486. The onset of the halo CME was observed in the SOHO/LASCO C2 field of view (Brueckner et al. 1995) at 19:54 UT. The CME propagated rather fast in the C2–C3 field of view, at about 2650 km s^{-1} linear speed (Fig. 1), and caught up with a preceding CME that originated at the same location at 12:54 UT but had much lower linear speed (about 600 km s^{-1}) and narrow sky-plane width (about 70°). The LASCO images show interaction of the two CMEs at about 25 $R_{\odot}$. It caused remarkably complex and intense radio emission (e.g., Gopalswamy et al. 2001).

2 Radio Spectra, Particle Fluxes, and Scintillation Images

Prior to the CME–CME interaction, intense type-III bursts were observed with space missions Wind, Cassini (Kliore et al. 2004), and Ulysses. In the WAVES spectrum

M. Lahkar (✉)
Centre for Radio Astronomy, Department of Physics, Cotton College, Guwahati, Assam, India
and
Radio Astronomy Centre, National Centre for Radio Astrophysics, Tata Institute of Fundamental Research, Udhagamandalam (Ooty), India

P.K. Manoharan, K. Mahalakshmi, K. Prabhu, G. Agalya, S.S. Begum, and P. Revathi
Radio Astronomy Centre, National Centre for Radio Astrophysics, Tata Institute of Fundamental Research, Udhagamandalam (Ooty), India

S.S. Hasan and R.J. Rutten (eds.), *Magnetic Coupling between the Interior and Atmosphere of the Sun*, Astrophysics and Space Science Proceedings,
DOI 10.1007/978-3-642-02859-5_64,

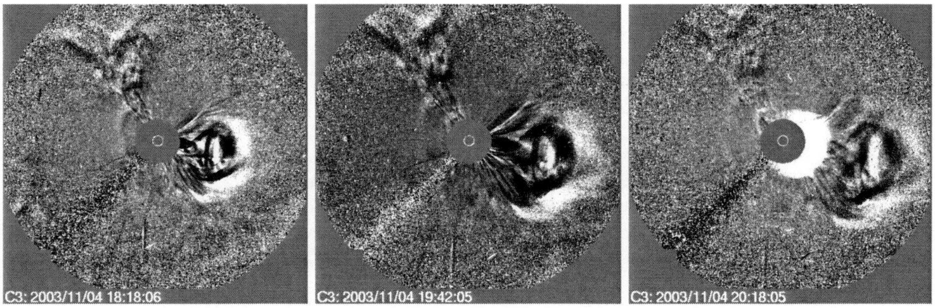

Fig. 1 LASCO images showing CME–CME interaction

(Bougeret et al. 1995), a fast-drifting intense type-II burst was observed during 19:50–21:00 UT at frequencies 1–11 MHz. It arrived at 1 MHz around 20:45 UT corresponding to $R = 10\text{–}15\,R_\odot$, in good agreement with the LASCO data.

The WAVES spectrum shows emission from the CME–CME interaction in the frequency range 1–3 MHz, higher than the type-II frequency at that time. Thus, this emission relates to about 15 times higher density than the typical ambient density at the interaction height (e.g., Gopalswamy et al. 2001). In the Cassini spectrum, an intense patch of emission occurred at frequencies below 1 MHz as the extension of the complex emission seen in WAVES. Thus, the interaction led to electron acceleration through the intense magnetic field and reconnection resulting from the interaction. The interaction and associated phenomena were observed more than 30 min on the above radio spectra (Fig. 2).

Type-IV emission was observed at frequencies >7 MHz during 20:10–21:00 UT (WAVES spectrum in Fig. 2), suggesting a plasmoid associated with the fast-moving CME (e.g., Manoharan and Kundu 2003). However, it disappeared just before the interaction.

The particle flux in the energy range 1–100 MeV showed no enhancement during the flare, but around 21:30 UT it increased in all energy bands to peak at about 06:00 UT the next day. Figure 3 shows that the increase was about 50 times above the pre-flare value at energies above 5 MeV. The sudden increase suggests that the CME–CME interaction favored development of magnetic connectivity with the Earth and particle acceleration (e.g., Gargate et al. 2006). The CME-produced shock arrived on 6 November, 19:20 UT (1-MeV profile in Fig. 3).

Figure 4 shows 3D tomographic interplanetary scintillation (IPS) images of the CME obtained with the Ooty Radio Telescope (e.g., Manoharan et al. 2001). They cover a range of $50\text{–}250\,R_\odot$. The interacting CMEs compressed the high-density and low-speed solar wind originating above a current sheet along the North-South direction. The IPS images obtained during 6 November 18:00–24:00 UT show the pushing and opening of the current sheet as viewed from the Earth. The central part of the CME compressed the fast solar wind belonging to a large coronal hole, which deflected the CME further away from the Sun–Earth line.

The solar wind parameters observed by near-Earth and with the Ulysses space missions are shown in Fig. 5. The shock arrival times are shown by vertical lines.

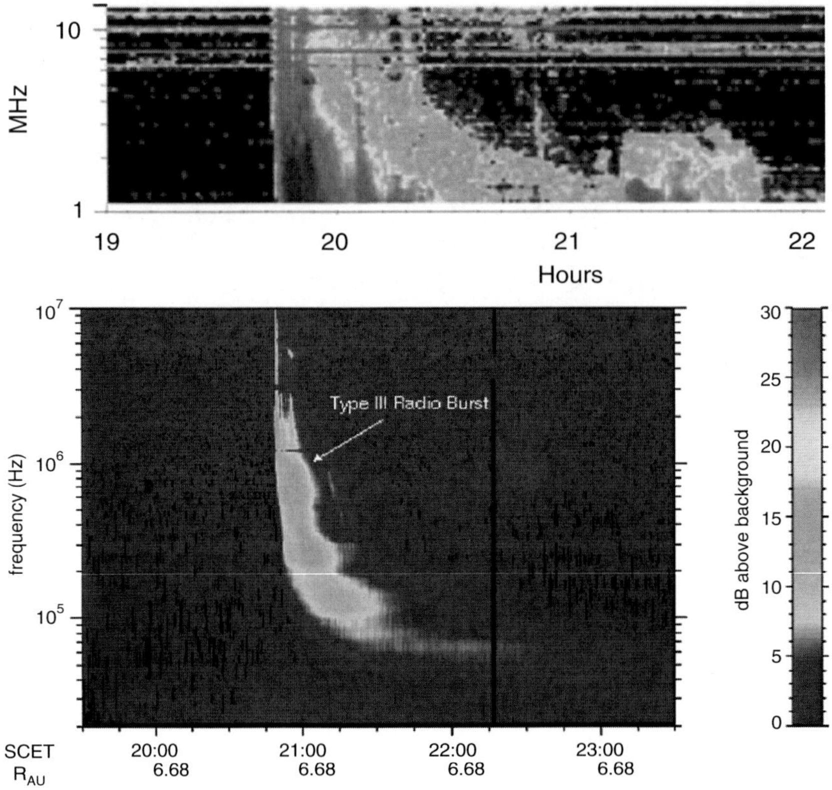

Fig. 2 Radio spectra from the WAVES and Cassini missions. Cassini was located at 8.7 AU

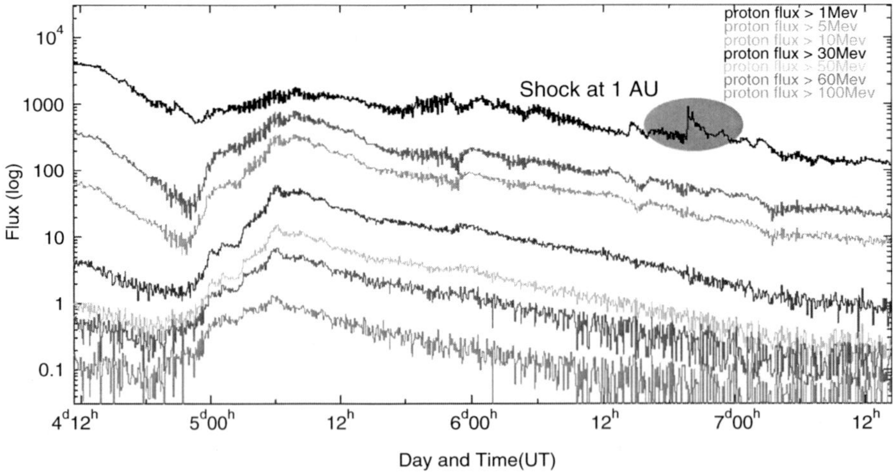

Fig. 3 Particle profiles in different energy bands

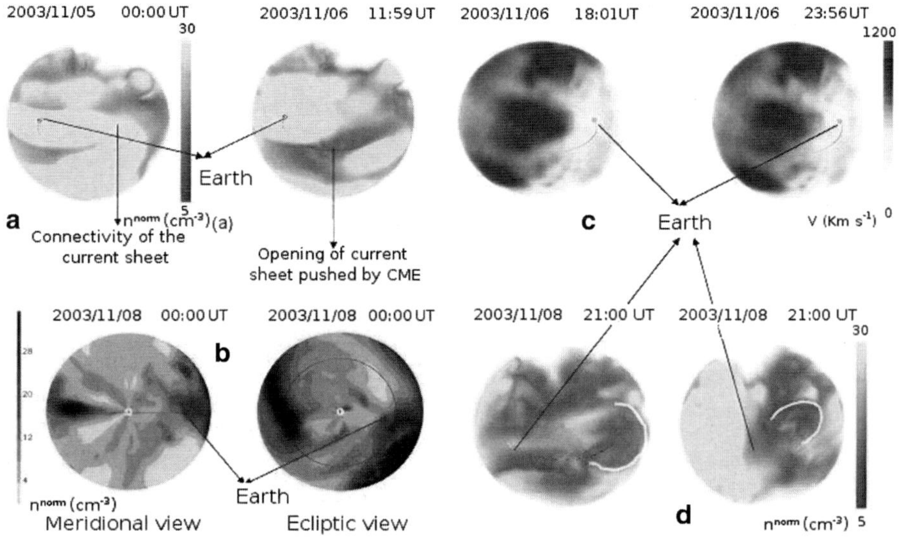

Fig. 4 Ooty IPS images showing the current sheet location (*top left*), CME deflection by the coronal hole (*top right*), CME compression of the solar wind (*bottom left*), and CME propagation (*bottom right*). The Sun is located at the center of each image. The two images at top right represent solar-wind speed; the others represent density

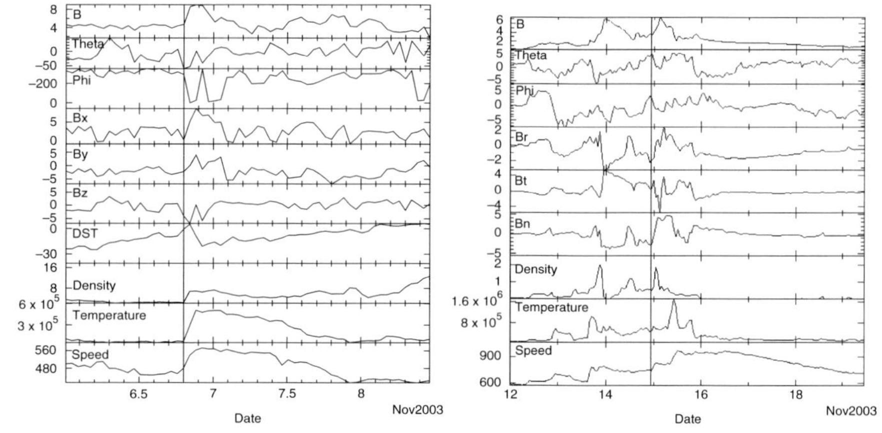

Fig. 5 1-AU and Ulysses hourly averages of solar wind parameters

As Ulysses was favorably located in the CME propagation direction, it could record the nose part of the CME and its shock, as indicated by a speed value of over 900 km s^{-1} at 5 AU. At Earth, the shock speed was below 600 km s^{-1}, suggesting that the eastern tail swept the Earth. From these measurements we infer a speed profile $V \sim R^{-0.4}$ to Earth. However, the deceleration $V \sim R^{-0.2}$ out to 5 AU near Ulysses implies gradual decline in speed along the CME propagation direction, which is in good agreement with the IPS measurements.

3 Conclusion

Our study shows the characteristics of a fast-moving CME and its interactions with transient and solar-wind structures at different distances from the Sun with good consistency between diverse diagnostics. The enhancement in radio emission and production of high-energy particles suggest that the magnetic field associated with the CME was strong. The gradual decline in CME speed suggests that the internal magnetic energy of the CME supported its propagation, including expansion in overcoming the aerodynamical drag imposed by the ambient solar wind (e.g., Manoharan 2006).

Acknowledgment We thank the Cassini, GOES, SOHO, TRACE, Ulysses, Wind, and OMNI-database teams for making their data available on the web. We also thank B. Jackson and the UCSD team for the IPS tomography analysis package. M. Lahkar thanks the National Centre for Radio Astrophysics (TIFR) for financial support. This work is partially supported by the CAWSES–India program sponsored by the Indian Space Research Organisation (ISRO).

References

Bougeret, J.-L., Kaiser, M. L., Kellogg, P. J., et al. 1995, Space Sci. Rev., 71, 231
Brueckner, G. E., Howard, R. A., Koomen, M. J., et al. 1995, Solar Phys., 162, 357
Gargate, L., Bingham, R., Fonseca, R. A., Silva, L. O. 2006, AGU Fall Meeting Abstracts, B1518
Gopalswamy, N., Yashiro, S., Kaiser, M. L., Howard, R. A., Bougeret, J.-L. 2001, ApJ, 548, L91
Kliore, A. J., Anderson, J. D., Armstrong, J. W., et al. 2004, Space Sci. Rev., 115, 1
Manoharan, P. K. 2006, Solar Phys., 235, 345
Manoharan, P. K., Kundu, M. R. 2003, ApJ, 592, 597
Manoharan, P. K., Tokumaru, M., Pick, M., et al. 2001, ApJ, 559, 1180

Solar System Resonances on Light-Travel Time Scales Set Up before Proto-Sun's Nuclear Ignition

M.H. Gokhale

Abstract A scenario is presented showing how solar-system resonances on time scales of light travel could have got set up before the onset of nuclear reactions in the proto-Sun. Such resonances may expedite the onset of nuclear ignition in the proto-Sun and the redistribution and loss of the proto-Sun's angular momentum.

1 Introduction

To ensure compatibility between models of solar variability phenomena and the standard model (SSM) of the Sun's mean structure and evolution, one must construct a hydrodynamic solar model (HDSM) whose mean structure equals the SSM and whose hydrodynamic state keeps producing acoustic waves and toroidal magnetic fields whose dissipation produces solar-like variability phenomena. The differential rotation that is needed to produce toroidal magnetic fields may be maintained by deposition of angular momentum by g-mode waves at loci of absorption. The maintenance of these waves (and of acoustic waves) needs maintenance of a spectrum of normal-mode oscillations of the HDSM's mass elements (i.e., oscillations with frequencies of the normal modes of the SSM).

I suggest that the power-input needed to maintain this spectrum may originate from gravitational energy–momentum exchanges of the HDSM's mass elements with the planets through resonances on time scales of planet-to-Sun speed-of-light travel time (e.g., about 43 min between the Sun and Jupiter). This suggestion is based on the facts that the frequencies of many solar acoustic modes lie in the $1/T_{\mathrm{P}}$ range for the inner planets, where T_{P} is the light-travel time per planet, and that the frequencies of many solar g-modes lie in the 360–410 μHz range perpetually traversed up and down by $1/T_{\mathrm{J}}$ as Jupiter moves in its elliptic orbit. This suggestion leads to the question how such resonances get set up initially. In this paper, I propose a mechanism setting up such resonances in the proto-solar system which

M.H. Gokhale (✉)
205 Sairang Aptts, New D. P. Road, Kothrud, Pune 311038, India

S.S. Hasan and R.J. Rutten (eds.), *Magnetic Coupling between the Interior and Atmosphere of the Sun*, Astrophysics and Space Science Proceedings,
DOI 10.1007/978-3-642-02859-5_65,

may also expedite the onset of nuclear ignition near the proto-Sun's center as well as redistribution and loss of the proto-Sun's angular momentum.

2 Fourier Frequencies in Momentum Transfer

Resonances on time scales of solar-system light travel are possible under the working hypothesis that the energy–momentum exchanges between the solar mass elements and the planets can be represented by waves with periodicities equal to the respective light-travel times and with amplitudes consistent with PPN expressions for the accelerations used in the standard ephemeris.

The standard theory of the origin of the solar system (cf. Shu et al. 1993; Boss 1998) says that the latter was formed by the break-up of a circum-solar parent disk into the proto-Sun and proto-planetary rings. Consider the turbulent gravitational dynamics of the parent-disk's earlier evolution that led to this break-up. Let ΔP_k, with $k = 1, 2, \ldots$, represent the disk mass elements that contributed to mass element ΔP of ring P, and let Δm_i represent a mass element of the proto-Sun. Throughout the evolution, small changes in the energy and momentum of ΔP_k at each instant of time t and the associated changes in the energy and momentum of Δm_i must both be spread over an interval of length $T = r(\Delta P_k, \Delta m_i, t)/c$ around t, with c the velocity of light.

Let $f(\Delta P \to \Delta m_i, t)$ represent the rate at which Δm_i receives gravitational momentum from any ΔP_k during the interval $(t - T/2, t + T/2)$. Along with each $r(\Delta P_k, \Delta m_i, t)$, the interval-length $T(\Delta P_k, \Delta m_i, t)$ and the light travel time profile (LTTP) of the rate f during this interval evolve both on longer time scales. Turbulence in the parent disk couples such LTTPs mutually during their evolution, so that the LTTP of every f during a given light-travel interval will contain ups and downs covering a wide range of frequencies, including $1/T(\Delta P_k, \Delta m_i, t)$ and depending on the locations of $\Delta P_1, \Delta P_2, \ldots$ relative to Δm_i. While different mass elements merge to form mass element ΔP in ring P and while all Δm_i converge to form the proto-Sun, the wide range of the Fourier frequencies of the LTTP of each f shrinks towards $\nu_\mathrm{p} = c/R_P$, where R_P is the average radius of the resulting ring P. Ultimately, the Fourier frequencies of the LTTP of each particular rate $f(\Delta P \to \Delta m_i, t)$ of momentum supply will lie in a band of small width, say $\Delta \nu_P$, around each respective $\nu_P = c/R_P$. This width will depend on the initial locations of the mass elements but will be much less than ν_P as the thickness of the resulting ring is much less than R_P.

Each term in the Fourier expansion of the resulting LTTP of each f over each light-travel interval T will be as if provided by a momentum wave of period T propagating from ΔP to Δm_i. Thus, the energy–momentum exchanges between the Sun's mass elements and the planetary rings under the resonances on time scales of light-travel.

3 Resonance Set-Up During the Approach to and Onset of Steady State

Before nuclear ignition, the only power source for the proto-Sun's luminosity on dynamical time scales would be the dissipation of its hydrodynamic and MHD flows. Hence, for the proto-Sun to approach a steady state with internal momentum equilibrium as well as internal power balance, the flows in the proto-Sun will have to be admissible under the approached mean structure (i.e., must be the normal modes of the approached mean structure), and also must be excitable by the momentum-input rates f. Therefore, the proto-Sun must approach a mean structure with normal mode frequencies within the Fourier bands of the f's. Thus, planetary-system light-travel resonances would be set up even before the onset of nuclear reactions.

Once the proto-Sun reaches steady state, the further evolution of its internal structure would imply slow evolution of its normal modes with coupled evolution of the proto-planetary rings in the presence of these resonances. The resonances will expedite the disposal of the proto-solar-system's gravitational energy through the decay of the normal mode waves at the boundaries of their propagation ranges within the proto-Sun. The resonant decay of the g-mode waves will expedite transport of angular momentum between the boundaries of their propagation ranges, leading to steep differential rotation at these boundaries and loss of angular momentum. The g-mode decay will also lead to enhanced heating at the inner boundaries, expediting the onset of nuclear ignition in the central region.

4 Conclusion

This scenario is qualitative only and is confined to the gravitational dynamics of the proto-solar-system, neglecting effects of rotation, magnetic fields, radiative processes, etc. The latter are known to be important during the formation of the parent disk (cf. Boss 1998), and also after the formation of the proto-solar-system in the Sun's outer parts. However, the rotation and magnetic fields will themselves be affected by the gravitational interactions between the Sun's mass elements and the planetary rings under the resonances on time scales of light-travel.

Acknowledgment The author thanks R.J. Rutten for much text improvement. The author also thanks LOC and IIA for travel support and hospitality during the meeting.

References

Boss, A. P. 1998, In: Origins, C. E. Woodward, J. M. Shull, H. A. Thronson (eds.), ASP Conf. Ser., vol. 128, p. 315

Shu, F., Najita, J., Galli, D., Ostriker, E., Lizano, S. 1993, Protostars and Planets III, 3

Part IV
Summaries of Presentations Published Elsewhere

Cycle Prediction from Dynamo Theory

A.R. Choudhuri

Abstract Many previous efforts in sunspot cycle prediction were based on various empirical correlations, most of which have limited statistical significance because they were inferred from observations of very few cycles. Perhaps the most successful of the empirical methods is to use the strength of the polar field in the previous sunspot minimum as a precursor for the next cycle. As the polar field at the present time is weak, Schatten [2005, Geophys. Res. Lett., 32, L21106] and Svalsgaard et al. [2005, Geophys. Res. Lett., 32, L01104] have predicted that cycle 24 will be a weak cycle.

The sunspot cycle is produced by a flux transport dynamo. One would like to make predictions of future cycles from dynamo models. By using the sunspot area data as the source of poloidal field, Dikpati and Gilman [2006, ApJ, 649, 498] predicted that cycle 24 will be very strong. Choudhuri et al. [2007, Phys. Rev. Lett., 98, 131103] pointed out that the Babcock–Leighton mechanism for producing the poloidal field involves randomness so that the sunspot area data cannot be taken as a deterministic source term. By feeding the polar field values occurring at sunspot minima in their code to account for the Babcock–Leighton process, Choudhuri et al. [2007, Phys. Rev. Lett., 98, 131103] found that cycle 24 should be weak. Dynamo models have been found to show good correlation between the polar field at the minimum and the strength of the next sunspot cycle when the turbulent diffusivity inside the convection zone is sufficiently high [Jiang et al. 2007, MNRAS, 381, 1527; Yeates et al. 2008, ApJ, 673, 544]. Goel and Choudhuri [2009, Res. Astron. Astrophys., 9, 115] analyze the historical faculae data to show that there is a good correlation between the hemispheric asymmetry of polar field at a minimum and the hemispheric asymmetry of the next cycle – again suggesting high diffusivity.

A.R. Choudhuri (✉)
Department of Physics, Indian Institute of Science, Bangalore, India

S.S. Hasan and R.J. Rutten (eds.), *Magnetic Coupling between the Interior and Atmosphere of the Sun*, Astrophysics and Space Science Proceedings,
DOI 10.1007/978-3-642-02859-5_66,

References

Choudhuri, A.R., Chatterjee, P., Jiang, J. 2007, Phys. Rev. Lett., 98, 131103
Dikpati, M., Gilman, P.A. 2006, ApJ, 649, 498
Goel, A., Choudhuri, A.R. 2009, Res. Astron. Astrophys., 9, 115
Jiang, J., Chatterjee, P., Choudhuri, A.R. 2007, MNRAS, 381, 1527
Schatten, K. 2005, Geophys. Res. Lett., 32, L21106
Svalgaard, L., Cliver, E.W., Kamide, Y. 2005, Geophys. Res. Lett., 32, L01104
Yeates, A.R., Nandy, D., Mackay, D.H. 2008, ApJ, 673, 544

Why Does the Torsional Oscillation Precede the Sunspot Cycle?

P. Chatterjee, S. Chakraborty, and A.R. Choudhuri

Abstract The Sun's rotation shows a periodic variation with the sunspot cycle, called torsional oscillations, the nature of which inside the solar convection zone has been determined from helioseismology. Several authors developed theoretical models of torsional oscillations by assuming that they are driven by the Lorentz force of the Sun's cyclically varying magnetic field. If this is true, then one would expect the torsional oscillations to follow the sunspot cycles. However, the torsional oscillations of a cycle begin a couple of years before the sunspots of that cycle appear and at a latitude higher than where the first sunspots are subsequently seen. Our aim in this paper is to provide an explanation for this seemingly causality defying phenomenon.

The sunspot cycle is produced by a flux transport dynamo (see Chatterjee, Nandy, Choudhuri 2004, A&A, 427, 1019). As the differential rotation is stronger at higher latitudes in the tachocline than at lower latitudes, the inclusion of solar-like rotation tends to produce a strong toroidal field at high latitudes rather than the latitudes where sunspots are seen. According to the Nandy and Choudhuri hypothesis (2002, Science, 296, 1671), the meridional circulation penetrates slightly below the bottom of the convection zone and the strong toroidal field produced at the high-latitude tachocline is pushed by this meridional circulation into stable layers below the convection zone, where magnetic buoyancy is suppressed and sunspots are not formed at high latitudes.

Presumably, the torsional oscillation gets initiated in the lower footpoints of intermittant vertical flux tubes inside the convection zone (Fig. 1 of Choudhuri 2003, Solar Phys., 215, 31), where the Lorentz force builds up due to the production of the azimuthal magnetic field. This perturbation then propagates upward along the vertical flux tubes at the Alfven speed.

P. Chatterjee (✉)
Department of Astronomy and Astrophysics, Tata Institute of Fundamental Research, Mumbai-400005, India

S. Chakraborty
Department of Theoretical Sciences, S. N. Bose Centre for Basic Sciences, Kolkata-700098, India

A.R. Choudhuri
Department of Physics, Indian Institute of Science, Bangalore-560012, India

S.S. Hasan and R.J. Rutten (eds.), *Magnetic Coupling between the Interior and Atmosphere of the Sun*, Astrophysics and Space Science Proceedings,
DOI 10.1007/978-3-642-02859-5_67,

The two completely novel aspects of our model are (1) the Nandy–Choudhuri hypothesis, which allows the formation of strong toroidal field in the high-latitude tachocline before the beginning of the sunspot cycle; and (2) the assumption that the perturbations propagate upward along flux tubes at the Alfvén speed. With these two assumptions incorporated, we find that our theoretical model readily explains most aspects of torsional oscillations.

The full paper will appear in Phys. Rev. Letters, vol. 102 (2009).

References

Chakraborty, S., Choudhuri, A.R., Chatterjee, P. 2009, Phys. Rev. Letters., 102, 041102
Chatterjee. P., Nandy, D., Choudhuri, A.R. 2004, Astron. Astrophys., 427, 1019
Choudhuri, A.R. 2003, Solar Phys., 215, 31
Nandy, D., Choudhuri, A.R. 2002, Science, 296, 1671

The Subsurface Magnetic Structure of Solar Active Regions

C.-H. Lin, S. Basu, and L. Li

Abstract The subsurface structures of active regions have often been inferred from helioseismic sound-speed inversions. However, pure acoustic waves are modified by magnetic fields when they propagate through solar active regions. Hence, the results of such inversions may not describe the pure sound speed. While changes in the sound speed are directly related to temperature variations, the wave speed of the modified wave does not have a simple relation with either the thermal or the magnetic structure.

In this work we first show that the "sound-speed" variation obtained from inversions is actually a combination of actual sound-speed variation and a magnetic component. Hence, the inversion result is not directly related to the thermal structure. Using solar models that include magnetic fields, we have developed a formulation to use inversion results to infer differences in magnetic and thermal structure between active regions and quiet regions. We then applied our technique to existing structure inversion results for different combinations of active and quiet regions and found that the effect of magnetic fields is strongest in a shallow region above $0.985\,R_\odot$ and that the strengths of magnetic-field effects at the surface and in the deeper ($r < 0.98\,R_\odot$) layers are inversely related, that is, the stronger the surface magnetic field, the smaller the magnetic effects in the deeper layers, and vice versa. We also found that the magnetic effects in the deeper layers are the strongest in the quiet regions, consistent with the fact that these are basically regions with weakest magnetic fields at the surface. Because the quiet regions were selected to precede or follow their companion active regions, the results could have implications about the evolution of magnetic fields under active regions.

This paper will be published in Solar Physics. Preprint: arXiv:0809.1427.

C.-H. Lin (✉)
Astrophysics Research Group, School of Physics, Trinity College Dublin, Ireland
and
Department of Astronomy, Yale University, U.S.A.

S. Basu and L. Li
Department of Astronomy, Yale University, U.S.A.

S.S. Hasan and R.J. Rutten (eds.), *Magnetic Coupling between the Interior and Atmosphere of the Sun*, Astrophysics and Space Science Proceedings,
DOI 10.1007/978-3-642-02859-5_68, © Springer-Verlag Berlin Heidelberg 2010

Acknowledgment This work utilizes data from the Solar Oscillations Investigation/Michelson Doppler Imager (SOI/MDI) on the Solar and Heliospheric Observatory (SOHO). SOHO is a project of international cooperation between ESA and NASA. MDI is supported by NASA grant NAG5-8878 to Stanford University. This work is partially supported by NSF grants ATM 0348837 and ATM 0737770 as well as NASA grant NNG06D13C. CHL is also supported by an ESA/PRODEX grant administered by Enterprise Ireland.

Sunspot Magnetometry from Kodaikanal

K. Nagaraju, K. Sankarasubramanian, and K.E. Rangarajan

Abstract Spectroscopic magnetic field measurements in sunspots have been carried out at Kodaikanal since such research was started by John Evershed (Evershed 1944). Subsequently, a Stokes polarimeter was built by Balasubramaniam and collaborators for the spectrograph at the Kodaikanal Tower Telescope (KTT), with the goal of measuring vector magnetic fields. Although the achieved accuracy is limited, the telescope model developed by them is very accurate (Balasubramaniam et al. 1985) and is still used even today to correct for instrumental polarization. A similar polarimeter was developed by Sankarasubramanian and collaborators, which is more modern in having motorized rotation of the polarization optics and in using CCD detectors instead of photographic registration on film (Sankarasubramanian et al. 2002). The accuracy in magnetic field measurement achieved with this instrument was quite high. Very recently, a dual-beam polarimeter has been installed, which uses an efficient and well-balanced modulation scheme. The calibration and characterization of this instrument are presented in Nagaraju et al. (2008b); results on magnetic structuring from the photosphere to the chromosphere are presented in Nagaraju et al. (2008a, 2009).

References

Balasubramaniam, K. S., Venkatakrishnan, P., Bhattacharyya, J. C. 1985, Solar Phys., 99, 333
Evershed, J. 1944, The Observatory, 65, 190
Nagaraju, K., Sankarasubramanian, K., Rangarajan, K. E. 2008a, ApJ, 678, 531
Nagaraju, K., Sankarasubramanian, K., Rangarajan, K. E., et al. 2008b, Bull. Astron. Soc. India, 36, 99
Nagaraju, K., Sankarasubramanian, K., Rangarajan, K. E. 2009, Solar Phys., submitted
Sankarasubramanian, R., Rangarajan, K. E., Ramesh, K. B. 2002, Bull. Astron. Soc. India, 30, 473

K. Nagaraju (✉) and K.E. Rangarajan
Indian Institute of Astrophysics, Bangalore, India

K. Sankarasubramanian
ISRO Satellite Center, Bangalore, India

S.S. Hasan and R.J. Rutten (eds.), *Magnetic Coupling between the Interior and Atmosphere of the Sun*, Astrophysics and Space Science Proceedings,
DOI 10.1007/978-3-642-02859-5_69,

Vector Magnetic Field in Emerging Flux Regions

B. Schmieder and E. Pariat

Abstract A crucial phase in magnetic flux emergence is the rise of magnetic flux tubes through the solar photosphere, which represents a severe transition between the very different environments of the solar interior and corona. Multi-wavelength observations with Flare Genesis, TRACE, SoHO, and more recently with the vector magnetographs at THEMIS and Hida (DST) led to the following conclusions. The fragmented magnetic field in the emergence region – with dipped field lines or bald patches – is directly related with Ellerman bombs, arch filament systems, and overlying coronal loops. Measurements of vector magnetic fields have given evidence that undulating "serpentine" fields are present while magnetic flux tubes cross the photosphere. See the sketch below, and for more detail see Pariat et al. (2004, 2007); Watanabe et al. (2008):

B. Schmieder (✉)
LESIA, Observatoire de Paris, Meudon, France

E. Pariat
NASA GSFC, Greenbelt MD and College of Science, GMU, Fairfax VA, USA

S.S. Hasan and R.J. Rutten (eds.), *Magnetic Coupling between the Interior and Atmosphere of the Sun*, Astrophysics and Space Science Proceedings,
DOI 10.1007/978-3-642-02859-5_70,

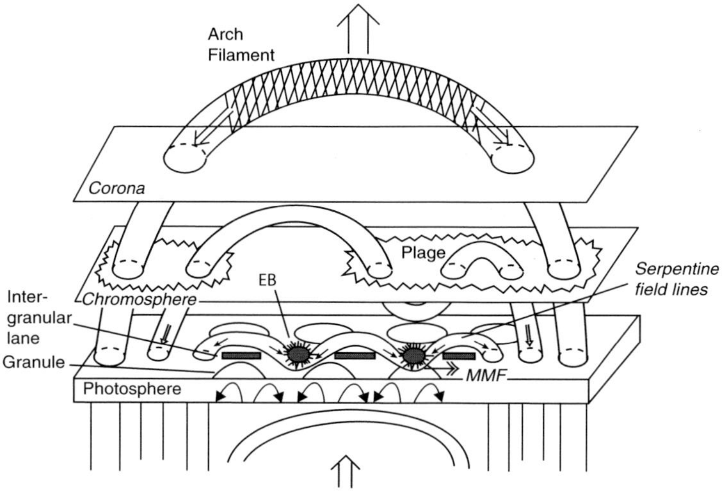

Fig. 1 Sketch of active-region flux emergence from photosphere to corona, from http://www.scholarpedia.org/article/Magnetic_flux_emergence

References

Pariat, E., Aulanier, G., Schmieder, B., et al., 2004, ApJ, 614, 1099
Pariat, E., Schmieder, B., Berlicki, A., et al., 2007, A&A 473, 279
Watanabe, H., Kitai, R., Okamoto, K., et al., 2008, ApJ, 684, 746

Evolution of Umbral Dots and Penumbral Grains

M. Sobotka and J. Jurčák

Abstract On 27 February 2007, Hinode SOT/SP acquired a time series of full-Stokes spectra of a regular, medium-sized sunspot NOAA 10944 located near the center of the solar disk. The inversion code SIR (Ruiz Cobo and del Toro Iniesta 1992, ApJ 398, 375) was applied to these data and a 3-h long time series of 34 spatial 3D maps of plasma parameters in the umbra and penumbra were computed. The temporal and spatial resolutions are 5.5 min and $0\overset{''}{.}32$, respectively. A simultaneous series of SOT/BFI G-band images was utilized for complementary measurements of horizontal motions and sizes of small-scale features.

The retrieved maps of plasma parameters show the spatial distribution of temperature, line-of-sight velocity, magnetic field strength, and inclination in two different ranges of optical depths corresponding to low and high photosphere. In these maps, the evolution of small-scale features – central and peripheral umbral dots and penumbral grains – was traced. Peripheral umbral dots and penumbral grains move toward the center of the umbra with a common speed of $390\,\mathrm{m\,s^{-1}}$. Penumbral grains often separate from tips of bright penumbral filaments and evolve into peripheral umbral dots. While central umbral dots do not show any excess of line-of-sight velocity nor magnetic field inclination with respect to the surrounding umbra, upflows of $420\,\mathrm{m\,s^{-1}}$ and a more horizontal magnetic field are detected in low photospheric layers of peripheral umbral dots. Penumbral grains have even stronger upflows (1–$2\,\mathrm{km\,s^{-1}}$) and magnetic field inclination in the low photosphere than peripheral umbral dots. The absolute values of these parameters decrease when penumbral grains evolve into peripheral umbral dots.

It seems that penumbral grains and peripheral umbral dots are of similar physical nature. Both types of feature appear in regions with a weaker and more horizontal magnetic field and their formation height reaches the low photosphere. On the other hand, central umbral dots appear in regions with stronger and more vertical

M. Sobotka (✉)
Astronomical Institute, Academy of Sciences of the Czech Republic, Czech Republic

J. Jurčák
Astronomical Institute, Academy of Sciences of the Czech Republic, Czech Republic
and
National Astronomical Observatory of Japan

S.S. Hasan and R.J. Rutten (eds.), *Magnetic Coupling between the Interior and Atmosphere of the Sun*, Astrophysics and Space Science Proceedings,
DOI 10.1007/978-3-642-02859-5_71,

magnetic field and are formed too deep to show upflows and changes in magnetic field inclination.

The full paper appeared in the meantime (ApJ, 694, 1080, 2009).

Acknowledgment This work has been supported by the grant IAA300030808 of the Grant Agency of the Academy of Sciences of the Czech Republic. J.J. was supported by the Japan Society for Promotion of Science. The work was partly carried out at the NAOJ Hinode Science Center.

Strong, Localized Downflows in a Sunspot Light Bridge

R.E. Louis, L.R. Bellot Rubio, S.K. Mathew, and P. Venkatakrishnan

Abstract We investigate spectropolarimetric observations of a sunspot light bridge in NOAA AR 10953 taken on 1 May 2007 with *Hinode* using the Fe I line pair at 630 nm. The SIR code (Ruiz Cobo and del Toro Iniesta 1992, ApJ, 398, 375) was used to invert the observed Stokes profiles, obtaining temperature stratifications and height-independent values of the magnetic field and Doppler velocity. The maps of the physical parameters show that the light bridge is a penumbral penetration into the umbra and has a relatively weak, inclined magnetic field. The highlight of our inversions is the presence of strong downflow patches in the light bridge, with line-of-sight velocities exceeding 4 km s^{-1}. The field azimuth also shows large rotation along a thin ridge close to one edge of the light bridge, essentially seen as a discontinuity in azimuth. Some of the downflow patches are also co-spatial with brightness enhancements in the Ca II H chromospheric filtergrams (Louis et al. 2008, Solar Phys., 252, 43). Inspection of the Stokes profiles for the downflow patches indicates doubly red-lobed Stokes-V signals. These profiles were also inverted with SIR using a two-component atmosphere with varying degrees of complexity. All such inversions indicate that the downflow patches consist of supersonic flows of about 10 km s^{-1}. Interestingly, the linear polarization also appears to be anomalous at the ridge demarcating the field azimuth change. We believe that the anomalous Q and U profiles result from mixing of the light bridge and the umbral magnetic field, which through reconnection may result in supersonic downflows in the photosphere and brightness enhancements in the chromosphere. The light bridge represents a strong inhomogeneity within a fairly regular sunspot. Our observations serve as useful inputs to future numerical models of light bridges.

The full analysis is to be submitted to ApJ.

R.E. Louis (✉), S.K. Mathew, and P. Venkatakrishnan
Udaipur Solar Observatory, Physical Research Laboratory, Udaipur, India

L.R. Bellot Rubio
Instituto de Astrofísica de Andalucía (CSIC), Granada, Spain

S.S. Hasan and R.J. Rutten (eds.), *Magnetic Coupling between the Interior and Atmosphere of the Sun*, Astrophysics and Space Science Proceedings,
DOI 10.1007/978-3-642-02859-5_72,

Small-Scale Velocities in Sunspot Penumbrae

M. Franz, R. Schlichenmaier, and W. Schmidt

Abstract To investigate the penumbral plasma flow at small scales, we used spectropolarimetric data of sunspots recorded by Hinode at 07:00 UT and 16:00 UT on 14 November 2008. We computed maps of apparent Doppler velocities by comparing the spectral position of the Fe I 630.15 nm line with the position of the line core of an average quiet Sun profile. We evaluated the bisector of the line wing to investigate the flow pattern in the deep photosphere.

Because of the nonzero heliocentric angle ($\theta = 8°$) of our data, not only the vertical but also the horizontal component of the Evershed flow contributes to the observed Dopplershifts. For the center-side penumbra, blueshifts are therefore to some extent due to projection effects. At equal azimuths, we find a sequence of elongated upflow patterns extending radially through the entire center-side penumbra. In these structures, strong upflows appear in concentrated patches alongside weaker upflows or even downflows. The strong upflows appear at the bright heads and the umbral side of the dark cores of the filaments, while the downflows are located at the penumbral side of the filaments. Projection effects lead to overall redshift of the limb-side penumbra, but the described pattern of up and downflows remains discernable.

This work will be submitted to Astronomy and Astrophysics.

M. Franz (✉), R. Schlichenmaier, and W. Schmidt
Kiepenheuer Institut für Sonnenphysik, Freiburg, Germany

S.S. Hasan and R.J. Rutten (eds.), *Magnetic Coupling between the Interior and Atmosphere of the Sun*, Astrophysics and Space Science Proceedings, DOI 10.1007/978-3-642-02859-5_73, © Springer-Verlag Berlin Heidelberg 2010

Photospheric Temperatures from Ca II H

V.M.J. Henriques, D. Kiselman, and M. van Noort

Abstract The temperature stratification in the upper photosphere can be extracted from Ca II H & K spectrograms following Shine and Linsky (1974) by assuming LTE, the Eddington–Barbier approximation, hydrostatic equilibrium, and that Ca II is mostly in the ground state. Rouppe van der Voort (2002) confirmed that these assumptions were solid for a wide range in the Ca II K wings and further developed the method including forward computation using MULTI (Carlsson 1986).

In this project we are adapting this technique to filtergrams in the Ca II H blue wing. Filtergrams imply many disadvantages such as having to deal with line blends (abundant in Ca II H) and reduced spectral resolution. However, our preliminary runs show that acceptable depth scaling is obtainable. With filtergrams the full resolution of the telescope (approximately $0.1''$ at Ca II H) is reached far more frequently and with much higher cadence than with spectra, as scanning with a slit over the same field of view takes over 20 min duration and the seeing evolves during that time (as does the photosphere itself). With the current setup available at the SST, and with the usual measurement procedure, we typically observe in seven filter positions sampling the blue wing of Ca II H. A simple 1.1 Å tiltable filter provides six of these positions; a fixed filter the seventh. Each full scan takes about 10 s and acquires 6 frames per position, which is enough for image restoration using MOMFBD (van Noort et al. 2005). This allows to collect time sequences of 3D snapshots at high resolution and high cadence. As the method uses only blue beam at the SST, complementary co-observing is possible using simultaneous spectropolarimetric imaging with the CRISP Fabry-Pérot instrument in the red beam.

Acknowledgment This project is supported by a Marie Curie Early-Stage Research Training Fellowship of the European Community's Sixth Framework Programme under contract MEST-CT-2005-020395 to the USO-SP Graduate School for Solar Physics.

V.M.J. Henriques (✉), D. Kiselman, and M. van Noort
Institute for Solar Physics, Royal Swedish Academy of Sciences, Stockholm, Sweden

S.S. Hasan and R.J. Rutten (eds.), *Magnetic Coupling between the Interior and Atmosphere of the Sun*, Astrophysics and Space Science Proceedings,
DOI 10.1007/978-3-642-02859-5_74,

References

Carlsson, M. 1986, Uppsala Astron. Obs. Report 33
Rouppe van der Voort, L. H. M. 2002, A&A, 389, 1020
Shine, R. A., Linsky, J. L. 1974, Solar Phys., 37, 145
van Noort, M., Rouppe van der Voort, L. H. M., Löfdahl, M. G. 2005, Solar Phys., 228, 191

Dual-Line Spectral Imaging of the Chromosphere

G. Cauzzi, K. Reardon, R.J. Rutten, A. Tritschler, and H. Uitenbroek

Abstract Hα filtergrams are notoriously difficult to interpret, "beautiful to view but not fit for analysis." We try to remedy this by using the IBIS bi-dimensional spectrometer at the Dunn Solar Telescope at NSO/Sacramento Peak to compare the quiet-sun chromosphere observed in Hα to what is observed simultaneously in Ca II 854.2 nm, sampling both lines with high angular and spectral resolution and extended coverage of space, time, and wavelength. Per (x, y, t) pixel we measured the intensity and Dopplershift of the minimum of each line's profile at that pixel, as well as the width of their inner chromospheric cores. A paper submitted to A&A (December 2008) compares these measurements in detail.

The figure below shows 1-h averages. The time averaging reduces the large modulation by repetitive 3-min chromospheric shocks seen everywhere in both lines. The figure shows remarkable dissimilarity between the time-averaged intensity scenes in the two lines and remarkable agreement between Ca II 854.2 nm intensity and Hα core width. The latter is a good indicator of chromospheric temperature through the low mass of the hydrogen atom, and so a principal Hα measure.

G. Cauzzi and K. Reardon
Osservatorio Astrofisico di Arcetri, Italy
and
National Solar Observatory/Sacramento Peak, USA

R.J. Rutten (✉)
National Solar Observatory/Sacramento Peak, USA
and
Sterrekundig Instituut Utrecht, The Netherlands
and
Institute of Theoretical Astrophysics Oslo, Norway

A. Tritschler and H. Uitenbroek
National Solar Observatory/Sacramento Peak, USA

S.S. Hasan and R.J. Rutten (eds.), *Magnetic Coupling between the Interior and Atmosphere of the Sun*, Astrophysics and Space Science Proceedings,
DOI 10.1007/978-3-642-02859-5_75,

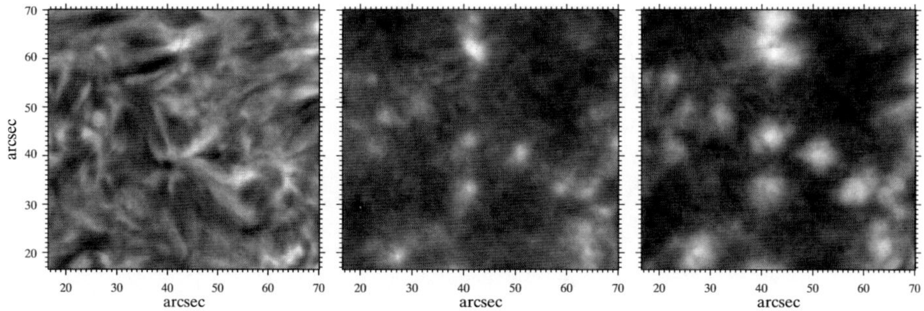

Fig. 1 Time-averaged IBIS measurements per pixel. *Left*: Hα profile-minimum intensity. *Middle*: Ca II 854.2 nm profile-minimum intensity. *Right*: Hα core width

Inversions of High-Cadence SOLIS-VSM Stokes Observations

C.E. Fischer, C.U. Keller, and F. Snik

Abstract We have processed full-Stokes observations made with the SOLIS-VSM using Fe I 630.15 and Fe I 630.25 nm. The data have high spectral and temporal resolution, moderate spatial resolution, and large polarimetric sensitivity and accuracy. We use the code LILIA, an LTE inversion code written by Socas-Navarro (2001), in order to invert the data in vector magnetic fields. The 180-degree ambiguity in magnetic field orientation is solved by using the Non-Potential Field Calculation (NPFC) method of Georgoulis (2005). The output product are maps of the full magnetic field vector at the photospheric level, as illustrated in Fig. 1. We performed such inversions for observations of active region NOAA 10808 taken during an X-class flare in September 2005. Details of the data processing and the first results are given in the proceedings of the Fifth Solar Polarization Workshop (ASP Conf. Ser., in press).

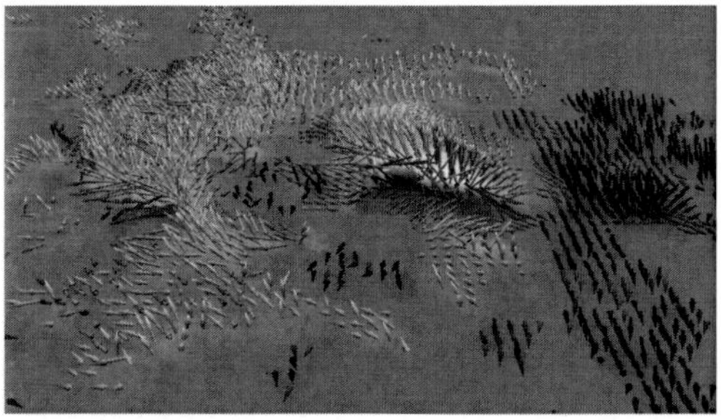

Fig. 1 After applying the data analysis steps, we obtain magnetic field vector maps of the umbra and surrounding of NOAA 10808 at optical depth $\tau_{500\,\mathrm{nm}} = -1$

References

Georgoulis, M. K. 2005, ApJ, 629, L69

Socas-Navarro, H. 2001, In: Advanced Solar Polarimetry – Theory, Observation, and Instrumentation, M. Sigwarth (ed.), ASP Conf. Ser., 236, 487

C.E. Fischer (✉), C.U. Keller, and F. Snik
Sterrekundig Instituut, Utrecht University, The Netherlands

S.S. Hasan and R.J. Rutten (eds.), *Magnetic Coupling between the Interior and Atmosphere of the Sun*, Astrophysics and Space Science Proceedings,
DOI 10.1007/978-3-642-02859-5_76, © Springer-Verlag Berlin Heidelberg 2010

Flows in Flaring and Dormant Active Regions

R.A. Maurya and A. Ambastha

Abstract During cycle 23, some active regions (ARs) produced extremely energetic flares and coronal mass ejections. These ARs are expected to be distinct from dormant ARs and quiet regions (QRs). It is of interest to identify whether the internal structure and dynamics of ARs is related to their outburst activity. For this investigation, we have obtained subsurface velocity flows in several ARs and QRs using ring diagram analysis, and derived the corresponding vorticities and kinetic helicity densities.

We find the following: (1) Flare-productive ARs show stronger internal zonal or meridional flows compared to QRs. (2) Strong downflows have better association with larger magnetic activity. (3) The sign change in the zonal or meridional velocity gradients observed at a depth of 3–5 Mm is associated with larger magnetic activity. (4) The steep meridional velocity gradient reported earlier by Ambastha et al. (2004) at $\approx$5 Mm depth in NOAA 10486 was further corroborated. (5) The correlation of this steep gradient with flare productivity was found to hold also in some other ARs. (6) The total unsigned kinetic helicity density is well correlated with the unsigned magnetic activity index. (7) The vertical vorticity changed sign for ARs between the North and South hemispheres. (8) The total unsigned kinetic helicity density of the ARs correlated remarkably well with the X-ray flux (Komm et al. 2004) and the Hα flare index. (9) Flaring ARs possessed larger kinetic helicity density in subsurface flows than corresponding QRs.

A detailed report of this work is to be submitted to Solar Physics.

Acknowledgment This work used data obtained by the Global Oscillation Network Group (GONG) program operated by AURA, Inc.

References

Ambastha, A., Basu, S., Antia, H. M., Bogart, R. S. 2004, Helio- and Astereoseismology: Towards a Golden Future, ESA-SP, 559, 293

Komm, R., Howe, R., González Hernández, I., Hill, F., Sudol, J., Toner, C. 2004, In: SOHO 14 GONG 2004, ESA-SP, 559, 158

R.A. Maurya (✉) and A. Ambastha
Udaipur Solar Observatory, Physical Research Laboratory, Udaipur, India

S.S. Hasan and R.J. Rutten (eds.), *Magnetic Coupling between the Interior and Atmosphere of the Sun*, Astrophysics and Space Science Proceedings,
DOI 10.1007/978-3-642-02859-5_77,

Magnetic and Velocity Field Changes Related to the Solar Flares of 28 and 29 October 2003

R.A. Maurya and A. Ambastha

Abstract Magnetic and velocity field measurements of solar active regions suffer from ambiguities caused by the change in spectral line profiles that occur during the impulsive phase of a major flare. This leads to difficulties in correct interpretation of any flare-related changes. Using magnetic and Doppler movies taken with GONG and MDI, we have detected transient, "moving" features around the peak phases of the X17.2/4B flare observed on 28 October 2003 and the X10/2B flare observed on 29 October 2003 in super-active region NOAA 10486. These features were located near the compact acoustic sources reported earlier by Donea and Lindsey (2005) and the seismic sources reported by Zharkova and Zharkov (2007). We find a moving feature, spatially and temporally associated with the flare ribbons, that separates away at speeds ranging from 30 to 50 km s^{-1} as observed in photospheric white light and in temperature-minimum (1600 Å), chromospheric (Hα), and transition-region (284 Å) intensities. We suggest that such moving features arise from the line-profile changes attributed to downward electron jets associated with the flare, and do not reflect real changes in the photospheric magnetic and velocity fields. However, abrupt and persistent changes in the pre- and post-flare phases were also found, which do not seem to be affected by line-profile changes. The detailed results have been appeared in Maurya and Ambastha (2009).

Acknowledgment This work utilizes data obtained by GONG (Global Oscillation Network Group), SOHO/MDI (Solar and Heliospheric Observatory, Michelson Doppler Imager), USO (Udaipur Solar Observatory), MLSO (Mauna Loa Solar Observatory), and TRACE (Transition Region and Coronal Explorer).

References

Donea, A. C., Lindsey, C. 2005, ApJ, 630, 1168
Maurya, R. A., Ambastha, A. 2009, Sol. phys., 258, 31
Zharkova, V. V., Zharkov, S. I. 2007, ApJ, 664, 573

R.A. Maurya (✉) and A. Ambastha
Udaipur Solar Observatory, Physical Research Laboratory, Udaipur, India

S.S. Hasan and R.J. Rutten (eds.), *Magnetic Coupling between the Interior and Atmosphere of the Sun*, Astrophysics and Space Science Proceedings,
DOI 10.1007/978-3-642-02859-5_78, © Springer-Verlag Berlin Heidelberg 2010

A Numerical Investigation of Unsheared Flux Cancelation

J.T. Karpen, S.K. Antiochos, C.R. DeVore, and M.G. Linton

Abstract Cancelation of magnetic flux in the solar photosphere and chromosphere has been linked observationally and theoretically to a broad range of solar activity phenomena, from filament channel formation to CME initiation. Because cancelation is typically measured at only a single layer in the atmosphere and only in the radial (line of sight) component of the magnetic field, the actual processes behind its observational signature are not fully understood. We have used our 3D MHD code with adaptive mesh refinement, *ARMS*, to investigate numerically the physics of flux cancelation, beginning with the simplest possible configuration: a subphotospheric Lundquist flux tube surrounded by a potential field in a gravitationally stratified atmosphere. Cancelation is driven by a two-cell circulation pattern imposed in the convection zone, in which the flows converge and form a downdraft at the polarity inversion line (PIL). We present and compare the results of 2D and 3D simulations of cancelation of initially unsheared flux – to our knowledge, these are the first such calculations in which the computational domain extends below the photosphere. The 2D simulation produces a flattened flux rope (plasmoid) whose axis remains centered along the PIL about 1650 km above the photosphere, without rising higher into the corona by the end of the run (10,000 s). Our calculations also show that 3D cancelation in an arcade geometry does not produce a fully disconnected flux tube in the corona, in contrast to the 2D results. Rather, most of the reconnected field stays rooted in the photosphere and is gradually submerged by the downdrafts at the PIL. An interchange-like instability develops above the region where the converging flows are driven, breaking the horizontal symmetry along the PIL. This generates an alternating pattern of magnetic shear (magnetic field component aligned with the PIL), which ultimately produces systematic footpoint shuffling through reconnection across the folds of the convoluted PIL. These simulations demonstrate the

J.T. Karpen (✉) and S.K. Antiochos
NASA Goddard Spaceflight Center, USA

C.R. DeVore and M.G. Linton
Naval Research Laboratory

S.S. Hasan and R.J. Rutten (eds.), *Magnetic Coupling between the Interior and Atmosphere of the Sun*, Astrophysics and Space Science Proceedings,
DOI 10.1007/978-3-642-02859-5_79,

importance of considering the effects of submergence, as well as the full 3D configuration of the magnetic field and atmosphere, in determining the physical processes behind flux cancelation on the Sun. A paper describing this work has been submitted to the Astrophysical Journal (January 2009).

Acknowledgment We thank Dr. Hasan, D. Banerjee, and their IIA colleagues for organizing and running a highly successful meeting. This research was funded by ONR and NASA.

Wave Heating of Coronal Loops with Steady Flows

V.S. Pandey, P. Venkatakrishnan, and A.S. Narayanan

Abstract We derive a general dispersion relation for hydromagnetic waves in inhomogeneous coronal loops with steady plasma flows. We model static loops by assuming the equilibrium magnetic and plasma configuration of solar coronal loops as slab geometry, with the z-axis of the slab along the uniform background magnetic field. We assume a density stratification perpendicular to the direction of the background magnetic field. Here we evaluate the extent to which hydromagnetic waves can be collisionally dissipated, that is, their dissipation by viscosity and heat conduction in the coronal loops when invoking steady flow, and we compare the results with the minimum that is required to satisfy the optically thin radiative cooling rate.

Our results are the following: (1) The differential equation is consistent with, for example, Roberts (1981a, b) in the absence of a steady flow. However, it is consistent with Nakariakov and Roberts (1995) when considering wave propagation only along the background magnetic field. (2) The steady flows in the loops may be either antiparallel or parallel to the direction of the longitudinal phase propagation of the waves. The antiparallel flows increase the wave period, while parallel ones decrease it. These results are consistent with the earlier findings of Nakariakov and Roberts (1995). (3) We found that flows also modify the wave heating rate. (4) The constraint that the wave period should be larger than the ion collision time places a limit on the maximum field strength for viscous damping. As the flow modifies the wave period, it also modifies the field strength at which waves damp sufficiently rapidly to balance radiative losses. (5) For both surface and body waves, we find that viscous heating exceeds thermal heating.

A detailed publication is to be submitted elsewhere.

References

Nakariakov, V. M., Roberts, B. 1995, Solar Phys., 159, 213
Roberts, B. 1981a, Solar Phys., 69, 27
Roberts, B. 1981b, Solar Phys., 69, 39

V.S. Pandey (✉) and P. Venkatakrishnan
Udaipur Solar Observatory, Physical Research Laboratory, Udaipur, India

A.S. Narayanan
Indian Institute of Astrophysics, Bangalore, India

S.S. Hasan and R.J. Rutten (eds.), *Magnetic Coupling between the Interior and Atmosphere of the Sun*, Astrophysics and Space Science Proceedings,
DOI 10.1007/978-3-642-02859-5_80,

Damping of Prominence Oscillations in Steady Equilibrium

K.A.P. Singh and B.N. Dwivedi

Abstract We have investigated the effect of field-aligned steady flows on the spatial damping of small-amplitude prominence oscillations. Non-adiabaticity was introduced through adding thermal conduction, radiation, and heating into the energy equation. Optically thick and thin radiative losses are mimicked, and various heating mechanisms are taken into account. A general dispersion relation is obtained by linearizing the MHD equations in the small-amplitude approximation.

A field-aligned steady flow breaks the symmetry between forward and backward propagating MHD waves in prominences. In the absence of non-adiabatic effects, the steady flow produces additional frequency shifts that are important for magneto-seismological diagnostics.

In the non-adiabatic case, the bulk motions contribute to wave damping. For a given wave period (5 min) and steady flow (5 km s^{-1}), the damping length for the longitudinal mode increases by a factor of two. Using the dispersion relation for the same characteristic steady flow of 5 km s^{-1} (Mach number $M_f = 0.4$) and wave number 10^6 m^{-1}, we obtained a wave period of 461 s for the thermal mode and 1,256 s for the slow mode, damping times of 1,207 s for the thermal mode and 26 s for the slow mode, and damping per wave period of 0.06 for the thermal mode and 7.64 for the slow mode. The thermal wave becomes a propagating wave in the presence of steady flow, making it observable in prominences. Observations of thermal wave modes in prominences will so help to understand the non-adiabatic mechanism that contributes wave damping of small-amplitude prominence oscillations.

Thus, the effect of steady flows on small-amplitude prominence oscillations is likely to play an important role in both wave detection and prominence seismology.

Acknowledgment We thank Robertus Erdélyi for helpful discussions.

K.A.P. Singh (✉)
Indian Institute of Astrophysics, Bangalore, India

B.N. Dwivedi
Department of Applied Physics, Banaras Hindu University, Varanasi, India

S.S. Hasan and R.J. Rutten (eds.), *Magnetic Coupling between the Interior and Atmosphere of the Sun*, Astrophysics and Space Science Proceedings,
DOI 10.1007/978-3-642-02859-5_81,

Variation of Network Contrast with Height

K.P. Raju

Abstract The bright Ca II H & K emission network known as "chromospheric network" outlines supergranulation cells. Convective motions in the supergranules sweep the magnetic flux tubes to their edges, resulting in field concentrations and enhanced emission at the cell boundaries, which form the network. Skylab observations in the 1970s have already shown the similarity of the Extreme UltraViolet (EUV) emission network originating from the transition region with the chromospheric network. It has been found that network emission is weakened in coronal holes as compared to quiet-Sun regions. This means that the intensity contrast of the network to the cell interiors is less in coronal holes than in the quiet Sun.

In the present study, the variation of intensity contrast of the network in the solar atmosphere has been examined. Intensities of 14 EUV emission lines in a coronal hole and in neighboring quiet-Sun areas have been obtained from observations with the Solar and Heliospheric Observatory (SOHO)/Coronal Diagnostic Spectrometer (CDS). The formation temperatures of the observed lines vary from 0.08 to 1.1 MK and hence represent increasing heights in the solar atmosphere, from the upper chromosphere and transition region to the low corona. The intensity contrast of the network with respect to the cell interior has been estimated for each line. Differences in the quiet-Sun areas and the coronal hole have also been noted. The network points and cell interiors were separated on the basis of the mean of the intensity distribution of the O V 629.73 Å line.

It is found that the contrast in lower transition-region lines is higher for quiet Sun than for a coronal hole. However, the contrasts become equal when one approaches the corona. For both regions, the contrast reaches a maximum at about $\log(T) = 5.4$.

A detailed paper will appear in Solar Physics.

K.P. Raju (✉)
Indian Institute of Astrophysics, Bangalore, India

S.S. Hasan and R.J. Rutten (eds.), *Magnetic Coupling between the Interior and Atmosphere of the Sun*, Astrophysics and Space Science Proceedings, DOI 10.1007/978-3-642-02859-5_82,

A Flaring Twisted Emerging Flux Region

R. Chandra, B. Schmieder, G. Aulanier, and J.M. Malherbe

Abstract We present signatures of the emergence of a twisted flux tube in decaying active region NOAA AR 10365 on 27 May 2003. When the magnetic flux tube is twisted, an asymmetry appears in the magnetogram because of the contribution of the azimuthal component to the observed vertical component of the field. In this case, the vertical component produces two "tongues" (Fig. 1, left). The twist of the flux tube is revealed by the photospheric longitudinal magnetic field pattern: diverging flows of opposite polarities, elongated polarities with a "tongue" shape. The asymmetry of the opposite polarities is interpreted as right-hand twist of the emerging flux tube.

The flare signatures in the chromosphere are ribbons observed in the MSDP Hα line (Fig. 1, right) and in TRACE 1,600 Å. The ribbons have a "J" shape and are shifted along the magnetic inversion line. The pattern of the ribbons indicates that the flare was triggered by coronal reconnection below the twisted flux tube of positive helicity. It is the first time that such a consistency between the signature of the emerging flux observed in the photosphere ("tongues") and flare ribbons in the chromosphere ('J' shape ribbons) is found. At the periphery of the active region, the enhanced brightening of network is interpreted as the stationary EIT wave component discussed by Delannée et al. (A&A 2007, 162, 603). Magnetic extrapolation using linear force-free-field modeling shows the existence of null points or a quasi-separatrix at the high altitude that may be responsible for this brightening (Chandra et al., Solar Phys., in press).

R. Chandra (✉), B. Schmieder, G. Aulanier, and J.M. Malherbe
Observatoire de Paris, LESIA, UMR8109 (CNRS), F-92195 Meudon, France

S.S. Hasan and R.J. Rutten (eds.), *Magnetic Coupling between the Interior and Atmosphere of the Sun*, Astrophysics and Space Science Proceedings,
DOI 10.1007/978-3-642-02859-5_83,

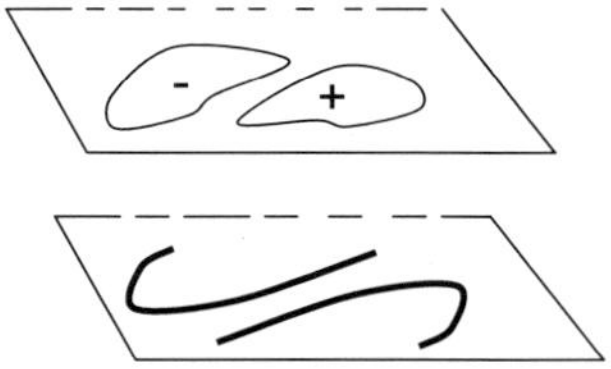

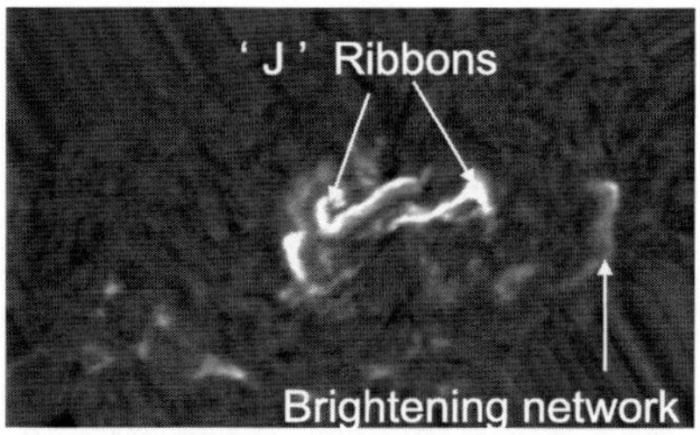

Fig. 1 Sketch of the signatures of positive magnetic helicity "tongues" (*left, top*), "J" shaped ribbons (*left, bottom*), and MSDP Hα image at 06:15 UT (*right*)

Reference

Chandra et al. Solar phys. 2009, 258, 53

Evidence of Magnetic Reconnection Outflows in a Flare seen by Hinode/EIS

P. Kumar, A.K. Srivastava, and W. Uddin

Abstract An X3.4-class flare occurred in AR NOAA 10930 on 13 December 2006 during 02:12–08:00 UT, peaking at 02:50 UT. We examine the flaring region during the late decay phase, analyzing raster data obtained with the EUV imaging spectrometer (EIS) onboard Hinode. The scanning started at 06:16:42 UT and ended at 06:20:56 UT. The 2″ wide slit was centered at $X = 272.59''$ and $Y = -130.19''$. The scan covered 800″ in X and 200″ in Y, with 1″x1″ binning. We applied standard EIS data reduction procedures and fitted spectral lines by single gaussian to obtain intensity maps, centroid Doppler velocity maps, and full width at half maximum (FWHM) maps for the strongest coronal line, Fe XII 195.12 Å, during the flare decay phase. The flaring region showed maximum downflow velocity $\approx$23 km s^{-1}, while the maximum upflow velocity was $\approx$59 km s^{-1}.

This flare is a long duration event (LDE) showing the presence of plasma downflows in the associated post-flare loop system during its decay phase. Strong upflows occurred approximately symmetrically around the downward moving plasma. Reconnection is an ongoing process throughout the entire duration of a solar flare. Hence, downflows and upflows may be closely related to reconnection. Such types of events often occur in the decay phase of LDE flares (cf., McKenzie, 2000, Sol. Phys., 195, 381). Using TRACE 195 Å images, Asai et al. (2004a, ApJ, 605, L77) have found the signature of outflows during both the impulsive and the decay phase. Asai et al. (2004b, ASPC, 325, 361) have observed the signature of reconnection-generated downflows above postflare loops, and also proposed that similar events may be observed with the Hinode instruments. We indeed find evidence of reconnection outflows (both up and down) in these high-resolution spectroscopic Hinode/EIS data during the late decay phase of the flare, about 3 h 30 min after the impulsive phase. Multi-wavelength high-resolution observations should be carried out to obtain more information. The detailed analysis will be submitted elsewhere.

P. Kumar (✉), A.K. Srivastava, and W. Uddin
Aryabhatta Research Institute of Observational Sciences (ARIES), Manora Peak, Nainital-263129, India

S.S. Hasan and R.J. Rutten (eds.), *Magnetic Coupling between the Interior and Atmosphere of the Sun*, Astrophysics and Space Science Proceedings,
DOI 10.1007/978-3-642-02859-5_84,

Does Coronal Rotation Period Depend on the Sunspot Number?

H.O. Vats, S. Chandra, and K.N. Iyer

Abstract The coronal rotation has been estimated using radio emissions at 2.8 GHz for the period 1947 to 2005. These measurements show a long term variability of the coronal rotation period. There are three components in this variability; irregular component, 11 year variation which is related to the sunspot number and 22 year component which may be related to solar magnetic field reversal or Hale effect.

1 Introduction

The solar rotation is known for very long time, but still there are many unsolved issues. There is variability on both spatial and temporal scales. The measurement of the rotation of Sun has been going on quite systematically for long time. Mainly three techniques have been used to estimate the rotation period. These can be summarized as below:

2 Feature Tracking

The features e.g. sunspots, filaments, prominences, faculae, bright points, edge of the sunspot groups, super granules, plages, magnetic features etc are very easily observable features in the solar images and have been very extensively used the estimate of solar rotation (Howard 1991, Sivaraman et al. 1993, Howard 1996 and references therein). The feature tracing method was largely used for the estimation of rotation of photosphere and chromosphere. Recently this method has

H.O. Vats (✉)
Physical Research Laboratory, Ahmedabad 380009, India

S. Chandra
PPN College, Kanpur, UP, India

K.N. Iyer
Physics Department, Saurashtra University, Rajkot, India

S.S. Hasan and R.J. Rutten (eds.), *Magnetic Coupling between the Interior and Atmosphere of the Sun*, Astrophysics and Space Science Proceedings,
DOI 10.1007/978-3-642-02859-5_85,

been successfully used for the X-ray bright points to determine coronal rotation (Kariyappa 2008) using images from Hinode/XRT and Yohkoh/SXT. The important findings from this study are; (1) corona rotates differentially like photosphere and chromosphere and (2) the rotation period does not depend on the phase of solar activity.

3 Spectral Measurements

The spectroscope marked the beginning of the modern era of solar and stellar studies. In 1977 W. W. Abney suggested that the effect of a stars rotation on its spectrum would be to broaden all the spectral lines. The line broadening actually gave the projected equatorial rotational velocity (v sin i) along the line of sight, where v and i are rotational velocity and the angle between rotation axis and line sight. The results of solar rotation by both these methods have been summarized by Beck 2000.

4 Rotation Measurements from Flux Modulation

This method has been used to estimate rotation period of several stars. We applied this method to solar radio flux and in this paper summarize the recent results.

5 Data Analysis and Results

Vats et al. 1998 used disk integrated radio flux at 2.8 GHz and derived sidereal rotation for one solar cycle i.e. 1985 to 1995. It was found that the persistence and period of rotational modulation on radio flux varies from year to year. The typical radio flux variation is as shown in Fig. 1 (left). This data was used by Vats et al. 2001 along with the data of other 10 radio frequencies and thus showed that there is a clear differential rotation as function altitude in the solar corona. The flux variation in Fig. 1 has an increasing trend due to the rising phase of solar activity. Here radio flux clearly shows a rotational modulation.

However, determining periodicity is no so easy in this. For the calculation of rotation period we obtained autocorrelation function using standard IDL subroutines and the autocorrelogram for this typical data set is shown in Fig. 1 (right panel). This figure clearly shows periodic modulation due to solar rotation, there are 15 cyclic up and down embossed on the slowly decreasing autocorrelation coefficient up to a large lag of 400 days. As 15 peaks is at 391 days this gives synodic rotation of about 26.1 day and sidereal rotation period as 24.1 days. Using this approach we estimated coronal rotation period from the radio flux measurements over a period of 1947 to 2005. This data set covers more than five solar activity cycles. These values

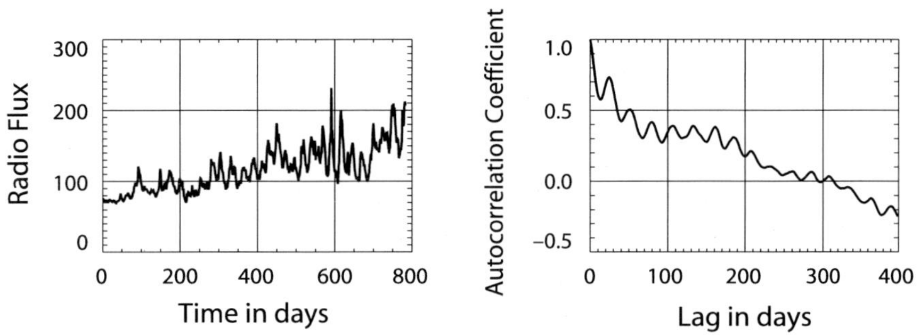

Fig. 1 Typical temporal variation of radio flux at 2.8 GHz starting from June 1, 1997. Figure on the right shows autocorrelation coefficient of radio flux as a function of lag in days

of sidereal rotation period show quite large variability. A minimum of 17 days in the year 1159 and a maximum of 31 in 1965, the average for whole period in 24.7 days with a standard deviation of 2.2 days. Further, analysis of the series of annual values of sidereal rotation period during 1947 to 2005 reveal to have three components, namely, 11 year period, 22 year period and an irregular component. The first two components namely, 11 and 22 year periodicities may be related to solar activity and Hale periodicities respectively. Mehta 2005 found no systematic relationship between solar rotation period and phase of the solar activity. In spite of the great care at all the observations and usage of modern methods, the results of rotation measurements derived by various techniques do not agree well. The difference may be due to the fact that features to a varying degree reflect the condition deeper inside the Sun. The multi-frequency radio measurements of Sun revealed that there is differential rotation in corona as a function of altitude (Vats et al. 2001). Thus corona has differential rotation both as a function of latitude and altitude.

Acknowledgment We have used data from SPIDER website for this work. We are grateful to the experimenters who contributed this excellent data to this website.

References

Beck, J. G. 2000, Solar Phys., 191, 47.
Howard, R. F. 1991, Solar Phys., 135, 327.
Howard, R. F. 1996, ARA&A, 34, 75.
Kariyappa, R. 2008, A&A, 488, 297.
Mehta, M. 2005, Bulletin of the Astronomical Society of India, 33, 323.
Sivaraman, K. R., Gupta, S. S., Howard, R. F., et al. 1993, Solar Phys., 146, 27.
Vats, H. O., Deshpande, M. R., Shah, C. R., Mehta, M., et al. 1998, Solar Phys., 118, 351.
Vats, H. O., Cecatto, J. R., Mehta, M., Sawant, H. S., Neri, J. A. C. F., et al. 2001, ApJ, 548, L87.

Coronal Magnetic Field from an Extreme Radio Burst

E. Ebenezer, K.H. Raveesha, and K.R. Subramanian

Abstract During the descending phase of cycle 23, active region NOAA 10720 produced multiple powerful flares including one on 20 January 2005, which started at 06:36 UT and peaked at 09:52 UT. The corresponding CME produced the largest ground-level particle event at Earth since 1956.

Dynamic radio spectra were obtained during the flare with the Gauribidanur Radio Array Solar Spectrograph (GRASS) and the BIRS spectrograph in Tasmania, where the ionospheric cutoff decreases to frequencies as low as 15–10 MHz. The figure below shows the Type-IV burst observed with GRASS, which lasted from 06:45 to 07:01 UT. There is no drift in the emission until 06:45 UT for the frequencies above 50 MHz sampled by GRASS. However, below 50 MHz, steep curvature occurred in the BIRS data. This cut-off can be attributed to the Razin–Tsytovitch effect, and therefore the coronal magnetic field strength B may be estimated from the relation $\nu_R = 20\, N_e/B$, using the observed cut-off frequency ν_R and the standard Newkirk model for the coronal electron densities N_e. We so derive decreasing field strength from 10.3 to 3.6 G during 274 s from 06:45 UT.

We intend to publish this work in more detail in a regular journal.

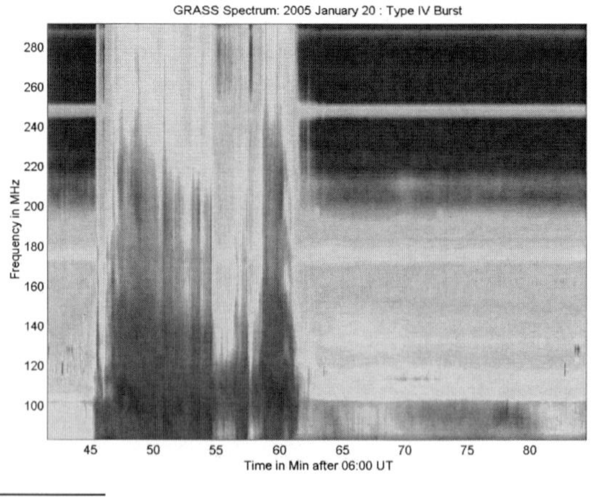

E. Ebenezer (✉) and K.R. Subramanian
Indian Institute of Astrophysics, Bangalore, India

K.H. Raveesha
Department of Physics, CMRIT, Bangalore, India

S.S. Hasan and R.J. Rutten (eds.), *Magnetic Coupling between the Interior and Atmosphere of the Sun*, Astrophysics and Space Science Proceedings,
DOI 10.1007/978-3-642-02859-5_86, © Springer-Verlag Berlin Heidelberg 2010

CME Kinematics and Dynamics

C.-H. Lin and P.T. Gallagher

Abstract The goal of this study is to investigate the driving mechanisms of CMEs and to infer the magnetic field properties at the onset of the instability. We use EIT 195 Å images and LASCO white-light coronagraph data of a CME event that occurred on 17 December 2006. It was a long-duration event, and was associated with an occulted C2.1 class flare. To determine the driving mechanism, we quantitatively and qualitatively compared the observationally obtained kinematic evolution with that predicted by three CME models: the breakout model (BO, see Antiochos et al. 1999; Lynch et al. 2008; DeVore and Antiochos 2008), the catastrophe model (CM, see Priest and Forbes 2000), and the toroidal instability model (TI, see Chen 1989; Kliem and Török 2006). Our results indicate that this CME is best represented by the CM model. We infer that, at the onset of the instability, the Alfvén speed is approximately 120 km s^{-1} and the height of the flux rope is roughly 100–200 Mm. These parameter values are related to the magnetic environment and the loop geometry and can be used to infer the magnetic condition at the onset of the eruption. We intend to submit the full analysis to A&A.

Acknowledgment This work is supported by an ESA/PRODEX grant administered by Enterprise Ireland.

References

Antiochos, S. K., DeVore, C. R., Klimchuk, J. A. 1999, ApJ, 510, 485
Chen, J. 1989, ApJ, 338, 453
DeVore, C. R., Antiochos, S. K. 2008, ApJ, 680, 740
Kliem, B., Török, T. 2006, Phys. Rev. Letters, 96, 255002
Lynch, B. J., Antiochos, S. K., DeVore, C. R., Luhmann, J. G., Zurbuchen, T. H. 2008, ApJ, 683, 1192
Priest, E., Forbes, T. (eds.), 2000, Magnetic Reconnection: MHD Theory and Applications, Cambridge University Press, Cambridge

C.-H. Lin (✉) and P.T. Gallagher
Astrophysics Research Group, School of Physics, Trinity College Dublin, Ireland

S.S. Hasan and R.J. Rutten (eds.), *Magnetic Coupling between the Interior and Atmosphere of the Sun*, Astrophysics and Space Science Proceedings,
DOI 10.1007/978-3-642-02859-5_87,

Solar Wind Monitoring with SWIM-SARA Onboard Chandrayaan-1

A. Bhardwaj, S. Barabash, R. Sridharan, M. Wieser, M.B. Dhanya, Y. Futaana, K. Asamura, Y. Kazama, D. McCann, S. Varier, E. Vijayakumar, S.V. Mohankumar, K.V. Raghavendra, T. Kurian, R.S. Thampi, H. Andersson, J. Svensson, S. Karlsson, J. Fischer, M. Holmstrom, P. Wurz, and R. Lundin

Abstract The SARA experiment aboard the Indian lunar mission Chandrayaan-1 consists of two instruments: Chandrayaan-1 Energetic Neutral Analyzer (CENA) and the Solar Wind Monitor (SWIM). CENA will provide measurements of low energy neutral atoms sputtered from lunar surface in the 0.01–3.3 keV energy range by the impact of solar wind ions. SWIM will monitor the solar wind flux precipitating onto the lunar surface and in the vicinity of moon. SWIM is basically an ion-mass analyzer providing energy-per-charge and number density of solar wind ions in the energy range 0.01–15 keV. It has sufficient mass resolution to resolve H^+ , He^{++}, He^+, O^{++}, O^+, and >20 amu, with energy resolution 7% and angular resolution $4.5° \times 22.5°$. The viewing angle of the instrument is $9° \times 180°$. Mechanically, SWIM consists of a sensor and an electronic board that includes high voltage supply and sensor electronics. The sensor part consists of an electrostatic deflector to analyze the arrival angle of the ions, cylindrical electrostatic analyzer for energy analysis, and the time-of-flight system for particle velocity determination. The total size of SWIM is slightly larger than a credit card and has a mass of 500 g.

The moon's orbit provides a unique location to observe solar wind continuously over 27 days period around the Earth. The measurement with SWIM will provide a unique database for near-Earth solar wind conditions over a 2 year period of the Chandrayaan-1 nominal mission life. With solar wind measurements being made at

A. Bhardwaj (✉), R. Sridharan, M.B. Dhanya, S.V. Mohankumar, and R.S. Thampi
Space Physics Laboratory, Vikram Sarabhai Space Centre, Trivandrum, India

S. Barabash, M. Wieser, Y. Futaana, Y. Kazama, D. McCann, H. Andersson, J. Svensson, S. Karlsson, M. Holmstrom, and R. Lundin
Swedish Institute of Space Physics, Box 812, 98128, Kiruna, Sweden

K. Asamura
ISAS, Japan Aerospace Exploration Agency, Sagamithara, Kanagawa, Japan

S. Varier, E. Vijayakumar, K.V. Raghavendra, and T. Kurian
Avionics Entity, Vikram Sarabhai Space Centre, Trivandrum, India

P. Wurz and J. Fischer
Physikalisches Institut, University of Bern, Bern, Switzerland

S.S. Hasan and R.J. Rutten (eds.), *Magnetic Coupling between the Interior and Atmosphere of the Sun*, Astrophysics and Space Science Proceedings,
DOI 10.1007/978-3-642-02859-5_88,

the L1 point (240 Re; Re = Earth's radius) and at around 6–10 Re, SWIM will not only help to study lunar–solar wind interactions but will also be of great importance to study changes in solar wind as it propagates from L1 to the lunar orbit (60 Re) to geostationary orbits.

Coupling of the Solar Wind and the Magnetosphere

Badruddin, V. Gupta, and Y.P. Singh

Abstract The field orientation in transient and co-rotating structures that are ejected from the Sun and evolve in interplanetary space is crucial in reconnection between the solar wind and the magnetosphere. However, fluctuations in the magnetic field and/or enhancements in plasma density may influence the coupling between the solar, interplanetary, and magnetospheric fields. Consequently, the geo-effectiveness of solar wind structures with enhanced field variance and density are likely to increase. We study the effects of both enhanced field fluctuations and enhanced plasma density on the magnetic reconnection/coupling efficiency. We consider solar plasma and field properties, particularly the field fluctuation and plasma density before, during, and after the interplanetary events with geospace consequences. Considering the geo-effectiveness of different magnitude as a measure of the coupling efficiency and reconnection rate, we study the influence of these enhancements on the solar wind–magnetosphere coupling efficiency, during southward field orientation within the interplanetary structures.

We have selected 212 events during the 1995–2006 period of varying geomagnetic activity. Scatterplots of the Dst index versus B_z/E_y show a number of events with almost the same $-B_z$ or E_y, but with quite different storm intensity index –Dst. Also, there are storms with the same amplitude produced by events of quite different $-B_z$ (E_y) intensity. This indicates an additional role of other parameter(s) in solar wind–magnetosphere coupling. To search for these, we have divided all selected events into six groups differing in geomagnetic activity level. We measure the average behavior of the solar plasma flow speed (V), density (N), dynamic pressure (P), magnetic field (B), its north south component (B_z) and its variance (σ_B), and the duskward electric field (E_y), and study their role in solar wind–magnetosphere coupling. For this purpose, we adopted the method of superposed epoch (SPE) analysis. Hourly solar wind plasma and field parameters (V, B, B_z, σ_B, N, P, E_y) together with the Dst index are analyzed using SPE analysis with respect to the start of geomagnetic activity, separately for all six groups of increasing activity.

Badruddin (✉) and V. Gupta
Department of Physics, Aligarh Muslim University, Aligarh, India

Y.P. Singh
Department of Physics, Mangalayatan University, Aligarh, India

S.S. Hasan and R.J. Rutten (eds.), *Magnetic Coupling between the Interior and Atmosphere of the Sun*, Astrophysics and Space Science Proceedings,
DOI 10.1007/978-3-642-02859-5_89,

The averaged plots obtained from our SPE method indicate that enhancement in the plasma density/pressure and in the filed fluctuation starting several hours before the storm or activity onset may play a role in enhancing the solar wind–magnetosphere coupling efficiency during southward B_z conditions.

The FIP Effect in RV Tauri Stars

B.E. Reddy

Abstract RV Tauri stars are evolved low-mass supergiants of spectral type F5 to K3 and are thought to be in the post-AGB (Asymptotic Giant Branch) phase evolving to planetary nebulae. They show deep and shallow minima with periods ranging from 40 to 150 days. While reviewing the abundances of CE Vir and EQ Cas (RV Tauri stars of semi-regular variable type D), we found that they show a clear trend with first ionization potential (FIP; Rao and Reddy 2005, MNRAS, 357, 235). This is an interesting twist to element depletion in stellar atmospheres. Our further search in a sample of 60 RV Tauri stars resulted in finding one more star with a clear FIP effect (Reddy and Rao, in preparation). At 3 out of 60, the phenomenon may not be uncommon among RV Tauri stars.

The FIP effect is familiar for solar coronal abundances, but in reverse: high-FIP elements such as Ne, O, etc. are found to be depleted compared to low-FIP elements such as Fe and Na. In the solar case, it is suggested that magnetic fields are the cause for these anomalies. The high-FIP elements are neutral and not trapped by magnetic fields but the charged low-FIP elements are. The separation of elements is thought to occur in the solar corona. The solar FIP phenomenon is also observed in a few active G and K giants. In the case of CE Vir and EQ Cas, the inverse FIP phenomenon is observed in their photospheres, which has obvious significance in interpreting such abundance anomalies.

As a first step towards establishing a possible connection between elemental separation and the presence of magnetic fields, it is necessary to measure the strengths of the fields in stars with a FIP effect, and also in counterparts without. Model computations with uniform radial magnetic field of about 1 KG over the whole surface suggest that infrared spectral lines with large Landé factor may Zeeman-broaden as much as $12\,\mathrm{km\,s^{-1}}$. Such broadening can easily be seen in high resolution ($R \approx 60,000$) IR spectrograms.

Understanding the FIP effect and its connection to stellar magnetism in stars as well as in the Sun is vital for many anomalous abundance trends observed in different stellar population including the first stars (Kim and Lambert 2008, AJ, 677, 572).

B.E. Reddy (✉)
Indian Institute of Astrophysics, Bangalore, India

S.S. Hasan and R.J. Rutten (eds.), *Magnetic Coupling between the Interior and Atmosphere of the Sun*, Astrophysics and Space Science Proceedings,
DOI 10.1007/978-3-642-02859-5_90,

Author Index

MIX
Papier aus verantwortungsvollen Quellen
Paper from responsible sources
FSC® C105338

Printed by Books on Demand, Germany